Methods in Enzymology

Volume 307
CONFOCAL MICROSCOPY

METHODS IN ENZYMOLOGY

EDITORS-IN-CHIEF

John N. Abelson   Melvin I. Simon

DIVISION OF BIOLOGY
CALIFORNIA INSTITUTE OF TECHNOLOGY
PASADENA, CALIFORNIA

FOUNDING EDITORS

Sidney P. Colowick and Nathan O. Kaplan

Methods in Enzymology

Volume 307

Confocal Microscopy

EDITED BY

P. Michael Conn

OREGON HEALTH SCIENCES UNIVERSITY
PORTLAND, OREGON, AND
OREGON REGIONAL PRIMATE RESEARCH CENTER
BEAVERTON, OREGON

ACADEMIC PRESS

San Diego   London   Boston   New York   Sydney   Tokyo   Toronto

Academic Press
A Harcourt Science and Technology Company
525 B Street, Suite 1900, San Diego, California 92101-4495, USA
http://www.academicpress.com

Academic Press Limited
24-28 Oval Road, London NW1 7DX, UK
http://www.hbuk.co.uk/ap/

International Standard Book Number: 0-12-182208-7

PRINTED IN THE UNITED STATES OF AMERICA
99 00 01 02 03 04 MM 9 8 7 6 5 4 3 2 1

Table of Contents

Section I. Theory and Practical Considerations

Section II. General Techniques

Section III. Measurement of Subcellular Relations and Volume Determinations

Section IV. Imaging of Processes

Section V. Imaging of Ions

Section VI. Imaging of Specialized Tissues

Section VII. Imaging Viruses and Fungi

Contributors to Volume 307

Article numbers are in parentheses following the names of contributors.
Affiliations listed are current.

JOHN H. ANDREWS (34), *Department of Plant Pathology, University of Wisconsin, Madison, Wisconsin 53706*

SILVIA M. ARRIBAS (15), *Departamento de Fisiología, Facultad de Medicina, Universidad Autónoma de Madrid, 28029 Madrid, Spain*

GEORGE F. BABCOCK (18), *Departments of Surgery and Cell Biology, University of Cincinnati College of Medicine, and Shriners Hospitals for Children, Cincinnati Burns Institute, Cincinnati, Ohio 45267-0558*

WERNER BASCHONG (11), *M. E. Müller Institute for Structural Biology and Department of Oral Surgery, Biozentrum of the University of Basel, CH-4056 Basel, Switzerland*

MIGUEL BERRIOS (4), *Department of Pharmacological Sciences and University Microscopy Imaging Center, University Hospital and Medical Center, State University of New York, Stony Brook, New York 11794-8088*

KANTI D. BHOOLA (22), *Department of Experimental and Clinical Pharmacology, Faculty of Medicine, University of Natal, Congella 4001, South Africa*

MIKE BIRCH (28), *Unit of Ophthalmology, Department of Medicine, University of Liverpool, Liverpool L69 3GA, United Kingdom*

GHASSAN BKAILY (8), *MRCC Group in Immuno-Cardiovascular Interactions, Department of Anatomy and Cell Biology, Faculty of Medicine, University of Sherbrooke, Sherbrooke, Québec, Canada J1H 5N4*

LOTHAR A. BLATTER (16), *Department of Physiology, Loyola University of Chicago, Maywood, Illinois 60153*

MATTHIAS BÖHNKE (30), *Department of Ophthalmology, University of Bern, 3010 Bern, Switzerland*

ALBERICO BORGHETTI (20), *Department of Clinical Medicine, Nephrology, and Health Sciences, University of Parma, 43100 Parma, Italy*

DAVID N. BOWSER (25), *Confocal and Fluorescence Imaging Group, Department of Physiology, The University of Melbourne, Parkville, Victoria 3052, Australia*

DANIEL BROTCHIE (28), *Unit of Ophthalmology, Department of Medicine, University of Liverpool, Liverpool L69 3GA, United Kingdom*

CHRISTOF BUEHLER (29), *Department of Mechanical Engineering, Massachusetts Institute of Technology, Cambridge, Massachusetts 02139*

NICK CALLAMARAS (10), *Department of Neurobiology and Behavior, University of California, Irvine, California 92697-4550*

SILVANO CAPITANI (12), *Institute of Human Anatomy, University of Ferrara, 44100 Ferrara, Italy*

H. DWIGHT CAVANAGH (14), *Department of Ophthalmology, University of Texas Southwestern Medical Center, Dallas, Texas 75235-9057*

CATERINA CINTI (12), *Institute of Citomorfologia Normale e Patologica, C.N.R., 66100 Chieti, Italy*

DAVID E. COLFLESH (4), *University Microscopy Imaging Center, University Hospital and Medical Center, State University of New York, Stony Brook, New York 11794-8088*

KIMBERLY A. CONLON (4), *Department of Pharmacological Sciences, University Hospital and Medical Center, State University of New York, Stony Brook, New York 11794-8651*

GUY COX (3), *Electron Microscope Unit, University of Sydney, Sydney, New South Wales 2006, Australia*

DANIEL CULLEN (34), *Forest Products Laboratory, U.S. Department of Agriculture Forest Service, Madison, Wisconsin 53705*

CRAIG J. DALY (15), *Autonomic Physiology Unit, University of Glasgow, Glasgow G12 8QQ, Scotland, United Kingdom*

MARKUS DÜRRENBERGER (11), *Interdivisional Electron Microscopy, Biozentrum of the University of Basel, CH-4056 Basel, Switzerland*

RALF ENGELMANN (31), *Leibniz-Institut for Neurobiology, D-39118 Magdeburg, Germany*

MARGHERITA FONTANA (27), *Oral Biology and Oral Health Research Institute, Indiana University School of Dentistry, Indianapolis, Indiana 46202*

RITA GATTI (20), *Institute of Histology and General Embryology, University of Parma, 43100 Parma, Italy*

GIAN CARLO GAZZOLA (20), *Institute of General Pathology, University of Parma, 43100 Parma, Italy*

CARLOS GONZÁLEZ-CABEZAS (27), *Oral Biology and Oral Health Research Institute, Indiana University School of Dentistry, Indianapolis, Indiana 46202*

ENRICO GRATTON (29), *Laboratory for Fluorescence Dynamics, Department of Physics, University of Illinois at Urbana–Champaign, Urbana, Illinois 61801*

IAN GRIERSON (28), *Unit of Ophthalmology, Department of Medicine, University of Liverpool, Liverpool L69 3GA, United Kingdom*

HISASHI HASHIMOTO (6), *Department of Anatomy, The Jikei University School of Medicine, Minato-ku, Tokyo 105-8461, Japan, and Center for Biogenic Resources, The Institute of Physical and Chemical Research (RIKEN), Tsukuba, Ibaraki 305-0074, Japan*

PENNY HOGG (28), *Unit of Ophthalmology, Department of Medicine, University of Liverpool, Liverpool L69 3GA, United Kingdom*

C. VYVYAN HOWARD (28), *Department of Fetal and Infant Toxico-Pathology, University of Liverpool, Liverpool L69 3GA, United Kingdom*

DAVID N. HOWELL (32), *Departments of Pathology and Laboratory Medicine Service, Veterans Affairs Medical Center, Durham, North Carolina 27705, and Duke University Medical Center, Durham, North Carolina 27710*

JON R. INGLEFIELD (26), *Neurotoxicology Division, National Health and Environmental Effects Research Laboratory, U.S. Environmental Protection Agency, Research Triangle Park, North Carolina 27711*

HIROSHI ISHIKAWA (6), *Department of Anatomy, The Jikei University School of Medicine, Minato-ku, Tokyo 105-8461, Japan*

DANIELLE JACQUES (8), *MRCC Group in Immuno-Cardiovascular Interactions, Department of Anatomy and Cell Biology, Faculty of Medicine, University of Sherbrooke, Sherbrooke, Québec, Canada J1H 5N4*

JAMES V. JESTER (14), *Department of Ophthalmology, University of Texas Southwestern Medical Center, Dallas, Texas 75235-9057*

MANABU KAGAYAMA (5), *Department of Anatomy, Tohoku University School of Dentistry, Sendai 980-8575, Japan*

KI HEAN KIM (29), *Department of Mechanical Engineering, Massachusetts Institute of Technology, Cambridge, Massachusetts 02139*

SUSAN M. KNOBEL (21), *Department of Molecular Physiology and Biophysics, Vanderbilt University, Nashville, Tennessee 37232*

SAMUEL KO (2), *Department of Biochemistry, The Chinese University of Hong Kong, Shatin, N.T., Hong Kong*

S. K. KONG (2), *Department of Biochemistry, The Chinese University of Hong Kong, Shatin, N.T., Hong Kong*

ULRICH KUBITSCHECK (13), *Institut für Medizinische Physik und Biophysik, Universität Münster, D-48149 Münster, Germany*

THORSTEN KUES (13), *Institut für Medizinische Physik und Biophysik, Universität Münster, D-48149 Münster, Germany*

MORIAKI KUSAKABE (6), *Center for Biogenic Resources, The Institute of Physical and Chemical Research (RIKEN), Tsukuba, Ibaraki 305-0074, Japan*

C. Y. LEE (2), *Department of Biochemistry, The Chinese University of Hong Kong, Shatin, N.T., Hong Kong*

P. Y. LUI (2), *Department of Biochemistry, The Chinese University of Hong Kong, Shatin, N.T., Hong Kong*

ANNA MANDINOVA (11), *M. E. Müller Institute for Structural Biology, Biozentrum of the University of Basel, CH-4056 Basel, Switzerland*

CARLOS B. MANTILLA (17), *Mayo Clinic, Rochester, Minnesota 55905*

FRANCESCO A. MANZOLI (12), *Institute of Human Anatomy, University of Bologna, 40126 Bologna, Italy*

NADIR M. MARALDI (12), *Institute of Citomorfologia Normale e Patologica, C.N.R., and Laboratory of Biologia Cellulare e Microscopia Elettronica, I.O.R., 40136 Bologna, Italy, and Institute of Human Anatomy, University of Bologna, 40123 Bologna, Italy*

BARRY R. MASTERS (29, 30), *Department of Mechanical Engineering, Massachusetts Institute of Technology, Cambridge, Massachusetts 02139, and Department of Ophthalmology, University of Bern, 3010 Bern, Switzerland*

JOHN C. MCGRATH (15), *Autonomic Physiology Unit, University of Glasgow, Glasgow G12 8QQ, Scotland, United Kingdom*

SARA E. MILLER (32), *Departments of Microbiology and Pathology, Duke University Medical Center, Durham, North Carolina 27710*

KAZUTAKA MOMOSE (24), *Department of Pharmacology, School of Pharmaceutical Sciences, Showa University, Shinagawa-ku, Tokyo 142-8555, Japan*

LUCA M. NERI (12), *Institute of Human Anatomy, University of Ferrara, 44100 Ferrara, Italy*

HISAYUKI OHATA (24), *Department of Pharmacology, School of Pharmaceutical Sciences, Showa University, Shinagawa-ku, Tokyo 142-8555, Japan*

GUIDO ORLANDINI (20), *Department of Clinical Medicine, Nephrology, and Health Sciences, University of Parma, 43100 Parma, Italy*

IAN PARKER (10), *Department of Neurobiology and Behavior, University of California, Irvine, California 92697-4550*

REINER PETERS (13), *Institut für Medizinische Physik und Biophysik, Universität Münster, D-48149 Münster, Germany*

W. MATTHEW PETROLL (14), *Department of Ophthalmology, University of Texas Southwestern Medical Center, Dallas, Texas 75235-9057*

STEVEN PETROU (25), *Confocal and Fluorescence Imaging Group, Department of Physiology, The University of Melbourne, Parkville, Victoria 3052, Australia*

DAVID W. PISTON (21), *Department of Molecular Physiology and Biophysics, Vanderbilt University, Nashville, Tennessee 37232*

TORSTEN PORWOL (7), *Max-Planck-Institut für Molekulare Physiologie, D-44202 Dortmund, Germany*

PIERRE POTHIER (8), *MRCC Group in Immuno-Cardiovascular Interactions, Department of Anatomy and Cell Biology, Faculty of Medicine, University of Sherbrooke, Sherbrooke, Québec, Canada J1H 5N4*

Y. S. PRAKASH (17), *Mayo Clinic, Rochester, Minnesota 55905*

DESHANDRA M. RAIDOO (22), *Department of Experimental and Clinical Pharmacology, Faculty of Medicine, University of Natal, Congella 4001, South Africa*

GOUSEI RIE (24), *Department of Pharmacology, School of Pharmaceutical Sciences, Showa University, Shinagawa-ku, Tokyo 142-8555, Japan*

NEIL ROBERTS (28), *Magnetic Resonance Research Centre, University of Liverpool, Liverpool L69 3GA, United Kingdom*

NICOLETTA RONDA (20), *Department of Clinical Medicine, Nephrology, and Health Sciences, University of Parma, 43100 Parma, Italy*

BERNHARD A. SABEL (31), *Institute of Medical Psychology, Otto-v.-Guericke University of Magdeburg, D-39120 Magdeburg, Germany*

SPARTACO SANTI (12), *Institute of Citomorfologia Normale e Patologica, C.N.R., 66100 Chieti, Italy*

YASUYUKI SASANO (5), *Department of Anatomy, Tohoku University School of Dentistry, Sendai 980-8575, Japan*

ROCHELLE D. SCHWARTZ-BLOOM (26), *Department of Pharmacology and Cancer Biology, Duke University Medical Center, Durham, North Carolina 27710*

AKIHISA SEGAWA (19), *Department of Anatomy, School of Medicine, Kitasato University, Sagamihara, Kanagawa 228-8555, Japan*

GARY C. SIECK (17), *Mayo Clinic, Rochester, Minnesota 55905*

GEOFFREY L. SMITH (33), *Sir William Dunn School of Pathology, University of Oxford, Oxford OX1 3RE, United Kingdom*

CELIA J. SNYMAN (22), *Department of Experimental and Clinical Pharmacology, Faculty of Medicine, University of Natal, Congella 4001, South Africa*

PETER T. C. SO (29), *Department of Mechanical Engineering, Massachusetts Institute of Technology, Cambridge, Massachusetts 02139*

RUSSELL N. SPEAR (34), *Department of Plant Pathology, University of Wisconsin, Madison, Wisconsin 53706*

EBERHARD SPIESS (7), *Biomedizinische Strukturforschung, Deutsches Krebsforschungszentrum, D-69009 Heidelberg, Germany*

STEFANO SQUARZONI (12), *Institute of Citomorfologia Normale e Patologica, C.N.R., 40136 Bologna, Italy*

GEORGE K. STOOKEY (27), *Oral Biology and Oral Health Research Institute, Indiana University School of Dentistry, Indianapolis, Indiana 46202*

ANJA-ROSE STROHMAIER (7), *Nikon GmbH, D-40472 Düsseldorf, Germany*

LIBORIO STUPPIA (12), *Institute of Biologia e Genetica, University "G. d'Annunzio" 66100 Chieti, Italy*

ROSMARIE SUETTERLIN (11), *M. E. Müller Institute for Structural Biology, Biozentrum of the University of Basel, CH-4056 Basel, Switzerland*

XUEJUN SUN (9), *Department of Oncology, University of Alberta, Cross Cancer Institute, Edmonton, Alberta T6G 1Z2, Canada*

YOSUKE UJIKE (24), *Department of Pharmacology, School of Pharmaceutical Sciences, Showa University, Shinagawa-ku, Tokyo 142-8555, Japan*

ALAIN VANDERPLASSCHEN (33), *Immunology–Vaccinology, Faculty of Veterinary Medicine, University of Liège, B-4000 Liège, Belgium*

ROBERT H. WEBB (1), *Schepens Eye Research Institute, and Wellman Laboratories of Photomedicine, Massachusetts General Hospital, Boston, Massachusetts 02114*

DAVID ALAN WILLIAMS (25), *Confocal and Fluorescence Imaging Group, Department of Physiology, The University of Melbourne, Parkville, Victoria 3052, Australia*

KAZUHIRO YAMAGUCHI (23), *Department of Medicine, School of Medicine, Keio University, Tokyo 160-8582, Japan*

MASAYUKI YAMAMOTO (24), *Department of Pharmacology, School of Pharmaceutical Sciences, Showa University, Shinagawa-ku, Tokyo 142-8555, Japan*

Preface

This *Methods in Enzymology* volume deals with the rapidly evolving topic of confocal microscopy. The OVID database (including MEDLINE, Current Contents, and other sources) lists 76 references to confocal microscopy for the five-year period 1985–1989. In contrast, for the four-year period 1995–1998, nearly 3600 references are listed.

This volume documents many diverse uses for confocal microscopy in disciplines that broadly span biology. The methods presented include shortcuts and conveniences not included in the sources from which they were taken. The techniques are described in a context that allows comparisons to other related methodologies. The authors were encouraged to do this in the belief that such comparisons are valuable to readers who must adapt extant procedures to new systems. Also, so far as possible, methodologies are presented in a manner that stresses their general applicability and potential limitations. Although for various reasons some topics are not covered, the volume provides a substantial and current overview of the extant methodology in the field and a view of its rapid development.

Particular thanks go to the authors for their attention to meeting deadlines and for maintaining high standards of quality, to the series editors for their encouragement, and to the staff of Academic Press for their help and timely publication of the volume.

P. MICHAEL CONN

METHODS IN ENZYMOLOGY

VOLUME 91. Enzyme Structure (Part I)
Edited by C. H. W. HIRS AND SERGE N. TIMASHEFF

VOLUME 92. Immunochemical Techniques (Part E: Monoclonal Antibodies and General Immunoassay Methods)
Edited by JOHN J. LANGONE AND HELEN VAN VUNAKIS

VOLUME 93. Immunochemical Techniques (Part F: Conventional Antibodies, Fc Receptors, and Cytotoxicity)
Edited by JOHN J. LANGONE AND HELEN VAN VUNAKIS

VOLUME 94. Polyamines
Edited by HERBERT TABOR AND CELIA WHITE TABOR

VOLUME 95. Cumulative Subject Index Volumes 61–74, 76–80
Edited by EDWARD A. DENNIS AND MARTHA G. DENNIS

VOLUME 96. Biomembranes [Part J: Membrane Biogenesis: Assembly and Targeting (General Methods; Eukaryotes)]
Edited by SIDNEY FLEISCHER AND BECCA FLEISCHER

VOLUME 97. Biomembranes [Part K: Membrane Biogenesis: Assembly and Targeting (Prokaryotes, Mitochondria, and Chloroplasts)]
Edited by SIDNEY FLEISCHER AND BECCA FLEISCHER

VOLUME 98. Biomembranes (Part L: Membrane Biogenesis: Processing and Recycling)
Edited by SIDNEY FLEISCHER AND BECCA FLEISCHER

VOLUME 99. Hormone Action (Part F: Protein Kinases)
Edited by JACKIE D. CORBIN AND JOEL G. HARDMAN

VOLUME 100. Recombinant DNA (Part B)
Edited by RAY WU, LAWRENCE GROSSMAN, AND KIVIE MOLDAVE

VOLUME 101. Recombinant DNA (Part C)
Edited by RAY WU, LAWRENCE GROSSMAN, AND KIVIE MOLDAVE

VOLUME 102. Hormone Action (Part G: Calmodulin and Calcium-Binding Proteins)
Edited by ANTHONY R. MEANS AND BERT W. O'MALLEY

VOLUME 103. Hormone Action (Part H: Neuroendocrine Peptides)
Edited by P. MICHAEL CONN

VOLUME 104. Enzyme Purification and Related Techniques (Part C)
Edited by WILLIAM B. JAKOBY

VOLUME 105. Oxygen Radicals in Biological Systems
Edited by LESTER PACKER

VOLUME 106. Posttranslational Modifications (Part A)
Edited by FINN WOLD AND KIVIE MOLDAVE

VOLUME 107. Posttranslational Modifications (Part B)
Edited by FINN WOLD AND KIVIE MOLDAVE

Section I

Theory and Practical Considerations

[1] Theoretical Basis of Confocal Microscopy

By ROBERT H. WEBB

A Simple View

A confocal microscope is most valuable in seeing clear images inside thick samples. To demonstrate this, I want to start with a conventional wide-field epifluorescence microscope shown in Fig. 1. The left diagram (Fig. 1) demonstrates the illumination light, and the right shows light collected from the sample. In the right diagram we see that a broad field of illumination is imaged into the thick sample. Although the illumination is focused at one plane of the sample, it lights up all of the sample. In the right diagram we see that the microscope objective has formed the image of the whole thick sample at the image plane of the microscope. If we put a film, charge-coupled device (CCD), or retina at the image plane, it will record the in-focus image of one plane within the thick sample, but it will also record all of those out-of-focus images of the other planes.

In Fig. 2 I show an alternative arrangement. Instead of a broad light source, I use a single point source of light and image it inside the thick sample. That focused light illuminates a single point inside the sample very brightly, but of course it also illuminates the rest of the sample at least weakly. On the right (Fig. 2) the image of the thick sample is very bright where the sample was brightly illuminated and dimmer where it was weakly illuminated. Since my intention is to look only at one point inside the thick sample, I will now put a pinhole in the image plane. The pinhole lets through only the light that is forming the bright part of the image. Behind the pinhole I put a detector, as shown in Fig. 3. That detector registers the brightness of the part of the thick sample that is illuminated by the focused light and ignores the rest of the sample. What we have here is a point source of light, a point focus of light inside the object or sample, and a pinhole detector, all three confocal with each other. That is a confocal microscope.[1-3]

This confocal microscope has all the features we need for looking at a point inside a thick sample. However, it is not very interesting to look at a single point. So we have to find a way to map out the whole sample point

[1] J. Pawley, ed., *in* "Handbook of Biological Confocal Microscopy," 3rd ed. Plenum, New York 1996.
[2] T. Wilson, ed., *in* "Confocal Microscopy." Academic Press, London, 1990.
[3] R. H. Webb, *Rep. Prog. Phys.* **59,** 427 1996.

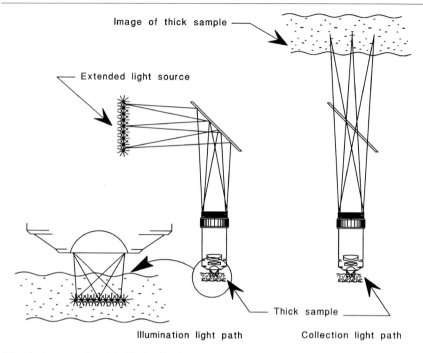

Image of thick sample

Extended light source

Thick sample

Illumination light path   Collection light path

FIG. 1. A conventional (wide-field) microscope for fluorescence in epitaxial configuration.

by point. Most laser-scanning confocal microscopes look at one point of the sample at a time. Other varieties look at many well-separated points at once, but locally they are imaging one point at a time.

The easiest way to look around in the sample is to move the sample, a technique called stage scanning. More complex scanning means allow the sample to be stationary while we move the illuminated spot(s) over the sample. But those are engineering details. Instead of concerning ourselves with them at the moment, let us assume that they are solved and investigate what properties this confocal microscope has.

Optical Sectioning

Our microscope discriminates against points near, but not in, the focal spot. When the unwanted points are beside the focal spot, the contrast has improved. However, this device also discriminates against points above and below the focus, a feature we call optical sectioning. Instead of using a microtome to slice a thin section out of a thick sample, we can now image that thin section inside the sample. Parts of the sample that are above the

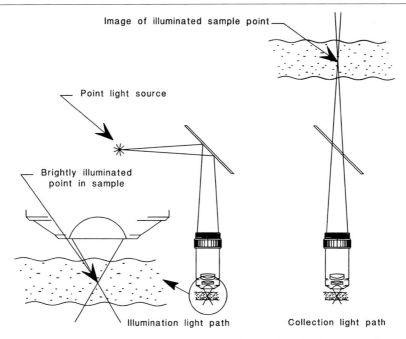

FIG. 2. The microscope of Fig. 1 with point illumination.

imaged point or below it will be illuminated weakly, and light from those parts will be mostly rejected by the pinhole. With scanning, this microscope can image a whole plane inside a thick sample and then be focused deeper into the sample to image a different layer, and those two images do not interfere with each other. With proper controls, the microscope can image a whole stack of optical sections, which can later be assembled into a three-dimensional display.[4,5]

Figure 4 shows an even more abstract sketch of a confocal microscope that emphasizes optical sectioning. An object in the sample that lies above the focal point is imaged above the pinhole. Light going toward that image is mostly blocked by the pinhole mask.

The confocal microscope also rejects light from points adjacent to the one illuminated. That increases the contrast, even for thin samples. Contrast enhancement is always desirable, particularly when we need to look at something dim next to something bright. This fact explains why confocal

[4] G. J. Brakenhoff et al., Scann. Microsc. **2,** 1831 (1988).
[5] F. E. Morgan et al., Scann. Microsc. **6,** 345 (1992).

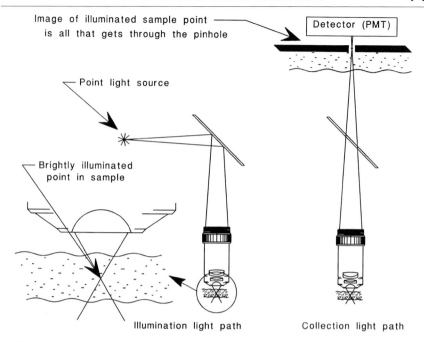

Image of illuminated sample point —
is all that gets through the pinhole

Detector (PMT)

— Point light source

— Brightly illuminated
point in sample

Illumination light path   Collection light path

FIG. 3. The microscope of Fig. 2 becomes confocal when a pinhole blocks light from all parts of the sample outside the focus.

microscopes are used so often for conventional (thin sections) fluorescence microscopy applications.

One thing our confocal microscope cannot do is look through walls. By that I mean that if a layer absorbs light, then deeper layers will be harder to see. That drop-off of visibility limits the sample thickness to about 50 μm in many cases, although there are many instances of looking 0.5 mm into tissue.

Point-Spread Function

Now I want to discuss the physics of the effects just observed. In Fig. 2 we saw that light from a point source is imaged inside the sample. In the sample, that light forms a double cone, as shown in Fig. 5a. Figure 5b uses gray scale to show where the light is most intense. The scale is logarithmic so that the peak is 10^5 times brighter than the darkest areas. On a linear scale, shown in Fig. 5c, only the peak has any intensity. The cross section of the cone, shown as lines in the gray scale images, represents a numerical

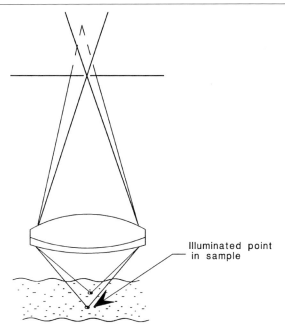

FIG. 4. Another schematic view of the confocal microscope. The point of interest is imaged in the pinhole, while light from the more proximate point is largely blocked by the pinhole. This organization is called "optical sectioning."

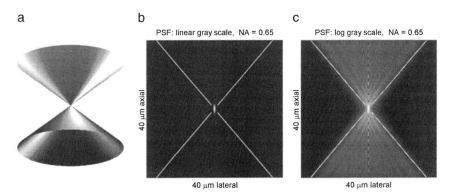

FIG. 5. (a) Light from an objective lens fills a (double) cone. (b) The actual light intensity is plotted as a linear gray scale. The same presentation, with a logarithmic gray scale, is shown (c), with the lightest value being 10^5 times the darkest.

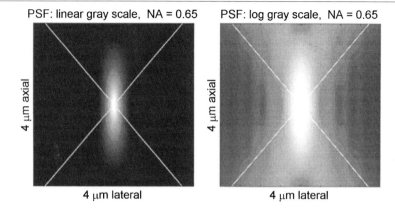

FIG. 6. The point-spread function close to the focus. These are the patterns of Fig. 5, magnified 10 times.

aperture (NA) of 0.65 for the sample in air or 0.86 if the index of the sample is 1.33 (water).

Looking close to the focal point, as shown in Fig. 6, there is structure to the intensity distribution. The lines show the geometric edges of the light seen in Fig. 5, but the pattern is much more complex. Physicists call this pattern the point-spread function. One way to think of the point-spread function is in terms of probability: the probability that a photon from the point source will reach some point ($\mathbf{q}$) is prob ($\mathbf{q}$). A photon is 10^5 times more likely to reach the focal point than some point far from it, still within the light cone. Keep this probability picture in mind because we will use it again to understand the confocal arrangement.

The complexity in the pattern in Fig. 6 is due to the effects of diffraction, a consequence of the wave nature of light.[6] It is the objective lens that causes the diffraction pattern and that pattern is displayed perfectly by using a point source of light. In the transverse plane (perpendicular to the symmetry axis of the lens), the central part of this pattern is called the Airy disk.[7] Figure 7 shows an Airy disk in linear and logarithmic gray scale. The radius of the inner dark ring in this pattern is a measure of the resolving power of the microscope. Such a resolution element (or resel) is nearly the same as the diameter of the disk at 50% intensity. Both are somewhat arbitrary measures of resolution.

It is important to understand that these point-spread functions (the patterns we have been looking at) are due to the lens. The smaller the lens

[6] M. Born and E. Wolf, "Principles of Optics." Pergamon Press, New York, 1991.
[7] E. Hecht, "Optics," 2nd ed. Addison-Wesley, Redding, MA, 1990.

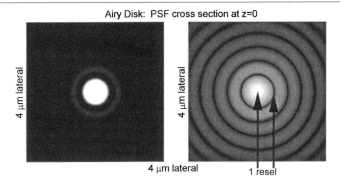

Fig. 7. The point-spread function in the focal plane: the pattern of Fig. 6 rotated about the *z* axis at *z* = 0. *Left:* Linear gray scale. *Right:* Logarithmic gray scale.

aperture, the wider the pattern (and the lower the resolution). A lens with a small aperture has a large point-spread function and a low resolution. A lens with a large aperture can resolve much smaller things because its point-spread function is smaller. The term numerical aperture (NA) refers to a measure of the lens aperture in angle and takes into account the smaller wavelengths in a refractive medium. (NA = $n \sin \vartheta$, and $\lambda = \lambda_0/n$.)

The image, our final result, is due to light returning from points in the sample. I will use the term "remittance" to include reflection, scattering, refraction, diffraction, and fluorescence—all the things that the sample does to redirect light. To analyze an image, we need to know how bright the illumination was at every point in the sample. For instance, a strong remitter illuminated by dim light will return the same amount of light as a weak remitter illuminated by bright light. Multiply the remittance by the illumination intensity at every point, which is what the point-spread function describes, and that describes the image brightness distribution in the image.

The image that we want is the image of the illuminated object, viewed through the objective lens. We know that the objective lens causes point sources of light to image as point-spread functions, so we should expect that to happen again. Here I am going to use a trick. I could take every point in the illuminated object, find its point-spread function at the image plane, and then add those all up—a process called convolution. A simpler way is to use the fact that the laws of optics work in both directions: it does not matter which way the light goes. So I will start with a (point) pinhole and know that its image in the sample is a point-spread function—in fact, the same point-spread function that we saw before. Because I only care about light that gets through the pinhole, I can use the distribution (the point-spread function) of the intensity of light that comes from the pinhole and use that to evaluate how much light goes to (and through) the pinhole. To distinguish the two, I will call the point-spread function for the illumination source PSF_S and the point-spread function for the collection pinhole PSF_P (S for source and P for pinhole).

Now imagine that the sample is featureless: its remittance is everywhere the same, as it would be in a fluorescent liquid. The detector is going to register light coming through the pinhole, from the part of the sample illuminated by PSF_S and sampled by PSF_P. Every point feels the influence of both point-spread functions, which means that the two should be multiplied. The product point-spread function is a point-spread function for the whole microscope, a confocal point-spread function.

$$PSF_{CF} = PSF_S \times PSF_P \qquad (1)$$

Every point on those gray scale plots is an intensity, and I need only to multiply the intensities at each point to find the mutual intensity—the intensity of light that came from the point source, was remitted by the sample, and passed through the pinhole.

One interpretation of the point-spread function is that of a probability. PSF_S is the pattern of prob $(\mathbf{q})$ for every point q in the sample, and PSF_P is another (independent) pattern of prob $(\mathbf{q})$ for every point q. The probability of detecting a photon is the probability that a photon goes from the point source to the sample and goes from the sample to and through the pinhole. These are independent probabilities, so the mutual probability is their product, as stated in Eq. (1).

Figure 8 shows how much sharper the confocal point-spread function is than either the source or the pinhole point-spread function. Subsidiary peaks that were 0.01 times the main peak become only 0.0001. That reduction is the source of the increased contrast and the optical sectioning.

There is, however, another hidden fact that helps with sectioning. Go back for a moment to Fig. 5, where we started with a double cone of light.

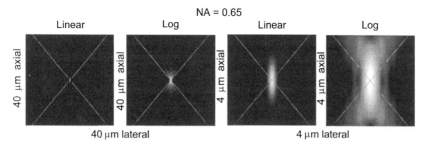

FIG. 8. The confocal point-spread function. Figures 5 and 6, for the wide-field microscope, show much larger point-spread functions.

Although the illuminated point in the sample gets concentrated light, the same total amount of light passes through each plane perpendicular to the axis. So we might worry that the out-of-focus remission could add up to a lot of light. That is just what happens in a wide-field microscope, where that extra returned light obscures the view of interior planes of the sample where the microscope is focused. However, with the point source and pinhole, those out-of-focus planes contribute so little light (see Fig. 4) that the interior planes are sharp and clear. Mathematically, we could predict that this is so by integrating over the confocal point-spread function, but Fig. 8 shows what to expect, with its drastic reduction of the intensity away from the central peak.

The formalism for this is:

$$\int_{\text{Plane}} \text{PSF} = \int_{\text{Plane}} \text{light} = \text{constant} \qquad (2)$$

for the wide-field (single) point-spread function.

For the confocal point-spread function, however, the integral is not of "light," it is of "light that reached a sample point *and* got back through the pinhole." Then the integral over the focal plane is much larger than the integral over any other plane.

$$\int_{\text{Focal plane}} \text{PSF}_{\text{CF}} \gg \int_{\text{Any other plane}} \text{PSF}_{\text{CF}} \qquad (3)$$

Better yet, the sum of all those integrals is still much less then the amount of intensity at the focus.

$$\int_{\text{Focal plane}} \text{PSF}_{\text{CF}} \gg \iint_{\text{All other planes}} \text{PSF}_{\text{CF}} \qquad (4)$$

That is just a complicated way of saying that the pinhole excludes almost all the light from anywhere but the focus. This is really a remarkable thing: the amount of light that gets through the pinhole from everywhere away from the focus is much less than what comes from the focus. The mathematics shows this and real confocal microscopes confirm it.

What makes a confocal microscope is a point-spread function that is the product of two individual point-spread functions.

Pinhole

Now I need to go back to the "point" pinhole and be a little more realistic.

What is a point source? A point is a mathematical fiction, so tiny that no light would come from it. A star, however, makes a pretty good real point source. That is because the image of the star is smaller than the point-spread function of the lens we use to observe it. The same is true of a point pinhole. If the extent of the pinhole is less than the point-spread function of the lens, then that is a point pinhole. It turns out that we can use a pinhole that is about three resels across and still get almost all of the confocal effects.[1,2] An even bigger pinhole will blur the point-spread function enough to degrade the optical sectioning and contrast enhancement. So my ideal confocal microscope will use a three resel pinhole, i.e., a pinhole that is three times the size of the Airy disk.

Magnification

There is a trick here that may be confusing. Every lens has two point-spread functions: one on each side. If the lens magnifies by a factor of 60, then the point-spread function on one side is small and that on the other side is roughly 60 times as big (roughly, because the diffraction peak details are not exactly images of each other and magnification is a concept from geometric optics). A 60× objective lens with NA = 0.85 forms a resel at the sample that is 0.4 μm across, but at the image plane of the microscope, the NA is 0.014 and the resel is 24 μm. So there is no need to make submicron pinholes; use one that is comfortably in the 50- to 100-μm range.

New objective lenses are becoming available that have high NA and low magnification, so pinholes need to be adjusted to compensate. We generally say that magnification is of no interest in confocal microscopy, as no image is ever really formed.[8] However, in this one case, it is important to adjust pinhole size to suit the objective lens.

As an example, the microscope I use has a 100/1.2 and a 40/1.2 objective lens. The two lenses have identical resolutions, but the 40× has a larger field of view, so it is the one I use. The 3 resel pinhole for the 100× would be 28 μm at the usual 150-mm image point, while for the 40× would be 9 μm there. Of course the actual pinhole location will probably not be a 150-mm point, as modern objective lenses work with an infinite conjugate ("infinity corrected"), and some extra magnification before the pinhole is used to make the physical device of manageable size. Also, I have control of the pinhole size, but I am not told how many resels it is, which would be useful information. These really are details, but it might be well to pay attention to them when running a confocal microscope.

[8] D. W. Piston, *Biol. Bull.* **195,** 1 (1998).

Complete Microscope

Figure 9 shows a complete confocal microscope in which the point-spread function is the product of two individual point-spread functions. Notice that the engineering details are still hidden in a box called "scanning engine." That box may have moving mirrors that sweep the laser beam over the sample or it might have a disk full of holes whose rotation sweeps many illumination spots over the sample. There are many varieties of scanning engine too, and the engineering details are in fact truly important. However, the theory of the confocal microscope does not require us to understand scanning engines. Rather, we should look at the two point-spread functions that go into making up the confocal point-spread function.

The microscope in Fig. 9 is usually used to detect fluorescence. The beam splitter is a dichroic mirror that reflects the fluorescent light and passes the excitation light. It can, however, be used equally well to detect light remitted without a change of wavelength by making the beam splitter a partially silvered mirror. If there is no wavelength change, the two point-spread functions are identical, except for the convolution on the pinhole. Even with the wavelength change due to fluorescence, there is not much change of point-spread function. Equation (5) gives the lateral resolution in terms of numerical aperture and wavelength for the confocal microscope. The quantity Δr is the full width at half-maximum intensity of the confocal point-spread function. I give this measure because there is some confusion in the literature as to how to use the Rayleigh criterion for resolution in the confocal situation. Equation (6) gives the axial resolution (the optical section).[1]

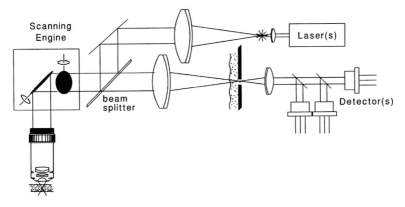

Fig. 9. A generic confocal microscope.

$$\Delta r = 0.32 \, \lambda/\text{NA} \qquad (5)$$
$$\Delta z = 1.26 \, n\lambda/\text{NA}^2 \qquad (6)$$

where NA is the numerical aperture, n is the index of refraction, and λ is the wavelength in vacuum (no medium). For fluorescence, λ should be replaced by the (geometric) mean of the two wavelengths.

As an example, suppose we use an objective of NA = 0.9 with fluorescein isothiocyanate (FITC). The excitation wavelength might be 488 nm, and the fluorescence will be centered around 530 nm. So the lateral resolution is 0.18 μm and the optical section is 1 μm.

Fluorescence may be very weak, so we probably want to use as big a pinhole as possible. Three (Rayleigh) resels would be 1 μm, at the sample, or somewhere around 80 μm in the first image plane of the objective lens. A bigger pinhole will let through even more light, but at the price of reducing the resolution and contrast.

There is another consequence of the wavelength shift of fluorescence. The confocal microscope requires that the pinhole be optically conjugate to the illuminated spot in the thick sample. That means that both the point source and the pinhole have to be imaged at the same place. Such a confocal arrangement is possible if the microscope objective is highly achromatic or if the pinhole position is adjusted to compensate for chromaticity. Some confocal microscopes use a single pinhole for all colors, so they need good apochromats or similar objectives. Other confocal microscopes separate their colors before the pinholes, so each pinhole can be adjusted separately and less expensive objectives can be used. This multipinhole design risks misalignment by a factor of the number of pinholes (usually three). There are trade-offs in both price and convenience.

Most confocal microscopists want the option of exciting fluorophores with two lasers at once. That demands good achromaticity so that both colors focus in the same plane. So do not try to save money on objective lenses!

Figure 9 sketches the generic confocal microscope for use primarily in fluorescence. The point source in this case is a laser that has been brought to a point focus before expansion to fill the objective lens with light. That point focus has to be less than the (magnified) point-spread function of the objective. The scanning engine in this design may be a pair of mirrors mounted on galvanometer motors, which are optically conjugate to the pupils of the objective lens. Notice that the scanning mirrors tilt the laser beam back and forth and untilt the remitted light back to a stationary beam. That stationary beam is then focused on a pinhole (here I have chosen the single pinhole design). After the pinhole, a series of dichroic beam splitters separate the various colors and send them to detectors appropriate to each color. Generally the detectors are photomultiplier tubes, although avalanche photodiodes are also used. For further discrimination against the excitation light, filters are placed in front of the detectors. Much of the cost of confocal microscopes has to do with changing those filters, the pinhole size, the choice of detector, the size of the scan (the field of view), and other parameters necessary to a useful picture.

Varieties of Confocal Microscope

One of the engineering details I have been ignoring is the scanning engine. Confocal microscopes come in two versions, O and P. The O version puts the scanning in an object plane, the P in a pupil plane.

In the P-confocal microscope (CM-P or variously CSLM, CLSM, and other permutations), the scanning occurs in a plane optically conjugate to the pupil of the objective lens. A deflection device, usually moving mirrors, changes the angle of a light beam, usually a laser beam, causing one or a few illumination spots to scan over the object. The same mirrors (usually) then descan the remitted beam to keep it stationary on a detection pinhole.[9] There are many variants to all this, but the theory of the confocal microscope applies to all.

The most common O-confocal microscope is the disk scanner or tandem-scanning microscope, in which a disk full of holes spins in a plane optically conjugate to the object, thus causing the images of those many holes to scan over the object. Then either the same set of holes or a different set on the same disk serve as detection pinholes.[10] The CM-O can use nonlaser light and provides a live image to the eye or a camera.

Multiphoton Microscope

There is another way to have two point-spread functions multiply so that the microscope becomes confocal.[11] A very intense light source, focused to a very small spot, can deliver two or more photons at once to an absorber. For instance, a single photon at 488 nm can excite fluorescence in fluorescein dye at around 530 nm. The same energy could also come from two photons at twice the wavelength (976 nm) if they arrive at very nearly the same time. They will arrive at very nearly the same time if the light is intense enough and the focus is tight enough. The position of each photon is controlled independently by the point-spread function of the objective

[9] R. H. Webb, *Appl. Opt.* **23**, 3680 (1984).
[10] G. S. Kino and T. R. Corle, *Phys. Today* **42**, 55 (1989).
[11] W. J. Denke *et al., Science* **248**, 73 (1990).

lens, so we have two identical point-spread functions. These are, again, independent probabilities, so they multiply to give a confocal point-spread function just like that of the confocal microscope.

There is an extra benefit to the multiphoton configuration. The illumination light is of much longer wavelength than the single photon excitation light, so it is less likely to damage the sample. Furthermore, the excitation light can only cause multiphoton processes in the very intense focus, so the light passing through out-of-focus planes does not bleach the fluorophore.

Finally, no pinhole is needed to achieve this confocal arrangement. Any light at the fluorescent wavelength has to originate in the focal volume, so any remitted light at that wavelength coming out of the objective lens will contribute to a good image. Figure 10 shows the generic multiphoton microscope that uses no pinhole. There are no alignment problems at the detector!

There is also a cost to the multiphoton configuration. First, it only works in fluorescence. Second, it is not simple to get all that light concentrated in space and time. Typically the source for this microscope is a pulsed laser pumped by some other big light source. That is both expensive and difficult to maintain.

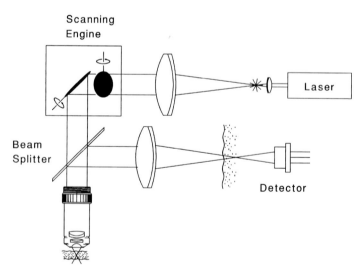

Fig. 10. A multiphoton microscope. The laser delivers all its energy in a very short pulse, so two photons can reach the sample nearly simultaneously. No pinhole is needed.

Light Sources

Laser

Real confocal microscopes have a lot of engineering details. In general, these are not appropriate for this discussion, but one of the details is—the light source. Confocal microscopes of the CM-P flavor use lasers as their source. Disk-scanning confocal microscopes (CM-O) have the advantage of being able to use almost any bright source. I will discuss briefly the properties of the lasers used in point-scanning microscopes and what the implications of the other light sources are.

Lasers are bright monochromatic sources whose light emerges in a tightly collimated coherent beam. What that means for a confocal microscope is that the point source is very nearly perfect. Lasers are also monochromatic and coherent, but the coherence is rather incidental for the use we make of the laser. Coherent light is light whose waves are all exactly in phase. That can be useful in confocal microscopy, but it is not necessary.

Speckle

In fact, coherent light makes very bad illumination for a broad field because it produces the phenomenon known as speckle. Speckle arises from the interference of light scattered from nearby points within the illumination field. That interference produces dark and bright spots next to each other in the image. Speckle does not arise in a confocal microscope because we look at only one illuminated point (one point-spread function) at a time. If we look more carefully, we find that speckle does affect the confocal image as out-of-focus planes scatter coherently into the pinhole. However, that is a second order effect that we need not worry about here and that does not occur in a fluorescent image.

Monochromaticity

A monochromatic light source is perfect for fluorescence imaging. A simple long-pass filter rejects the excitation light and allows the fluorescence through.

Incoherent Sources

Nonlaser sources are generally grouped under the heading incoherent. Both monochromaticity and good collimation are necessary for coherence. Incoherent sources include arc lamps and other bright sources that are too big to act as points. Often we limit the color with narrow band filters so that only useful colors fall on the sample. The result of such limiting is to reduce the intensity of the light at the sample. Sometimes that is a good idea if we are having trouble with bleaching, but generally nonlaser sources are light starved.

For a disk-scanning confocal microscope, we need to use a light source that covers the object field (the sample) with uniform light. Köhler illumination accomplishes just that, but requires some care.[12] Thus, although a nonlaser source may seem simpler, it often requires as much attention as a laser.

One of the useful aspects of disk-scanning confocal microscopes is that one may obtain a true-color image. If we do not know what is in a sample it is often useful to illuminate it with white light. That also may give us a more familiar looking image, closer to what the eye sees in a simple bench microscope.

Bleaching

One of the major uses of confocal microscopy is for fluorescence imaging. The fluorophores used tend to bleach out when exposed to too much light. Bleaching is generally thought to be proportional to light dose, although there are some examples of nonlinearities, both favorable and unfavorable to the high intensities of the laser-scanning confocal microscopes.[1] Fluorophore saturation certainly occurs, and that limits the useful intensity, but bleaching is permanent, as is photodamage to biological structures. There really is not much one can do about a process proportional to light dose, as the fluorescence has the same dependence, but one can avoid exposure to light that is not giving a fluorescent signal. That means not exposing the sample to light during "fly-back"—the time when the scanning engine is merely returning to zero. It also means avoiding exposure to one laser while another is being used. These are just points that a careful microscope builder pays attention to.

A major benefit of the multiphoton microscope is that only light in the focal volume participates in the bleaching. All single photon microscopes expose the layers above and below the focal plane, even though they image only the focal plane. The advantage of the multiphoton microscope is clear and is one of the major reasons for using it.

---

[12] S. Inoue and K. R. Spring, "Video Microscopy, the Fundamentals," p. 22. Plenum, New York, 1997.

Colocality

It is often desirable to test whether two fluorophores are attached to the exact same point on a cell, i.e., the same molecule. To do this the microscope displays an image in each of the two colors, and the observer checks whether the redder image exactly overlies the bluer. This condition will be met if the two pinholes (if there are two) are imaged at exactly the same point of the sample and also if two laser sources (if there are two) are imaged at exactly the same point of the sample. To check, use a small particle such as a polystyrene bead (used in flow cytometry) that fluoresces in two colors.[13] The two color images should be coincident in position and size.

Numerical Aperture

In a conventional microscope, numerical aperture is well defined. It depends on the diameter of the objective lens' pupil, the focal length, and the wavelength in the sample medium. In the cone of Fig. 5a, these quantities are unambiguous, and even when diffraction is included, the term NA is clear. We use it to describe the resolution obtainable with a microscope, as used in Eqs. (5) and (6).

$$NA = n \sin \vartheta \qquad (7)$$

Snell's law tells us that $n \sin \vartheta$ does not change across interfaces, so NA is a handy quantity. But resolution really depends on the angle ϑ, the most extreme ray in the light cone of Fig. 5a, and the index of refraction only enters because the reduced wavelength λ/n is the proper distance metric. The term aperture enters because $\sin \vartheta$ is governed by the aperture of the lens: the pupil. A big aperture is needed for a big NA, but you do not get the benefit of a big aperture if you are sending a small beam of light through it.

Whether we use NA or $\sin \vartheta$ as a synonym for resolution, we assume that the objective lens pupil is uniformly filled with light. In a confocal microscope, that might well not be true! The profile of a laser beam is generally gaussian, so there is less light at the pupil edge. The profile of the beams in a CM-O will be a diffraction pattern, so the same is true. The microscope manufacturer will make a choice as to how much light can be wasted by overfilling the pupil, and that will affect the resolution of the microscope. I do not see much that any of us can do about this situation, but it might be a good idea to test the microscope resolution.

---

[13] Molecular Probes, Inc. http://www.probes.com/.

Summary

A confocal microscope forms its image by recording light primarily from a small focal volume, largely ignoring points to the side or above or below. That volume, described as a point-spread function, is the product of two similar functions that are generated by the objective lens. Because of that multiplication, the recorded light is greater than even the integrated total of the light from all other points in a thick sample. Some of the implications of implementing this theory are reflected in the choices available to users of confocal microscopes.

Acknowledgment

This work was supported in part by DE-FG02-91ER61229 from the Office of Health and Environmental Research of the Department of Energy.

[2] Practical Considerations in Acquiring Biological Signals from Confocal Microscope

By S. K. KONG, S. KO, C. Y. LEE, and P. Y. LUI

Introduction

With the development of fluorescent indicators and recombinant proteins such as fluo-3[1] and green fluorescent protein-tagged chimeras,[2] fluorescence microscopic imaging (FMI) offers unparalleled opportunities to study biochemical events in living cells with a minimum of perturbation.[3,4] In the conventional wide-view FMI, not only is a sharp image generated from an in-focus area, signals above and below the focal plane are also acquired as out-of-focus blurs that distort and degrade the contrast and sharpness of the final image. However, in a confocal microscope, the excitation light generated by a laser is focused to a discrete point of the specimen to reduce the wide-view illumination and a pinhole is put in front of the detector to prevent the passage of signals coming from planes other than

[1] A. Minta, J. P. Kao, and R. Y. Tsien, *J. Biol. Chem.* **264,** 8171 (1989).
[2] R. Rizzuto, M. Brini, P. Pizzo, M. Murgia, and T. Pozzan, *Curr. Biol.* **5,** 635 (1995).
[3] R. Y. Tsien, *Am. J. Physiol.* **263,** C723 (1992).
[4] R. Y. Tsien and A. Miyawaki, *Science* **280,** 1954 (1998).

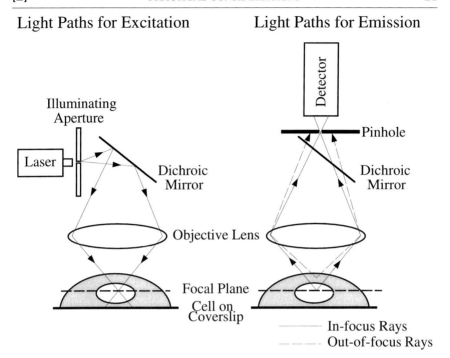

Light Paths for Excitation   Light Paths for Emission

FIG. 1. The confocal principle in epifluorescence scanning confocal microscopy. Light from laser source passes through the illuminating aperture, is reflected by the dichroic mirror, and is focused on one point in the specimen (left). After excitation, fluorescence emissions from the focal point and the out-of-focus illuminating cones are collected by the objectives. Both in-focus and out-of-focus rays can pass through the dichroic mirror, but only the emissions from the focal point are able to pass through the pinhole to the detector (right). By tilting the dichroic mirror and oscillating other scanning mirrors (not shown), scanning of the entire specimen becomes possible.

the confocal section (Fig. 1). As a consequence, only in-focus signals are obtained while all of the out-of focus fluorescence is eliminated. By moving the laser spot across the specimen with scanning mirrors, a sharp image from the confocal plane can be generated with a sensitive photodetector. The image so generated can be stored into a computer for data analysis. With these technical advancements, a number of cellular activities have been observed and their significance elucidated. For example, the uncoupling of the rise of nuclear and cytosolic Ca^{2+} and the release of Ca^{2+} from the nuclear envelope have been discovered.[5,6] However, potential problems

[5] P. P. Lui, S. K. Kong, K. P. Fung, and C. Y. Lee, *Pflug. Arch.* **436,** 371 (1998).
[6] P. P. Lui, S. K. Kong, D. Tsang, and C. Y. Lee, *Pflug. Arch.* **435,** 357 (1998).

and flaws in the use of confocal microscopy have seldom been documented. This article presents several potential problems that we have encountered that either reduced the efficiency of the experiment process or gave rise to erroneous interpretations.

Laser-Induced Rise of Fluo-3 Fluorescence Signals

The development of the lipophilic membrane-permeant acetoxymethyl (AM) ester form of fluorescent indicators provides a nondestructive method of introducing probes into the cytoplasm for various biochemical studies without perforating the cells with micropipettes and injectors.[7] After crossing the cell membrane, the AM form of the indicator is hydrolyzed by esterases and converted back to its hydrophilic form. Practically, the esterified indicator is first dissolved in organic solvents such as dimethyl sulfoxide (DMSO) before cell loading. A technical difficulty encountered is that indicators in some cell types have been shown to be compartmentalized in endosomes as a consequence of endocytosis[8] or in other organelles such as mitochondria and endoplasmic reticulum.[9] As the probe compartmentalization is a temperature-dependent process, the temperature for dye loading should be low enough to slow down endocytosis but high enough to hydrolyze all intracellular indicators. The compartmentalization problem can be circumvented further by dispersing indicators with detergents such as Pluronic F-127 (Molecular Probes, Eugene, OR) to improve the dye solubility and avoid the accumulation of dye on the cell surface.[10]

In the course of measuring fluo-3 fluorescence, we sometimes observed a spontaneous and sustained increase in fluorescence signal intensity. In Fig. 2a (see color insert) HeLa cells were loaded with fluo-3/AM (1.5 μM) for 1 hr at room temperature and were then exposed to laser scanning without the addition of stimuli. Interestingly, fluo-3 signals increased progressively with time. It was suspected initially that the spontaneous cellular activity had led to the fluorescence burst. However, when it was discovered that only illuminated cells were affected in this way, while all other cells outside of the area exposed to the laser light had only background fluorescence (Fig. 2b, see color insert), we concluded that laser irradiation might have converted residual nonfluorescent fluo-3/AM molecules into its fluorescent form by heat generated by laser irradiation.

In our experimental condition, the lipophilic membrane-permeant form

[7] R. Y. Tsien, *Nature* **290**, 527 (1981).

[8] A. Malgaroli, D, Milani, J. Meldolesi, and T. Pozzan, *J. Cell Biol.* **105**, 2145 (1987).

[9] F. Di Virgilio, H. Steinberg, and S. C. Silverstein, *Cell Calcium* **11**, 57 (1990).

[10] R. Haugland, *in* "Handbook of Fluorescent Probes and Research Chemicals," 6th ed., p. 544. Molecular Probes, Eugene, OR, 1996.

of fluo-3/AM was admitted into the cell by simple diffusion. As mentioned earlier, esterases remove the AM group and convert the indicator into the hydrophilic and fluorescent form, which is trapped intracellularly. We surmised that the hydrolytic reaction might need time for completion and, should there be sufficient unconverted fluo-3/AM in the cytoplasm when the experiment began, the heat generated by laser irradiation or fluorescent energy loss would promote the enzymatic reaction and lead to the apparently spontaneous fluorescence emission. The confirmation of this hypothesis would be possible with a ratiometric indicator by measuring the fluorescence emission at the isosbestic point (a wavelength at which fluorescence varies only with the concentration of indicator but not the parameter being measured). However, in the present case with fluo-3, we could only test the postulate by observing that the averaged fluo-3 fluorescence of cells tended toward a maximum. Figure 2c supports this notion and indicates that prolonged irradiation of fluo-3 loaded cells with laser led to a maximal

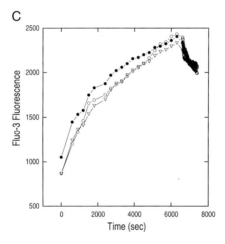

Fig. 2. HeLa cells were loaded with fluo-3 (1.5 μM) for 1 hr in physiological buffer with shaking at room temperature. Sequential images of the same optical section in the x–y plane (256 × 256 pixels) were obtained at 2.5-sec intervals at room temperature by a confocal imaging system (Molecular Dynamics, Sunnyvale, CA) with Nikon objective ×60, NA 1.4. For fluorescence measurement, excitation and emission were 488 and above 510 nm, respectively. No stimulant has been added. (a, see color insert) Pseudocolor images of four single cells recorded at the times indicated. (b, see color insert) Fluorescence images (1024 × 1024 pixels) of other HeLa cells cultured on the same coverslip as those shown in (a). The scale bar in white indicates dimension in micrometers. The palette below the images illustrates the degree of fluorescence. L929 cells were loaded with fluo-3 as mentioned earlier. Sequential images of the same optical section in the x–y plane were obtained at room temperature initially at 5 min and then at 10-sec intervals as indicated. No stimulant has been added. The changes in fluo-3 fluorescence of three cells against time were determined (c). Reproduced from P. P. Lui, M. M. Lee, S. Ko, C. Y. Lee, and S. K. Kong, *Biol. Signals* **6,** 45 (1997), with permission from Karger, Basel.

fluorescence emission. A different cell system, L929 cells, has been used in this experiment, which further shows that the erroneous signal arises from the hardware setup rather than that of cells.

To prevent the problem of unwanted fluorescence, we might increase the temperature during dye loading to speed up conversion, but this may enhance indicator compartmentation. Alternatively, we recommend that cells under study should first be illuminated until the fluorescence has reached a stable level, indicating that the unhydrolyzed dye has been exhausted before beginning the experiment. Care should of course also be taken to prevent photobleaching because of overstrong excitation. Figure 2c is a typical example that too frequent excitation provokes photobleaching.

Therefore, it is better to test a few dye loading concentrations and loading times before starting experiments. Longer loading time will create the problem mentioned earlier. Normally, dye concentrations of 5–10 μM are used, and loading times can be in the range of 30 min to 1 hr, depending on the cell type. An ideal method for studying cellular activity is to employ dual emission probes such as indo-1/AM for Ca^{2+} with a UV laser (Coherent, Santa Clara, CA) for excitation. Emissions at two different wavelengths can then be used ratiometrically. The advantage of using this semiquantitative ratiometric determination is that information obtained in this way is independent of the intracellular dye concentration. The limitation of this suggestion is that it requires UV excitation, and the UV light source is not as widely available as visible laser lines in commercial setups.

Water Content of Dimethyl Sulfoxide and Efficiency of Dye Loading

In our experience, the dryness of DMSO also affects the dye loading efficiency. Figure 3 (see color insert) illustrates the confocal images of cells from several experiments in which Calcium Green-1 (Molecular Probes), a Ca^{2+}-sensitive fluorescent indicator, was dissolved in dry DMSO or DMSO containing 10% (v/v) or 30% (v/v) H_2O. As shown in Fig. 3a, when cells were loaded with Calcium Green-1 that had been dissolved in dry DMSO at room temperature, a relatively homogeneous cytosolic fluorescence was observed. In contrast, when cells were incubated with Calcium Green-1 dissolved in non-dry DMSO, localized bright spots were seen in the optical sections. Comparison of Fig. 3b with Fig. 3c reveals that the higher the water content in DMSO, the more pronounced that anomaly. We surmised that low dye solubility in water-containing DMSO gives rise to insoluble indicators, which produced bright spots after endocytosis.

The results in Fig. 3 clearly indicate that the dryness of DMSO used is important for a successful experiment. With wet DMSO, the hot spots can saturate the detector, and the wide range of fluorescence intensity creates

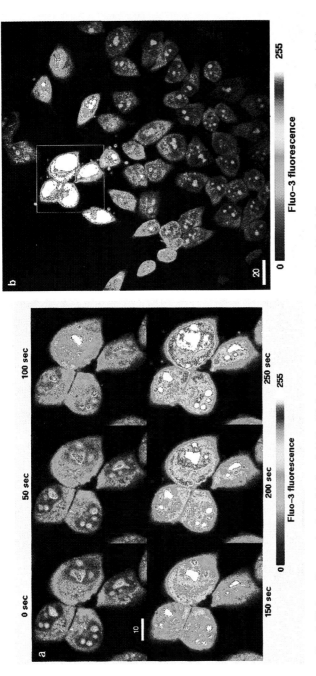

FIG. 2. HeLa cells were loaded with fluo-3 (1.5 μM) for 1 hr in physiological buffer with shaking at room temperature. Sequential images of the same optical section in the x–y plane (256 × 256 pixels) were obtained at 2.5-sec intervals at room temperature by a confocal imaging system (Molecular Dynamics) with Nikon objective ×60, NA 1.4. For fluorescence measurement, excitation and emission were 488 and above 510 nm, respectively. No stimulant has been added. (a) Pseudocolor images of four single cells recorded at the times indicated. (b) Fluorescence images (1024 × 1024 pixels) of other HeLa cells cultured on the same coverslip as those shown in (a). The scale bar in white indicates dimension in micrometers. The palette below the images illustrates the degree of fluorescence. L929 cells were loaded with fluo-3 as mentioned earlier. Sequential images of the same optical section in the x–y plane were obtained at room temperature initially at 5 min and then at 10-sec intervals as indicated. No stimulant has been added. The changes in fluo-3 fluorescence of three cells against time were determined (c, see black and white version in text). Reproduced from P. P. Lui, M. M. Lee, S. Ko, C. Y. Lee, and S. K. Kong, *Biol. Signals* **6**, 45 (1997), with permission from Karger, Basel.

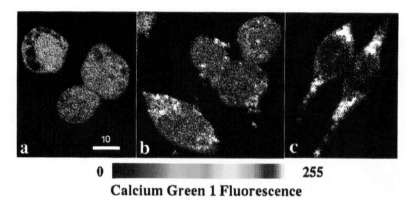

Calcium Green 1 Fluorescence

FIG. 3. Confocal fluorescence pseudocolor images of PU5-1.8 cells loaded with Calcium Green-1 dissolved in dry DMSO (a), DMSO with 10% (v/v) H_2) (b), or 30% (v/v) H_2O (c) at room temperature. The scale bar in white shows the dimension in micrometers. The palette below the images illustrates the degree of fluorescence. Reprinted from M. F. Lee, S. K. Kong, K. P. Fung, C. P. Lui, and C. Y. Lee, *Biol. Signals* **5,** 291 (1996), with permission from Karger, Basel.

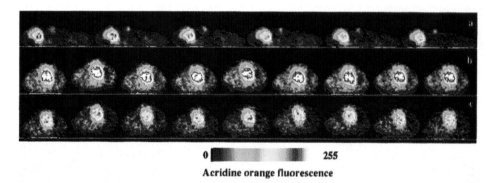

Acridine orange fluorescence

FIG. 4. U87MG cells were fixed and stained with acridine orange (10 ng/ml). Sequential images along the Z axis (vertical scannings) of the cells were obtained at 24° (a), 37° (b), or 43° (c) at 2.5-sec intervals. The palette below the images illustrates the degree of fluorescence. Note the changes in the position of the growth substratum. Reprinted from M. F. Lee, S. K. Kong, K. P. Fung, C. P. Lui, and C. Y. Lee, *Biol. Signals* **5,** 291 (1996), with permission from Karger, Basel.

difficulties in data acquisition and image analysis, e.g., background subtraction. Most importantly, the artifact hot spots might be misinterpreted and mask genuine local responses.

To obtain dry DMSO, molecular sieves (Sigma, St. Louis, MO) were first prepared by leaving the particles in an oven at 150° overnight. DMSO from the supplier was then incubated with the dry molecular sieve in an air-tight bottle for at least 2 hr. These procedures can remove the water in wet DMSO. However, the addition of dry molecular sieves back to the stock of indicator contaminated with H_2O did not improve the efficiency of dye loading (data not shown).

Effect of Temperature on Focus

Temperature is one of the most important physical parameters that control physiological processes. To mimic the *in vivo* system, the temperature for experiments is usually set to the body temperature of the host animal, which is normally higher than the room temperature. In single-cell studies, cells are seeded on coverslips that are mounted in a chamber or a holder with physiological buffer.[11] One approach to controlling temperature is to heat up the holder to the desired temperature using a commercially available temperature control unit. Unfortunately, many temperature control units usually rely on a negative feedback system to keep the temperature at a narrow range near the desired temperature rather than keeping the temperature precisely and constantly at a fixed point. In the case of conventional confocal microscopic studies, these temperature fluctuations affect the refractive index of the immersion oil between the objectives and the coverglass. As a result of the fluctuations of the refractive index of the immersion oil, different optical sections would have come into focus, even if the microscope is trained on a fixed x–y plane. For example, with the use of a low fluorescence immersion oil [refractive index at 23° ($\eta^{23°}$) = 1.515 and the change of refractive index/° ($\Delta \eta$) = 0.385 × 10^{-3}/°], a temperature change of 1° could cause a focal plane shift of 0.25 μm.[12] In relation to cellular dimension, this introduces a very significant error in confocal studies. In Fig. 4 (color plate) U87MG cells were fixed in methanol/acetic acid (3 : 1, v/v) and stained with acridine orange, a marker for DNA and RNA, for fluorescence determination. Figure 4a shows that at room temperature (24°), there were no periodic fluctuations in fluorescence against time, but at 37° (Fig. 4b) and 43° (Fig. 4c), oscillations in terms of fluorescence and the position of growth substratum were clearly observed.

---

[11] C. Ince, J. T. van Dissel, and M. M. Diesselhof, *Pflug. Arch.* **403,** 240 (1985).
[12] M. F. Lee, S. K. Kong, K. P. Fung, C. P. Lui, and C. Y. Lee, *Biol. Signals* **5,** 291 (1996).

In view of the serious error that can result from even very small temperature fluctuations reported here, it is imperative that the temperature be controlled rigidly. In confocal microscopy work, a delicate instrument, which provides an accurate temperature control not only in the bath holder but also in the coverslip and microscopic objectives (Bioptechs, Butler, PA), is available commercially to meet this purpose. However, one should check with the supplier whether the objectives used could be heated. Alternatively, one can perform experiments in a thermostatically controlled laboratory at a fixed temperature to avoid temperature leak, between the cell holder and the microscope and frequent temperature fluctuations.

Acknowledgment

Work described in this article was supported by grants from the Faculty of Science, the Chinese University of Hong Kong.

Section II

General Techniques

[3] Equipment for Mass Storage and Processing of Data

By GUY COX

Confocal microscopes do not produce images that are patterns of light: their "images" are numerical data in the memory of a computer. Such an image has many advantages; it can be transmitted and copied simply, with 100% fidelity, but its longevity depends entirely on the medium on which it is stored. Initially, it is probably stored in the random access memory (RAM) of the computer, a very insecure location from which it will be lost if the computer is turned off or if that area of memory is overwritten. To store that image with all its information intact, we must write it in digital form on some less transient medium; a reproduction on paper or film of the image encoded by these data, however good, cannot contain all the information of the original.

An image represented by a series of numbers in a computer is not directly accessible to human senses. When the image is planar—a series of values or pixels representing intensities in a two-dimensional array—it is normal to display it as corresponding intensities on a monitor. Even this may be less than trivial. Most confocal microscopes, or CCD cameras, can store 12- or 16-bit images, capable of storing 4096 or 65,536 intensity levels, respectively. Monitors or, more accurately, the display boards driving them typically can display only 256 levels in each color channel. The human eye, however, can distinguish no more than 64 different intensity levels. Much information can be lost in the transition from numerical data to human perception.

A confocal microscope does not produce a planar image, it produces a three-dimensional array of sample values or voxels, which is even harder to present to the eye in a meaningful form. Much of the image processing associated with confocal microscopy is devoted to overcoming the problem; it is visualization software, dedicated to making electronic data comprehensible to the human eye and brain. A further step is to take advantage of the numerical nature of the information to extract measurements about our sample, and here we are entering the realm of image analysis or morphometry rather than image processing.

This article reviews some of the available solutions to these two problems. This is a rapidly changing area, and new alternatives will doubtless become available almost as soon as this is printed. Computers become more powerful daily, and tasks that require supercomputers one year are simple on personal computers the next. As well as assessing currently available technologies, therefore, this article will try to provide enough background information to enable users to assess what the future has to offer in an informed and rational way.

Mass Storage

Confocal images demand substantial storage capacity. The smallest image one is likely to acquire is a single-plane, single-channel image measuring 512×512 pixels. If it is saved as an 8-bit image (1 byte per pixel), each point in the image can have one of 256 possible gray levels. This image will require one-quarter of a megabyte (Mbyte) to store, with an additional overhead to store basic information about the picture, in a "header" at the start of the file, at the end, or even in a separate file. Confocal microscopes will capture much larger images than this, and the image may contain more than one channel. A three-channel, 1024×1024 pixel image would be typical of most current systems, and most also offer the capability of storing 16 bits per pixel (65,536 gray levels). We are therefore storing 3 Mbytes per plane at 8-bit precision or 6 Mbytes per plane at 16-bit. A three-dimensional image data set could contain 50 planes or more, thus possibly requiring 300 Mbytes of storage space. At the time of this writing, personal computers typically have 4-gigabyte (Gbyte) hard disks, although the entry level may well be twice that by the time this appears in print. However, even a 30-Gbyte disk will only hold about 100 such images. Permanent image storage is always going to require some form of removable storage media.

Removable Storage Media

The key requirement for image storage is generally permanence. Silver-based black and white film negatives provide an archival medium proved to last longer than a century, whereas dye-based color photographic media are less durable but are still demonstrably capable of surviving for decades away from daylight. One would hope for at least equal permanence from digital storage, but only one of the storage media available to a present-day computer user has a history longer than 20 years. That medium is magnetic tape, and the experience from its relatively long history is that its longevity cannot match photographic film. Estimates of durability of other types of media are based on accelerated aging tests; at best these can only be an indication.

The permanence of different media types is only part of the story, however, as a digital record is of no use without devices to read it. Hence it is important to assess the likelihood of the survival of appropriate hardware in a functional form. Popularity can be relevant; popular hardware is more likely to survive than something that is sold in small quantities, regardless of technical merit. Popularity brings the additional benefit that images saved on a popular medium are readable at home and in other laboratories. Eight-inch diskettes were the standard for laboratory computers in the early 1980s, but the author's laboratory now no longer has a single drive in working order; any data from old diskettes that has not been transcribed to more modern media will, at the very least, be inconvenient to recover. Even $5\frac{1}{4}$-inch floppy drives are becoming scarce, and some varieties, such as those written by Apple II computers, are effectively unreadable. For years Unix and OS-9 systems used $\frac{1}{4}$-inch tape in DC6000 format cassettes for backup and software distribution,[1] but the author's laboratory now has just one drive left to read these tapes, on an old computer saved from the scrap heap for just this reason. These examples are all of reasonably popular media types, but the need for high data capacities has often driven the confocal microscope user to "high-end" devices that have a much higher price and smaller user base. Various optical disk systems with a $5\frac{1}{4}$-inch form factor were popular as recently as the mid 1990s,[1] but, as many plaintive requests on the Internet show, drives to read some of these disks are now scarce. This factor, as well as the cost per megabyte, needs to be taken into account when choosing a storage system.

Sequential Devices

Magnetic tapes are referred to as sequential devices because they can be read from, and written to, only by moving sequentially from one end. Although they are erasable and rewritable, this means that one cannot generally replace a single file or group of files. New data can only be added to the end of the tape; if something is to be erased, the entire tape, or at least one writing session, must be erased and rewritten. Costwise, tape is one of the least expensive bulk storage media available, but as an image storage system it suffers from the time taken to locate and recover any one file. Also, although it is rewritable, it will not stand an infinite number of uses. The tape surface has a much harder life than the surface of a disk: it comes into direct contact with the recording heads and capstans and is coiled and uncoiled each time. Even reading files repeatedly wears the tape, yet if tapes are left untouched for long periods, a different problem is encountered: "print through," where the magnetic signal on one coil of

---

[1] G. Cox, in "Handbook of Biological Confocal Microscopy" (James Pawley, ed.), p. 535. Plenum Press, New York, 1995.

tape affects the layer above or below it. The long-term archival potential of tape must be regarded as dubious.

Tape comes into its own as a backup medium, where a whole file system is recorded at one time, and can therefore be recovered in the event of a system crash. Modern backup software provides powerful tools for managing backups, including user-friendly interfaces for finding individual files to restore (invaluable in a confocal facility where files always seem to get deleted accidentally) and incremental backups so that only files that have changed are added to the tape. There is a bewildering variety of tape drives on the market, based on a range of technologies, although two formats are clear market leaders at present.

Data cassette drives, based on audio cassette standards, are a format dating back to the very early days of small computers, and current examples are all based on a TEAC mechanism. Capacities ranging from 150 to 600 Mbyte, while hugely larger than early examples offered, are too small to give them a place in the confocal laboratory.

Quarter-inch tape drives are still on the market, having been for many years the "standard" tape format, especially on Unix-based systems. They use static recording heads, which limits both their speed and the data density on the tape. Two cassette formats are used: DC2000 or QIC (aimed mainly at the personal computer market) and DC6000 (the Unix workstation format). DC2000 drives can store up to 2 Gbyte, whereas DC6000 will store up to 5 Gbyte. The cost per megabyte is unlikely to match that of more modern systems, and the main reason for buying a quarter-inch system would be backward compatibility with a collection of older tapes. However, because many of the new versions are not fully backward compatible, make sure that the new drive will read your existing tapes.

The principal current tape backup systems all use rotating heads to increase the amount of data on a finite length of tape. Because tape speeds are also a limitation, this also increases the speed of reading and writing. They are small cassettes, similar in size to audio cassettes, and the major contenders are based either on the 4-mm digital audio tape (DAT) format or on the 8-mm video-8 system. DAT tapes come in three standard types: DDS 1 (2 Gbyte), DDS2 (4 Gbyte), and DDS3 (12 Gbyte). Because the format is quite standard, drives from one manufacturer will usually read tapes written by other drives. Despite the name, audio tapes are not suitable; the tapes are special data quality versions. Because of their low media cost and reasonable purchase price, DAT tapes have become one of the most popular backup systems.

Eight-millimeter systems, derived from video-8 videotape, have a substantially higher purchase price but a commensurately higher capacity, storing up to 24 Gbyte of data; media costs are very reasonable and, in fact, very similar to DAT at less than 1c per megabyte. A proprietary (and incompatible) development of the 8-mm system is Sony's 25 Gbyte AIC (Advanced Intelligent Tape), which gains its "intelligence" from a 16-Kbyte chip on the cassette itself, which stores the table of contents. Very high data transfer rates are claimed and indeed achieved, although as always in this field, independent tests give much lower figures than manufacturers' claims!

One of the latest tape drive developments has abandoned the helical scan system and reverted to linear storage. Digital linear tapes (DLT) offer capacities of up to 20 Gbyte with sustained transfer up to 100 Mbyte/sec. This is again a proprietary system based on mechanisms made by Quantum. The performance of these systems is critically dependent on the interface; if the data stream cannot keep pace with the speed of the tape, the tape must repeatedly stop and reposition, which degrades performance dramatically.

Many manufacturers quote "compressed" capacities for their tape drives based on the compression ratio they expect to achieve with their archiving software, usually a notional 2× factor. However, such compression ratios cannot be achieved with image files, and all figures quoted here refer to the actual uncompressed capacity of the system. When evaluating competing systems, it is important to compare actual uncompressed storage capacities, as this is much closer to the achievable figure with microscope images.

Random Access Devices

Random access devices are disks of one type or another and, as their name implies, can move their reading heads to any part of the disk surface to read a particular file without reference to whatever else is stored on the disk. In most cases, this is also true when recording, but CD-based recording systems must record sequentially and are only random access on playback. There are both erasable and nonerasable random access devices; in the past this meant also a distinction between magnetic and optical media, but that line is now distinctly blurred.

The cost per megabyte of storing data on removable discs varies enormously (Table I), but the lower cost options usually offer considerable sacrifices in speed or convenience. For the small computer user, purchase cost is often also an important consideration, but this is unlikely to be a factor in the confocal laboratory. The cost of even an expensive drive is a small part of the price of a confocal microscope and the amount of data written ensures that low cost per megabyte is always the more important factor on the bottom line.

TABLE I
Costs and Characteristics of Various Bulk Storage Media[a]

Media	Capacity	Sustained speed	Cost per megabyte
Tape			
Teac data cassette	155–600 Mb		5–12c
DC2000 QIC	2 Gb		2c
DC6000	1–5 Gb		1–5c
DAT	1.2–12 Gb		0.5c
8 mm	1.2–24 Gb	1.5 Mb/sec	0.3–0.9c
DLT	10–20 Gb	1.6 Mb/sec	0.4c
Disk			
Superdisk	120 Mb	7 Mb/sec	14c
Zip	100 Mb	0.7 Mb/sec	15c
Jaz	1–2 Gb	2.3 Mb/sec	6–12c
Syquest	0.25–1.5 Gb	0.7–1.7 Mb/sec	6–24c
MO	0.23–2.6 Gb	0.9 Mb/sec	10c
PD	650 Mb		9c
CD-R	650 Mb	0.6 Mb/sec (4×)	0.3c

[a] Speeds quoted are approximate only and may not match real-world results; they are, however, taken from published tests, not manufacturers' claims. The testers noted[2] that the very high speeds obtained by the relatively low-capacity Superdisks were probably affected by cacheing. Prices are only very approximate; they are quoted in U.S. cents but are based on figures from the United States,[3] the United Kingdom,[2] Australia (prices quoted to author); in all cases, prices free of any taxes are given. At the time of compilation, an approximate conversion would be 1p (U.K.) = 2c (U.S.) = 3c (Australia). Where a range of prices is given, the highest price is always for the lowest capacity media. "Consumer" devices such as CD-R and Zip disks are likely to be subject to more discounting than "professional" media.

Because confocal images are always large, the speed of the device can be an important factor, particularly its sustained transfer rate. Speeds of individual systems are considered later, but one also needs to keep in mind the speed of the connection to the computer. Options here are typically SCSI (the fastest and available on a wide range of platforms), IDE (the traditional PC hard-disk interface and therefore fast and always available without adding a special SCSI board), and parallel port. The latter uses a printer port and is therefore very convenient if a drive is to be shared between different computers, but is the slowest option by far. PC card (formerly called PCMCIA) interfaces are another option: they are fast and easily interchanged but are designed for notebook computers, which are not often connected to confocal microscopes.

Floppy Diskettes and "Super Floppies"

The oldest and simplest of removable media, rewritable, random-access systems is the humble floppy diskette. Standard floppies have a capacity of a meager 1.4 Mbyte so are of little use for the routine storage of confocal images but still have a role to play in data transport because just about every computer has a $3\frac{1}{2}$-inch floppy drive. "Super floppies" with greater capacity have a long history; a 2.88-Mbyte version was announced as the "new standard" some years back but failed to make any impression on the marketplace.

Much more substantial increases in capacity were obtained from "floptical" technology. These diskettes record data magnetically, just as on a conventional floppy disk, but the disk also has an optical track encoded on it. The drive has a laser to read the optical track, which is used to guide the magnetic heads with much greater precision than a normal floppy drive can achieve. Typically the drives have been able also to read and write standard $3\frac{1}{2}$-inch floppies, but not necessarily other flavors of "flopticals." The current incarnation of this technology is the LS-120 Superdrive, available from Imation, Nexus, and possibly other suppliers.[2,3] It holds 120 Mbyte per disk, and the drives remain compatible with conventional floppies. This means that an IDE-interface drive could simply be installed in place of the standard floppy drive with minimal inconvenience.

Iomega made their name with a "Bernouilli disk" technology that "flew" a flexible disk on a cushion of air, and in the 1980s, many confocal microscope laboratories depended on these systems for image storage. To the best of the author's knowledge, these drives are no longer made, but their descendant is the 100-Mbyte Zip drive, which, thanks to a low purchase cost and reasonably priced media, has become hugely popular. They are available with SCSI, IDE, or parallel interfaces; the parallel interface limits performance severely but is convenient for domestic use because drives can be transferred easily between systems. Zip disks are neither as fast nor as high in capacity as their closest rivals, but their cost per megabyte is comparable, and because the drives are inexpensive and versatile, they are very widely used. This feature alone makes them well worth considering for the confocal laboratory. It is very probable that microscope users will have Zip drives both in their offices and at home so data interchange becomes extremely easy. Also, the sheer size of the installed base makes it probable that compatible drives will continue to be available for many years to come.

Hard Disk Cartridges

Removable cartridges based on conventional hard disk (Winchester) technology have been available for many years, although early versions earned themselves a reputation for fragility. The removable cassette con-

---

[2] R. Gann, *Person. Comp. World,* August, 116 (1998).
[3] Dantz software (Retrospect Remote) home page www.dantz.com.

tains the platter and the remainder of the mechanism is fixed. SyQuest is the pioneer of this technology and still holds a major position in the marketplace with their EZFlyer (250 Mbyte), Sparq (1 Gbyte), and SyJet (1.5 Gbyte) drives. Iomega is also competitive with their Jaz 1 and 2 Gbyte drives, and various other manufacturers offer similar products. These are usually reasonably fast, and most brands come with a range of alternative interfaces; only the parallel port is likely to affect the overall speed as the others exceed the speed of media. Cost per megabyte is very reasonable (down to as little as 6c per megabyte) in the larger capacities but is considerably higher in the smaller sizes. (Iomega Jaz disks are the same price in 1 and 2 Gbyte capacities, for example). Many confocal and other imaging laboratories find the larger capacity (1 Gbyte plus) versions of these drives invaluable, but they have made little headway in the domestic market as a consequence of the relatively high up-front cost of drives and media. This means that the long-term availability of drives might become a problem, although it must be said that the current generation of these drives have a vastly wider availability than their predecessors.

Magneto-optical Media

Magneto-optical (MO) drives are rewritable media that are based around a hybrid of Winchester and CD technology. They use a laser to heat a reflective, magnetic coating to above its Curie point; the temperature at which a metal loses any existing magnetism. A magnetic head fixes the orientation of the coating as it cools. Different magnetic polarities reflect polarized light differently so that the laser (on a lower setting) can then read the data. Writing data to an MO drive is a relatively slow, two-pass process; first an erase pass to set the entire area to a uniform polarity, then a write pass to set the appropriate bits to the opposite polarity. Reading is much faster. This is probably not too much of a disadvantage for confocal images, as files are likely to be written once and read many times. Although they are in one sense magnetic media, they should not be affected by external magnetic fields (so long as the temperature is below 150°), which improves their archival stability. Manufacturers quote archival lives of 30 years and 10 million read–write cycles. Nevertheless, there have been reports of MO disks becoming unreadable. MO disks in both $5\frac{1}{4}$- and $3\frac{1}{2}$-inch form factors are still on the market, but are no longer as attractive an option as they were a few years ago. They used to offer far higher capacities than removable magnetic media and consequently a lower cost per Mbyte. Now magnetic media have caught up and costs are comparable. Three and a half-inch drives have always offered a degree of standardization, which has made them the more popular option for most users. Their capacity now reaches up to 640 Mbytes and drives will generally read the older 128 and 230 Mbyte disks. Five and a quarter-inch drives have never been standardized so closely, and backward compatibility can generally only be assured by sticking to one manufacturer. The trade-off has been much higher capacities, currently up to 2.6 Gbyte.

A related technology to MO is PD or phase change, a proprietary technology offered by Panasonic. The disk has a coating that can be changed by the laser beam between crystalline and amorphous states. These reflect light differently, allowing data on the disk to be read. The drives will also read conventional CD-ROM disks, but will not write CD-R disks. The capacity of the PD or phase-change disk is 650 Mbyte, which offers a reasonable storage cost of around 9c per Mbyte. The advantage of this system is that it can simply replace the CD drive of the host computer, but the PD disks cannot be read by a conventional CD-ROM drive.

WORM Drives

Write Once Read Many (WORM) drives are totally optical drives, using a technology similar to, but not compatible with, CD-R drives. A laser modifies the coated surface of the disk to record data. To read the information back the laser is operated at a lower power, and the changes in reflectivity provide the signal. Unlike audio compact disks, the disk itself is permanently contained in a cartridge and many designs are able to record on both sides of the disk. Some drives are compatible with the same manufacturer's MO disks, but not with any other media. The drives are typically rather slow and have suffered from virtually total incompatibility between systems from different manufacturers. Once common in confocal laboratories, this technology must now be regarded as obsolescent. The newer technology of packet CD does everything a WORM drive will do, at an equally slow speed, but at a fraction of the cost. Sony, one of the manufacturers, suggests that users of their MO drives might wish to sometimes make nonerasable archive copies of important files using a WORM disk in the same drive, which is probably the only current use for this technology.

CD-ROM Systems

CD-ROM disks, as read-only media, have been available for many years. CD-ROM readers are virtually universal on all types of computers and the format is standardized; because the disks are physically identical to audio compact disks, compatibility is not often an issue. Most software is now distributed on CD-ROM. Recordable CDs (CD-R) are newer but have still been on the market for several years. Initially the drives were expensive, temperamental, and slow, but now they are inexpensive, reasonably foolproof, and faster, although still slow relative to other media. Part of the reason for the improvement is the increase in processor and bus speeds; in the past, computers often had difficulty keeping pace with the continuous data flow required to write a CD, especially at faster rates than "single speed" (the speed of an audio CD). "Single speed" is a rather pedestrian 150 Kbyte/sec, but reading speeds are now typically 24 times that and writing speeds of at least 600 Kbyte/sec are commonplace.

The disks have a dye layer (which is bleached by the writing laser) in front of a reflective layer of evaporated metal. The metal in turn is protected only by a thin layer of laquer; it is not advisable to write on this with spirit-based pens or to affix adhesive labels, either of which could damage the protection offered by the laquer. Archival stability for up to a century is claimed by the makers, and the wide-spread availability of CD readers should ensure that hardware will still be extant well into the next century. The nature of the metal used has also been a topic of some discussion, as gold, used in the more expensive blank disks, is claimed (plausibly enough) to be less likely to suffer from corrosion than the aluminum of less expensive disks. Originally, CDs could only be written in a single session—the entire disk had to be written at once—but multisession systems have been common for many years. Because there is substantial overhead in writing multisession CDs, many users still prefer to accumulate a full 650 Mbyte and write their disks in a single session. Either way the cost per megabyte is trivial, less than 1c per megabyte; CD-R is the only random access medium to rival tape in cost, although of course, unlike tape, it is not rewritable.

The inconvenience of the "streaming" write process has been overcome to some extent by a new format known as "packet CD." This enables a CD-R to emulate a hard disk and data to be written whenever it comes to hand. Packet CD is a software standard and most new CD-R drives can support it once appropriate drivers have been installed. However, a packet CD is not readable on a conventional CD-ROM drive until it has been "fixed" by the packet software, and thereafter it cannot be altered.

In parallel with the availability of these new formats is the emergence of CDs that can be erased and rewritten (CD-RW). These currently suffer from the twin disadvantages that they cannot be read by conventional CD-ROM drives and that the erasing process is slow. However, the drives can also write conventional CD-R media and many laboratories may find the versatility worth having. One strategy might be to accumulate data to packet CD-RW media and, when a disk is full, transfer it to conventional CD-R. The CD-RW can then be reformatted and used for a fresh stock of images.

Digital Video Disk

Digital video disks (DVD) are a development of compact disk technology that makes it possible to store an entire movie on a single digital disk. Because the data are digital, it is clear that the technology is immediately adaptable to the computer domain. DVD-ROM drives are already available and can store 4.7 Gbyte on one side of a DVD disk. The disks can be double sided and, because the optics that read the disk are confocal, they can carry two separate layers on each side, bringing the total capacity to a theoretical 17 Gbyte. These are, of course, read-only devices, but DVD-R (write-once DVD disks) are now becoming available and promise to offer enormous potential for image storage. DVD-R disks are compatible with DVD-ROM drives. However, when moving into the realm of rewritable media, the situation becomes more complex. Two standards are battling for acceptance. DVD-RAM offers 2.6 Gbyte per side, whereas DVD-RW offers 3 Gbyte per side. The systems are not compatible with each other and neither is compatible with DVD-ROM. At this stage, both are single-layer systems.

It is clear that DVD offers the potential of huge capacities and, at least if the base technology becomes popular for its prime purpose of video, the likelihood of low costs. However, at the time of this writing it is too early to know if this promise will be fulfilled.

Data Storage Conclusions

Tape drives have a place primarily as backup devices. In this role they provide fast and cost-effective data storage. If a laboratory is buying a new system, it should be either DAT, if the systems to be backed up run only to single-figure or low double-figure gigabytes, or 8 mm if the systems are large enough to warrant the extra up-front costs. Newer, incompatible systems such as AIT and DLT offer a clear performance advantage (especially on systems with high-speed bus transfer) but at the price of being locked into proprietary systems. Older system such as DC6000 are only of interest if backward compatibility is important.

For long-term archival storage of confocal images, CD-R is a clear winner because it offers very low cost, good anticipated archival integrity, and universal readability. It is possible that DVD-R may take over this mantle in the future. However, optical technology is too inefficient in its writing arrangements to be the sole image storage system; for effective management of a confocal laboratory, a fast, rewritable storage medium is also needed and Zip, Jaz, or Syquest disks can fill this need. The author's choice is Zip because of its ubiquity; most laboratories and homes have a

Zip drive. However, larger-capacity media undeniably offer more bangs (or bytes) for the buck.

Image Processing

Data Compression

Having looked at the considerable data storage problem presented by confocal images, a natural question to ask is "Can we reduce the size of the problem? Is it possible to compress the image data, to make it smaller?" There are many general-purpose data compression systems around, but most of these work on some variation or combination of three well-known algorithms. Run-length encoding (RLE) looks for sequences of identical values and replaces them with one copy of the value and a multiplier. Lempel-Ziv-Welch (LZW) and Huffman encoding look for repeated sequences, assign each a token, and then replace each occurrence by its token. None of these works well with real images so that if you compress confocal images using one of the popular systems such as PKzip or WinZip you will find that gains are very modest—at best the files will shrink by 10–20%, which does little to alleviate the problem.

Compression techniques designed for real-world images are always "lossy": part of the data is discarded. The philosophy behind this approach is that not all of the data in an image are relevant to the impression it makes on the human eye. In terms of a confocal image, for example, there will always be noise in the background caused by statistical fluctuations in the numbers of photons arriving at the detector and random release of electrons in the photomultiplier and amplifier. The eye will typically not even see these and certainly will not be perturbed if they are missing. Using this approach, very large file compressions can be achieved with losses that are undetectable to the eye.

The almost universal compression system in use today is the Joint Photographic Experts' Group (JPEG) compression protocol,[4,5] which is supported by many painting and image manipulation programs. This breaks the image into blocks of 8 × 8 pixels, each of which is then processed through a discrete cosine transform. This is similar to a Fourier transform, but much faster to implement, and gives an 8 × 8 array in frequency space. The frequency components with the lowest information content are then eliminated, discarding high-frequency information (fine detail) preferen-

[4] L. F. Anson, *Byte* **19**(11), 195 (1993).
[5] W. B. Pennebaker and J. Mitchell, "JPEG Still Image Compression Standard." Van Nostrand Rheinhold, New York, 1993.

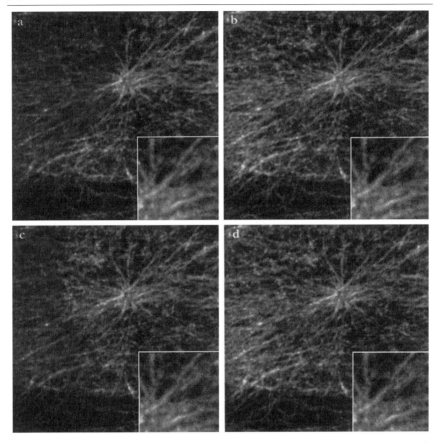

FIG. 1. Effects of image compression on a confocal fluorescence image of a rat aortic smooth muscle cell labeled with FITC antitubulin. (a) The original, uncompressed image, (b) JPEG compressed to 50% of the original size, (c) JPEG compressed to 20% of the original size, and (d) compressed to 10% of the original size. Insets show a small part of the image at a 2× magnification.

tially, and the remaining components are stored (using Huffman encoding) in the compressed image. The amount to be discarded in frequency space can be specified, which gives the user control over the trade-off between image quality and degree of compression. Typically monochrome images can be compressed down to one-fifth or less of their original size with no visible loss of quality[6] (Fig. 1). Color images can be compressed further,

[6] G. B. Avinash, "Image Compression and Data Integrity in Confocal Microscopy," p. 206. Proc. 51st Annual Meeting, Microscopical Society of America, 1993.

as luminance (brightness) and chrominance (color) are treated separately, and the eye can tolerate a greater loss of information in the chrominance signal. Compression and decompression are similar operations and require similar amounts of computer time. A consequence of the separation of chrominance and luminance information is that JPEG can only operate on gray scale or true color (RGB) images. If (as is common in confocal microscopy) a false-color palette is added to a gray-scale image, it would have to be converted to a full color image, tripling its size, before compression could start.

Rivals to JPEG include Portable Net Graphics (PNG), a format designed to compress images with particular features designed to aid decompression over the Internet. (The JPEG standard promptly added similar features, termed Progressive JPG.) PNG has not to date won widespread acceptance, largely because, unlike JPEG, full details of the compression technique have not been made freely available. It also gives poorer compression performance than JPG. Fractal compression, a proprietary technique developed by Iterative Systems Inc.,[4,7] takes a totally different approach: it creates mathematical expressions, which, when iterated, recreate the original image. It can give spectacular levels of compression. Unlike JPEG compression, creating the compressed image is a very time-consuming process but decompression is very quick. This has made it most useful for such items as CD-ROM encyclopedias, where the images are compressed once for mass distribution and frequent decompression. It does not seem to have gained much acceptance in more general markets.

The effectiveness of, and the sacrifices in, JPEG compression of a confocal image is explored in Fig. 1. The original image, of microtubules in a smooth muscle cell, was 512 × 512 pixels, so image data occupied 262,144 bytes. As a TIFF file, the file size was 263,866 bytes, with the increase being the size of the TIFF header. A compressed TIFF file using lossless LZW encoding reduced the file size to 225,844 bytes, 85% of the original. With RLE encoding, the file size actually increased to 275,848 bytes. Compressing to half the original size with JPEG, the loss in quality is barely noticeable (Fig. 1b). At 20% of the original size (Fig. 1c) the difference is becoming apparent but is not obtrusive, and even reducing the file to only 10% of the original size (Fig. 1d) gives an image that is at least usable.

JPEG compression is a viable approach to reducing the size of confocal files, giving high compression ratios before the loss of quality becomes visible. In some cases, e.g., where it is necessary to send confocal images on floppy diskettes, even compression to a level where some loss in quality is visible may be acceptable. Confocal manufacturers are now

[7] M. F. Barnsley and L. P. Hurd, "Fractal Image Compression." A. K. Peters, Wellelsey, 1993.

recognizing this. Noran offers hardware-assisted, high-speed JPEG compression with their Odyssey video-rate confocal microscope, which uses a Silicon Graphics workstation as its host computer. Bio-Rad has introduced JPEG compression in the latest releases of their companion software Confocal Assistant.

Three-Dimensional Rendering

Confocal microscopes collect images in three dimensions, which should make them appear more realistic than the two-dimensional images generated by conventional fluorescence microscopes. Achieving that goal is far from easy. We believe that we see the world in three dimensions, but this is not really so; we *interpret* the world in three dimensions. The difference in parallax between the views seen by our two eyes is one factor in this, and it can be the major factor, but in everyday life it is but one among many. Experiments have shown that people can tell one object is in front of another even when the parallax is below the resolution of the eye. Other cues we use are motion parallax when we move our head, convergence and focus of our eyes, perspective, concealment of one object by another, and our knowledge of the size and shape of everyday things. All together they are extremely effective; you need only watch a game of tennis to realize just how good our spatial perception is.

The last item on this list, our intuitive knowledge of the size and shape of everyday objects, is useless in the microscopic world. In the everyday world we are also not used to looking at totally transparent objects. In the microscope, when looking at wide-field or confocal fluorescence images, or even at a histological section, everything is transparent. Another of our essential visual cues, objects obscuring their background, is therefore lost.

It is quite straightforward to generate binocular parallax, or at least an approximation to it, from our data set. Stereoscopy involves producing one view for each eye from our series of slices. In a stereo pair, depth is converted into displacement between two images. Distant objects appear at the same place in both pictures, whereas near objects appear displaced toward the center line between the pictures. There is a limit to how much displacement the eye can tolerate; in a real specimen we converge and focus our eyes differently when looking at near and far objects. In a stereo pair picture, everything is actually at the same plane, so we cannot do this. If a data set extends for some considerable depth into a sample, we may therefore have to restrict the displacement artificially, i.e., foreshorten the depth, for the sake of comfortable viewing. The resulting stereoscopic image will still suffer from the disadvantage that all objects are transparent and the lack of motion parallax. If we move our head when looking at a stereo pair the entire image appears to swing. Nevertheless it can be a dramatically effective way of presenting confocal images, and confocal microscopes all offer stereoscopic views of confocal data sets without requiring additional software.

Stereo pairs are generally produced in confocal microscopy by projecting the series of images onto a plane, displacing each plane by one pixel from its predecessor. This is computationally simple but has the disadvantage that the perceived depth is essentially arbitrary, a function of pixel size and section thickness. An alternative approach is to rotate the data set to generate the two views, and if we were generating stereo pairs as part of a full three-dimensional reconstruction, this tactic would usually be employed. Neither method is geometrically ideal but in practice both are extremely effective.

The resulting stereo pair can be presented on screen as an "anaglyph," with one image in red and the other in green or cyan, and viewed through appropriately colored glasses. This means that the original images must be monochrome-single channel without false-color palettes. This is rather limiting, and special viewers are available for viewing stereo pairs placed side by side on the computer screen, thereby avoiding these limitations. A more sophisticated approach requires the viewer to wear special glasses containing liquid-crystal shutters linked to the computer by an infrared receptor. The left and right images are alternated rapidly on the monitor while the glasses switch between the two eyes in synchronism. This is a catalogue option on most Silicon Graphics computer systems and is available as a third-party addition to some others, but it is an expensive item.

To bring in more of the visual cues into the perception of three-dimensional data sets we need to produce a series of computer-generated views of our specimen, mimicking the views we would get if we looked all around it as a real object. This is computer three-dimensional reconstruction or volume rendering: the application of formulas or algorithms that can be implemented on a computer to create three-dimensional projections from data sets. If several projections are created from different viewpoints, these can be played in quick succession to form rotations or animations. This brings in the element of motion parallax, which often gives us a clearer understanding of the three-dimensional structure of the sample.

Rendering Techniques

There are several approaches to rendering volume data sets on a computer. In general, there is always a trade-off between speed and preservation of accuracy and detail. Because the amount of data that must be manipulated is very large, both processor speed and memory are important.

Surface Extraction

The original confocal data sets consist of a set or grid of xyz coordinates of points in space with an intensity value for each point. We can reduce this to a simpler case by deciding which intensity values actually correspond to the "object" or objects we wish to render and then extracting a reduced data set containing just the x, y, and z coordinates of our objects. By geometric rendering, fitting tiny polygons to the points defining the surface of the "object," we can produce a view of the surfaces of the objects we have extracted. This approach is fast, but we can only see the surface of our object, as internal detail is lost. It can give an excellent view of a specimen if it is something that can be represented just by a surface and presents it to the eye in a way that is easy to comprehend as transparency is no longer an issue. Directional "lighting" effects are often added to enhance the "realism."

Simple Projections

In many cases, especially in cell biology, we are concerned with more complex relationships than can be represented with surface extraction. In this case our reconstructions must make use of all the information in the original data set. The simplest approach is to project the three-dimensional array, from a given angle, onto a plane. If we take the average of all the intensity values (voxels) along each line, the end result will be an average brightness projection (Fig. 2a). Alternatively, if we find and record the maximum brightness along each line rather than taking the average, we will form a maximum brightness projection (Fig. 2b). Which is the more effective will depend on the sample, but in many samples of interest to the biologist the end results are surprisingly similar.

With these projections the computation is simplified because we do not need to calculate the z coordinates as we rotate the data set; it is enough to know where each point is in x and y. Typically we would keep either x or y fixed and rotate about the other axis so that we are tracking translations in one direction only. This approach has the advantage that all internal detail contributes to the final image, but suffers from the major problem that there is no distinction between front and back: both projections are identical. This can be alleviated if we create a stereo pair rather than a single image at each viewpoint. Nevertheless, if the image is complex,

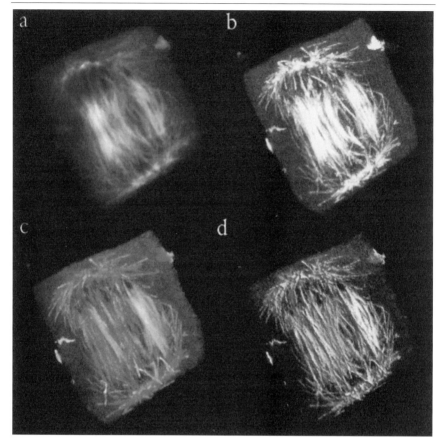

FIG. 2. Different rendering methods applied to a confocal data set of a dividing wheat root cell. The specimen is stained with FITC antitubulin and shows the spindle microtubules at late anaphase. The data set is $256 \times 256 \times 88$ (extracted from a larger field) and was acquired on a Bio-Rad Micro-Radiance system. In each case the projection is a top view of data looking straight down the stack of sections with no rotation. All projections were scaled to a uniform contrast range. (a) An average brightness projection and (b) a maximum brightness projection, both created with Confocal Assistant from Bio-Rad. The average brightness projection gives a very good impression of the overall distribution of microtubules but with very little impression of depth; in fact, a view from the opposite direction would be identical. The maximum brightness projection gives much more prominence to individual microtubules and small groups. However, noise is much more apparent because there has been no averaging. (c) A weighted reconstruction created in Voxel View; the tonal range gives a much clearer indication of what is near the top and what is further down the stack. (d) Also created in Voxel View, with a lighting model applied. Individual microtubules (or bundles of microtubules) are much easier to distinguish thanks to the "shadows" cast.

the problem of visualizing totally transparent objects may make it hard to comprehend.

Weighted Projection: Alpha Blending

In this technique a projection is again created by summing all the voxels along each line of sight, but with the added refinement that the voxels in front are made to contribute more than those behind. In order to do this the z position of each voxel must also be calculated at each rotation angle, which makes it the most computationally demanding technique. In this way, at each viewpoint the front is distinct from the back, yet internal detail can still be seen (Fig. 2c). The level of transparency allocated to a voxel will be based on its brightness, but the exact relationship is under the user's control and, in practice, needs considerable experimentation to give the best reconstruction (Fig. 3). The end result is inevitably a compromise between seeing internal detail and giving a clear front/back differentiation, but it can be a very effective compromise.

Most software that offers weighted projection will also be able to add a lighting model to the reconstruction. In this case each voxel is treated not as an intensity value but as a reflective point onto which light "shines" from a source that the user can position. This generates shadows that can make surface detail much clearer as well as helping with front/back relationships (Fig. 2d). It will also strongly emphasize noise, which will appear as granular detail on the surface; it is important not to mistake this for true surface structure. If noise is obtrusive, it may be beneficial to preprocess data with a smoothing filter (see later). Applying a lighting model adds hugely to the processing requirements and will be a slow task even on a fast computer.

When rotating a data set there will be some positions in which the one sampling line will "hit" a much higher number of voxels than its neighbor. The reconstruction will then show alternating dark and bright stripes in a parallel or criss-cross pattern (Fig. 4a). This problem is known as aliasing and can be avoided by treating each voxel as a cube in space rather than as a point (Fig. 4b). This is obviously a better approximation to the truth, but equally presents another large increase in the amount of processing required and so will slow down the process of generating the reconstruction. In general the user will first experiment to obtain all the correct parameters for the reconstruction and only then turn on antialiasing to generate the final view or series of views.

If we create a movie from a series of views generated by a weighted projection algorithm we have recovered the two key visual cues of motion parallax and objects in front obscuring those behind. If we create the movie

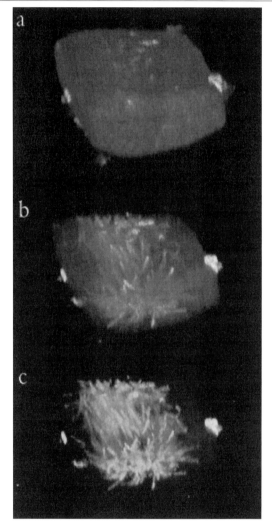

FIG. 3. The control of transparency. The data set is the same as in Fig. 2, but is now reconstructed in an oblique view. If we keep even midlevel voxels fairly opaque (a) we see mostly the cytoplasm of the cell, looking like a suitcase with little inkling of the contents. Bringing the opacity curve closer to linear (b) we begin to see the spindle through the cytoplasm. Setting the opacity so that mid- to low-intensity voxels have little effect shows the spindle almost exclusively (c). Projections are generated in Voxel View.

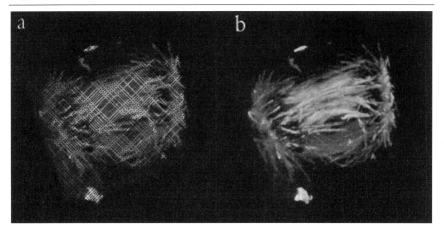

Fig. 4. Aliasing. (a) At certain orientations, aliasing can be extremely apparent, not to say distracting, and information about the sample can be almost totally obscured. (b) Anti-aliasing does a very good job of rectifying the situation at the expense of a substantial increase in rendering time. Projections are generated in Voxel View.

as a series of stereoscopic views we also bring in binocular parallax. Top-end systems have also started to offer perspective, introducing a scaling difference between objects at the front and the back. So far this still needs a very powerful computer to handle it.

Software and Hardware

The choice of software for three-dimensional reconstruction (see Appendix) goes hand in hand with the choice of hardware. In general the platforms used will either be PCs, running Windows or OS2, Macintosh computers, or Unix workstations. PCs and Macintoshes tend to be close in performance with any gain by one camp soon matched by the other. Unix workstations still generally maintain a substantial performance lead over both. Workstations benefit from more than sheer processor speed because many of them have very advanced graphics processing subsystems, which can be extremely valuable to rendering software. However, to take full advantage of these the software must be designed to interact specifically with them. This means that most rendering software for the Unix environment is written for a specific marque. Silicon Graphics workstations have always specialized in offering the most advanced graphic subsystems available and have therefore become brand leaders in this market segment. Accordingly, Unix three-dimensional reconstruction software is most often Silicon Graphics specific. Software that will run on other workstations is generally written for a generic X windows (the common Unix windowing interface) system and will not therefore achieve the performance of more specialized programs optimized for Silicon Graphics hardware.

The question of software is complicated by an extraneous factor. A company holds a patent in some countries (although not in Australia, where this article is written) on the combination of confocal microscopy and three-dimensional reconstruction. Although other manufacturers of confocal microscopes have not accepted the validity of this patent (which has not yet been tested in the courts) and all include bundled three-dimensional software with their microscopes, the company has threatened third-party vendors of reconstruction software with lawsuits. Many of these companies have therefore ceased to mention confocal microscopy in their product specification and market their software as suited to computer-deconvolved wide-field microscopy, serial-section light and electron microscopy, and computerized tomography. The software products nevertheless remain eminently suitable for confocal applications, but some companies have therefore requested that their products not be mentioned here. Any suggestion in this article that a product is suitable for reconstruction of confocal data sets should be taken as applying only to Australia and other countries where the patent issue does not apply.

Software for simple maximum brightness and average brightness projections and for producing pixel-displacement stereo pairs is bundled with, probably, every confocal microscope on the market, certainly all those from major manufacturers. Some include considerably more than this minimum. When purchasing third-party software, therefore, one is generally looking for something more sophisticated, something capable of producing weighted (alpha-blended) projections, with antialiasing, lighting, and maybe surface extraction. Measurement may also be a requirement, but this is covered separately later.

VoxBlast, from Vaytek Systems, is probably the brand leader in the PC market segment. It is also available for Macintosh and Unix workstation platforms, but unlike other Unix products, it is not tailored closely to the Silicon Graphics environment and is also available for Sun and Hewlett-Packard systems. All versions have the same specification. It offers a very full suite of features: a wide range of rendering options, including alpha blending, surface extraction and simple projections, the ability to render two- and three-channel images, free rotation in all planes, freely placeable cutting planes, and a scripting language to automate complex series of operations; these are only a sample of the list. The impressive specification, at a very reasonable price, comes at the cost of a fairly steep learning curve and a distinctly complex user interface. Other systems for the PC platform will not be discussed here but mention must be made of the three-dimensional extension to the Carl Zeiss Imaging (formerly Kontron) KS series of image analysis software. This is a much more expensive option but does come as part of a full-scale image analysis/morphometry package so that if the laboratory also needs image analysis capabilities, one has the advantage of a single product with a single user interface and single line of support.

In the Silicon Graphics arena, prices automatically become much higher because of the more limited market. One of the oldest and best established systems, and certainly the most expensive, is Voxel View, from Vital Images. It offers magnificent performance, thanks to its extensive tailoring to the Silicon Graphics imaging hardware, and has a very straightforward and elegant user interface once one becomes accustomed to its quirky use of the right mouse button for many selections. It is exclusively an alpha-blending rendering system; simple projections and surface extraction are not offered. One feature it does offer is perspective, but only on the most powerful systems (those with a R10,000 processor or better). Its companion programs Voxel Math and Voxel Animator provide useful extensions to its functionality, although many of the features they add are standard in competing products. Voxel View is also offered for the Macintosh but with a much reduced feature set.

Another Silicon Graphics-only option is Imaris from Bitplane AG, Zurich. It is targeted closely at the confocal market and offers a range of special options for that purpose. These include compensation for fall off in intensity with depth, cross-talk reduction, and compensation for misalignment between channels. It also offers a rich set of visualization modes from simple projections to lighted reconstructions, including surface extraction.

Most of the packages mentioned here also include facilities for making three-dimensional morphometric measurements, but this capability properly belongs in a later section.

Image Filtering and Preprocessing

Confocal data sets tend to have much worse signal/noise ratios than wide-field images. This is a matter of simple statistics. If one acquires a wide-field, 512×512 pixel, image with a CCD camera and a 1-sec accumulated exposure, then each point in the image is being sampled for one second. If one acquires a confocal image of the same size over 10 sec, each point in the image is sampled for just 0.04 msec. This means that a noise-reducing filter can often be applied advantageously to our data, particularly if it is to be used for further processing or visualization.

A convolution filter moves through a data set pixel by pixel or voxel by voxel, modifying each value by reference to its immediate neighbors. Simple smoothing filters carry out a weighted average, averaging (for example) nine pixels but giving the center one more weight and then replacing the original value of the central pixel with the new average. Thus a pixel that is much brighter than its neighbors will become darker and one that is darker will become lighter. The new value is placed in the output image; original data are unchanged as the filter moves onto the next pixel and its neighbors. Smoothing filters of this type are not computationally demanding and give effective noise reduction with the penalty of some loss of detail.

A median filter is similar but typically gives a better compromise between noise reduction and preservation of detail. It is more computer-intensive and therefore slower. It looks at all the pixels within, typically, a 3×3 region. The central pixel is then replaced (in the output image) by the median value of these nine pixels. The diagram shows this in action:

127	136	56		127	136	56
100	75	90	becomes	100	**100**	90
101	105	97		101	105	97

Ranking our original nine pixels, we have 56, 75, 90, 97, **100,** 101, 105, 127, 136. The new value for pixel being processed is 100, the median of the list with four values above it and four below.

Most confocal microscopes offer smoothing, and often median, filters, but typically only in two dimensions. Some three-dimensional reconstruction and deconvolution software (e.g., Imaris) offers filters that operate in three dimensions, which is definitely preferable.

Confocal microscopes do not have equal resolution in all three dimensions. The lateral resolution has a linear relationship with the numerical aperture (NA) of the lens, whereas the depth resolution varies as the square of the NA. With the highest NA lenses (NA of 1.3–1.4) the lateral resolution will be around 200 nm, whereas the depth resolution will be about 500 nm. With lower NA lenses the difference becomes progressively larger. This can spoil the appearance of three-dimensional reconstructions, and taking too small a number of optical sections in the interests of conserving disk space will make matters worse. To fully exploit the depth resolution of a particular objective we should take at least two slices within the minimum resolved distance (the Nyquist criterion). As three-dimensional reconstructions become more powerful, the problem of poor z resolution in confocal sections becomes more serious: images become fuzzier as they rotate toward a side-on view. We may be able to improve the situation by preprocessing our image data.

The most thorough (and time-consuming) approach is to deconvolve the images with the point-spread function of the confocal microscope. In principle this could offer good results, as deconvolution of confocal images is a less intractable problem than deconvolution of wide-field ones. Most deconvolution packages—Huyghens, from Bitplane; AutoDeblur from AutoQuant; MicroTome from Vaytek; and the freeware XCOSM from Washington University—offer the ability to handle confocal data sets as well as wide-field. However, the relatively poor signal/noise ratio of confocal images limits the application of deconvolution algorithms. Whether for this reason or because of the cost or the (very substantial) time required, this approach has not achieved any great popularity.

A much simpler alternative was proposed by Cox and Sheppard,[8] who showed that a simple transform based on an edge-enhancing algorithm, applied in the z direction only, was enough to improve the appearance of three-dimensional confocal data sets dramatically. No commercial implementation has yet appeared but it would be simple for any programmer to implement the algorithm.

Three-Dimensional Morphometry

Morphometry, segmentation and measurement of two-dimensional features in images, is a routine part of many imaging laboratories and many two-dimensional measurement features are included among the standard software of most confocal microscopes. Three-dimensional measurement facilities are less common, but are, in the author's experience, often in demand. Surface area and volume of three-dimensional structures are fundamental questions for anatomists and histologists to ask. To some extent, stereological techniques, which lie beyond the scope of this article, can provide the answer, but tools for the direct measurement of morphometric parameters still have an important part to play.

Volume is one of the simplest measurements to implement because a simple count of the voxels within the object is all that is needed. With some labor it could even be done, slice by slice, in a two-dimensional image analysis program. However, it is provided by the built-in software in some recent confocal microscopes, including the Lasersharp software, which controls Bio-Rad microscopes, and by three-dimensional reconstruction software such as Imaris, VoxBlast, and VoxelView. The typical approach is that the user will "seed" a volume by selecting a point within the volume of interest and then select the range of gray values that segment out the desired volume. More advanced systems will automatically find the edges of a structure by looking for abrupt changes of contrast. The voxels within the volume are then counted and converted to absolute units.

Surface area is less simple because a simple count of surface voxels does not give the correct result. Only high-end packages are currently likely

[8] G. C. Cox and C. Sheppard, *Bio-Imaging* **1,** 82 (1993).

to offer this feature at present, but it is available in VoxelView, VoxBlast, and Imaris. Once a volume has been segmented, the area will be calculated, but it will take longer than the simple count required for volume. On VoxBlast, this feature is turned off by default because of the processing time required.

Another measurement feature implemented by most high-end three-dimensional packages is line length, the length of a line in three-dimensional space. While this is a less commonly required measurement than area and volume, it would be difficult to obtain any other way. A rather specialist development of this, but invaluable to the neurobiologist, is the measurement of neuron length and branching, and specialist packages such as Bitplane's Neuron Tracer are available for this purpose.

The range and scope of volume measurements offered in current systems still fall far short of the list of measurement parameters offered in two-dimensional image analysis packages. However, in either case a small list of core features meets most needs, and that core functionality is now available in three dimensions.

Image Processing Conclusions

Image compression will also have a growing part to play, especially as presentations from lap-top computers take over from slides as the standard format for conference presentations. In this area the JPEG standard shows no sign of being superseded. JPEG compression is offered as a standard with some microscopes, but all imaging software also supports the format.

The prime need in the confocal laboratory is for rendering systems for three-dimensional data sets. High-end three-dimension rendering systems offer genuinely worthwhile advantages over the simple projections offered with all confocal systems. Most of these will also offer core three-dimensional measurement parameters. The range of facilities bundled with a confocal microscope will vary greatly, depending on the microscope chosen. It is worthwhile taking time to learn and test the capabilities of the bundled software first before purchasing additional rendering and analysis software. One is then in a better position to choose the packages that will best fill the additional needs of the laboratory.

The key decision to be made when moving to a high-end solution is whether to stay with low-cost PC/Mac hardware or move up to Silicon Graphics. The latter is a quantum leap in cost, because realistically it is not worth doing without also committing to a maintenance contract. The extra cost is mirrored in the higher cost of software to run on it. A Unix system also needs an experienced person to administer it. However, the results in terms of speed, graphic quality, and smoothness of animation are still definitely superior to the personal computer platforms. It is likely to be the solution for the large, multipurpose facility, whereas smaller laboratories are likely to find that the cost and complexity are not worthwhile.

Appendix

Manufacturers of Software Mentioned in Text

VoxBlast and MicroTome: Vaytek Inc., 305 West Lowe St., Fairfield, IA 52556, www.vaytek.com vaytek@vaytek.com

VoxelView: Vital Images Inc., 505 North Fourth Street, Fairfield, IA 52556, www.vitalimages.com

Imaris, Huyghens: Bitplane AG, Technoparkstrasse 1, CH-8005, Zürich, Switzerland, www.bitplane.com

KS400: Carl Zeiss Vision GmbH, Oskar-von-Miller-Strasse 1, 85386 Eching-bei-Munchen, Germany

XCOSM: Biomedical Computer Laboratory, Washington University, 700 S. Euclid Avenue, St. Louis, MO 63110, www.ibc.wustl.edu/bcl/xcosm/xcosm.html

AutoDeblur: Autoquant Imaging Inc., 877 25th Street, Watervliet, NY, 12189, www.aqi.com sales@aqi.com

[4] Antifading Agents for Confocal Fluorescence Microscopy

By Miguel Berrios, Kimberly A. Conlon, and David E. Colflesh

I. Introduction

Advances in fluorescence probe chemistry, more economical and powerful lasers, and improvements in confocal and image acquisition systems, together with faster and inexpensive computers, have made the confocal laser scanning microscope (CLSM) available to a wider range of biological applications and users. Currently, single- and dual-labeling immunofluorescence,[1–3] imaging of green fluorescent pro-

---

[1] V. H. Meller, P. A. Fisher, and M. Berrios, *Chromosome Res.* **3**, 255 (1995).
[2] I. S. Nathke, C. L. Adams, P. Polakis, J. H. Sellin, and W. J. Nelson, *J. Cell Biol.* **134**, 165 (1996).
[3] M. Furuse, K. Fujita, T. Hiiragi, K. Fujimoto, and S. Tsukita, *J. Cell Biol.* **141**, 1539 (1998).

tein,[4-10] and fluorescent *in situ* hybridization (FISH)[11-14] are the most widely used applications of the CLSM. Single- or dual-labeling immunofluorescence may involve either a direct or an indirect method.[15,16] The greatest advantage of the CLSM is its ability to discriminate between signals originating from in-focus and out-of-focus optical planes to produce digital images that can be merged pixel by pixel to generate perfectly registered two- and three-dimensional renditions.

To obtain the maximum information from these renditions, however, each optical section must provide a good quality image. Optical aberrations may be reduced when all objects in the light path of the microscope, including the objective lens, coverslip, and immersion oil, are used following manufacturer's specifications. In CLSM immunofluorescence, the quality of signal from fluorochrome-conjugated antibodies is dependent not only on the properties of the confocal system, the objective lens of the microscope, and the image acquisition device itself, but on the condition of the specimen. Specimens should be prepared with care to avoid introducing artifacts that, aside from some other deleterious effects, may increase background fluorescence levels. Important considerations in this regard are the fixation procedure, antibody probes and incubation conditions, and ultimately the method used for mounting the specimen between slide and coverslip. In fact, immunofluorescence and coverslip mounting protocols suitable for the wide-field microscope may prove inadequate for the CLSM.

[4] M. Chalfie, *Photochem. Photobiol.* **62**, 651 (1995).
[5] A. B. Cubitt, R. Heim, S. R. Adams, A. E. Boyd, L. A. Gross, and R. Y. Tsien, *Trends Biochem. Sci.* **20**, 448 (1995).
[6] J. W. Larrick, R. F. Balint, and D. C. Youvan, *Immunotechnology* **1**, 83 (1995).
[7] D. C. Prasher, *Trends Genet.* **11**, 320 (1995).
[8] H. H. Gerdes and C. Kaether, *FEBS Lett.* **389**, 44 (1996).
[9] T. Darsow, C. G. Burd, and S. D. Emr, *J. Cell Biol.* **142**, 913 (1998).
[10] P. M. Roberts and D. S. Goldfarb, *in* "Methods in Cell Biology" (M. Berrios, ed.), p. 545. Academic Press, San Diego, 1998.
[11] P. B. Moens and R. E. Pearlman, *in* "Methods in Cell Biology" (B. A. Hamkalo and S. G. Elgin, eds.), p. 101. Academic Press, San Diego, 1991.
[12] N. Lemieux, B. Dutrillaux, and E. Viegas-Pequignot, *Cytogenet. Cell Genet.* **59**, 311 (1992).
[13] A. F. Dernburg and J. W. Sedat, *in* "Methods in Cell Biology" (M. Berrios, ed.), p. 187. Academic Press, San Diego, 1998.
[14] J. Loidl, F. Klein, and J. Engebrecht, *in* "Methods in Cell Biology" (M. Berrios, ed.), p. 257. Academic Press, New York, 1998.
[15] A. Kawamura, "Fluorescent Antibody Techniques and Their Applications." University of Tokyo Press, Tokyo, 1977.
[16] D. R. Springall and J. M. Polak, *in* "Image Analysis in Histology: Conventional Confocal Microscopy" (R. Wooton, D. R. Springall, and J. M. Polak, eds.), p. 123. Cambridge Univ. Press, New York, 1995.

In addition, a pervasive phenomenon restricting image acquisition is the deterioration of fluorescence emitted by fluorochrome-conjugated antibodies during laser scanning of optical planes. Fluorescence fading may be very pronounced for fluorescein isothiocyanate (FITC)-conjugated antibodies.[17] This fading phenomenon is particularly striking during the imaging of dual-labeled specimens because of the differential deterioration of fluorescence emitted by several fluorochromes [e.g., FITC vs tetramethylrhodamine isothiocyanate (TRITC)]. Increasing the concentration of fluorochrome-conjugated antibodies is not a viable solution, as this usually results in higher background fluorescence. In most situations, mounting biological specimens with antifading agents can solve fluorescence fading problems.

This article discusses strategies to reduce specimen-associated background fluorescence and the properties and performance of several antifading agents during the acquisition of multiple optical sections. Although discussions are focused on indirect immunofluorescence, most sections are applicable to other *in situ* localization methods. In our hands, several antifading agents, including the commercial preparations Vectashield, SlowFade, and ProLong have fluorochrome antifading properties that alleviate the problems mentioned earlier, making acquisition by the CLSM of fluorescent optical Z-series from a multilabeled specimen reproducible.

II. Specimen Preparation

In our experience, several aspects of specimen preparation should receive special consideration when attempting to obtain fluorescent images with low backgrounds and reduced fluorochrome fading. In general, we place as much effort into reducing fluorochrome fluorescence fading as we place into avoiding artifacts that may result in background fluorescence. Attention to fixation and antibody probing procedures and, when necessary, introduction of procedures directed at reducing naturally occurring background fluorescence during specimen preparation has helped us obtain good quality fluorescent images (Fig. 1). These three aspects of specimen preparation directly affect capturing images with low backgrounds and reduced fluorochrome fading and are discussed in the following sections.

A. Fixation

There are at least as many fixation procedures available in the literature as there are immunofluorescence protocols. All of them, however, are at

[17] G. Bock, M. Hilchenbach, K. Schauenstein, and G. Wick, *J. Histochem. Cytochem.* **33**, 669 (1985).

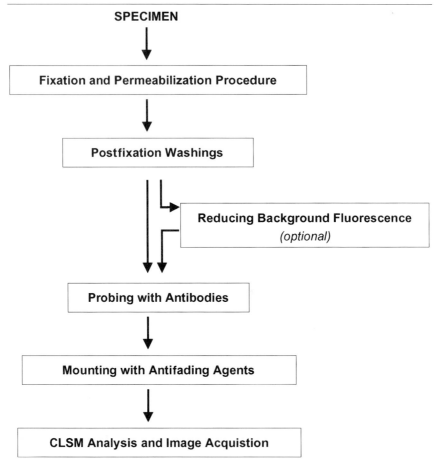

FIG. 1. Flow chart showing the main steps during the preparation of specimens for CLSM immunofluorescence. Reduction of specimen-associated background fluorescence is optional.

best a compromise between preservation of structure and preservation of antigenic determinants. Examples of fixation procedures for immunofluorescence can be found in comprehensive reviews by Kawamura[15] and Springall and Polak[16] and in articles by Perez et al.,[18] Poot et al.,[19] Dernburg

[18] J. L. Perez, M. De Ona, J. Niubo, H. Villar, S. Melon, A. Garcia, and R. Martin, *J. Clin. Microbiol.* **33,** 1646 (1995).
[19] M. Poot, Y. Z. Zhang, J. A. Kramer, K. S. Wells, L. J. Jones, D. K. Hanzel, A. G. Lugade, V. L. Singer, and R. P. Haugland, *J. Histochem. Cytochem.* **44,** 1363 (1996).

TABLE I
PREPARATION OF AN 8% (w/v) PARAFORMALDEHYDE 154 mM PIPES, pH 7.5, STOCK[a]

1. Weigh 2 g of paraformaldehyde powder
2. Dissolve powder in 15 ml of 256 mM PIPES, pH 7.5, while stirring in water bath at 70° (approximately 20 min)
3. Once in solution, complete volume to 25 ml with cold water
4. Clarify resulting solution by centrifugation at 5000g for 10 min at room temperature
5. Store supernatant in a capped glass bottle at room temperature[b]

[a] From M. Berrios and D. E. Colflesh, *Biotech. Histochem.* **70**, 40 (1995).
[b] Discard stock after 6 weeks.

and Sedat,[13] and Solari.[20] Although fixatives containing aldehydes have a tendency to increase background fluorescence, we have found that these compounds provide superior morphological preservation over, for example, fixation with either ethanol or methanol. Fixatives based on buffered salt solutions containing 3 to 4% (w/v) paraformaldehyde or combinations of paraformaldehyde and 0.1% (v/v) glutaraldehyde work well for most confocal applications.[1,21,22] In fact, buffered paraformaldehyde alone may be an adequate fixative for most confocal immunofluorescence applications. To prepare these fixatives at an appropriate pH, we prefer to dissolve paraformaldehyde powder in a buffer rather than the more traditional resuspension of this reagent in hot water and pH titration with 6.25 N NaOH. A procedure that uses 1,4-piperazinediethanesulfonic acid (PIPES) buffer instead of sodium hydroxide to produce an 8% (w/v) paraformaldehyde, pH 7.5, stock solution is shown in Table I.[21] This stock solution should be kept at room temperature and is stable for many weeks. Another key step in reducing background fluorescence staining is the inclusion of one or two postfixation washes, preferably with buffered solutions containing primary amines [e.g., Tris (tris[hydroxymethyl]aminomethane), Tris–glycine, Tris–Cl, glycine, 2,2′,2″-nitrilotriethanol (triethanolamine) buffer solutions]. These postfixation washes reduce the risk of cross-linking antibodies by quenching available aldehyde groups.

B. Reducing Specimen-Associated Background Fluorescence

The detection of antigens by confocal immunofluorescence has been limited by two phenomena: (1) fluorochrome fading and (2) specimen-

[20] A. F. Solari, *in* "Methods in Cell Biology" (M. Berrios, ed.), p. 236. Academic Press, San Diego, 1998.
[21] M. Berrios and D. E. Colflesh, *Biotech. Histochem.* **70**, 40 (1995).
[22] D. E. Colflesh, K. A. Conlon, and M. Berrios, *J. Histotech.* **22**, 23 (1999).

associated background fluorescence (also known as autofluorescence). Both phenomena act together to reduce the effectiveness of this technique, particularly when antigens are found in low abundance. As discussed in Sections III and IV, fluorochrome fading may be reduced effectively using a combination of antifading agents and rapid scanning lasers (short dwelling time).[21] Much less has been done, however, to reduce background fluorescence associated with the specimen. This phenomenon still remains as an unsurmountable threshold to the limit of detection of immunofluorescent assays. We have devised a procedure in which reducing agents are used to lower specimen-associated background fluorescence before the detection of subcellular antigens by confocal immunofluorescence microscopy.[22]

The reduction of specimen-associated background fluorescence is performed immediately after paraformaldehyde fixation of the sample (see Table II). We found that it is best to prepare the solution of reducing agent fresh just before use.[22] The reducing agent (sodium borohydride or sodium cyanoborohydride) is dissolved in a defined buffered saline solution, such as 140 mM NaCl, 10 mM phosphate buffer, pH 7.4 (PBS). These solutions must be discarded after use because their ability to reduce background fluorescence diminishes with time. The final or working concentration of the reducing agent is defined by the amount of specimen-associated background fluorescence of the sample and is determined by incubating the specimen in increasing concentrations of reducing agent. We have successfully used concentrations ranging from 10 μM to 100 mM of sodium borohydride with no detectable deterioration of specimen morphology or its antigenic determinants.[22] At higher concentrations (e.g., 10 to 100 mM), the pH of a solution of sodium borohydride in PBS increases from 7.2 to about 9.0–9.5. If this increase in pH is a concern, sodium cyanoborohydride can be used instead. We found that the pH of solutions of sodium cyanoborohydride increased to about 8.5. To maximize reproducibility from experiment to

TABLE II
BUFFERED SOLUTIONS AND REAGENTS FOR
REDUCING SPECIMEN-ASSOCIATED
BACKGROUND FLUORESCENCE

Buffered solutions and reagents
MSM-PIPES
MSM-PIPES containing nonionic detergents
4% (w/v) paraformaldehyde in MSM-PIPES
PBS
Sodium borohydride or sodium cyanoborohydride[a]

[a] Fresh bottle (or store powder dry in a desiccator).

experiment, the specimen is fixed, if necessary, specimen-associated background fluorescence is reduced, and antibody incubations and washes are performed in 5.4 × 20-mm-diameter wells of two-well immunofluorescence chambers.[23,24] These chambers (Electron Microscopy Sciences, Fort Washington, PA) are assembled around the microscope slide without any bonding substance and are taken apart before microscopical examination. Previous experience using hydrophobic substances such as nail polish, bee's wax, dental wax, rubber cement, or silicone glue to confine solutions over specimens during incubations proved difficult to reproduce and, in general, these substances had the tendency to increase fluorescence backgrounds. Other devices, such as chamber slides (Lab-Tek, Nalge Nunc International, Naperville, IL), have a tendency to leave adhesive residues and their use precludes examination of specimens such as tissue smears and sections. Observations described in this article were obtained using these immunofluorescence chambers to fix specimens, to reduce specimen-associated background fluorescence, if necessary, and to probe specimens with antibodies.

Typically, a general procedure to reduce specimen-associated background fluorescence is as follows: after fixation, the sample is washed twice with a buffered defined salt solution [e.g., 18 mM $MgSO_4$, 5 mM $CaCl_2$, 40 mM KCl, 24 mM NaCl, 0.5% (v/v) Triton X-100, 0.5% (v/v) Nonidet P-40 (NP-40), and 5 mM PIPES, pH 6.8 (MSM-PIPES)][25] and then incubated with 400 μl of a solution containing the desired reducing agent at 37° for 10 min in a humidified chamber [a humidified chamber can be built using a plastic box 110 × 160 × 40 mm deep (e.g., Electron Microscopy Sciences) containing an 8- to 10-mm-thick water-soaked plastic sponge insert at the bottom]. These conditions are optimum in reducing specimen-associated background fluorescence. When the incubation is performed at 23°, no reduction of specimen fluorescence is observed. Incubation is followed by two washes with a buffered defined salt solution (e.g., MSM-PIPES) to completely remove all traces of the reducing agent. A general procedure is outlined next.

Reduction of Background Autofluorescence

1. Place slide containing tissue culture cells or frozen section into a two-well immunofluorescence chamber.[23,24]

2. Fill each well with freshly prepared 4% (w/v) paraformaldehyde in MSM-PIPES.[21,22]

---

[23] M. Berrios, *Med. Lab. Sci.* **46,** 276 (1989).

[24] P. A. Fisher and M. Berrios, *in* "Methods in Cell Biology" (M. Berrios, ed.), p. 397. Academic Press, San Diego, 1998.

[25] D. E. Smith, Y. Gruenbaum, M. Berrios, and P. A. Fisher, *J. Cell Biol.* **105,** 771 (1987).

3. Incubate at room temperature with gentle shaking for 3 min.

4. Remove fixative.

5. Fix again by adding 400 µl per well of 4% (w/v) paraformaldehyde in MSM-PIPES.[21,25]

6. Incubate at room temperature with gentle shaking for 3 min.

7. During second fixation, weigh out reducing agent (sodium borohydride or sodium cyanoborohydride) and dilute to desired concentration with phosphate-buffered saline (PBS).

8. Remove fixative and wash twice (1 min each time, with gentle shaking) with 800 µl MSM-PIPES containing nonionic detergents.

9. Remove second wash and add 400 µl per well of reducing agent in PBS solution.

10. Incubate for 10 min at 37° in a humidified chamber.

11. Remove reducing agent and wash twice (1 min each time, with gentle shaking) with 800 µl MSM-PIPES containing nonionic detergents.

We have tested several strategies for their ability to reduce naturally occurring background fluorescence associated with insect (*Drosophila* female abdomen) and mammalian (mouse liver) 7-µm-thick cryosections.[22] Freshly prepared cryosections were incubated with either sodium borohydride or sodium cyanoborohydride (Table II). Figure 2 contains results of one of these experiments. Sodium cyanoborohydride was slightly more effective in the reduction of autofluorescence than sodium borohydride, possibly because the former is less reactive and consequently more stable in aqueous solutions (compare Fig. 2A with Fig. 2B). When mouse liver cryosections were incubated in 10 mM sodium borohydride for 10 min at 37°, no reduction of tissue autofluorescence was observed (Fig. 3A). Mouse liver tissue autofluorescence was reduced by about 48% when cryosections were incubated in 100 mM sodium borohydride for 10 min at 37° (Fig. 3B). The inability of 10 mM sodium borohydride to quench autofluorescence in this tissue may be a result of the high fluorescence intensity of the tissue. A comparison of specimen-associated background fluorescence between *Drosophila* adult female abdomen and that of mouse liver sections showed that the fluorescence intensity of the former is lower than that of the latter (insets in Figs. 2 and 3).

C. Probing with Antibodies

Once the specificity of an antibody has been determined using enzyme-linked immunosorbent assay (ELISA), competitive ELISA, immunoblots, and so on, focus should be placed on identifying an appropriate fluorochrome-conjugated secondary antibody. In our experience, using fluorochrome-conjugated secondary antibodies for the visualization of first spe-

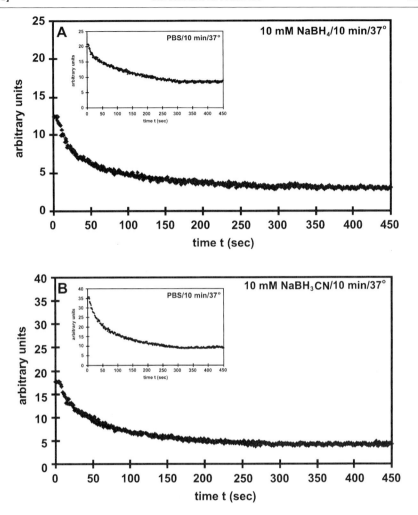

FIG. 2. Autofluorescence of *Drosophila* cryosections after incubation in either sodium borohydride or sodium cyanoborohydride. Graphs represent actual recordings of autofluorescence intensity measured at 1-sec intervals during 450 sec of continuous laser illumination. Each time point presents measurements from three random fields. Samples were incubated for 10 min at 37° in PBS (insets), freshly prepared 10 m*M* sodium borohydride (A), or 10 m*M* sodium cyanoborohydride (B). *X* axis, laser exposure time (sec); *Y* axis, autofluorescence intensity expressed in arbitrary units. Adapted from D. E. Colflesh *et al. J. Histotech.* **22,** 23 (1999), reproduced with permission.

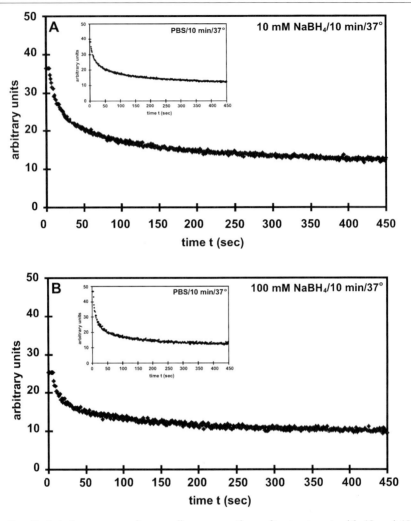

FIG. 3. Autofluorescence of mouse liver cryosections after treatment with 10 and 100 m*M* sodium borohydride. Graphs represent actual recordings of autofluorescence intensity measured at 1-sec intervals during 450 sec of continuous laser illumination. Each time point represents measurements from three random fields. Samples incubated for 10 min at 37° in PBS (insets), freshly prepared 10 m*M* sodium borohydride (A), or 100 m*M* sodium borohydride (B). *X* axis, laser exposure time (sec); *Y* axis, fluorescence intensity in arbitrary units. Adapted from D. E. Colflesh *et al.*, *J. Histotech.* **22,** 23 (1999), reproduced with permission.

cific antibody–antigen complexes has several practical advantages over the direct labeling of specific (primary) antibodies. Secondary antibodies are widely available commercially, they are relatively inexpensive, and they are available as IgG fractions from different hosts (e.g., chicken, donkey, goat, mouse, and rabbit) conjugated to a wide range of fluorochromes. In addition, the use of secondary antibodies allows for signal amplification and eliminates possible detrimental effects of the chemical conjugation procedure on specific antibodies.

Fluorochrome-conjugated secondary antibodies are available most often as specific anti-IgG (antiheavy and/or antilight chains) affinity-purified antibody fractions. A complete list of fluorochrome-conjugated antibody fractions is beyond the scope of this article and may be found at http://www1-78.dev-com.com/kshreder/onlinecomp.html or in well-organized charts included in catalogs from several commercial vendors (see Table III).

We have had success performing confocal immunofluorescence using commercially available fluorochrome-conjugated secondary antibodies (affinity-purified fractions) containing approximately 3.6 mol of fluorochrome per mole of immunoglobulin G (IgG). To reduce nonspecific fluorescence, we have found that it is best to obtain affinity-purified antibody fractions and to use them at relatively low working concentrations, usually at concentrations that are lower than those recommended by the commercial supplier. Most often we obtain optimal staining using working dilutions ranging anywhere from 1:1000 to 1:10,000 (1.5 μg IgG/ml to 0.2 μg IgG/ml) of the original stock. We also prefer to keep fluorochrome-conjugated secondary antibodies as 20-μl stocks in 50% (v/v) glycerol at $-20°$. Under these conditions, these reagents do not show signs of deterioration for up to 4 years. These antibodies are diluted into buffered salt solutions (e.g., PBS) supplemented with nonionic detergents [e.g., 0.5% (v/v) of Triton X-100, Nonidet P-40, or Tween 20] just before use, discarding the remainder. As stated earlier, to maximize staining reproducibility from experiment to experiment, we perform antibody incubations and washes in wells of immunofluorescence chambers.[23,24]

III. Antifading Agents

A. Mounting Specimens with Antifading Agents

This is a simple procedure, but in our experience some care must be taken to avoid shearing the specimen, introducing air bubbles and/or fluorescent contaminants, and/or leaving too much mounting media between the slide and the coverslip. An excess of mounting media between the slide and the coverslip does not improve the antifading properties of the agent

TABLE III
SUPPLIERS OF FLUOROCHROME-CONJUGATED ANTIBODIES[a]

Name	Address
Amersham Corporation	2936 South Clearbrook Drive
	Arlington Heights, IL 60005
	Phone: 800-323-9750
Aves Labs, Inc.	16200 South West Pacific Highway #146
	Tigard, OR 97224
	Phone: 503-245-1858
	Fax: 503-245-8784
	http://www.aveslab.com
Boehringer Mannheim Corporation	9115 Hague Road
	P.O. Box 50414
	Indianapolis, IN 46250
	Phone: 800-262-1640
	Fax: 800-428-2883
ICN Biomedical	3300 Hyland Avenue
	Costa Mesa, CA 92626
	Phone: 800-854-0530
	Fax: 800-334-6999
E. Y. Laboratories	P.O. Box 1787
	San Mateo, CA 94401
	Phone: 800-821-0044/650-342-3296
	Fax: 650-342-2648
	http://www.eylabs.com/
Jackson ImmunoResearch Laboratories	P.O. Box 9
	West Grove, PA 19390
	Phone: 800-367-5296/610-869-4024
	Fax: 610-869-0171
	curserjaxn@aol.com/
Molecular Probes	P.O. Box 22010
	Eugene, OR 97402
	Phone: 541-465-8300
	Fax: 541-344-6504
	http://www.molecularprobes.com/

[a] Partial list; in the United States.

and may hamper the gathering of confocal slices when using objective lenses with short working distances. To maintain better control over this operation, we prefer to use individual glass coverslips over each specimen (e.g., $18 \times 18 \times 0.13$–0.17 mm thick) instead of one large coverslip over the whole slide or most of it. When using the immunofluorescence chambers,[23,24] we generate two samples per slide; this leaves sufficient space for placing an adequate amount of seal (e.g., colorless nail polish) around each coverslip. We recommend adding, over the wet specimen, approximately

0.03 μl of mounting media per square micrometer of coverslip (i.e., 10 μl for an 18-mm^2 coverslip). Alternatively, to help standardize this procedure, excess mounting media may be squeezed out by gently pressing the coverslip against the slide for 3 min using the attraction force of two small permanent magnets (e.g., Edmund Scientific Co., Barrington, NJ).[21]

Excess mounting media may be removed using a 200-μl ultramicropipette tip, a yellow pipette tip (Eppendorf, Germany), or a similar disposable plastic pipette tip attached to a low vacuum line (e.g., -10 psi). Care must be taken to avoid the drying of specimens before mounting the coverslip. In our experience, the antifading properties of mounting agents are not affected by wet specimens. Each coverslip may be sealed with clear nail polish and kept in the refrigerator and in the dark until use to reduce evaporation.

B. Agents and Their Antifading Properties

Several antifading preparations are reported in the literature, and a number of these are available commercially under various trade names (Table IV). Antifading preparations are used most often as coverslip mounting media to reduce the fading of fluorescence after UV light or laser excitation of fluorochromes. Although it is not our goal to evaluate the antifading properties of all mounting media available, we have surveyed,

TABLE IV
COMMERCIAL ANTIFADING AGENTS[a]

Trade name	Supplier
ProLong	Molecular Probes, Inc.
SlowFade	4849 Pitchford Avenue
SlowFade Light	Eugene, OR 97402
	Phone: 541-465-8300
	Fax: 541-344-6504
	http://www.probes.com/
Vectashield (H-1000)	Vector Laboratories
	30 Inbold Road
	Burlingame, CA 94010
	Phone: 650-697-3600
	Fax: 415-697-0339
Citifluor	Stephens Scientific
	107 Riverdale Road
	Riverdale, NJ 07457
	Phone: 201-831-9800

[a] Partial list.

using a CLSM, the properties of several common antifading agents reported in the literature and available commercially.

To evaluate the antifading properties of several mounting media, we selected nuclei from *Drosophila melanogaster* adult male accessory glands[26] and histone proteins as targets. Cells in this *Drosophila* tissue are arrested at interphase, and nuclei are homogeneous in size and morphology. The histones are an abundant and widely distributed group of nuclear proteins. Briefly, *Drosophila* accessory glands are dissected out and transferred to a sodium dodecyl sulfate-cleansed microscope slide, and nuclei are extruded from cells exactly as described.[27] Indirect immunofluorescence is as described previously.[21] Cell nuclei are probed with a mouse monoclonal antibody (MAb052; Chemicon International, Temecula, CA) directed against histones (H1 and core proteins H2a, H2b, H3, and H4) followed by FITC-conjugated donkey antimouse IgG antibodies, TRITC-conjugated donkey antimouse IgG antibodies, or Texas Red sulfonyl chloride (TRSC)-conjugated donkey antimouse IgG antibodies. Stained nuclei are mounted with either MSM-PIPES without detergents[25,28] as a control or this defined buffered salt solution supplemented with one of the following antifading mounting media: 0.1% (w/v) *p*-phenylenediamine,[29] 0.3% (w/v) gallic acid *n*-propyl ester (*n*-propyl gallate),[30] and the commercial preparations Vectashield (Vector Laboratories, Burlingame, CA), SlowFade, or ProLong (Molecular Probes, Eugene, OR). The pH of mounting solutions is kept between pH 6.8 and 7.2, and the initial nuclear fluorescence staining is similar for all mounting media.

Fluorescence signal intensity data from each nuclei are collected simultaneously at 60-sec intervals during 15 min of continuous laser epi-illumination using an Odyssey confocal 200-mW argon ion multiline laser scanning system (Noran Instruments, Middleton, WI) attached to a Nikon Diaphot inverted microscope (Nikon, Melville, NY) equipped with a Nikon 60×/1.4 CFN planapochromat oil immersion objective lens.[21] The objective lens is focused near the diameter of 12 randomly selected nuclei. On going through this objective lens, the laser scans an area of about 630 μm.[2] The laser scanning rate is 30 frame/sec and the dwell time is 0.1 μsec/pixel. To detect fluorescence emission from FITC, specimens are exposed to a laser excitation wavelength of 488 nm, and images are captured through a 500-nm barrier filter. To detect fluorescence emission from either TRITC or TRSC,

[26] T. Schmidt, P. S. Chen, and M. Pellagrini, *J. Biol. Chem.* **260,** 7645 (1985).
[27] M. Berrios, *Biotech Histochem.* **69,** 78 (1994).
[28] M. Berrios and P. A. Fisher, *J. Cell Biol.* **103,** 711 (1986).
[29] G. D. Johnson and G. M. Araujo, *J. Immunol. Methods* **43,** 349 (1981).
[30] H. Giloh and J. W. Sedat, *Science* **217,** 1252 (1982).

TABLE V
LASER SCANNING[a] POWER MEASURED[b] AT SPECIMEN

	Excitation wavelength	
Nominal laser power[c]	488 nm (FITC)	529 nm (TRITC/TRSC)
20%	90 nW/μm^2	20 nW/μm^2
50%	500 nW/μm^2	120 nW/μm^2

[a] Laser scanning rate was 30 frames/sec.
[b] Measurements were made with an 840C Optical Power Meter (Newport Corporation, Irvine, CA).
[c] Power settings were regulated through an acousto-optic deflector driven by the Odyssey software package (Noran Instruments, Middleton, WI). Adapted from M. Berrios and D. E. Colflesh, *Biotech. Histotech.* **70**, 40 (1995), reproduced with permission.

specimens are exposed to a laser excitation wavelength of 529 nm and images are captured through a 550-nm barrier filter. The confocal detector aperture is adjusted to 15 μm, and laser power output is set at either 20 or 50% (see Table V). Photomultiplier gain is set at 88% (1.1 kV), a value at which no signal is detected from specimens probed in the absence of first specific antibodies.[21,22]

FITC fluorescence fades rapidly when specimens are mounted without an antifading agent and exposed to CLSM (Fig. 4). The fluorescence signal from FITC is reduced to about 5% after 1 min of exposure to 50% laser power (Fig. 4A) and to about 50% after 1 min of exposure to 20% laser power (Fig. 4B). In contrast, TRITC fluorescence fades to about 55% and TRSC fluorescence fades to about 80% after 15 min of exposure to 50% laser power (Fig. 4A). At 20% laser power, TRITC and TRSC fluorescence remains nearly unchanged (Fig. 4B).

When specimens are mounted in either *p*-phenylenediamine or *n*-propyl gallate, two widely used antifading agents,[12,17,31–35] FITC fluorescence fades to about 10% after 15 min of exposure to 50% (Figs. 5A and 6A) or 20% (Figs. 5B and 6B) laser power. TRITC fluorescence in the presence of either *p*-phenylenediamine or *n*-propyl gallate is reduced to about 70% after 15 min of exposure to 50% laser power (Figs. 5A and 6A). Although TRITC fluorescence in the presence of *p*-phenylenediamine remains nearly

[31] A. M. Soliman, *Arch. Oto-Rhino-Laryngol.* **245,** 28 (1988).
[32] A. Scheynius and P. Lundahl, *Arch. Dermatol. Res.* **281,** 521 (1990).
[33] M. Battaglia, D. Pozzi, S. Grimaldi, and T. Parasassi, *Biotech. Histochem.* **69,** 152 (1994).
[34] H. Jensen, N. Broholm, and B. Norrilb, *J. Histochem. Cytochem.* **43,** 507 (1995).
[35] D. E. Colflesh and M. Berrios, unpublished.

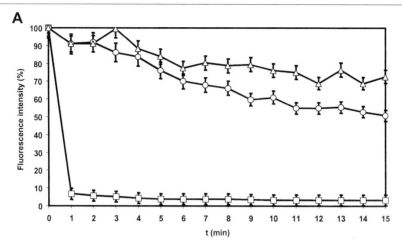

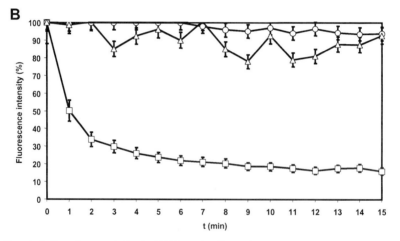

FIG. 4. Fading characteristics for FITC-, TRITC-, and TRSC-conjugated IgG exposed to CLSM without antifading agents. The X axis is laser exposure time (min) and the Y axis is fluorescence intensity expressed as a percentage relative to the initial signal (i.e., $t = 0$ min fluorescence intensity is 100%). Each point is an average of 20 measurements at 1-min intervals. Fading during exposure to 50% laser power (A) and 20% laser power (B). FITC-conjugated IgG (□), TRITC-conjugated IgG (○), and TRSC-conjugated IgG (△). Unpublished results of D. E. Colflesh and M. Berrios and adapted from M. Berrios and D. E. Colflesh, *Biotech. Histochem.* **70,** 40 (1995), reproduced with permission.

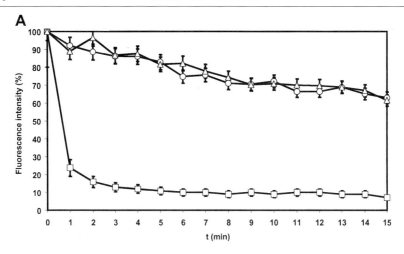

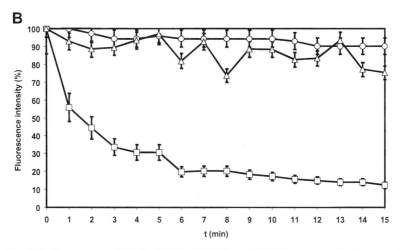

FIG. 5. Fading characteristics for FITC-, TRITC-, and TRSC-conjugated IgG exposed to CLSM in the presence of 0.1% (w/v) p-phenylenediamine. The X axis is laser exposure time (min) and the Y axis is fluorescence intensity expressed as a percentage relative to the initial signal (i.e., t = 0 min fluorescence intensity is 100%). Each point is an average of 20 measurements at 1-min intervals. Fading during exposure to 50% laser power (A) and 20% laser power (B). FITC-conjugated IgG (□), TRITC-conjugated IgG (○), and TRSC-conjugated IgG (△). Unpublished results of D. E. Colflesh and M. Berrios and adapted from M. Berrios and D. E. Colflesh, *Biotech. Histochem.* **70,** 40 (1995), reproduced with permission.

A

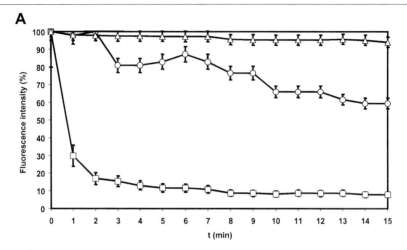

B

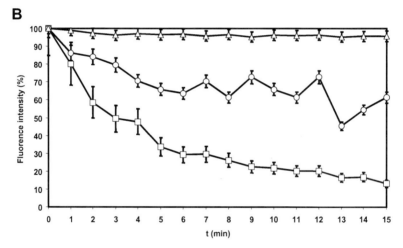

Fig. 6. Fading characteristics for FITC-, TRITC-, and TRSC-conjugated IgG exposed to CLSM in the presence of 0.3% (w/v) n-propyl gallate. The X axis is laser exposure time (min) and the Y axis is fluorescence intensity expressed as a percentage relative to the initial signal (i.e., $t = 0$ min fluorescence intensity is 100%). Each point is an average of 20 measurements at 1-min intervals. Fading during exposure to 50% laser power (A) and 20% laser power (B). FITC-conjugated IgG (□), TRITC-conjugated IgG (○), and TRSC-conjugated IgG (△). Unpublished results of D. E. Colflesh and M. Berrios and adapted from M. Berrios and D. E. Colflesh, *Biotech. Histochem.* **70,** 40 (1995), reproduced with permission.

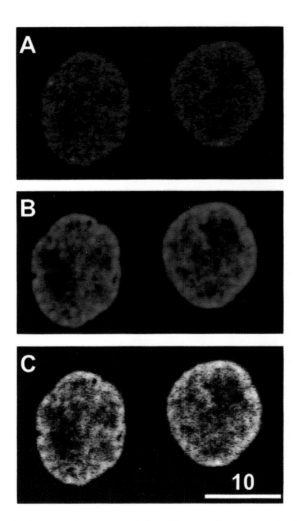

Fig. 10. Confocal immunofluorescent staining of *Drosophila* accessory gland nuclei probed with anti-CRP1 and antihistone antibodies. Optical sections through the diameter of two accessory gland nuclei stained with a mixture containing an antihistone mouse monoclonal antibody and affinity-purified rabbit anti-CRP1 antibodies followed by a mixture containing FITC-conjugated donkey antimouse IgG and TRSC-conjugated donkey antirabbit IgG. Specimen-associated autofluorescence was reduced with sodium borohydride. Specimens were mounted with ProLong. (A) TRSC fluorescence signal (CRP1). (B) FITC fluorescence signal (histones). (C) Pixel-by-pixel superimposition of (A) and (B). Bar: 10 μm. Adapted from D. E. Colflesh *et al., J. Histotech.,* in press (1999), reproduced with permission.

unchanged after 15 min of exposure to 20% laser power (Fig. 5B), TRITC fluorescence is reduced to 68% in the presence of n-propyl gallate (Fig. 6B). TRSC fluorescence in the presence of p-phenylenediamine fades to 70 and 80% after 15 min of exposure to 50 and 20% laser power, respectively (Fig. 5). In contrast, TRSC fluorescence in the presence of n-propyl gallate remains nearly unchanged after 15 min of exposure to either 50 or 20% laser power (Fig. 6).

The commercial antifading preparations, Vectashield, SlowFade, and ProLong, were also tested for their ability to quench the fading of fluorescence signals under the previous conditions. FITC fluorescence is reduced between 20 and 40% when specimens are mounted in Vectashield, SlowFade, or ProLong after 15 min of exposure to 50% laser power (Figs. 7A, 8A, and 9A). However, when specimens are mounted in either Vectashield or SlowFade and exposed to 20% laser power, about 50 to 70% of the FITC fluorescence signal remains after 15 min of exposure (Figs. 7B and 8B). FITC fluorescence is reduced to 20% under 20% laser power when using ProLong (Fig. 9B). In the presence of ProLong, TRITC and TRSC fluorescence is reduced to about 88 and 70%, respectively, after 15 min of exposure to 50% laser power (Fig. 9A) and to 95 and 90%, respectively, after 15 min of exposure to 20% laser power (Fig. 9B).

As shown earlier, 0.1% (w/v) p-phenylenediamine and 0.3% (w/v) n-propyl gallate provide a level of protection to fluorochrome fluorescence fading over mounting specimens in PBS or MSM-PIPES.[25,28] Commercially available preparations such as the ones tested here, however, are superior in reducing fluorochrome fluorescence fading and provide protection for longer periods of time. In our hands, antifading solutions containing 0.1% (w/v) p-phenylenediamine[29] or 0.3% (w/v) n-propyl gallate[30] had shorter shelf-lives than commercial preparations and some had the tendency to generate nonspecific fluorescence as the preparation aged. Nonspecific fluorescence staining of cell nuclei was a common observation when using p-phenylenediamine for example.[36]

IV. Acquisition of Fading Fluorescent Images

Attempts to circumvent problems associated with differential deterioration of fluorescence emitted by separate fluorochromes by reducing laser dwelling time and laser power output[37,38] severely limit the ability to capture comparable image pairs from, for example, FITC and TRITC or TRSC

[36] R. Kittelberger, P. F. Davis, and W. E. Stehbens, *Acta Histochem.* **86,** 137 (1989).
[37] G. I. Kaufman, J. F. Nester, and D. E. Wasserman, *J. Histochem. Cytochem.* **19,** 469 (1971).
[38] J. P. Rigaut and J. Vassy, *Anal. Quant. Cytol. Histol.* **13,** 223 (1991).

A

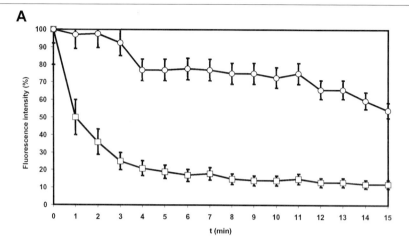

B

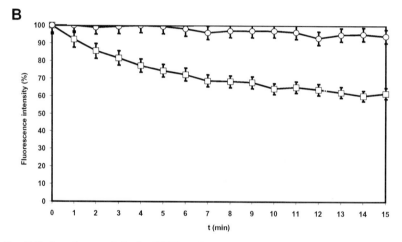

Fɪɢ. 7. Fading characteristics for FITC- and TRITC-conjugated IgG exposed to CLSM in the presence of Vectashield. The *X* axis is laser exposure time (min) and the *Y* axis is fluorescence intensity expressed as a percentage relative to the initial signal (i.e., $t = 0$ min fluorescence intensity is 100%). Each point is an average of 20 measurements at 1-min intervals. Fading during exposure to 50% laser power (A) and 20% laser power (B). FITC-conjugated IgG (□) and TRITC-conjugated IgG (○). Adapted from M. Berrios and D. E. Colflesh, *Biotech. Histochem.* **70,** 40 (1995), reproduced with permission.

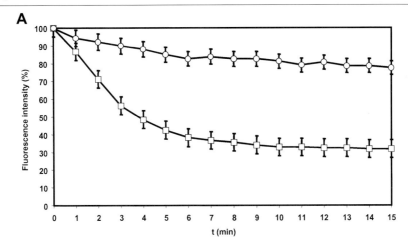

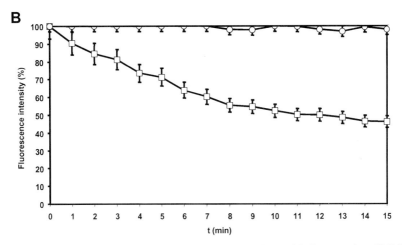

FIG. 8. Fading characteristics for FITC- and TRITC-conjugated IgG exposed to CLSM in the presence of SlowFade. The X axis is laser exposure time (min) and the Y axis is fluorescence intensity expressed as a percentage relative to the initial signal (i.e., $t = 0$ min fluorescence intensity is 100%). Each point is an average of 20 measurements at 1-min intervals. Fading during exposure to 50% laser power (A) and 20% laser power (B). FITC-conjugated IgG (□) and TRITC-conjugated IgG (○). Adapted from M. Berrios and D. E. Colflesh, *Biotech. Histochem.* **70,** 40 (1995), reproduced with permission.

A

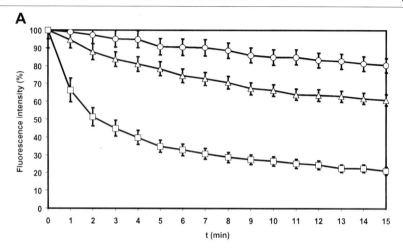

B

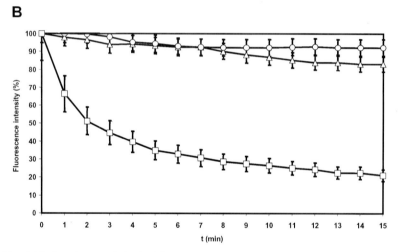

Fig. 9. Fading characteristics for FITC-, TRITC-, and TRSC-conjugated IgG exposed to CLSM in the presence of ProLong. The X axis is laser exposure time (min) and the Y axis is fluorescence intensity expressed as a percentage relative to the initial signal (i.e., $t = 0$ min fluorescence intensity is 100%). Each point is an average of 20 measurements at 1-min intervals. Fading during exposure to 50% laser power (A) and 20% laser power (B). FITC-conjugated IgG (□), TRITC-conjugated IgG (○), and TRSC-conjugated IgG (△). Adapted from D. E. Colflesh *et al., J. Histotech.* **22,** 23 (1999), reproduced with permission.

fluorescence. Although reducing laser power and dwelling time has not proven practical in our hands, these strategies applied in conjunction with reducing autofluorescence and using antifading agents are effective in limiting the deterioration of fluorescence signals. Furthermore, as a consequence of the CLSM selectiveness, the signal intensity at each plane of focus tends to be lower than that observed with the wide-field fluorescence microscope. This reduction in brightness must also be taken into consideration during specimen preparation, antibody probing, and image capturing. In addition, prior to image capturing, the acquisition system should be adjusted to employ the full dynamic range of the gray scale (i.e., 256 gray levels). This precaution only improves the signal-to-noise ratio but also prevents pixel saturation. More detailed discussions on this and related subjects may be found in works edited by Shotton[39] and Wootton et al.[40] and in comprehensive reviews by Majlof and Forsgren,[41] Brelje et al.,[42] and Paddy.[43]

We found these strategies essential to acquire comparable image pairs from *Drosophila* accessory gland nuclei to show the distribution of *Drosophila* chromatin remodeling protein 1 (CRP1)[44,45] and chromatin. We took advantage of the reduction of tissue autofluorescence after treatment with sodium borohydride and the antifading properties exhibited by ProLong to perform dual optical Z-series on *Drosophila* accessory gland nuclei[45] immunostained with mixtures containing anti-CRP1 antibodies and antihistone antibodies.[22] *Drosophila* adult male accessory glands were prepared, fixed, incubated with sodium borohydride, and probed with antibodies for indirect immunofluorescence as described previously[22] and in this article. To colocalize CRP1 with histones, nuclei were probed with a mixture containing antihistone mouse monoclonal antibody (MAb052; Chemicon International, El Segundo, CA) and affinity-purified rabbit anti-CRP1 antibodies.[45] Specimens were mounted with ProLong and examined by CLSM.

[39] D. Shotton, "Electronic Light Microscopy: Techniques in Modern Biomedical Microscopy." Wiley-Liss, New York, 1993.
[40] R. Wootton, D. R. Springall, and J. M. Polak, "Image Analysis in Histology: Conventional Confocal Microscopy." Cambridge Univ. Press, New York, 1995.
[41] L. Majlof and P. Forsgren, in "Methods in Cell Biology" (B. Matsumoto, ed.), p. 79. Academic Press, San Diego, 1993.
[42] T. C. Brelje, M. W. Wessendorf, and R. L. Sorenson, in "Methods in Cell Biology" (B. Matsumoto, ed.), p. 97. Academic Press, San Diego, 1993.
[43] M. R. Paddy, in "Methods in Cell Biology" (M. Berrios, ed.), p. 49. Academic Press, San Diego, 1998.
[44] K. Kawasaki, A. Philpott, A. A. Avilion, M. Berrios, and P. A. Fisher, *J. Biol. Chem.* **269,** 10169 (1994).
[45] G. Crevel, H. Huikeshoven, S. Cotterill, M. Simon, J. Wall, A. Philpott, R. A. Laskey, M. McConnell, P.A. Fisher, and M. Berrios, *J. Struct. Biol.* **118,** 9 (1997).

To minimize fading, images derived from FITC and TRSC wavelengths were obtained using 20 and 50% laser power, respectively. Staining with antihistone monoclonal antibody MAb052 followed by FITC-conjugated secondary antibodies was assigned the color green. Staining with anti-CRP1 antibodies followed by TRSC-conjugated secondary antibodies was assigned the color red. Figure 10 (see color insert) shows two sets each containing three immunofluorescent optical sections from two accessory gland nuclei. Figure 10A shows the TRSC image, Fig. 10B shows the corresponding FITC image, and Fig. 10C shows a pixel-by-pixel superimposition of images shown in Figs. 10A and 10B. Yellow indicates areas of colocalization (Fig. 10C). Anti-CRP1 antibodies and antihistone antibodies stained the nuclear interior of accessory gland nuclei with different patterns (compare Fig. 10A with Fig. 10B). Antihistone antibodies showed areas of dense chromatin around the nuclear periphery (Fig. 10A). In contrast, CRP1 was distributed as a diffused network that spanned the nuclear interior, including the nucleolus (Fig. 10B). In these nuclei, anti-CRP1 antibodies showed areas of overlap with antihistone antibody staining, particularly at or near the peripheral chromatin (Fig. 10C).

Appendix

Vendors

Materials and reagents included in this article for which the supplier was not specified may be purchased from the following sources.

Aldrich Chemical Company, Inc.
1001 West Saint Paul Avenue
Milwaukee, WI 53233
Phone: (800) 558-9160/(414) 273-3850
Fax: (800) 962-9591/(414) 273-4979
Telex 26843 Aldrich MI

Chemicon International, Inc.
100 Lomita Street
El Segundo, CA 90245
Phone (800) 437-7500
Fax: (213) 322-1086
TELEX 18-2079

Electron Microscopy Sciences
321 Morris Road
Fort Washington, PA 19034
Phone: (800) 523-5874
Fax: (215) 646-8931
http://www.emsdiasum.com/ems/

Fisher Scientific
711 Forbes Avenue
Pittsburgh, PA 15219-4785
Phone: (800) 766-7000
Fax: (800) 926-1166

Lancaster Synthesis, Inc.
P.O. Box 1000
Windham, NH 03087-9977
Phone: (800) 238-2324
Fax: (603) 889-3326

Mallinckrodt Chemical Inc.
16305 Swingley Ridge Drive
Chesterfield, MO 63017
Phone: (314) 530-2000
Fax: (314) 539-1110

Sigma Chemical Company
P.O. Box 14508
St. Louis, MO 63178-9916
Phone: (800) 325-3010
Fax: (800) 325-5052
http://www.sigma.sial.com

VWR Scientific
P.O. Box 626
Bridgeport, NJ 08014
Phone: (800) 932-5000
Fax: (609) 467-3336
http://www.vwrsp.com

Thomas Scientific
Swedesboro, NJ 08085-6099
Phone: (609) 467-2000
Fax: (609) 467-3087
Telex: 6851166 (WUI)
Cable: BALANCE Swedesboro

Acknowledgments

We thank Drs. Colin Dingwall and Farshid Guilak for valuable suggestions and critical reading of the original work. We also thank Wen Hui Feng, Laura Rosenberger, and Noran Instruments' engineers Dennis Maier and Adam Myerov for expert technical assistance and Joan Martin for help in assembling the manuscript. This work was supported by Research Grants PO1CA47995 from the National Cancer Institute, DCB-8615969 from the National Science Foundation, and an American Cancer Society institutional grant.

[5] Mounting Techniques for Confocal Microscopy

By Manabu Kagayama and Yasuyuki Sasano

Introduction

The confocal microscope is a valuable research tool for imaging biological specimens labeled with fluorochromes. Images produced by confocal microscopy are different from those by usual light microscopy in that the former is cut at the confocal plane of the objective lens. This offers a higher resolution of the image. Further, it is possible to cut optically serial sections and reconstruct three-dimensional structure in thick biological specimens using a confocal microscope.[1,2]

The quality of images taken by the confocal microscope is dependent on the specimen preparation procedure, such as fixation, embedding, cutting, staining, and mounting. Fluorescein isothiocyanate (FITC) and tetraethylrhodamine isothiocyanate (TRITC) are the most commonly used fluorochromes in immunohistochemistry and *in situ* hybridization. Fluorescence decay is rapid for these fluorochromes when glycerin or buffer solution is

[1] J. R. Salisbury, *Histol. Histopathol.* **9**, 773 (1994).
[2] S. W. Paddock, *Curr. Biol.* **7**, R182 (1997).

used as the mounting medium. *p*-Phenylenediamine and 1,4-diazobi-cyclo[2.2.2]octane are well-known additives that retard fading consider-ably,[3,4] Commercial mounting media of unknown chemical composition are available to preserve the intensity of fluorescence. These mounting media work well in thin sections. When applying these mounting media to thick sections, air bubbles occur frequently between coverslips and glass slides. This results in an increase in background noises by confocal microscopy.

Using a confocal microscope, a series of optical sections can be produced from a thick specimen labeled with fluorochromes. The number of optical sections collected is dependent on the characteristics of the specimens and is a function of objective lenses. A confocal microscope is usually equipped with oil immersion-type objective lenses for maximum resolution. Because the working distance of these lenses is very short (0.14–0.16 mm), the thickness of the coverslip impedes the observation of deeper parts of thick sections.

This article describes mounting methods for confocal microscopy.

Materials and Methods

Fixation

Tissues from rats are used for observations. Under pentobarbital anes-thesia, the tissues are fixed by perfusion through arteries with 4% (w/v) paraformaldehyde solution adjusted at pH 7.4 with 0.1 M phosphate buffer. For the vital staining of calcified tissues, some of the animals are injected with calcein (2 mg/kg body weight) and alizarin red (12 mg/kg) before fixation. After perfusion fixation, tissues are dissected and fixed in the same fixative for 10–20 hr and stored in phosphate-buffered saline (PBS) at 4°. Because most of the tissues used are calcified, part of the tissues are decalci-fied in 10% (w/v) EDTA at 4° for 1–3 months.

Methods for Thin Sections

After decalcification, tissues are dehydrated, immersed in xylene, and embedded in paraffin. Sections (3–5 μm) cut with a sliding microtome are placed on slide glasses coated with adhesive (APS slide glass, Matunami Glass Co., Oosaka, Japan) and are allowed to dry on a hot plate. The sections are stored at 4°. For frozen sections, the tissues are immersed in 10–20% (w/v) sucrose solution and are frozen by a mixture of dry ice and

---

[3] J. L. Platt and A. F. Michael, *J. Histochem. Cytochem.* **31**, 840 (1983).
[4] K. D. Krenik, G. M. Kephart, K. P. Offord, S. L. Dunette, and G. J. Gleich, *J. Immunol. Methods* **117**, 91 (1989).

acetone. Sections are cut with a cryostat microtome at a thickness of 10–30 μm, attached on the slide glasses, and allowed to dry.

The sections are stained with FITC- or TRITC-conjugated phalloidin (Sigma Chemical Co., St. Louis, MO) diluted to 50 μg/ml by PBS containing 0.05% (v/v) Triton X-100 at 4° for 1–3 hr. After staining, the sections are rinsed with PBS and mounted. To mount the sections, one to two drops of mounting medium are dispensed, and coverslips are placed onto the sections. The mounted specimens are stored at 4° in the dark. A mounting medium containing polyvinyl alcohol (PVA), Permafluor (Lipshaw, Pittsburgh, PA), and Vectashield (Vector Laboratories, Inc., Burlingame, CA) is used.

The PVA solution is prepared by adding 20 g of PVA (average molecular weight 30,000–70,000, Sigma Chemical Co.) to 80 ml of Tris–HCl buffer (65 mM) with continuous stirring to form a slurry. Then 15 ml of fluorescence-free glycerol (Merck, Darmstadt, Germany), 100 mg of chlorobutanol, and 100 mg of p-phenylenediamine are added to this solution. The solution is adjusted to pH 8.2 with 1 M HCl. The slurry is stirred and heated on a hot plate to the boiling point for about 15 min until the particles are dissolved. The mixture is degassed with a rotary vacuum pump and stored at 4° in the dark.

Methods for Thick Sections Cut with Microslicer

After fixation and decalcification, the tissue blocks are affixed with cyanoacrylate resin to a metal block holder, and sections are cut at the thickness of 100–200 μm within PBS using a microslicer (Dosaka, Kyoto, Japan) with continuous visual monitoring. The sections are transferred onto the slide glasses with a small brush and allowed to dry. The sections are stained with FITC- or TRITC-conjugated phalloidin (Sigma Chemical Co.) diluted to 50 μg/ml by PBS containing 0.05% Triton X-100 at 4° for 1–3 days, depending on the section thickness. After staining, the sections are rinsed thoroughly with PBS for 6–10 hr and mounted as described earlier.

Methods for Whole Mount Specimens and Thick Sections Cut with Saw Microtome

After fixation, the tissues stained vitally with calcein and alizarin red are dehydrated, immersed in propylene oxide, and embedded in Spurr's epoxy resin (Taab Laboratories Equipment, Aldermaston, Berkshire, England). The tissues are sliced with a saw microtome (SP1600, Leica, Heidelberg, Germany) at a thickness of 100–200 μm. As whole mount specimens, calvaria of 8-day-old rats are stained directly with FITC- or TRITC-conjugated phalloidin, rinsed, dehydrated, and embedded in epoxy resin as described earlier.

Thick sections and whole mount specimens are ground with sanding paper to expose the desired surface. After washing with distilled water, these specimens are dried and attached on the slides with cyanoacrylate. Specimens embedded with resins can be mounted temporally by immersion oil and covered with glasses for the observation with dry-type objective lenses. If the structure of interest is present at a deeper part than the working distance of oil immersion objectives, the surface of specimens can be ground again.

Confocal Microscopy

Confocal microscopy is performed using a microscope (CLSM, Leica, Heidelberg, Germany) equipped with objective lenses ×10, ×25, ×40, ×63, and ×100 (all were oil immersion type, except ×10, which was dry type) with aperture settings of 0.25, 0.75, 1.30, 1.40, and 1.32, respectively. The working distance of oil immersion objectives is 0.14–0.16 mm. The epifluorescence mode is used for the observation. To obtain optimum images of three-dimensional structures, images are cut serially and reconstructed with the computer of CLSM. The wavelength of excitation light used is 488 nm for FITC and calcein and 512 nm for TRITC and alizarin red.

Results and Discussions

Mounting methods using water-soluble media for thin sections worked well for confocal microscopy. Fluorescence of the specimens mounted with Permafluor and PVA was preserved for more than 1 year (Figs. 1 and 2).

As in the thin sections, images obtained from thick sections mounted with water-soluble media were very clear, even in objects deep within the thick sections. However, in most of the specimens mounted with Permafluor, air bubbles occurred near the perimeter of coverslips after prolonged storage. No bubbles occurred in the sections mounted with PVA and stored more than 1 year. Air bubbles may be formed possibly by the shrinkage of the hardened medium, because shrinkage of Permafluor was larger than that of PVA when mounting media were placed onto the glasses and allowed to dry. We have not used Vectashield until recently and performance of the thick sections mounted with it is not known. The manufacturer of Vectashield recommends sealing the coverslips with clear nail polish or a plastic sealant for prolonged storage.

In sections thicker than 150 μm, the deeper part of the specimens could not be focused because of the short working distance of oil immersion objective lenses. One of the possible methods to overcome this difficulty is the use of a thinner coverslip (0.04–0.06) instead of a standard one (0.17 mm).

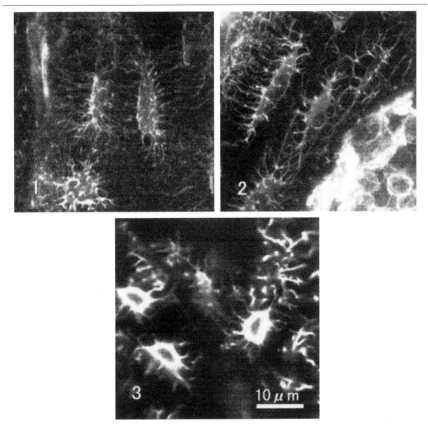

FIGS. 1–3. Confocal micrographs were reconstructed from 30 serial images taken by a 0.3-μm step using ×100 objective lens. Scale of the micrographs as indicated in Fig. 3.

FIGS. 1 and 2. Confocal micrographs of osteocytes taken using frozen sections stained with FITC-conjugated phalloidin, mounted with Permafluor, and stored for 1.5 years (Fig. 1) and 2 weeks (Fig. 2). Note that fine cytoplasmic processes of osteocytes are similarly visible in Figs. 1 and 2.

FIG. 3. Confocal micrographs of osteonal lacunae and canaliculi stained vitally with calcein in thick sections embedded in epoxy resin.

Tissues embedded in epoxy resin were transparent to the light used, and labeling lines of calcein and alizarin red were clearly visible at depths greater than 200 μm below the surface of the preparation. Figure 3 is a confocal micrograph of osteonal lacunae and canaliculi stained vitally with calcein in the thick section embedded by epoxy resin. Although the staining of osteonal lacunae and canaliculi was less intense than that of the surface of bone at the time of fluorochrome injection (calcification front), confocal microscopy can reveal those structures very clearly. As reported previously,[5] the rate of fading appeared to be less in specimens mounted with the resin than with PVA. An advantage of this method is that whole mount specimens can be examined by confocal microscopy. Without the use of a coverslip, there was no interface in which air bubbles could occur, and deeper parts of thick specimens could be analyzed than in the specimen covered with glasses.

[5] M. Kagayama, Y. Sasano, M. Hirata, I. Mizoguchi, and I. Takahashi, *Biotech. Histochem.* **71,** 231 (1996).

[6] Preparation of Whole Mounts and Thick Sections for Confocal Microscopy

By Hisashi Hashimoto, Hiroshi Ishikawa, and Moriaki Kusakabe

Introduction

Numerous researchers have attempted to reveal a three-dimensional relationship in tissue and cellular architectures because it is fundamental to a proper understanding of the biological system. It is very difficult to examine the distribution of nerve fibers and capillary nets in a tissue, the morphological changes of the epithelial tissue in a developing organ, and the shape and appearance of a cell from tissue sections. Distribution of a molecule in a tissue and interrelationships between tissue components may not be revealed until a three-dimensional reconstruction is made. One three-dimensional image may give us far more information than hundreds of tissue sections.

Three-dimensional analysis has mainly been performed by reconstructing from serial tissue sections, but there are technical difficulties and problems during the process due to damage or artifacts caused by the sectioning process. The confocal laser scanning microscope (CLSM) has been used for three-dimensional analysis. The CLSM possesses some attractive features: it enables us to get a sharp and vivid optical tomogram that is free from glaring fluorescence in out-of-focus planes and to digitalize and store the image in a permanent form. Consequently, serial optical images from thick specimens can be obtained easily by relocating focus planes at an arbitrary interval and can be reconstructed into a three-dimensional image with a high-powered personal computer.

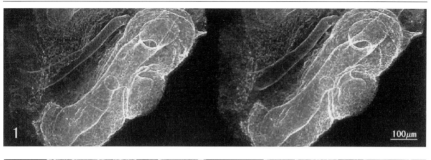

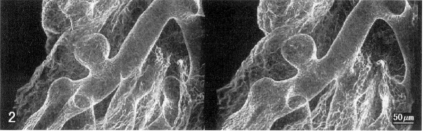

FIGS. 1 AND 2. Stereo pair images of laminin in a fetal mouse lung on day 11 (Fig. 1) and day 12 (Fig. 2) of gestation. Fetal mouse lungs were removed, fixed, pretreated, and immunostained for laminin as described in Procedure I. The antibodies used were rabbit antimouse laminin IgG (AT-2404, E.Y. Labs. Inc., San Mateo, CA) diluted 1 : 500 and FITC-conjugated goat antirabbit IgG antibody (No. 454, MBL Co., Nagoya, Japan) diluted 1 : 100. Three-dimensional stereo pair images were reconstructed with optional software for Carl Zeiss LSM-510 from 25 images taken at a 5-μm interval (Fig. 1) and from 43 images taken at a 3-μm interval (Fig. 2) with a Carl Zeiss LSM-10 and a Plan NeoFLUAR 20× objective lens (NA 0.5).

Specimens subjected to three-dimensional investigation with CLSM must satisfy certain criteria: they should have little autofluorescence and little absorbance of light for excitation and emission and they should not have obstacles for penetration of a fluorochrome tracer. This article presents a method (Procedure 1, see Appendix) that was developed primarily in order to observe the three-dimensional distribution of the extracellular matrixes: laminin and tenascin.[1] Laminin is a noncollagenous glycoprotein and is one of the main constituents of basement membranes in both fetal and adult organs (Figs. 1–4, 16 and 18). Three-dimensional visualization revealed branching in the fetal lung (Figs. 1 and 2) and the development of the pituitary gland.[2] The expression of tenascin is restricted spatiotempo-rally and its heterogeneous distribution is well indicated by three-dimen-

[1] H. Hashimoto and M. Kusakabe, *Connect. Tissue Res.* **36,** 63 (1997).
[2] H. Hashimoto, H. Ishikawa, and M. Kusakabe, *Dev. Dyn.* **212,** 157 (1998).

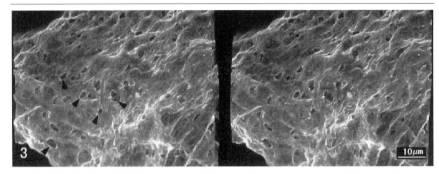

FIG. 3. Stereo pair images of a laminin sheet in the basement membrane zone of an adult mouse small intestinal villus. A proximal small intestine was removed from an adult mouse, fixed, sectioned at 150 μm, pretreated, and immunostained for laminin as described in Procedure I. The specimen was dehydrated and mounted in Rigolac as described in Procedure IV. The antibodies used were rabbit antimouse laminin IgG (AT-2404, E.Y. Labs. Inc.) diluted 1:3000 and BodipyFL-conjugated goat antirabbit IgG antibody (B-2766, Molecular Probes, Inc., Eugene, OR) diluted 1:200. Three-dimensional stereo pair images were reconstructed from 74 images taken at a 0.45-μm interval with a LSM-510 and a Plan NeoFLUAR 100× objective lens (NA 1.3). Arrowheads indicate pores in the laminin sheet.

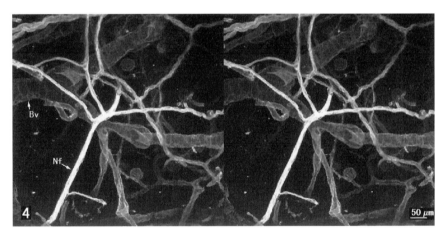

FIG. 4. Stereo pair images of laminin in subcutaneous tissue of an adult mouse ear. An adult mouse ear was fixed as usual and treated with 0.1 M disodium ethylenediaminetetraacetic acid (EDTA) in 0.1 M sodium phosphate buffer at 4° for 1 week. The skin and subcutaneous tissue of the ear were mechanically peeled off and immunostained for laminin after pretreatment with sodium deoxycholate. The antibodies used were rabbit antimouse laminin IgG (AT-2404, E.Y. Labs. Inc.) diluted 1:500 and FITC-conjugated goat antirabbit IgG antibody (No. 454, MBL Co.) diluted 1:100. Three-dimensional stereo pair images were reconstructed with a LSM-510 from 12 images taken at a 3-μm interval with a LSM-310 and a 20× objective lens. In the subcutaneous tissue of the mouse ear, immunofluorescence of laminin localized at the basement membrane indicated the three-dimensional distribution of numerous blood vessels (Bv) and nerve fibers (Nf).

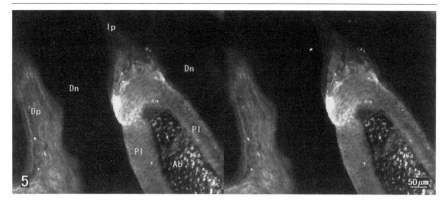

FIG. 5. Stereo pair images of tenascin in the periodontal ligament and tooth pulp of a mouse. A molar portion in the mandible of a mouse was fixed as usual and decalcified with 0.1 M disodium ethylenediaminetetraacetic acid (EDTA) in 0.1 M sodium phosphate buffer at 4° for 1 week. The specimen was sliced at 150 μm, pretreated with sodium deoxycholate, and immunostained for tenascin. The antibodies used were a rat antihuman tenascin monoclonal antibody (developed in our laboratory, RIKEN) at a concentration of 10 μg/ml and FITC-conjugated goat antirat IgG antibody (No. 6270, TAGO, Inc., Burlingame, CA) diluted 1:100. Three-dimensional images were reconstructed with a LSM-510 from 25 images taken at a 3-μm interval with a LSM-310 and a 20× objective lens. Tenascin is observed in the periodontal ligament, the odontoblast layer in the dental pulp (Dp), and around the osteocyte. Prominent heterogeneity in tenascin distribution is found in the periodontal ligament (Pl). Dense tenascin is localized at the periodontal ligament adjoining the dental cervix. Dn, dentine; Ab, alveolar bone; Ip, interdental papilla.

sional images (Figs. 5, 6 and 16). Our method is composed simply of fixation, sectioning if necessary, pretreatment, and immunostaining as usual, but each step was reexamined. This method is now applicable to other tissue components, such as neurotransmitters, including vasoactive intestinal peptide (VIP, Fig. 17),[3] somatostatin (Figs. 7 and 8), and substance P, peptide hormones, neurofilaments, and interleukins. In addition, this article also introduces a novel method for the three-dimensional investigation of vascular nets (Figs. 12–14 and 17–20).[2,4,5]

Fixation

The strongly fixed tissue is not suited for three-dimensional observation with CLSM. Various fixatives, such as formaldehyde, picric acid, mercuric

[3] L. Lee, H. Hashimoto, and M. Kusakabe, *Acta Histochem.* **99**, 101 (1997).
[4] H. Hashimoto, H. Ishikawa, and M. Kusakabe, *Anat. Rec.* **250**, 488 (1998).
[5] H. Hashimoto, H. Ishikawa, and M. Kusakabe, *Microvasc. Res.* **55**, 179 (1998).

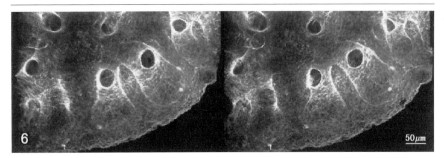

FIG. 6. Stereo pair images of tenascin in a fetal mouse lung at day 14 of gestation. Fetal mouse lungs were removed, fixed, pretreated, and immunostained for tenascin as described in Procedure I. The antibodies used were a rat antihuman tenascin monoclonal antibody (developed in our laboratory, RIKEN) at a concentration of 10 μg/ml and FITC-conjugated goat antirat IgG antibody (No. 6270, TAGO, Inc.) diluted 1 : 100. Three-dimensional images were reconstructed with a LSM-510 from 19 images taken at a 3-μm interval with a LSM-10 and a 20$\times$ objective lens. Tenascin is found encompassing the latest branching portion of bronchia.

chloride, ethanol, methanol, and acetone, have been used for preparing microscopic sections. However, each fixative more or less induces autofluorescence in biological samples by the denaturation or chemical modification of macromolecules. Strong fixatives, such as formaldehyde solution, surpass in preserving tissue and cellular architectures and immobilizing antigenic molecules, but coincidentally induce intense autofluorescence. In a preliminary study, the autofluorescence of rat liver and kidney fixed by calcium–formalin solution, 4% paraformaldehyde in sodium phosphate buffer, Zamboni's fixative, Carnoy's fixative, or Methacarn fixative was examined. Although each tissue showed intense autofluorescence in the cytoplasm, fluorescence from the liver fixed by Methacarn fixative was the brightest. The autofluorescence in these fixed tissues was not reduced by pretreatment with Triton X-100 or deoxycholic acid. Pretreatment with pepsin reduced the autofluorescence of the paraformaldehyde-fixed specimen and improved the penetration of antibodies, but this pretreatment was not appropriate for a whole mount specimen, as stated in the following section.

We scanned previous reports for immunostaining against laminin and found a report by Belford et al.[6] Belford et al.[6] immunostained laminin in a whole mount retina of rats and mice fixed by 0.5% paraformaldehyde in 15% saturated picric acid solution for 1–2 hr. They mentioned that fixation with formaldehyde concentrations greater than 0.5% reduced specific immunofluorescence and increased background fluorescence in whole mount

[6] D. A. Belford, G. A. Gole, and R. A. Rush, *Invest. Ophthalmol. Vis. Sci.* **28,** 1761 (1987).

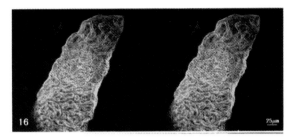

FIG. 16. Stereo pair images of mouse small intestinal villi immunostained for laminin (red) and tenascin (green). The mouse small intestine was processed as in Procedure I with rabbit antimouse laminin antibody, RITC-conjugated goat antirabbit IgG antibody, rat antihuman tenascin antibody, and FITC-conjugated goat antirat IgG antibody. Serial red-colored RITC and green-colored FITC images were taken simultaneously at a 1.5-μm interval with a LSM-310 and a Plan NEOFLUOR 40× objective lens (NA 0.75). Three-dimensional images were reconstructed with a LSM-510 from 20 images. By superimposing red-colored RITC and green-colored FITC stereo pair images, the interrelationship of the localization of laminin and tenascin is indicated clearly.

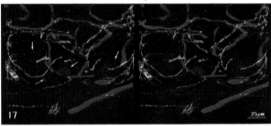

FIG. 17. Stereo pair images of vascular nets and VIP positive nerve fibers at the crypt of the mouse small intestine. The small intestine from a mouse perfused with RITC-labeled gelatin as described in Procedure III was sliced and immunostained for VIP with a rabbit antisynthetic porcine VIP serum (provided by Dr. Yanaihara, Shizuoka College of Pharmacology) and BodipyFL-conjugated goat antirabbit IgG antibody. The specimen was mounted in Rigolac as mentioned in Procedure IV. Serial red-colored RITC and green-colored BodipyFl images were obtained simultaneously at a 0.7-μm interval with a LSM-510 and a 40× objective lens, and stereo pair images were reconstructed from 90 images. The crypt is surrounded with capillary nets at its base and with VIP-positive nerve fibers at various postions. Red fluorescence found in granules of Paneth cell is nonspecific (arrows).

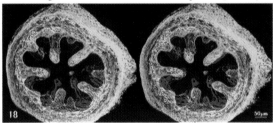

FIG. 18. Stereo pair images of vascular nets and laminin in a mouse small intestine at day 15 of gestation. A mouse fetus was perfused with RITC-labeled gelatin as in Procedure III and the small intestine was obtained. The tissue was sliced at a thickness of 150 μm, pretreated with sodium deoxycholate, and immunostained for laminin with a rabbit antimouse laminin antibody and a FITC-conjugated goat antirabbit IgG antibody. Stereo pair images were reconstructed from 55 serial red-colored RITC and green-colored FITC images obtained simultaneously at a 1.99-μm interval with a LSM-510 and a 20× objective lens. In a fetal small intestine, the interrelationship between developing villi and capillary nets is shown clearly.

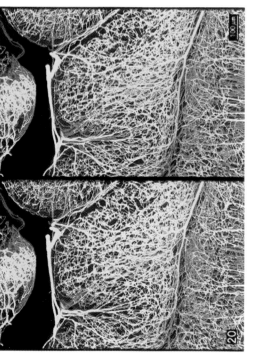

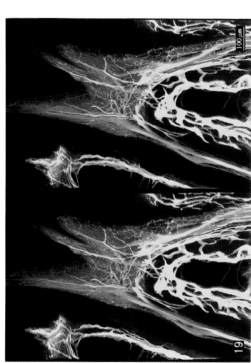

FIGS. 19 AND 20. Stereo pair images of vascular nets in a tooth (Fig. 19) and cerebellum (Fig. 20) of the mouse. A mouse was perfused with RITC-labeled gelatin and injected with 0.11 ml (net) of FITC-labeled gelatin just before the end of perfusion. The mandible was removed and decalcified with EDTA; the cerebellum was also removed. The specimens were sliced at 500 μm, treated with sodium deoxycholate, and mounted in Rigolac. Stereo pair images were reconstructed from 35 images taken at a 5.54-μm interval (Fig. 19) and from 38 images at a 5.81-μm interval (Fig. 20) with a LSM-510 and a Plan NeoFLUOR 10× objective lens (NA 0.3). The perfusion of the circulatory system from the left ventricle with a fluorochrome-labeled gelatin followed by a small amount (0.1–0.6 ml) of gelatin labeled with another colored fluorochrome shows that arterial vessels are distinguished from the venous vessels by the fluorescence from gelatin filling them.[5] In these figures, venous blood vessels are filled only with the RITC-labeled gelatin and emit red fluorescence, whereas most of the RITC-labeled gelatin in the arterial blood vessels, resulting in green to yellowish fluorescence. Vascular nets in hard tissue such as tooth and bone can be visualized three dimensionally after decalcification by our methods. In Fig. 19, arteries running through an accessory canal are seen.

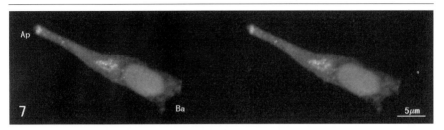

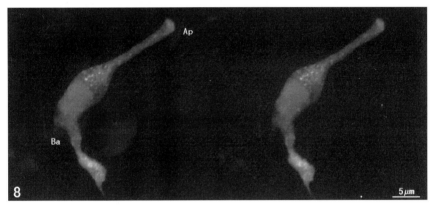

FIGS. 7 AND 8. Stereo pair images of somatostatin containing cells in the mouse small intestinal villus. A proximal portion of an adult mouse small intestine was fixed, sectioned at 150 μm, pretreated, and immunostained for somatostatin as described in Procedure I. The specimen was dehydrated and mounted in Rigolac as described in Procedure IV. The antibodies used were rabbit antisomatostatin serum (JIMRO, Gunma, Japan) and BodipyFL-conjugated goat antirabbit IgG antibody (B-2766) diluted 1:200. Three-dimensional stereo pair images were reconstructed from 35 images taken at a 0.25-μm interval (Fig. 7) and from 58 images taken at a 0.27-μm (Fig. 8) with a LSM-510 and a 100× objective lens. These three-dimensional images clearly indicate shapes and forms of the somatostatin cell situated among epithelial cells in the villus. The apex (Ap) of the somatostatin cell is exposed to the lumen of the small intestine and its base (Ba) lies on the basement membrane. Some somatostatin cells have a basal process traveling along the base of the epithelium as reported previously by Larsson *et al.*[31] and Sjölund *et al.*[32]

preparations. Picric acid is one of the chief constituents in Bouin's fixative and Zamboni's fixative[7] and is thought to produce intermolecular salt links by inducing polarity in the amino acid side chains of proteins.[8] Since the introduction of a buffered picric acid–formaldehyde mixture by Stefani *et al.*,[7] some researchers applied this mixture to the fixation of peptide-

[7] M. Stefani, C. de Martino, and L. Zamboni, *Nature* **216,** 173 (1967).

[8] A. G. E. Pearse, "Histochemistry: Theoretical and Applied," Vol. 1. Churchill Livingstone, Edinburgh, 1980.

containing endocrine cells and nerves. Costa et al.[9] stated that the picric acid–formaldehyde mixture preserved the structure of the tissue and the antigenicity of peptide-containing structures better than the other fixative tested. Somogyi and Takagi[10] presumed that the use of picric acid not only allows the use of low concentration glutaraldehyde, but also improves fine structural detail and, through the preservation of immunogenicity, provides better immunostaining than aldehydes alone. Therefore, we tried a 0.1 M sodium phosphate-buffered (SPB, pH 7.0) mixture of 0.5% paraformaldehyde and 15% (v/v) saturated picric acid solution for a whole mount preparation. The specimen fixed by this fixative showed little autofluorescence and preserved immunoreactivity against various molecules, including laminin, tenascin, type IV collagen, glial fibrillary acidic protein (GFAP), neurofilaments, neurotransmitters, and peptide hormones. This fixative has a comparatively mild action on tissue, but electron microscopy revealed that basic tissue architectures were well preserved after pretreatment with a sodium deoxycholate solution[3] (Fig. 10).

Sectioning

Early fetal organs and even the adult intestinal wall of a mouse can be handled as a whole organ. However, more developed and massive organs have a more complicated structure and are not practical as specimens for CLSM. In a specimen prepared by vascular perfusion with fluorochrome-labeled gelatin, emission and excitation light can reach more than 500 μm. Such thick specimens are not practical when fluorescence staining is administered after fixation.

For making a section 100 to 500 μm thick, a Sartorius-type microtome system, a sliding microtome equipped with a cryostage, is feasible for sectioning without deformation and fracture of a tissue. A vibratome, which is generally utilized for cutting thick sections, is suited for a well-fixed hardened tissue, but it is difficult to prepare useable sections from a weak-fixed soft tissue. It is also difficult for a cryostat, or a cold microtome, to make a thick section because such sections are liable to fragment due to freezing.

Prior to sectioning, the specimen is infiltrated with the cryoprotectant solution, 20% (w/v) sucrose and 10% (v/v) glycerol in phosphate-buffered saline (PBS, pH 7.2) through 5 and 10% (w/v) sucrose in PBS. The specimen on a cryostage should not be so cold that a slice may break before it thaws. The best sections can be obtained when the section thaws on the knife as soon as it is cut and then transferred to PBS with a slender brush.

[9] M. Costa, R. Buffa, J. B. Furness, and E. Solcia, Histochemistry 65, 157 (1980).
[10] P. Somogyi and H. Takagi, Neuroscience 7, 1779 (1982).

Pretreatment

Pretreatment of the specimen has two purposes: (1) to remove obstacles in order that tracers, such as fluorochromes and antibodies, can penetrate deeply and evenly into the specimen and (2) to reduce unexpected autofluorescence. Theoretically, the best image would be obtained if structural components and a target molecule are left intact and any other constituents, including plasma membrane and cytosolic proteins, are removed.

Originally, a detergent was utilized for improving the penetration of antibodies and for reducing nonspecific binding of antibodies in the immunoenzyme method for thick sections. Hartman et al.[11] first introduced Triton X-100, a nonionic detergent, in the diluting buffer for a primary antibody and in the subsequent washing buffer to reduce nonspecific, low-affinity protein–protein binding. Grzanna et al.[12] also added Triton X-100 to the diluting buffer and to the washing buffer. Variations in the concentration of Triton X-100 in the first incubation step produced variations in the intensity of staining, no staining of dopamine β-hydroxylase-containing processes was observed when Triton was omitted. Dehydration, clearing, and rehydration of the whole mount preparations before incubation with antibodies were attempted by Costa et al.[9] and the results were compared with those obtained by freezing and thawing or by using Triton X-100 in the incubation medium. As they reported, freezing and thawing or the use of Triton X-100 in the incubation media did not improve penetration and did not reduce the background fluorescence; indeed the positive nerve fibers appeared disrupted and fragmented. Attempts made to facilitate penetration by allowing the whole mount preparations to air dry or by freezing and thawing or by using detergents failed to solve the problem. The process of dehydrating, clearing, and rehydrating whole mount preparations before incubating with the antibodies improved the penetration of the antibodies dramatically. This treatment seems to have extracted some of the tissue and cell barriers to penetration of large protein molecules. Franklin and Martin[13] modified a procedure described by Grzanna et al.[12] and incubated tissue slices in PBS with 0.4% Triton X-100 for 15 min at room temperature as a pretreatment. They also tried to improve the permeation of antibodies with absolute methanol at −20° to −30° and then brought to PBS through an ethanol series at 4°. They found that there was a slight difference between tissue that had been permeabilized by absolute methanol at −20° and by Triton X-100 treatment; the latter tissue seemed to give a slightly weaker fluorescence than the former. Therefore, they selected

[11] B. K. Hartman, D. Zide, and S. Udenfriend, *Proc. Natl. Acad. Sci. U.S.A.* **69,** 2722 (1972).
[12] R. Grzanna, M. E. Molliver, and J. T. Coyle, *Proc. Natl. Acad. Sci. U.S.A.* **75,** 2502 (1978).
[13] R. M. Franklin and M. T. Martin, *Histochemistry* **72,** 173 (1981).

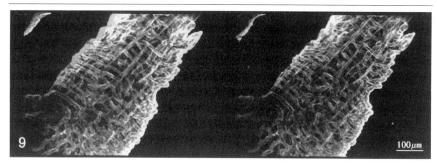

FIG. 9. An example of pepsin pretreatment (immunofluorescence staining for laminin in the small intestinal villus). The adult mouse small intestine was fixed with 4% paraformaldehyde in 0.1 M sodium phosphate buffer at 4° overnight. After several rinses with phosphate buffer, the small intestine was sliced between villi with razor blades under a dissection microscope to obtain intact villi. The specimen was pretreated with 3% sodium deoxycholate solution at room temperature for 4 hr and with 0.1% pepsin in 0.01 N HCl at room temperature for 4 hr. The specimen was then immunostained for laminin with rabbit antimouse laminin antibody (AT-2404) diluted 1:500 and FITC-conjugated goat antirabbit IgG antibody (N. 454) diluted 1:100. Three-dimensional images were reconstructed with a LSM-10 from eight images taken at a 3-μm interval. The pepsin pretreatment digested laminin in the epithelial basement membrane, whereas intense immunofluorescence of laminin clearly indicate the capillary basement membrane in lamina propria.

methanol permeation, although it entails more processing steps. Somogyi and Takagi[10] simply stated that penetration is enhanced greatly by the freeze–thaw treatment and although this affects preservation adversely, the fine structural details are still fairly good.

As mentioned earlier, the method and effect of pretreatment differ from researcher to researcher. These differences are derived from the diversity in target organs, antigens, the method for fixation, the specificity of antibodies, etc. Therefore, we cannot select the method of choice for any organ and any antigen.

In our early trials, we have performed pepsin treatment on a 4% paraformaldehyde-fixed specimen for the three-dimensional observation of laminin (Fig. 9). The enzymatic digestion by proteinase, including pepsin, is usually carried out prior to immunostaining in order to restore antigenicity.[14,15] However, in the enzymatic digestion of a whole mount specimen or a thick slice, the enzyme acts on the surface of the specimen and gradually on the deeper parts. The surface of the specimen may be digested fully

[14] J. C. W. Finley and P. Petrusz, in "Techniques in Immunocytochemistry" (G. R. Bullock and P. Petrusz, eds.), Vol. 1, p. 239. Academic Press, London, 1982.
[15] H. Hashimoto and K. Hoshino, Acta Histochem. Cytochem. 21, 125 (1988).

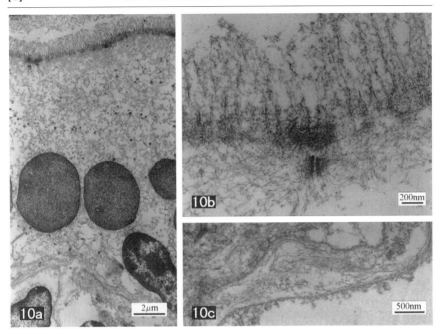

FIG. 10. Electron micrographs of the small intestine pretreated with sodium deoxycholate solution. The mouse small intestine was processed as described in Procedure I with an anti-mouse laminin antibody, observed with a LSM-10, and then postfixed with glutaraldehyde and osmium tetroxide, dehydrated with a graded series of ethanol, and embedded in Epon 812 through propylene oxide. The ultrathin section was stained with uranium acetate and lead citrate. Original magnification: ×4000 (a), ×20,000 (b), and ×8000 (c). The tissue and cellular architecture is well preserved, although the pretreatment with sodium deoxycholate was performed on a mildly fixed tissue. Filamentous structures in a cytoplasm, microvilli, and desmosome are observed in the enterocyte (a and b). In the basement membrane zone, filamentous structures are also present and collagenous fibers are found within the lamina propria of the villus (c). Adopted from L. Lee, H. Hashimoto, and M. Kusakabe, *Acta Histochem.* **99,** 101 (1997).

before its deeper part is digested optimally. In case of weak digestion, the surface of the specimen may show intense immunoreactivity whereas little immunoreaction will occur in its deeper part. As the enzymatic digestion proceeds, immunoreaction in the deeper parts becomes intense whereas that on the surface is diminished. Moreover, enzymatic digestion is affected by various conditions, such as fixation of the tissue, pH, ionic strength and temperature of the enzyme solution, enzyme concentration, and reaction time. Therefore, the enzymatic pretreatment is not suitable for a thick specimen except when the treatment is needed in order to restore specific antigenicity.

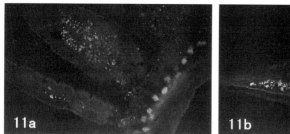
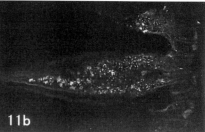

Fig. 11. Examples of nonspecific binding of two kinds of FITC-conjugated antibodies in the small intestinal villi. The mouse small intestine was fixed and pretreated as in Procedure I. The specimen was incubated with FITC-conjugated antirat IgG antibodies without incubation with a primary antibody. Nonspecific fluorescence is found at some cells in the lamina propria and the granules of the Paneth cell. The specificity and nonspecific binding of a secondary antibody should be checked carefully in addition to those of a primary antibody. Modified from L. Lee, H. Hashimoto, and M. Kusakabe, *Acta Histochem.* **99,** 101 (1997).

After changing the fixative to 0.5% paraformaldehyde and 15% (v/v) saturated picric acid solution, we attempted pretreatment with a detergent instead of pepsin because the detergent had been utilized to isolate a basement membrane. Von Bruchhausen and Merker[16] introduced sodium deoxycholate to purify the basement membrane of rat renal cortex for the first time. They found that the fine structure of the isolated basement membrane is identical to that of the corresponding structure in the renal cortex. Welling and Grantham[17] isolated the basement membrane by perfusing renal tubules with sodium deoxycholate or Triton X-100 and stated that several observations support the view that deoxycholate does not alter the physical properties of the basement membrane appreciably. They also mentioned that prolonged incubation of a basement membrane in deoxycholate had no effect on hydraulic conductivity and that a basement membrane isolated in sodium deoxycholate appeared morphologically indistinguishable from untreated membranes when examined by electron microscopy and stained positively with the periodic acid–Schiff reagent, indicating preservation of the carbohydrate moieties in the membrane. They had difficulty obtaining consistently clean basement membranes using Triton X-100, although the resulting basement membranes in successful studies appeared identical to those obtained with sodium deoxycholate. Meezan *et al.*[18] reported a method for the isolation of ultrastructurally and chemically pure basement membranes from kidney glomeruli and tubules

---

[16] F. Von Bruchhausen and H. J. Merker, *Histochemistry* **8,** 90 (1967).
[17] L. W. Welling and J. J. Grantham, *J. Clin. Invest.* **51,** 1063 (1972).
[18] E. Meezan, J. T. Hjelle, K. Brendel, and E. Carlson, *Life Sci.* **17,** 1721 (1975).

and brain and retinal microvessels by the sequential solubilization of cell membranes, intracellular protein, and plasma proteins by Triton X-100 and sodium deoxycholate. They mentioned that direct treatment of the pure organ subfractions with sodium deoxycholate was adequate for the preparation of basement membranes from a few milligrams of tissue, but with larger amounts of tissue, graded solubilization and deoxyribonuclease treatment were necessary to avoid the formation of a gel-like aggregate of DNA with the basement membrane, which was then difficult to disperse.

These previous reports utilized sodium deoxycholate to isolate pure basement membranes from an unfixed tissue and indicated that the ultrastructural and physical properties of the isolated basement membranes are identical to those seen in the intact tissue. We have attempted to apply the pretreatment with Triton X-100 or sodium deoxycholate on a fixed tissue. Pretreatment with sodium deoxycholate reduced autofluorescence from a mouse liver and renal cortex fixed by a phosphate-buffered mixture of 0.5% paraformaldehyde and 15% saturated picric acid solution, 2% paraformaldehyde in sodium phosphate buffer, or 4% paraformaldehyde in sodium phosphate buffer, but that with Triton X-100 had a slight effect. The transparency of the light ranging from 400 to 800 nm in a 100-μm slice of liver and renal cortex fixed with the mixture of paraformaldehyde and picric acid was improved by pretreatment with sodium deoxycholate, but not with Triton X-100. Light and electron microscopic examination of a tissue fixed with a mixture of paraformaldehyde and picric acid and incubated in sodium deoxycholate indicated that the tissue and cellular architecture was not altered; membranous structures and cytosolic macromolecules in a cell were lost[3] (Fig. 10). The fluorescence immunostaining of the tissue for laminin and tenascin showed intense fluorescence even deep in the tissue. Prolonged pretreatment with sodium deoxycholate caused no effect on the antigenicity of laminin and tenascin or on the histoarchitecture.

We now routinely utilize a 3% sodium deoxycholate solution as pretreatment. However, sodium deoxycholate cannot be used in all cases because a target molecule may be extracted by this reagent. Various kinds of detergents are now available commercially. Each detergent has different properties, such as critical micelle concentration, and exerts different actions on lipids and proteins. If a detergent is not successful with a molecule, it is advisable to try another kind of detergent with different electrical properties.

Fluorescent Staining

The most widely used fluorescence staining for CLSM would be the fluorochrome-labeled antibody method. Considering the purpose of three-dimensional analysis, one should select the most specific staining method and avoid the possibility of unexpected background fluorescence. Conventional staining methods, including PAS reaction and Feulgen reaction with fluorescent Schiff-type reagents, may be performed. However, these conventional methods are likely to label so many structures that the three-dimensional image becomes complicated in analyzing their distribution.

On the topic of fluorescent staining, there are two points to which attention should be paid: (1) uniformity in staining and (2) the prevention of nonspecific reaction. In order to attain uniform staining of a specimen with a specific antibody, the antibody must be able to penetrate evenly and deep into it. Pretreatment of the specimen is needed for this purpose. The concentration of the antibody and the duration of incubation influence the results deeply and must be determined carefully by the researcher. In our experiment, a more desirable result was obtained at a lower concentration and longer incubation time than those used for staining of the conventional section. In the case of the high concentration of the antibody, the antibody bound with the antigen situated at the surface area of the specimen may aggregate and interfere with the unbound antibody in penetration into a deeper area of the specimen. The surface area of the specimen may show intense immunofluorescence, while the deeper area emits fluorescence weakly.

The specificity of the primary antibody is indisputably important for staining the specimen as well. In addition, care must be taken regarding the specificity of the secondary fluorochrome-labeled antibody, as the weak-fixed and pretreated specimen is likely to show fluorescence by the nonspecific binding of antibodies, especially by the secondary antibody. We usually utilize a fluorochrome-labeled antibody that has been purified by affinity chromatography and absorbed by the serum of a target animal. Some antibodies, however, showed a nonspecific reaction (Fig. 11).

Vascular Cast with Fluorochrome-Labeled Gelatin:
  Procedures II and III

In a series of three-dimensional investigations of fetal organs with the antilaminin antibody, it was difficult to distinguish the basement membrane of vessels from that of the epithelium. In developing organs, vascular nets seem to play a crucial role in the morphogenesis (Figs. 12–14 and 18).

Vascular nets in a organ have been investigated by the intravascular injection of India ink before. Since the introduction of acrylic monomer as

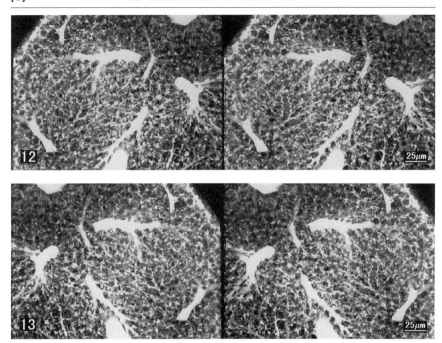

FIGS. 12. AND 13. Vascular nets in the mouse liver at postnatal day 5. A mouse was perfused with FITC-labeled gelatin and processed as described in Procedure III. The liver was removed, sectioned at 150 μm, and observed with a LSM-310. Three-dimensional images were reconstructed with a LSM-510 from 40 images taken at a 2-μm interval from the front side (Fig. 12) and the far side (Fig. 13). Sinusoidal capillary nets in the young mouse liver are well observed without leakage of FITC-labeled gelatin. As the three-dimensional reconstruction can be performed from both the front and the far side and with an arbitrary number of serial images, vascular nets of interest can be examined easily.

an injection medium by Batson,[19] plastic resins, such as prepolymerized methylmethacrylate and Mercox, have been applied and injection replica of vascular nets were produced.[20,21] After the removal of all soft and hard tissues, the injection replica has been principally observed by a scanning electron microscope. However, it is impossible to reveal the interrelationship between vascular nets and the other tissue components because this method requires removing soft tissues for observation.

[19] V. Batson, *Anat. Rec.* **121,** 425 (1955).
[20] T. Murakami, *Arch. Histol. Jap.* **33,** 179 (1971).
[21] K. C. Hodde and J. A. Nowell, *in* "SEM/1980/Part II," p. 89. SEM Inc., AMF O'Hare, Chicago, 1980.

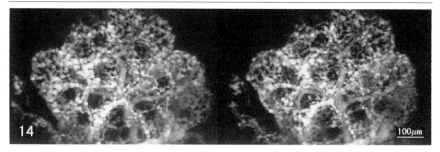

FIG. 14. Vascular nets in a fetal mouse lung at day 15 of gestation. A mouse fetus was perfused with RITC-labeled gelatin as mentioned in Procedure III. The lung was removed and observed with a LSM-10. Three-dimensional images were reconstructed with a LSM-510 from 28 images taken at a 3-μm interval. Numerous vascular nets surrounding the developing bronchia are found.

There are some reports concerning the observation of vascular nets with light or fluorescence microscopes.[22–25] In these reports, antibodies against a vascular endothelial cell-specific antigen such as factor VIII and angiotensin-converting enzyme II or lectins specific for a vascular endothelial cell such as *Ulex europaeus* agglutinin I (UAE-I) and *Ricinus communis* agglutinin I (RCA) were utilized to detect endothelial cells. However, some species and strain specificity of these antibodies and lectins may be present and it is not certain whether these tracers react uniformly in each fetal immature to adult mature endothelial cells. Therefore, we attempted to make an injection replica of vascular nets with gelatin labeled with a fluorochrome in order to examine with CLSM.[4,5] (see Appendix). Gelatin has some advantageous characteristics. An aqueous gelatin solution does not solidify in a vital warmth. It is solidified not by polymerization, but by cooling. This is particularly favorable for the perfusion of a fetus because no time limit is set for injection as long as the fetus and the gelatin solution are kept warm. There are many reactive residues, including amine in a gelatin molecule. These residues bind covalently with an isothiocyanate derivative of fluorochrome, such as fluorescein isothiocyanate, and are cross-linked easily to each other with fixatives. Once fixed, the solidified gelatin rarely dissolves by warming.

[22] S. E. Connolly, T. A. Hores, L. E. H. Smith, and P. A. D'Amore, *Microvasc. Res.* **36,** 275 (1988).
[23] H. Holthöfer, I. Virtanen, A.-L. Kariniemi, M. Hormia, E. Linder, and A. Miettinen, *Lab. Invest.* **47,** 60 (1982).
[24] R. D'Amato, E. Wesolowski, and L. E. H. Smith, *Microvasc. Res.* **46,** 135 (1993).
[25] V. Rummelt, L. M. G. Gardner, R. Folberg, S. Beck, B. Knosp, T. O. Moninger, and K. C. Moore, *J. Histochem. Cytochem.* **42,** 681 (1994).

This method is easily applicable to a fetus because no time limit is set for injection and cardiac contractions are maintained during perfusion (Figs. 14 and 18). In addition, a specimen prepared by vascular perfusion of fluorochrome-labeled gelatin can be stained again by the immunofluorescence method (Figs. 17 and 18). This means that the interrelationship between vascular nets and other tissue components can be examined three dimensionally. This method will aid research on the role of vascular nets.

Mounting

The preparation for immunofluorescence observation is usually mounted in an aqueous mounting medium that is a mixture of glycerol and a buffer solution. The most widely used buffer is 0.1 M sodium phosphate buffer (pH 7.4) or 0.01 M sodium phosphate-buffered saline (pH 7.2). In many cases, an antifading agent is added to the mounting medium.

Fading of fluorescence by continuous excitation is an inevitable problem in fluorescence microscopy, including CLSM. In order to make a three-dimensional observation, dozens of images must be taken from the same region, although the focal plane is different at each image. Fluorochrome located in a different focal plane is not a little excited by being exposed to a laser light source and quenching of the fluorescence is caused. Several chemicals are reported to prevent fading of the fluorochrome by the addition to mounting media.[26-28] These are sodium azide, sodium iodide, polyvinyl pyrrolidone (PVP), polyvinyl alcohol (PVA), 1,4-diazobicyclo[2.2.2]octane (DABCO), p-phenylenediamine, n-propyl gallate, and sodium dithionite. Giloh et al.[26] recommends 0.1 to 0.25 M of n-propyl gallate in glycerol to reduce the photobleaching of tetramethylrhodamine and fluorescein. Johnson et al.[27] mentions that p-phenylenediamine is the most efficient retarding agent and that DABCO, which is an extremely stable compound and is nonionizing, inexpensive, and readily available, appears to be a suitable alternative. Contrary to these previous reports, Böck et al.[28] states that sodium azide and sodium iodide increase fluorescence intensity and inhibit bleaching but that the other compounds reported previously to exert beneficial effects did not lead to an increase in fluorescence and had questionable effects on bleaching.

[26] H. Giloh and J. W. Sedat, *Science* **217,** 1252 (1982).
[27] G. D. Johnson, R. S. Davidson, K. C. McNamee, G. Russel, D. Goodwin, and E. J. Holborow, *J. Immunol. Methods* **55,** 231 (1982).
[28] G. Böck, M. Hilchenbach, K. Schauenstein, and G. Wick, *J. Histochem. Cytochem.* **33,** 699 (1985).

We usually use 0.05 M Tris–HCl-buffered saline (pH 8.0) containing 90% (v/v) glycerol and 10 mg/ml of DABCO as a mounting medium. A mixture of glycerol and PBS (9 : 1) containing n-propyl gallate had been used previously, but this mounting medium seemed to induce autofluorescence in a nucleus and the fluorescence of fluorescein was not stable. The Tris–HCl buffer was adopted because the fluorescence intensity of fluorescein derivatives increases and becomes stable as the pH of the solution increases in the physiological range. In our experience, the addition of DABCO reduces the photobleaching of fluorescein and rhodamine, although Böck et al.[26] report that the effect of DABCO on photobleaching is questionable. When a vulnerable fetal organ is to be mounted, it is desirable to increase gradually the concentration of glycerol (30%, 50%, and 70%) in order not to cause shrinkage or deformation of the specimen. This mounting medium is stable over several months. The specimen mounted in this mounting media can be stored for a few weeks at 4°, but longer storage increases autofluorescence in cytoplasm and reduces the specific fluorescence of fluorescein and rhodamine.

For mounting a whole organ or a thick section, spacers should be placed between the glass slide and the coverslip in order not to crush it. In our laboratories, two glass strips made from a coverslip are attached to a glass slide as a spacer and a specimen is mounted between the strips. We can obtain coverslips having thicknesses ranging from 0.13 to 0.60 mm. The thickness of the spacer should be approximately equal to that of the specimen.

As mentioned earlier, the specimen for fluorescence microscopy is usually mounted in an aqueous mounting medium. The specimen in the medium is pinched lightly between the glasses and is not anchored firmly. This brings up a problem that the specimen may shift when the coverslip is pushed by an oil-immersion objective lens to take serial images. The shift of specimen causes the deviation of images, resulting in a deformation of the three-dimensional image. A specimen mounted in an aqueous mounting medium increases background fluorescence gradually, even if it is returned to PBS. For these reasons, we have attempted to mount the specimen in a resin to make a semipermanent preparation. The tissue section is mounted conventionally in a balsam or a synthetic styrene resin dissolved in an organic solvent such as xylene through dehydration with a graded series of ethanol. These resins are hardened by the evaporation of solvent. If a whole mount specimen or a thick section is mounted in these resins, it will take a long time to harden completely and shrinkage of the resins during hardening may cause deformation. Franklin et al.[13] attempted to embed immunofluorescence-stained tissue slices in hydroxypropyl methacrylate to obtain high-resolution staining of antigens in 1-μm tissue sections. However,

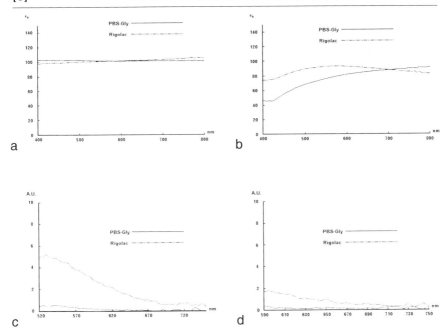

FIG. 15. (a) Transparency spectra of PBS–glycerol (1 : 9) and polymerized Rigolac, a polyester resin. Rigolac was polymerized as in Procedure IV in a space 170 μm thick between the glass slide and the coverslip. Transparency spectra of the polymerized resin and PBS–glycerol in the space between the glass slide and the coverslip were examined against distilled water with a Carl Zeiss UMSP microspectrophotometer. (b) Transparency spectra of a liver mounted in PBS–glycerol (1 : 9) and Rigolac. The mouse liver was removed, fixed, and sliced as in Procedure I. The liver slice was mounted in PBS–glycerol or Rigolac. The transparency spectrum of the mounted liver slice was recorded against each mounting media. (c and d) Fluorescence spectra of a liver mounted in PBS–glycerol (1 : 9) and Rigolac. The liver slice was prepared as in (b). Autofluorescence spectrum of the mounted liver slice was investigated under blue excitation (c) or green excitation (d). By mounting in Rigolac, the transparency of the liver slice is improved but autofluorescence of the liver is increased, especially under blue excitation.

a 170-μm-thick polymerized hydroxypropyl methacrylate resin showed autofluorescence as well as epoxy resin, Epon 812, in our experiment. An unsaturated polyester resin, Rigolac, which is one of the acrylic resins and is utilized as an embedding medium for transmission electron microscopy, was found to have a high transparency of visible light and no autofluorescence (Fig. 15a). The refractive index to the polymerized polyester resin [a mixture of Rigolac 2004 and Rigolac 70F (7 : 3)] is 1.5292,[29] a little higher

[29] Sato, Showa Kobunshi Co. Ltd., personal communication, 1997.

than that of glass. The transparency of the light through a slice of liver mounted in the polyester resin is higher than that mounted in glycerol/ PBS (9:1) (Fig. 15b). When a specimen is mounted in polyester resin between two coverslips spaced with glass strips, the specimen can be observed from both the front and the far side. In the case of observation with an oil-immersion lens, a coverslip is removed from the resin and an immersion oil is dropped directly on the resin so that the short working distance of the oil-immersion lens may be compensated. However, polyester resin has some problems. The autofluorescence in the cytoplasm increases especially under blue excitation (Figs. 15c and 15d). The fluorescence intensity of fluorochromes conjugated to antibodies may decrease due to polymerization. In the fluorochrome tested, the fluorescence of antibodies labeled with fluorescein isothiocyanate, Oregon Green 488, and tetramethylrhodamine isothiocyanate was reduced, whereas that of BodipyFL was preserved and was observable with a fluorescence microscope and CLSM (Figs. 7, 8, 17, 19, see color insert and Fig. 20, see color insert). The fluorescence of fluorescein isothiocyanate and rhodamine isothicyanate conjugated to gelatin was hardly affected. Injection replica of vascular nets by fluorochrome-labeled gelatin was clearly indicated in a specimen 500 μm thick mounted in this polyester resin. The application of resin mounting is still limited, but the specimen mounted in the resin is easy to handle and maintain and would be semipermanently observable if stored in the dark.

Observation

We have been utilizing the CLSMs of Carl Zeiss, LSM-10, LSM-310, LSM-410, and LSM-510. Because the operation of the CLSM depends mainly on the specific control software for it, the comments in this section may not be applicable for every CLSM.

Prior to the observation with CLSM, the specimen should be examined with a conventional fluorescence microscope, especially on the specific and background fluorescence. A confocal image is a computer graphic constructed from ten thousands or millions of pixels whose brightness corresponds with the integrated fluorescence intensity of a limited spectrum in a focal point. If a background fluorescence has a broad or a similar spectrum with a specific fluorescence, it is impossible to distinguish specific fluorescence from background on a confocal image. Although the pseudo-colored image may assist it, the real color of specific and background florescence is not reproduced accurately on the pseudo-colored image.

The interval between images should be determined from the numerical aperture (NA) of the objective lens used and the size of the pinhole used for obtaining a confocal image. A confocal image is not the image of a mathematical plane, but it includes fluorescence from an area with some thickness. The thickness is conventionally referred to as full-width at half-maximum. The interval between images is principally determined by reference to the value of full-width at half-maximum. If the interval is fairly large, some structures may be lost in the reconstructed three-dimensional image. However, if it is short, so many similar images are collected and it takes a large memory to store and a long time to reconstruct. The latest control software for LSM-510 automatically calculates full-width at half-maximum at the current setting of the objective lens, the size of confocal pinhole, and the wavelength of laser beam and suggests half of it as the optimal interval between images.

One of the most important and intrinsic issues found on a confocal image is that the fluorescence intensity from a structure oriented in parallel with the optical axis becomes high and that from a structure perpendicular to the optical axis becomes relatively low.[30] That is, when a tubular structure emits a fluorescence of uniform intensity, confocal images show high intensity on the side walls and low intensity on the top and bottom walls (refer to Figs. 1 and 2). This phenomenon is more conspicuous at low magnification than at high magnification because the NA of the lens determines full-width at half-maximum or the thickness of a confocal image.

Three-dimensional images reconstructed with a CLSM will afford meritorious information on the biological system to researchers.

Appendix

Procedure I: Standard Procedure for Sample Preparation

1. Remove a tissue, slice into small pieces with razor blades, and fix by immersion in a fixative consisting of 0.5% paraformaldehyde and 15% (v/v) saturated picric acid in 0.1 M sodium phosphate buffer (SPB, pH 7.0) for 2 hr at room temperature or overnight at 4°. Fixation by perfusion with the fixative is desirable, if possible. After perfusion, the tissue is treated in the same manner as described earlier.

2. Rinse the tissue with 0.1 M SPB (pH 7.4) several times until picric acid is no longer found in the buffer.

[30] H. T. M. van der Voort and G. J. Brakenhoff, *J. Microsco.* **158,** 43 (1990).
[31] L.-I. Larsson, N. Goltermann, L. De Magistris, J. F. Rehfeld, and T. W. Schwartz, *Science* **205,** 1393 (1979).
[32] K. Sjölund, G. Sandén, R. Håkanson, and F. Sundler, *Gastroenterology* **85,** 1120 (1983)

3. Sectioning (optional): Immerse the tissue in 0.01 M sodium phosphate-buffered saline (PBS,[33] pH 7.2) containing 20% sucrose and 10% glycerol through PBS[33] containing 5 and 10% sucrose. Freeze the tissue on a cryostage and section at 100 to 200 μm. Collect sections in PBS.
4. In the following steps, a 48-well multiwell plate is used to incubate the specimen under mild agitation. Incubate the specimen in a 3% sodium deoxycholate solution for 4 hr at room temperature or overnight at 4°. The incubation time may be prolonged, if necessary.
5. Rinse the specimen with distilled water two times and then with PBS for three times 1 hr each.
6. Incubate the specimen with 10% preimmune serum appropriate for the secondary antibody, such as normal goat serum, overnight at 4°.
7. Apply the first antibody, such as antilaminin rabbit serum, for 1 or 2 days at 4°. The optimal dilution of the antibody should be determined previously.[34]
8. Rinse the specimen with PBS several times 1 hr each.
9. Apply the secondary antibody, such as fluorescein isothiocyanate (FITC)-labeled goat antirabbit IgG antibody, for 1 day at 4°.[34]
10. Rinse the specimen as in step 8.[35]
11. Mount the specimen in 0.05 M Tris–HCl-buffered saline (pH 8.0) containing 90% glycerol and 10 mg/ml of DABCO.

Procedure II: Preparation of Fluorochrome-Labeled Gelatin

1. Swell 8 g of gelatin (from bovine skin, approximately 225 bloom, Sigma, St. Louis, MO) in 30 ml of distilled water in an incubator at 37°.
2. Dissolve the gelatin completely in an incubator at 60° after swollen.
3. Adjust the total volume of the gelatin solution to 40 ml (the final gelatin concentration is 20%).
4. Raise the pH of the gelatin solution to 11 by adding a 1 N NaOH solution (it requires about 1.2 ml).
5. Dissolve 20–50 mg of isothiocyanate derivative of fluorochrome (fluorescein isothiocyanate, rhodamine isothiocyanate, or tetramethylrhodamine isothiocyanate) in 1 ml of absolute dimethyl sulfoxide (DMSO).
6. Gently pour the fluorochrome solution into the warm gelatin solution. The final weight ratio of fluorochrome to gelatin is 1:160 to 1:400.

[33] Sodium azide is added to PBS at the concentration of 0.05%.
[34] For diluting serum or an antibody, PBS containing 1% bovine serum albumin and 0.1% sodium azide is used.
[35] For double staining, repeat steps 6–10.

7. Place the mixture in a dark incubator at 37° overnight and agitate mildly to progress the labeling of the gelatin with the fluorochrome.
8. Transfer the reacted mixture into a seamless dialyzing tube and dialyze against 0.01 M phosphate-buffered saline containing 0.01% sodium azide in a dark incubator at 37°.
9. Continue the dialysis for 7–10 days; change the dialyzing buffer every day.
10. End the dialysis when no fluorochrome is found in the dialyzing buffer.
11. Place the dialyzed mixture in a dark refrigerator to gel the mixture.
12. Transfer the gelled mixture to a tightly stoppered glass bottle and store in a dark refrigerator.

Procedure III: Perfusion with Fluorochrome-Labeled Gelatin

1. Dissolve the fluorochrome-labeled gelatin in a warm water bath.
2. Adjust the final concentration of fluorochrome-labeled gelatin approximately 10% with 0.01 M PBS.
3. Load the fluorochrome-labeled gelatin into a plastic syringe equipped with a plastic tube.
4. Keep the syringe with the plastic tube and needle in a warm water bath.
5. Perfusion of an animal

(a) Whole Body Perfusion for a Small Animal[36,37]

Anesthetize an animal with intraperitoneal injection of sodium pentobarbital solution.
Perform a thoracotomy and expose the heart.
Insert the needle into the left ventricle and clamp with a Kocher clamp. Care must be taken to keep the pinpoint of the needle in the left ventricle.
Sever the right atrium or auricle and drain blood from it.
Inject the fluorochrome-labeled gelatin solution slowly.
Continue the perfusion until no more blood cells flow.

(b) Perfusion of a Fetus[36]

Kill a pregnant animal by excess inhalation of ether or by cervical dislocation.

---

[36] Heparinization prior to the perfusion with fluorochrome-labeled gelatin solution may be recommended if blood coagulation precludes the perfusion in capillaries.
[37] Perfusion with a fixative solution before the fluorochrome-labeled gelatin solution may improve tissue fixation.

Perform a laparotomy.

Remove fetuses with their fetal membrane intact and transfer to ice-cold PBS.

Transfer a fetus to warm PBS and cut off its fetal membrane in order to expose the fetus, the placenta, and its umbilical cord under a dissection microscope.

Wait until the umbilical artery and vein expand and the pulsation of the umbilical artery is obvious.

Insert a 27- to 30-gauge needle or a glass needle connected to a syringe or microinjector via a silicone tube into the umbilical vein. A micromanipulator may help this process.

Sever the umbilical artery.

Inject the fluorochrome-labeled gelatin solution slowly and drain blood from the umbilical artery. Care must be taken to keep the heart pulsating. A high pressure of the injection may cause the disruption of the capillary in the liver and the peritoneal cavity may fill with the leaked fluorochrome-labeled gelatin solution.

After a while the fluorochrome-labeled gelatin solution flows from the umbilical artery with blood, but continue the perfusion until no blood cells flow.

(c) In case the governing artery of the target organ is obvious and a needle can be inserted into it, only the organ can be perfused.[36,37]

6. Clamp the base of the heart or the vessels proximal to the inserted point and the severed portion with a Kocher clamp after perfusion and immerse the whole body immediately in ice-cold fixative in order to solidify the fluorochrome-labeled gelatin. The fixative consists of 0.5% paraformaldehyde and 15% (v/v) of a saturated picric acid solution in sodium phosphate buffer as mentioned in Procedure I.

7. After solidification of the fluorochrome-labeled gelatin, remove the target organ and cut it into small pieces and fix further in the same fixative overnight at 4° in the dark.

8. Rinse the specimen with ice-cold SPB several times.

9. Section the specimen as stated earlier if necessary.

10. Perform pretreatment with chilled 3% sodium deoxycholate solution.

11. Rinse the specimen twice with chilled distilled water for 1 hr each.[38]

---

[38] The specimen perfused with the fluorochrome-labeled gelatin can be stained further by the immunofluorescence method. Immunofluorescence staining is performed between steps 11 and 12. Each procedure should be carried out at 4°.

12. Perform postfixation with 4% paraformaldehyde in SPB if the fluorochrome-labeled gelatin is fixed imperfectly and dissolves during observation.
13. Mount the specimen as in Procedure I.

Procedure IV: Mounting the Specimen in Polyester Resin

1. Prepare the specimen according to steps 1 to 10 in Procedure I or steps 1 to 11 in Procedure III.
2. Dehydrate the specimen through a graded series of ethanol to 100% ethanol in the dark.
3. Transfer the specimen to pure styrene monomer twice in the dark.
4. Prepare the polyester resin by mixing Rigolac 70F and Rigolac 2004 (3:7) and dissolve 0.05% of benzoylmethyl ether in the mixture. The resin mixture can be stored at 4° in the dark for a few months.
5. Immerse the specimen in the resin mixture in the dark.
6. Mount the specimen in the resin mixture between coverslips spaced with glass strips. Seal each edge of the coverslips with paraffin wax.
7. Polymerize the resin mixture under UV (360 nm) irradiation for 30 min at room temperature.

Acknowledgments

We thank Dr. D. C. Herbert (The University of Texas Health Science Center at San Antonio, Texas) for help during the preparation of this manuscript. We also thank Mr. Fumiyoshi Ishidate (Carl Zeiss Co., Ltd.) for technical advice and Carl Zeiss Co., Ltd., for providing us with opportunities to utilize LSM-310 and LSM-410 confocal laser scanning microscopes. The original work described in this manuscript was supported in part by a grant-in-aid from the Ministry of Education, Science, Sports and Culture of Japan (Nos. 08670038 and 09670032), the Special Coordination Funds of the Science and Technology Agency of the Japanese Government, the CASIO Science Promotion Foundation, and the Foundation for Advancement of International Science.

[7] Cytotomography

By Torsten Porwol, Anja-Rose Strohmaier, and Eberhard Spiess

Introduction

Our understanding of the three-dimensional organization of microscopical organisms and their constituents is based on two-dimensional images. Using serial slices, it is now possible by computer-assisted techniques to reconstruct from such series images that generate the impression of the external and internal three-dimensional morphology and interrelations of cellular constituents.

Methods to achieve such goals are subsumed under the term "tomography," which is derived from the Greek words $\tau o\mu\eta$ (tome) and $\gamma\rho\alpha\phi\epsilon\iota\nu$ (grafein), which means "to cut" and "to write or "to record." In medical science, a number of physical methods have been applied to obtain insight into whole-body organization. Computer tomography is well known based on the use of X rays. At the other end of the biological size scale, electron microscopy is used to obtain information on the three-dimensional organization of macromolecules and small organelles as summarized by Kostner et al.[1] The development of confocal microscopy and/or image deconvolution techniques now makes it possible to also get insight into the three-dimensional organization of cells.

A tomogram can be recorded by two totally distinct methods: (i) Series of optical sections are taken by moving the focus of the irradiation stepwise through the object (through focusing) and (ii) the object is irradiated stepwise from different directions, ideally in a 4π geometry. The first approach was used initially in medical tomography (conventional tomography) and is applicable for the tomography of cells. The second approach is used in modern medical tomography (CT) and in molecular electron tomography. Some authors[1] are inclined to use "tomography" exclusively for the second approach. However, we see no need for this restriction of the term; but, to avoid conflicts, we propose to use the term "cytotomography" for the through-focus technique rendering a three-dimensional impression from cells and their interior by light microscopical image acquisition. In electron microscopy the tomography is performed on living objects. Common to all these procedures is that the sectioning is a noninvasive and nondestructive process leaving the integrity of the object undamaged.

---

[1] A. J. Koster, R. Grimm, D. Typke, R. Hegerl, A. Stoschek, J. Walz, and W. Baumeister, J. Struct. Biol. 120, 276 (1997).

Three-dimensional impressions of microscopical objects have so far been obtained by scanning electron microscopy and transmission electron microscopy of embedded and dissected objects or by taking images from tilted objects. The use of these methods, however, is limited in application and in the results obtained. Using cytotomography, a more general comprehension will be possible as applications in morphology[2-6] and physiology[7-11] of cells clearly reveal. Practical instructions on this matter are given in the following sections.

Specimen Preparation

Staining

Microscopical imaging depends on contrast/intensity differences within the object. A number of different details of an object in a space volume such as a cell can therefore only be distinguished and visualized by their distinct optical properties. In addition to differentiation on a purely physical basis, e.g., interference contrast or phase contrast, the differential staining of details by dyes is necessary. The transmission detectors of commercial confocal light microscopes are not set up in a confocal optical arrangement. Furthermore, a confocal set up would be difficult to calibrate for high numerical aperture lenses and would eventually lead to low photon efficiency and contrast. These restrictions do not hold for fluorescence microscopy. The invention of a great variety of fluorochromes, which are per se specific probes or can be linked to such probes, was therefore the prerequisite and is the basis for the elucidation of cellular architecture. Using fluorescent dyes, two different detection modes are possible: The most common

[2] M. Messerli, M. E. Eppenberger-Eberhardt, B. M. Rutishauser, P. Schwarb, P. von Arx, S. Koch-Schneidemann, H. M. Eppenberger, and J.-C. Perriard, *Histochemistry* **100,** 193 (1993).
[3] M. Messerli and J.-C. Perriard, *Microsc. Res. Techniq.* **30,** 521 (1995).
[4] M. Fricker, M. Hollinshead, N. White, and D. Vaux, *J. Cell Biol.* **136,** 531 (1997).
[5] A.-R. Strohmaier, T. Porwol, H. Acker, and E. Spiess, *J. Histochem. Cytochem.* **45,** 975 (1997).
[6] J. B. Dictenberg, W. Zimmermannn, C. A. Sparks, A. Young, C. Vidair, Y. Zheng, W. Carrington, F. S. Fay, and S. J. Doxsey, *J. Cell Biol.* **141,** 163 (1998).
[7] T. Porwol, E. Merten, N. Opitz, and H. Acker, *Acta Anat.* **157,** 116 (1996).
[8] T. Porwol, E. W. Ehleben, K. Zierold, J. Fandrey, and H. Acker, *Eur. J. Biochem.* **256,** 16 (1998).
[9] W. Ehleben, T. Porwol, J- Fandrey, W. Kummer, and H. Acker, *Kidney Int.* **51,** 483 (1997).
[10] W. F. Graier, J. Paltauf-Doburzynska, B. J. F. Hill, E. Fleischhacker, B. G. Hoebel, G. M. Kostner, and M. Sturek, *J. Physiol.* **506,** 109 (1998).
[11] P. A. Negulescu and T. E. Macken, *Methods Enzymol.* **192,** 38 (1990).

is the intensity measurement of photons using appropriate filters to separate the wavelength of interest. A new approach is the measurement of the lifetime of relaxation. In such measurements at least two well-defined time windows are used (boxcar) and the total amount of photons is determined for each window. In the case of known functional relationships, the fluorescence lifetime can be calculated for the individual dye at its specific location in the sample.

Fixed Specimen. The first choice in targeting organelles are specific antibodies coupled to fluorochromes. Generally, protocols for antibody labeling include fixation and permeabilization of the object. Fixation is achieved by methanol/acetone at $-20°$, by formaldehyde (up to 3% in phosphate-buffered saline, PBS), or by a combination of both techniques. Permeabilization is achieved additionally by, e.g., Triton X-100, Nonidet, or sodium dodecyl sulfate in PBS at concentrations of up to 0.2%. Of course, such protocols produce an artificial situation; careful handling may reduce extraction of antigens, shrinkage, and distortions. It is self-evident that the specificity of the primary and secondary antibodies for their targets has to be proven (Western blotting; competitive incubation). It is essential to minimize background staining by unspecific bindings or cross-reactivity and unspecific staining. The following precautions are recommended: (i) preabsorption of secondary antibodies on the target cells; (ii) preincubation of the primary antibody-tagged cells with preimmune serum of the secondary antibody species; and (iii) strongest possible dilution of the secondary antibody; it is even advisable to reuse secondary antibody preparations. Specimens may be embedded in water, glycerol (50–90% in PBS), or Mowiol in glycerol–PBS. The addition of antifading agents such as 1,4-diazobicyclo(2.2.2)octane (DABCO) or *p*-phenylendiamine (2%, w/v) to the embedding media is dispensable for most new generation dyes. The use of spacers (fragments from coverslips) between slide and coverslip is an advantage. The most positive effect of these protocols is the high stability of the specimen which facilitates microscopy.

Live Specimen. The construction of vital fluorogenic dyes specific for organelles,[12] such as the nucleus, mitochondria, the acidic compartment, or the endoplasmic reticulum, for physiological situations such as pH indication, ion detection,[11] and biochemical reactions, e.g., enzymatic activities[13] or oxygen sensing,[9] offers the possibility of investigating the live cell situation. However, it has to be kept in mind that the application of these "vital"

---

[12] R. P. Haugland and M. T. Z. Spence, eds., "Handbook of Fluorescent Probes and Research Chemicals." Molecular Probes Inc., Eugene, OR, 1996.

[13] C. J. F. van Noorden, T. G. N. Jonges, J. van Marle, E. R. Bissell, P. Griffini, M. Jans, J. Snel, and R. E. Smith, *Clin. Exp. Metast.* **16,** 159 (1998).

dyes also has limitations as dyes might interfere with cellular processes, be metabolized, or displaced over time. However, the risk of imaging an artificially deformed cell architecture is reduced. Therefore, such approaches should be the ultimate goal of researchers. The application of dyes is delicate because it is concentration dependent; a variation in stain concentration might alter the predominant target. This compels users to punctiliously elaborate staining protocols for the cell type under investigation and to achieve a particularly well-balanced staining procedure for different constituents.

The discovery and cloning of the green fluorescent protein (GFP) as opened a completely new avenue to label cell constituents in live and fixed cells.[14,15] By the construction of chimeras containing the protein of interest and the functionally neutral fluorescent tag, an almost ideal labeling technique was achieved. The potential of this approach was further extended by the mutation of GFP, which led to a variety of tags different in their spectral properties and in stability.[16] Limitations of the technique arise from the relatively large size of the GFP (30 kDa), which may influence protein processing, folding, trafficking, and stability and thus protein function. The strength of GFPs is in live cell observation. The very recently developed "fluorescence arsenic helix binder technique" (FLASH)[17] reduces these disadvantages of GFP as only a short peptide sequence has to be attached for tagging and visualization. As GFPs also have a limited stability after formaldehyde fixation, it is possible to combine this labeling technique with antibody labeling.

Living specimens have to be observed under cell culture conditions. This means controlled temperature, atmosphere, and medium composition. To achieve this, either an air-conditioned box surrounding the stage of the microscope or a closed, temperature-controlled microscopical culture device is necessary. The use of sophisticated culture devices,[18] simpler homemade chambers,[19] or devices such as TC chamber slides or slips (Nunc A/S, Kampstrup, DK) are indispensable when inverted microscopes are used. Only such devices provide a firmly fixed cell support that cannot move in the z axis during microscopy. It is further recommended using an indicator-free culture media to avoid additional background formation.

Lowering the temperature in native specimens, e.g., to 20°, over a short

[14] D. C. Prasher, *Trends Genet.* **11,** 320 (1995).
[15] S. R. Kain, M. Adams, A. Kondepudi, T.-T. Yang, W. W. Ward, and P. Kitts, *BioTechniques* **19,** 650 (1995).
[16] Living Colors, Clontech Lab. Inc, Palo Alto, CA, 1997.
[17] B. A. Griffin, S. R. Adams, and R. Y. Tsien, *Science* **281,** 269 (1998).
[18] S. Pentz and H. Hörler, *J. Microsc.* **167,** 97 (1992).
[19] A.-R. Strohmaier, H. Spring, and E. Spiess, *Histochem. Cell Biol.* **105,** 179 (1996).

period of time is a means of reducing fast cellular and intracellular movements without influencing the vitality of cells. This procedure will reduce an object introduced blur in imaging.

Cell Surface. To obtain a proper reference for intracellular localization of cell constituents and to clearly delineate cell shape it is imperative to stain the cell surface. We propose the use of the carbocyanine dye 3,3'-dihexyloxocarbocyanine iodide [DiOC$_6$(3)] in a concentration of 1 μg/ml in PBS for 5 min followed by a 5-min washing step in PBS.[6] This staining is the last step in the staining sequence and is followed by immediate microscopy.

It is obvious that the combination of all these different staining techniques not only allows us to elucidate the three-dimensional architecture of a cell, but also introduces a fourth dimension, which is time to visualize rearrangements. Finally, a fifth dimension, which is the intensity of the signals, will allow following physiological and/or biochemical processes.

In multicolor applications, the possible mutual interaction/cross-talk between dyes has to be considered; e.g., energy transfer might quench signals of a certain kind and obscure structural details and positions. Careful selection of stains and controls addressing these possibilities are necessary before the final imaging of objects.

Physical Properties of Specimens

For accurate determination of the section thickness of the specimen all parameters influencing the optical properties of the imaging system have to be determined. These parameters are the (i) diffraction index of media in which the specimen is embedded, (ii) diffraction index of the immersion medium between the front lens and the coverslip, and (iii) diffraction index and thickness of coverslip. The ideal situation would be equal indices.

A correction for the discrepancies between the nominal section thickness indicated by the system and the real section thickness has to be performed according to Eq. (1):[20]

$$\Delta f = \tan[\arcsin(\mathrm{NA}/n_{\mathrm{i.m.}})]/\tan[\arcsin(\mathrm{NA}/n_{\mathrm{m.m.}})] \qquad (1)$$

(NA, numerical aperture; $n_{\mathrm{i.m.}}$ diffraction index of immersion fluid; and $n_{\mathrm{m.m.}}$, diffraction index of medium surrounding the object).

Filters

The success of application of the various dyes depends strongly on the availability of suitable light sources and specific filter systems. Therefore, great care should be taken to select the optimal dye–filter combinations.

[20] T. D. Visser, J. L. Oud, and G. J. Brakenhoff, *Optik* **90,** 17 (1992).

Optical Setup

In principle, any high-performance confocal system can be used. However, several pitfalls have to be avoided in order to optimize such a demanding system for cytotomography. The following aspects are especially critical: (i) calibration of magnification, (ii) scanner stability, and (iii) imaging properties.

Calibration of Magnification

The calibration of the magnification is of importance due to the fact that the computer system needs information on the magnification of the specific objective or in general the analog technology (see Appendix). The software must be supplied with the correct information on the physical characteristics in order to obtain the proper coordinate system.

Scanner Stability

Scanner stability is a function of temperature and mechanical vibrations, as well as other electronic or mechanical design-dependent features. Only the two first parameters can be influenced by the experimentalist planning the setup. Movement along the optical axis influences the resolution of the optical sectioning capabilities. Optimal performance can be achieved via the piezoelectrically driven movement of the stage or objective with minimal hysteresis.

Several authors have shown that lens aberrations play a major role in accurate imaging. To minimize such aberrations, we strongly recommend the use of water-immersion objectives. Using an inverted microscope assumes matching coverslip thickness to the objective. The thickness of the individual coverslip can be measured easily on a confocal laser scanning microscol (CLSM) as described in the Appendix.

Imaging Properties

The optical characterization of the mount and immersion medium is another key feature for accurate imaging. An important rule of thumb suggests minimizing the difference of the mount and immersion medium in order to reduce spherical aberration. Otherwise artifacts can seriously alter the imaging properties of the optical system.[7]

Fluorescence microscopy is impaired by chromatic and spherical aberration of the objective. Chromatic aberration causes light of different wavelengths to be brought to a focus at different points along the optical axis. In addition to this axial chromatic aberration, the system may also show a lateral aberration. In addition the optical system will show spherical aberration due to the fact that parallel rays passing through the edge of the lens travel a shorter optical length than the ones close to the optical axes. A detailed analysis shows that two kinds of spherical aberrations, longitudinal and transversal, are distinguishable. In practice the ideal focal point is blurred and shows a three-dimensional intensity distribution. To avoid image distortions in reconstruction it is therefore necessary to use objectives that are optimally corrected for these aberrations.

All these effects build up the three-dimensional (in space) and wavelength-dependent imaging property of the optical system. However, it should be kept in mind that even an Apochromat, a high-quality objective, is calibrated only for three wavelengths (Fraunhofer lines). This three discrete wavelength calibration covers the visible window of the spectrum, but the objective has a usable transmission range of about 600 nm. Chromatic aberration-free objectives are available commercially that do not contain refractive optical elements. They are based on reflective materials, i.e., metal coatings optimized for a specific wavelength. As discussed earlier by Porwol et al.,[7] they are of limited use in optical sectioning due to the spherical aberration in living tissue. In multidimensional microscopy the imaging properties of the optical setup are important for high-resolution studies. The imaging property of an infinitely small point below the resolution limit is named the "point-spread function" (PSF) and can be calculated for an ideal optical system or measured using test objects of appropriate size and shape. It should be mentioned that the PSF is related to the optical transfer function (OTF) by a mathematical operation. Here we recommend well-characterized monodisperse polymer particles (e.g., from Microparticles GmbH, Berlin, Germany) with diameters of 199 nm that can be double stained with appropriate dyes, e.g., fluorescein isothiocyanate (FITC) and rhodamine B. The images have to be sampled at the Nyquist frequency of the microscope used.[5,21] The Nyquist rate for the confocal laser scanning microscope can be calculated perpendicular to the optical axis using Eq. (2) and along the optical axis by Eq. (3):

$$\Delta x = \frac{1}{2} \left\{ \frac{\lambda_{\text{emission}}}{4 n_{\text{m.m}} \sin \left[\arcsin \left(\frac{\text{NA}}{n_{\text{m.m.}}} \right) \right]} \right\} \qquad (2)$$

$$\Delta z = \frac{1}{2} \left\{ \frac{\lambda_{\text{emission}}}{2 n_{\text{m.m}} \left[1 - \cos \left(\frac{\text{NA}}{n_{\text{m.m.}}} \right) \right]} \right\} \qquad (3)$$

[21] P. Shaw, *Histochem. J.* **26,** 687 (1994).

NA is the numerical aperture and $n_{m.m.}$ is the index of refraction of the mount medium. The emission wavelength ($\lambda_{emission}$) can be obtained from the spectral characteristics of the dye and the filter settings used for the emission channel.

The measurements of the point-spread function have been done for the optical setup used previously[6] in order to compare the resolution of the measurement. The point-spread function of the CLSM has been calculated on an UNIX-based workstation (SGI) using the commercial software package Huygens2 (Scientific Volume Imaging BV, Hilversum, the Netherlands). Details on the measurement of the three-dimensional point-spread function can be found in the literature, e.g., van der Voort and K. C. Strasters.[22]

The exact knowledge of the PSF of the optical system allows an increase of the resolution using deconvolution procedures.[23] We use the maximum likelihood estimation (MLE) as described by Shepp and Vardi.[24] It has been shown[22] that a lateral resolution of nearly one-fourth of the wavelength is feasible and that the resolution can be enhanced further by a factor of 3–4, mathematically. Therefore, the combination of confocal microscopy and deconvolution techniques now has the potential of resolving biomedical samples in the 50- to 100-nm range by nondestructive methods. The appropriate three-dimensional PSF for the wavelength has been used to perform the iterative deconvolution of the individual channels. In our measurements, up to three different channels of the same sample have been recorded and processed on the workstation.[5,8]

The total number of optical sections depends on a variety of specimen-inherent parameters (e.g., stability of object and staining), which are limited by the physical resolution and scanner stability. Scanning parameters for specimens always represent a compromise between those limitations. Large numbers (up to 80) of optical sections have been recorded.[5,8,9] In these studies, the individual sections were scanned with a resolution of up to 512×512 pixels. Because of the high scanning resolution of the data set, several days were needed to perform the MLE deconvolution on an R5000 SGI workstation with 256 MB of random access memory (RAM).

Data Visualization

Multidimensional microscopy shows its full potential only in combination with state of the art data visualization approaches. Data visualization is characterized by the use of sophisticated computer graphics to gain

[22] H. T. M. van der Voort, and K. C. Strasters, *J. Microsc.* **178**, 165 (1995).
[23] M. Schrader, S. W. Hell, and H. T. M. van der Voort, *Appl. Phys. Lett.* **69**, 3644 (1996).
[24] L. A. Shepp and Y. Vardi, *IEEE Trans. Med. Imag. MT* **1**, 113 (1982).

insight into complex data sets. For reconstitution work we used the software packages "Application Visualization System" (AVS) and AVS/Express, which were modified to meet the requirements of confocal scanning laser microscopy. This approach is convenient due to the high flexibility provided by the programming environment of AVS software for visualization in microscopy.[25] AVS is a collection of software components called modules, which can be combined into executable flow networks or executables. All variables of the components can be modified interactively, allowing flexible data visualization. Due to the fact that a few modules have been written for nonspecific application, the execution time is not optimal for a specific data visualization. However, the interface to computer languages allows the experienced user to write the optimal routine for one's own purpose. The individual data sets as given in the example show a rather complex three-dimensional structure. The data set obtained from the collagen gel had to be averaged down by a factor of four perpendicular to the optical axis. However, even with this reduced resolution a total number of about 6,000,000 triangles has been obtained with the marching cube algorithm. For several reasons it is helpful to have an "internal view" of the volumetric data sets generated by the CLSM. Here, we have alternatively altered the transparency of the different data sets in order to enhance the data visualization. Calculation of the contours of some optical sections using the full resolution in the plane perpendicular to the optical axis shows the shape of the object. We found not only a better understanding of the spatial arrangement of cell morphology and intracellular compartments but also an increase in sensitivity. Signals that easily escaped detection in individual sections became more prominent in the reconstruction.

In addition, quantitative determinations of volumes are possible. The full potential of modern data visualization approaches can be obtained via automatic calculation of a large amount of images showing variations of distinct parameters. The variable parameters can be of different origin. The parameter might be the value of the isosurface of the different channels or the transparency of the isosurface in order to mention a few. The variation of the isosurface level allows the detection of the volume of a specific stained area in the sample as a function of the fluorescence level. The technique has been used in our work on the effect of nickel and cobalt on putative members of the oxygen-sensing pathway.[7,8] The multiparameter analysis allows the study of complex multidimensional interrelations.

Multimedia technology improves the results greatly, i.e., we used a computer system to produce computer animation based on CLSM data visualization approaches. Real time video sequences have been produced

[25] B. Sheehan, S. D. Fuller, M. E. Pique, and M. Yeager, *J. Struct. Biol.* **116,** 99 (1996).

using the cosmo motion compression JPEG (Joint Photographic Expert Group) card from SGI and the INDY video card. The video card is connected to a S-VHS recorder. The hardware JPEG card allows the production of real time video animation using the two analog broadcast standards. Due to the resolution and number of frames per second (30 frames/sec NTSC), the production is rather resource demanding. Several compression schemes are known to provide those sequences via the Internet or on a compact disk as well as other digital storage media.

Example

The migration of a human tumor cell (SB 3) into a reconstituted collagen matrix was studied to demonstrate the possible contribution of the lysosomal cysteine protease cathepsin B to this process. Only by tomography and three-dimensional image reconstruction can the realistic expansion of the cell in the gel become conceivable and it can be demonstrated unambiguously that fractions of the enzyme were transported into the invadopodia and eventually appear at the cell surface. Staining of the cell surface (blue) was obtained by $DiOC_6(3)$; cathepsin B (red) was detected by indirect immunocytochemistry using a secondary antibody labeled by the fluorochrome Cy3; and the collagen gel (white) in which the cell was penetrating was visualized by reflection microscopy.[5] In Fig. 1 (see color insert) different visualization possibilities are shown. In Figs. 1A and 1B the cell is shown with and without the gel, respectively. The surface of the gel is oblique. The degree of penetration and the scattered vesicles become visible in the matrix-stripped representation (Fig. 1B). The collagen is omitted and the cell surface is attenuated to open the view to the inside (Figs. 1C–1F), revealing cathepsin B-containing organelles. In Figs. 1C and 1D the cell surface is invisible and only a couple of contour lines trace the extension of the cell. Figure 1C shows the superposition of two visualization approaches with special emphasis on the localization of the cathepsin B. Here the volume-rendering visualization has been superimposed on the surface-rendered isosurface. In Fig. 1D only the isosurface visualization approach is shown. The comparison of Figs. 1C and 1D clearly reveals that the threshold chosen sufficiently represents the enzyme localization within the cell. In Figs. 1E and 1F the surface of the cell is shown in a transparent manner. Figure 1E shows the isosurface that has been calculated for an appropriate threshold. Figure 1F represents the volume-rendered data set. A comparison of the graphs published previously by Strohmaier et al.[5] clearly reveals the increase of resolution. The different visualization procedures of the deconvoluted data sets allow the identification of the distribution of cathepsin B in an up to now unknown quality.

Appendix

Calibration of Magnification

A CLSM has to be calibrated for x, y and z axes extensions. The magnification perpendicular to the optical axis can be checked easily using a calibration carrier glass provided by the manufacturers of microscopes. On the CLSM the reflection mode is the method of choice using ordinary calibration glass. The scanner software needs the appropriate magnification of the objective lens to calculate the distance between two scanned points. The calibration and verification of the extents along the optical axis can be done in reflection mode using a coverslip of defined thickness (for details, see determination of cover glass thickness) or the fluorescence mode using monodisperse polymer particles of a defined size and shape. The coverslip is, however, a far more stable calibration object for an initial check of the instrumentation.

Determination of Coverslip Thickness

The thickness of individual coverslips has to be determined. This can be done either by a high precision microcaliper or by optical means. A line scan along the optical axis using the reflection mode will allow the high precision measurement of the coverslip thickness. The reflection mode measurement will result in two Gaussian shape-like intensity maximums, which can be used for resolution measurements, at the top and bottom of the coverslip. The total difference between the two maximums is equal to the coverslip thickness.

Acknowledgments

We are indebted to Professor Dr. Helmut Acker, Max Planck Institute for Molecular Physiology, Dortmund, Germany, for continuous support and helpful discussions. A.-R.S. and E.S. acknowledge a grant from NCDR-DKFZ Israeli-German Cooperation in Cancer Research. T.P. acknowledges financial support by Deutsche Forschungsgemeinschaft, Deutsche Krebshilfe, and Den Norsk Kreftforening.

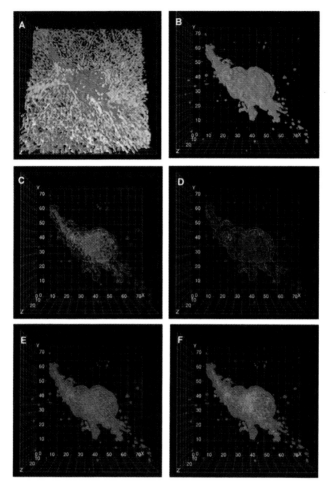

Fig. 1. Multidimensional deconvolution and subsequent data visualization of a tumor cell invading into a collagen matrix. The specimen was fixed by methanol, and cells were stained by $DiOC_6(3)$ and immunolabeled by anti-cathepsin B antibody detected by Cy3-labeled second antibodies. Individual data sets were recorded sequentially and have been deconvoluted using the maximum likelihood estimation method. The multidimensional reconstruction of a cubus from a series of 60 light optical sections is shown. The gel, recorded in the reflection mode, is in white, the cell is given in blue, and cathepsin B in red. Contour lines emphasize the surface topography. The perspective is unchanged in A to F. (A and B) The cell is shown with and without the gel, respectively. The surface of the gel is oblique. The degree of penetration and scattered vesicles become visible in the matrix-stripped representation (B). The collagen is omitted and the cell surface is attenuated to open the view to the inside (C–F), revealing cathepsin B-containing organelles. (C and D) The cell surface is invisible and only a couple of contour lines outline the extents of the cell. (C) The superposition of two visualization approaches with special emphasis on the localization of the cathepsin B. Here the volume-rendering visualization has been superimposed on the surface-rendered isosurface. (D) Only the isosurface is shown parva componere magnis. The comparison of C and D clearly reveals that the threshold chosen sufficiently represents the enzyme localization within the single cell. (E and F) The surface of the cell is shown in a transparent manner. (E) The isosurface calculated for an appropriate threshold. (F) The volume-rendered data set. The axes are indicated, along (z) and perpendicular (x, y) to the optical axes, respectively. The scales for all axes are shown in micrometers.

[8] Use of Confocal Microscopy to Investigate Cell Structure and Function

By GHASSAN BKAILY, DANIELLE JACQUES, and PIERRE POTHIER

Introduction

Initially, fluorescent probes were commonly used to reveal cell morphology and biochemistry in static conditions. With time, the need for a technique to probe functional elements in living cells pushed scientists to develop fluorescent probes that could be both monitored and quantified such as calcium dyes. However, measurement of a functional process occurring in living cells using fluorometry and fluorescent Ca^{2+} probes did not permit visualization of the inner workings of the cell, which was overcome by the development of real time two-dimension fluorescence microscopic digital imaging. Although this technique shed some light on the so-called "black box" of working cells, it did raise several additional questions. Are we actually measuring cytosolic free ion? Is the dye limited to the cytosol and, if not, what is the contribution of other intracellular compartments? Is the probe taken up by other intracellular organelles and, if yes, is it distributed homogeneously? What is the contribution of organelles to the observed changes in the cytosol? At what subcellular level does a particular drug produce its effect? Where are the ionic channels and receptors localized and are they distributed homogeneously? Is voltage distributed homogeneously across the membrane of cells and organelles? Does cell-to-cell contact play a role in cell function and its response to drugs? What are the effects of photobleaching and/or leak of the fluorescent probe?

Until the development of three-dimensional (3D) imaging, such questions were raised but never fully answered because of the absence of a technique enabling the visualization of dye distribution at the 3D level. The recent development and use of confocal microscopy coupled to high-performance hardware and software systems have provided the scientist with the capability of overcoming some of the limitations of standard microscopic imaging measurement. Up until a few years ago, access to confocal microscopy was fairly limited. Today, a majority of laboratories dealing with molecular and cell biology, pharmacology, biophysics, biochemistry, and so on are or in the process of being equipped with this powerful scientific tool. This increase in the need for confocal microscopy has been highly supported by a growing industry for developing new fluorescent probes. However, in order to better optimize this technique, scientists need to be familiar with the basic approaches and limitations of confocal microscopy. This article discusses sample preparations to be used, labeling of structures and functional probes, parameter settings, and the development of specific ligand probes, as well as measurements and limitations of the technique. For further information, the reader should refer to other volumes in *Methods in Enzymology* as well as to other key references in the literature.[1-4]

Preparations Used in Confocal Microscopy

To date, several cell types have been used routinely for confocal microscopy studies, including cells from heart, vascular smooth muscle, vascular endothelium, nerve, bone, T lymphocytes, liver, *Xenopus* oocytes, fibroblasts, sea urchin eggs, and basophilic leukemia cells. Most of the work has been done using freshly isolated and/or cultured single cells either in primary culture or from cell lines. There is no doubt that the choice of preparation is crucial for obtaining sastisfactory results. For example, in order to obtain sufficient details from 3D reconstructions, the following criteria should be taken into account: (1) isolated cells should be plated on glass coverslip (plastic should be avoided); (2) cells should be attached in order to avoid displacement during serial sectioning; (3) serial sectioning should be taken at rest or when the steady-state effect of a compound is reached; (4) in order to eliminate electrical and chemical coupling between cells, cells should be grown at low density; and (5) cells should be relatively thick and not flattened in order to allow a minimum of 75 to 150 serial sections (according to cell type). This could be done by using the cells as soon as they are attached and/or by lowering serum concentration. (6) When cell lines or proliferative cells are used, it is recommended to replace the culture medium overnight by a medium free of serum and growth factors prior to starting the experiments. This will allow the cells to settle in a latent phase of mitosis and will increase the accessibility of receptors and channels for agonist and antagonist actions by washing out the effects of compounds present in the culture medium. (7) Using specific fluorescent markers, the origin, purity, and phenotype of the cells should be checked continually. Because ions and other free functional components of the cells are not distributed homogeneously and because the increase or decrease of the

[1] G. Bkaily, P. Pothier, P. D'Orléans-Juste, M. Simaan, D. Jacques, D. Jaalouk, F. Belzile, G. Hassan, C. Boutin, G. Haddad, and W. Neugebauer, *Mol. Cell. Biochem.* **172,** 171 (1997).
[2] G. Bkaily, P. D'Orléans-Juste, P. Pothier, J. B. Calixto, and R. Yunes, *Drug Dev. Res.* **42,** 211 (1997).
[3] S. W. Paddock, *Proc. Soc. Exp. Med.* **213,** 24 (1996).
[4] R. Rizzuto, W. Carrington, and R. A. Tuft, *Trends Cell. Biol.* **8,** 288 (1998).

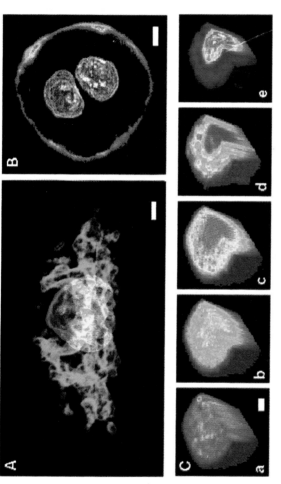

FIG. 1. (A) Mitofluor Green-stained mitochondria (green) followed by Syto 11 staining of the nucleus (red) in an embryonic chick heart myocyte. Stained mitochondria form a ribbon-like mass surrounding the nucleus. Images were generated from 40 serial sections (0.08 μm) at a step size of 0.49 μm. Bar: 2 μm. (B) Di-8-ANEPPS potentiometric dye labeling of plasma membrane potential in an embryonic chick myocyte (pseudo-colored) followed by Syto 11 staining of the bilobed nucleus (gray scale). Transmembrane voltage staining appears distributed unevenly around the cell surface with no internalization after 30 min. Bar: 2 μm. (C) Sagittal and cross-sectional view of a 3D-reconstructed perforated plasma membrane ventricular myocyte illustrating intracellular calcium distribution in response to increasing concentrations of extranuclear calcium. Coverslips were mounted in the bath chamber and, after loading with 13.5 μM Fluo-3 AM, cells were bathed in a buffered saline solution containing 100 nM of Ca^{2+}. Following rapid perforation of the cell with ionomycin (1 to 2 min), the cells were then washed quickly and stabilized for 5 min in buffer solution without Ca^{2+}. Image acquisitions were performed at the following settings: 488-nm excitation and 510-nm emission wavelengths, laser power at 9 mV, 3% attenuation filter, and PMT set at 700. Z-axis serial sections (0.08 μm) were performed at a step size of 0.9 μm. Nine identical vertical serial series were obtained 2 min after the incremental addition of 100 (a), 700 (b), and 1500 (c) nM extranuclear Ca^{2+}. Images are shown as pseudo-colored representations according to an intensity scale from 0 to 255 (bottom). At the end of the last concentration of the ion, mitochondria were stained with Mitofluor Green (d) followed by Syto 11 staining of the nucleus (e) for 8–10 min and scanned under identical conditions. The reconstructed images depict a gradual increase of [Ca^{2+}]$_i$ within the mitochondria and the nucleus with increasing concentrations of external calcium. Bar: 2 μm.

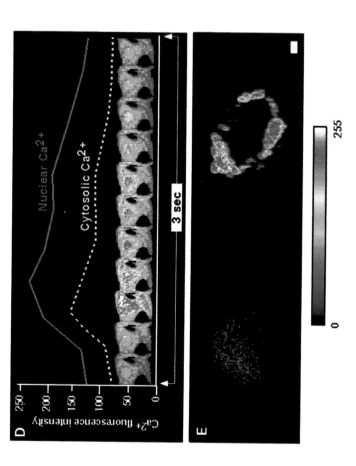

Fig. 1. (*continued*). (D) Time-lapse rapid scans and graphic representations of cytosolic and nuclear-free Ca^{2+} variations of spontaneously contracting 10-day-old chick embryonic ventricular myocytes loaded with 13.5 μM Fluo-3 AM. Whole cell images were generated continuously every 330 msec for 3 min at a resolution of 512 × 512 pixels and 0.17 μm pixel size. Total time span: 3 sec. Images are shown as pseudo-colored representations according to an intensity scale from 0 to 255. (Left) (E) Cultured left ventricular endocardial endothelial cells were incubated with 10^{-7} M fluorescein-conjugated human PYY following a 20-min preincubation with a 100-fold excess of cold human PYY. The image represents a sequential rendering of 12 optical sections (step size: 0.52 μm; pixel size: 0.08 μm). Only very little diffuse nonspecific labeling is observed. (Right) Three-dimensional reconstruction of a labeled ventricular cultured endocardial endothelial cell following a 2-hr incubation with fluorescein-conjugated human PYY at room temperature. The image represents a slightly angled 3D rendering of a cell generated from eight optical sections (step size: 0.6 μm; pixel: 0.17 μm) and viewed as a look-through projection. Note the distribution of PYY labeling along the circumference of the cell.

probed free component seems to be faster than computer sampling and acquisition, it is difficult to assess these fast occurring phenomena at the 3D level. Hence, large rapid scans of the area under interest could be done.

Settings of Confocal Microscope

There is no doubt that one of the major elements in confocal microscopy is the setting of optimal parameter conditions. This should be determined prior to initiating any serious experiments and should be captured and maintained for each specific fluorescent probe. This will allow a realistic comparison between results obtained using the same fluorescent probe. Table I shows typical probe data and optical settings for a Molecular Dynamics (Sunnyvale, CA) Multi Probe 2001 confocal argon laser scanning (CSLM) system equipped with a Nikon Diaphot epifluorescence inverted microscope and a 60× (1.4 NA) Nikon Oil Plan achromat objective. These optical settings are also valid with other confocal microscopes. When the argon laser is used, the 488-nm (9.0–12.0 mV) or 514-nm laser line (15–20 mV) is directed to the sample via the 510- or 535-nm primary dichroic filter and attenuated with a 1 to 3% neutral density filter to reduce photobleaching of the fluorescent probe. The pinhole size could be set at 50 nm for structural determination and 100 μm for ion-specific dyes. In most cases, the image size is set at 512 × 512 pixels with pixel size of 0.08 μm for small ovoid-shaped cells and 0.034 μm for elongated cells. In order to optimize and validate fluorescence intensity measurements, laser line intensity, photometric gain, photomultiplier tube (PMT) settings, and filter attenuation should be determined and kept rigorously constant throughout the duration of the experimental procedures.

Organelle Fluorescent Probes

Several commercially available dyes permit the localization and delimitation of several organelles and membranes such as the plasma membrane (Di-8-ANEPPS), mitochondrial membrane (Mitofluor Green), endoplasmic reticulum DiOC6(3), Golgi apparatus (NBDC$_6$-ceramide), and the nucleus (live nucleic acid Syto stains). Once the cultured cell type is available, these probes should be used initially in order to be familiar with their distribution and localization. Once this is done, these structure delimitation probes could be used at the end of each experiment. Other dyes can also be used to delimit organelle structure such as mitochondrial (JC-1) and plasma membrane potential dyes. The choice of the structural delimiting probe depends on the excitation and emission of the fluorescent probe used to study the functional element of the cells. Double and triple labeling

TABLE I
PROBE DATA TABLE AND OPTICAL SETTINGS FOR ARGON-BASED CLSM

Probe/dye	Application	EX_{max}	EM_{max}	Argon laser line	Primary beam splitter	Secondary beam splitter	Detector 1 or barrier[a] filter	Detector 2 filter
Fluo-3, AM	Ca^{2+} ion indicator	464	526	488	510	None	510	None
Fura-Red, AM	Ca^{2+} ion indicator	458	600	488	535	None	595	None
Calcium Orange, AM	Ca^{2+} ion indicator	550	567	514	535	None	570	None
Sodium Green	Na^{2+} ion indicator	507	532	488	510	None	510	None
Amiloride	Na$^+$/H$^+$ pump inhibitor	488	420	488	510	None	510	None
SNARF-1, AM	pH indicator	488–530	Dual 587/636	488/514	535	595 DRLP	600	540DF30
DiOC$_6$(3)	Endoplasmic reticulum dye	484	501	488	510	None	510	None
MitoFluor Green	Mitochondrion probe	490	516	488	510	None	510	None
BODIPY FL C$_5$-ceramide	Golgi dye	505	511	488	510	None	510	None
JC-1	Mitochondrial membrane potential dye	490	Dual 527/590	488	510	565 DRLP	570	530DF30
Di-8-ANEPPS	Cell membrane	481	605	488	510	None	>570	None
Syto 11	Live nucleic acid stain	500–510	525–531	488	510	None	510	None
Angiotensin II-FITC	Angiotensin-binding sites	494	520	488	510	None	510	None
AngII-TRITC[b]	Angiotensin-binding sites	ND	ND	514	535	None	535	None
PYY-FITC	PYY-binding sites	ND	ND	488	510	None	510	None

[a] If the secondary beam splitter is absent, then value indicated applies to barrier filter.
[b] Custom-prepared.

of functional fluorescent probes can be performed if the two or three probes are excited and emit at different wavelengths. Also, certain functional probes such as membrane potential sensitive dyes could be used simultaneously both as delimiting (structural) and functional probes in conjunction with other functional probes. Furthermore, plasma membrane potential sensitive probes could be used to study cell coupling as well as voltage distribution and cell shortening.[1]

Volume Rendering and Nuclear Calcium, Sodium,
  and pH Measurements

Once the scanned images are obtained, they can be transferred onto a workstation similar to a Silicon Graphics Indy 4000 workstation equipped with Imagespace analysis and volume workbench software modules (Molecular Dynamics). Reconstruction of 3D images of ion-selective dyes are then performed on unfiltered serial sections and are represented as closest intensity projections for calcium, sodium, or pH distribution and look-through extended focus projections for both nucleus and calcium (or Na^+ or pH) colocalization studies. Three-dimensional reconstructions of cell organelles can be performed on Gaussian- or median-filtered serial sections. The look-through method depicts a specimen as if it were transparent while being in focus throughout its depth. It is useful for studying both surface and interior features and their relative positions. Closest intensity projections, however, produce a stack of opaque images depicting actual voxel intensities. Calcium, sodium, or pH (or any functional dye) images can be represented as pseudo-colored representations according to an intensity scale of 0–255 with lowest intensity in black and highest intensity in white. As a rule, pseudo-scales should be included with all images illustrating pseudo-colored representations of intensity levels.

Measurement of uptake of calcium, sodium, or any free element of the cell within the nucleus can be performed on both 2D images (individual sections) and 3D reconstructs (section series). The nuclear area following Syto 11 staining can be isolated from the rest of the cell by setting a lower intensity threshold filter to confine relevant pixels. A 3D binary image series of the nuclear volume can then be generated for each cell using the exact same x, y, and z set planes as those used, for example, during calcium ion uptake (Figs. 1A, 1C, and 1D, see color insert). By applying these binary image patterns of the nucleus to the same cell initially labeled for calcium, Na^+, or pH (the binary image serves as a "cookie cutter"), a new 3D projection is created depicting fluorescence intensity levels exclusively within the nucleus. Hence, by "removing" the nucleus from the surrounding cytoplasm, we are then able to measure mean calcium, Na^+, or pH intensity

values in the entire nuclear volume while eliminating the possible contribution of the perinuclear space to the measurements.

Rapid Scan Imaging

Rapid line scan imaging can be used to monitor temporal oscillations
of a functional probe such as Ca^{2+} dye during the spontaneous contraction
or cytosolic, nuclear, and mitochondrial Ca^{2+} oscillation and/or release.
Contracting cells or cytosolic and nuclear Ca^{2+} oscillations can be recorded
easily at a rate of 330 msec per scan (three scans/sec) for a total of up to
1000 frames. Each frame, which consists of 32 lines/scan (512 pixels) and
a pixel size of 0.42–0.64 μm, enables the visualization of the entire cell
throughout the contractile process with good spatial and temporal resolution. However, if faster acquisition times are needed, line scanning can be
performed at a rate of up to 10 msec per image using the Multi Probe
CSLM (Molecular Dynamics). Figure 1D shows an example.

Coculture of Cells

The reconstitution of interaction between two cell types is an excellent
preparation for confocal microscopic studies. For example, single vascular
endothelial and smooth muscle cells can be used in coculture conditions.
Because vascular endothelial cells grow less rapidly in culture than vascular
smooth muscle cells, vascular endothelial cells are first cultured in the
middle of a sterile coverslip using a custom-made watertight compartment
9 mm in diameter. Once the endothelial cells have attached and reached
confluence (or subconfluence), the compartment can be removed, leaving
a confined area of endothelial cells on the glass lamella. Then, vascular
smooth muscle cells can be added in an area adjacent to that of the endothelial cells. On reaching confluence, the vascular smooth muscle cells establish
side-to-side and/or overlapping contact with endothelial cells. Once this is
achieved, the cocultures can be loaded with Fluo-3 (or other fluorescent
probes) for 3D Ca^{2+} measurement using confocal microscopy. An alternate
culture method can also be used. This method consists of plating a drop
of high-density endothelial cells in the middle of the lamella and allowing
the cells to attach overnight. The next day, the culture dish is washed gently
and fresh culture medium containing a high density of vascular smooth
muscle cells is added to the culture dish containing the attached endothelial
cells. This simple technique allows for a better study of the development
of cell-to-cell contact between vascular endothelial and smooth muscle cells
as a function of time and can be used within an hour or more. Although
this technique has a disadvantage in that cell contacts are generated at

random, it may, however, more adequately represent a nonhomogeneous organizational contact between two different cell types, such as what we observed in certain pathological conditions and healing processes following surgery or vascular bypass. Also, an interaction between cells of the immune system and endothelial system or at any other cell type could be done in a similar fashion with the exception that the immune cells in suspension should be loaded with the dye, collected by centrifugation, and then added to the attached preloaded monolayer cells.

Loading of Fluorescent Probes for Confocal Microscopy

Cytosolic and Nuclear Ca^{2+} Measurements by Confocal Microscopy: Fluo-3

Freshly cultured single cells and cells in primary cultures are excellent preparations for studying cytosolic and nuclear-free Ca^{2+} using 3D confocal microscopy. Several visible wavelength Ca^{2+} dyes were initially tested in order to select the ion indicator best suited for laser-based confocal microscopy. Of the various Ca^{2+} dyes available, we tested Fluo-3, Fura Red, Calcium Green-1, and Calcium Orange (Molecular Probes). Calcium Orange was the least successful with respect to loading and calcium fluorescence. The dye was very difficult to load in all cell preparations tested regardless of temperature, time of incubation, or dye concentration. Moreover, the addition of Pluronic F-127, a dispersing agent used to aid solubilization, did not improve loading of the dye. Calcium Green-1 loaded reasonably well and displayed good baseline fluorescence. However, in most preparations tested, very little or no Ca^{2+} response was observed in the presence of a number of pharmacological agents known to increase intracellular Ca^{2+}. Fura Red, a long wavelength analog of Fura-2 with a high emission spectrum, was tested equally. Loading was performed at room temperature for 45 min at a concentration of 20 μM. At baseline calcium levels, labeling was relatively uniform in the cytoplasm and nucleus in heart cells but appeared slightly higher in the nucleus of vascular endothelial and smooth muscle cells. Probe fluorescence appeared relatively stable and showed little photobleaching during our experiments. However, one of the singular characteristics of Fura Red is that its fluorescence emission decreases on Ca^{2+} binding in contrast to other indicators where emission intensity increases with Ca^{2+} binding. This particular property of Fura Red allows its use in combinations with Fluo-3 for dual emission ratiometric measurements, thus enabling the expression of fluorescence intensity in terms of Ca^{2+} concentration. There are some drawbacks, however: this dye, being much weaker than other visible wavelength calcium indicators, must be loaded at higher concentrations. Moreover, the intracellular distribution of both indicators must be identical for ratiometric measurements to be valid. Finally, in some of our experiments involving dual labeling with Fura Red and Fluo-3, both fluorescent dyes increased in response to depolarization of the cell membrane or pharmacological stimulation, making Fura Red unreliable, at least with some cell preparations. Among all Ca^{2+} dyes tested, Fluo-3 provided the best overall loading and fluorescence features in all our cell preparations. Fluo-3 was loaded easily, and was distributed homogeneously within the cytosol and the nucleus, with low photobleaching at a laser attenuation setting of 3%. More importantly, calcium responses to both electrical and pharmacological stimuli were comparable to those using Fura-2 in classical Ca^{2+} imaging studies. Hence, this dye was used in the majority of our calcium experiments with the confocal microscope in intact (Fig. 1D) and perforated plasma membrane (Fig. 1C) of many cell types.[1]

Single cells from heart, aortic vascular smooth muscle (VSM), T lymphocytes, osteoblasts, or endothelium can be cultured on 25-mm-diameter glass coverslips that fit a 1-ml bath chamber. The cells are then washed three times with 2 ml of a balanced Tyrode's salt solution containing 5 mM HEPES, 136 mM NaCl, 2.7 mM KCl, 1 mM MgCl$_2$, 1.9 mM CaCl$_2$, and 5.6 mM glucose, buffered to pH 7.4 with Tris base, and supplemented with 0.1% bovine serum albumin (BSA). The osmolarity of the buffer solution with or without BSA should be adjusted with sucrose to 310 mOsm. The Fluo-3/AM probe (Molecular Probes, Eugene, OR) or other Ca^{2+} probes should be diluted in Tyrode–BSA from frozen 1 mM stocks in dimethyl sulfoxide (DMSO) to a final concentration of 6.5 or 13.6 μM. Cells are then loaded by placing the coverslips cell side down on a 50- to 100-μl drop of diluted probe on a sheet of parafilm stretched over a glass plate and incubated for 45–60 min in the dark at room temperature. The inverted coverslip method offers considerable savings with regard to the amount of probe used, especially at high concentrations. More importantly, the use of smaller aliquots of concentrated stock solution needed to prepare the final probe concentration reduces the percentage of DMSO content in the incubation medium. We found that cell blebbing and photobleaching were encountered commonly with final DMSO concentrations greater than 0.6%. Loading should also be performed in an humidified environment when inverting coverslips on less than 100 μl in order to avoid evaporation problems. After the loading period, coverslips should be recovered carefully and the cells washed twice with Tyrode–BSA buffer and twice in Tyrode buffer alone. Loaded cells should then be left to hydrolyze for an additional 15 min to ensure complete hydrolysis of acetoxymethyl ester groups. Optical settings for Fluo-3 AM are shown in Table I.

Cytosolic and Nuclear Na$^+$ Measurement by Confocal Microscopy:
  Sodium Green/AM

Sodium Green AM is the only argon laser excitable Na$^+$ indicator available commercially. This sodium ion probe exhibits a 41-fold selectivity for Na$^+$ over K$^+$ and, on binding sodium, increases its fluorescence emission intensity. Loading of Sodium Green into heart and vascular cells is difficult unless coupled with Pluronic F-127 to increase aqueous solubility. When well loaded, the dye is relatively stable, although slightly more subject to photobleaching than Fluo-3. This can be minimized by setting the laser attenuation filter down to 1%.

Single cells can be loaded with 13.5 μM Sodium Green (Molecular Probes, Eugene, OR) similarly to that described earlier for Fluo-3 with one modification: the Na$^+$-sensitive dye should be loaded in the presence of Pluronic F-127 (20%, w/v). When used alone, Sodium Green loaded poorly into cells. However, when reconstituted initially in equal volumes of DMSO and Pluronic acid and then diluted into the loading buffer (final concentration of 0.1% detergent), the probe offered good fluorescence intensity. Moreover, it was found that the loading temperatures should not exceed 19–29° as loading tended to be inconsistent and often compartmentalized. Optical settings for Sodium Green/AM are given in Table I.

After loading the ion probe for 30 min at 19°, and if heart cells are to be used, one can see that the nuclear Na$^+$ level is similar to that in the cytosol. However, if resting aortic vascular endothelial and smooth muscle cells are to be used, one can see that the level of cytosolic Na$^+$ is slightly greater in the immediate cytosolic zone surrounding the nucleus. No information is available concerning a possible role of Na$^+$ in nuclear function or whether the nucleus may play a role in cytosolic Na$^+$ buffering. The mechanism by which Na$^+$ crosses the nuclear membrane, as well as its physiological role in the nucleus, if any, is not known. However, it has been demonstrated that confocal microscopy can be highly useful in determining the physiological role of Na$^+$ and the modulation of this ion at the cytosolic and nuclear levels as well as in subcellular organelles.[2]

Intracellular pH Measurement by Confocal Microscopy:
  Carboxy SNARF-1/AM

Several long wavelength pH fluorescent dyes have been developed and are available commercially such as BCECF, SNARF-1, SNARL, DM-NERF, CL-NERF, and HPTS. The use of an argon-based confocal microscope limits the use of all these pH indicators. This limitation can be overcome by modifying the laser capabilities into a multiline argon/UV laser or an argon/krypton laser.

Carboxy SNARF-1, however, is well suited for all types of laser instrumentation and is excited efficiently by both 488- and 514-nm argon laser lines. The emission spectrum of the dye undergoes a pH-dependent wavelength shift, thus allowing the ratioing of fluorescence intensities at two emission wavelengths, typically 580 and 640 nm, and hence can be used for more accurate determinations of pH. This pH dye has been applied extensively in many cell types.[5–8] The dye has a pK_a of approximately 7.5, making it well suited to cytoplasmic pH measurements.[6] It is reported to be less susceptible to photobleaching and exhibits a good pH-dependent shift of its fluorescence emission spectrum.[6] Using confocal microscopy, studies on single cells loaded with SNARF-1 suggest that subsarcolemmal and nuclear areas have a pH value near 7.2, but for regions corresponding to the distribution of mitochondria, the pH is 7.8–8.2.[5,7] This raises the possibility that cytosolic pH is not homogeneous and is a function of the location of organelles within the cytosol. It is also possible that pH$_i$ measurements near the plasma (or nuclear and mitochondrial) membrane may lead to erroneously low pH$_i$ values due to the pronounced spectral contribution of the enriched protonated indicator component at its associated apparent alkaline pK_i shift.[7] Moreover, pH$_i$ measurement with H$^+$-sensitive dye may depend on the cell type used, isolation, and/or culture conditions, as well as the degree of metabolic activity. In order to evaluate intracellular pH changes, cells can be loaded with the ratiometric pH indicator Carboxy-SNARF-1/AM, using the inverted coverslip method. Cells should be incubated in darkness for 30 min at ambient temperature with 5 μM of freshly prepared indicator dye in Tyrode–BSA buffer from 1 mM stock solutions reconstituted in DMSO. Following the loading period, cells should be washed and left to hydrolyze for 15 min as described for the Ca^{2+} dyes. Coverslips are then placed in a 25-mm (1 ml) bath chamber for visulization with the confocal microscope. Optical settings for SNARF-1/AM are given in Table I. The use of SNARF-1 dye in chick embryo heart cells revealed a sometimes heterogeneous, sometimes homogeneous pH$_i$ distribution in the cytosol and the nucleus. In some instances, labeling appeared to be more intense at the point of attachment of the cell to the coverslip than

[5] E. Chacon, J. M. Reece, A. L. Nieminen, G. Zahrebelski, B. Herman, and J. J. Lemasters, *Biophys. J.* **66,** 942 (1994).

[6] K. W. Dunn, S. Mayor, J. N. Myers, and F. R. Maxfield, *FASEB J.* **8,** 573 (1994).

[7] J. J. Lemasters, E. Chacon, G. Zahrebelski, J. M. Reece, and A. L. Nieminen, in "Optical Microscopy: Emerging Methods and Applications" (B. Herman and J. J. Lemasters, eds.), p. 339. Academic Press, New York, 1993.

[8] N. Opitz, E. Merten, and H. Acker, *Pflug. Arch.* **427,** 332 (1994).

at other cell periphery. In heart and vascular cells, the indicator responded well to changes in extracellular pH from 6.5 to 8.5. The dye reached an isosbestic point at an external pH of 6.0 where no difference could be seen between acidic and basic measurements. Also, the distribution of pH levels in heart cells is relatively homogeneous within the cytosol and the nucleus. However, in human vascular aortic smooth muscle cells, the distribution of pH level in the cytosol is less homogeneous and appears to be much more basic within and in the immediate vicinity of the nucleus as compared to cytosolic pH. This was particularly evident when lowering extracellular pH. We should caution that expressing pH_i by quantitative values may lead to significant errors in comparison to qualitative (fluorescence intensity) or ratiometric measurements.[8] Care should also be taken even when expressing pH_i in ratiometric terms because of possible intracellular redistribution of the indicator between cytosol and lipophilic cell compartments.[8]

Endoplasmic Reticulum: Carbocyanine $DiOC_6(3)$ for Endoplasmic Reticulum Staining

Several short-chain carbocyanine dyes such as $DiOC_6(3)$ and $DiOC_5(3)$ and long-chain carbocyanines such as $DIIC_{18}(3)$ have been widely used to visualize the endoplasmic/sarcoplasmic reticulum.[9,10] These probes pass through the plasma membrane easily, and while some carbocyanine probes have been reported to stain several other intracellular membranes, such as Golgi and mitochondria, endoplasmic reticulum (ER) membranes are labeled preferentially when used at higher concentrations and are easily distinguishable by their characteristic morphology.

Sarcoplasmic/endoplasmic reticulum (SR) can be visualized using the short-chain carbocyanine $DiOC_6(3)$ dye from Molecular Probes (Eugene, OR). Cells grown on coverslips can be placed in a 25-mm bath chamber and bathed in 1 ml of Tyrode buffer solution. Freshly prepared ER dye should be added at a final concentration of 50 nM obtained after serial dilutions in Tyrode–BSA loading medium from 10 mM stock solutions in DMSO. Labeling is very rapid (within 3–5 min). In some instances, ER staining can be followed by labeling of the nucleus with Syto 11 as described later. This stepwise dual-labeling technique enables the localization of both cell organelles using the same excitation wavelengths. Optical settings for $DiOC_6(3)$ are given in Table I.

The $DiOC_5$ and $DiOC_6$ probes were tested in isolated vascular endothelial and smooth muscle cells as well as in embryonic heart myocytes. The sarcoplasmic reticulum can be recognized easily by its filament-like mor-

[9] M. Terasaki, J. Song, J. R. Wong, M. J. Weiss, and L. B. Chen, Cell 38, 101 (1984).
[10] R. M. Wadkins and P. J. Houghton, Biochemistry 34, 3858 (1995).

phology. In three-dimensional reconstructions, the SR appears as leaf-like undulations in heart cells or tubular structures in VSM cells. $DiOC_6$ staining in these cell preparations, when used at concentrations of 10 mM, is more consistent and more intense than that of its equivalent $DiOC_5$. This ER staining can also be used jointly with plasma membrane markers or nuclear stains to evaluate intracellular distribution of the organelle or its spatial relationship with the nucleus. Unfortunately, the use of these carbocyanine probes, apart from being toxic at high concentrations, cannot be used simultaneously (in double-labeling experiments) with standard long wave-length Ca^{2+} or Na^+ indicators such as Fluo-3, Calcium Green, or Sodium Green because of similar excitatory and emission wavelengths (488 excitation, 500–530 emission range). If one wishes to associate cell function or ionic responses with ER membranes using these particular probes, one approach is to first complete calcium or sodium experiments and then label the organelle with $DiOC_6$ stain to visualize ER localization. Several washes in low calcium buffer to reduce Ca^{2+} content followed by ER staining with 10–50 mM $DiOC_6$ stain produces a signal sufficiently strong, allowing to reduce the gain (2 to 1×), attenuation filter (3–1%), and/or PMT voltage to levels below Ca^{2+} fluorescence detection. The latter staining, being much stronger than the ionic markers, thus enables subsequent correlation between structure and function.

Mitochondria: MitoFluor/AM

The new MitoFluor mitochondrion-selective probes developed by scientists at Molecular Probes are novel mitochondrial stains that appear to accumulate preferentially in the organelle regardless of its membrane potential (Molecular Probes Handbook). A 1 mM MitoFluor stock solution in DMSO should be freshly diluted to a final concentration of 50–100 nM in Tyrode–BSA buffer. Labeling can be performed directly in the bath chamber at room temperature in darkness. Staining becomes evident within 5 min after addition of the probe. Excess dye in the surrounding medium should be washed out after 5 min. As in the case of ER/SR staining, mitochondrial staining can be followed by nuclear labeling with Syto 11 and Fig. 1A shows an example. Optical settings for MitoFluor Green are given in Table I.

Figure 1 demonstrates an example of mitochondrial labeling in embryonic chick ventricular myocytes. As evidenced in Fig. 1, fetal heart cells are rich in mitochondria surrounding the nucleus. These organelles may contribute substantially to total measured intracellular Ca^{2+} and their contribution to cytosolic Ca^{2+} buffering should be taken into consideration.

Measurement of Membrane Potential Using Confocal Microscopy:
Di-8-ANEPPS and JC-1 Probes

Since the late 1960s, the measurement of fluctuations in membrane potential using fluorescent indicators and the search for specific voltage-sensitive dyes with an acceptable signal-to-noise ratio has sparked the development of several commercially available dyes such as thiazole orange, di-0-Cn(3), tetramethylrhodamine ethyl ester, bisaxonal, RH2g2, di-8-ANEPPS, and the mitochondrial membrane dye JC-1.[11,12] The di-4-ANEPPS and di-8-ANEPPS indicators belong to the class of fast potentiometric dyes of the styryl type. This family of indicators was found to be extremely effective optically and less phototoxic transducers of membrane potential.[11]

Optical imaging of plasma membrane potential can be performed in cells using the fast voltage-dependent potentiometric dye, di-8-ANEPPS (Molecular Probes). After washings in Tyrode–BSA buffer, coverslip-plated cells are then loaded with the freshly prepared styryl dye (13.5 μM final concentration in Tyrode–BSA buffer) from DMSO-reconstituted stock solutions. Loading of the dye should be performed at room temperature in darkness. Labeling is very rapid, and strong fluorescence intensity is usually obtained within 3–5 min. Figures 1B and 1C show examples.

The monomeric mitochondrial voltage-sensitive dye JC-1 can also be used to evaluate intracellular membrane potential responses to high extracellular KCl depolarization. Evaporated stocks of cationic JC-1 (Molecular Probes) can be reconstituted in 100% ethanol and further diluted to a working solution of 10 μg/ml in HEPES–NaCl buffer. Loading should be performed directly into the experimental bath chamber at room temperature in darkness. Good fluorescence is achieved within 10 min after addition of the probe. Optical settings for Di-8-ANEPPS and JC-1 are given in Table I.

Di-8-ANEPPS is a hydrophobic compound that seems to anchor in the cell membrane and is more stable than other membrane voltage dyes with less leakage.[11] This dye was reported to offer better time recordings of transmembrane voltage changes (about 30 min) in cultured neonatal rat heart cells than other fluorescent voltage probes, before leakage into the cytosol. In experiments with embryonic chick heart single cells as well as human adult aortic vascular endothelial and vascular smooth muscle cells, di-8-ANEPPS staining is stable for up to 1 hr with only a diffuse halo of

---

[11] S. Rohr and B. M. Salzberg, *Biophys. J.* **67,** 1301 (1994).
[12] J. C. Smith, *Biochim. Biophys. Acta* **1016,** 1 (1990).

staining observed in a few cells after 30–60 min, but never in the perinuclear region. The quality of di-8-ANEPPS membrane potential staining could be influenced by several factors, including cell type as well as species origin.[11] For example, the distribution of transsarcolemmal membrane potential in heart, endothelium, and VSM cells did not appear homogeneous. This irregular distribution of membrane potential in single cells could be due to scattered protein distribution on the sarcolemmal membrane, as was suggested for ion channels. This nonhomogeneity of labeling can be more apparent when single cells are depolarized with high extracellular K^+. Sustained depolarization of the human VSM cell with 30 mM $[K]_o$ rapidly (10 sec) induced a nonhomogeneous depolarization of the cell membrane accompanied by cell contracture. This latter contracture remained as long as the cell membrane can be depolarized with 30 mM $[K]_o$.

J-aggregate formation has also been used to visualize mitochondria in a variety of cells.[11,13] The mitochondrial voltage-sensitive probe JC-1, a specific energy potential-dependent mitochondrial dye, was reported to display a fairly narrow red peak that was sensitive to a variety of mitochondrial membrane potential modulating agents.[11] In preparations such as embryonic chick heart cells as well as in adult human aortic vascular endothelial and smooth muscle cells, mitochondrial membrane potentials are well labeled with the JC-1 indicator. Epifluorescence visualization reveals a decrease in green fluorescence (monomer) compared to orange-red fluorescence (JC-1 aggregate) on plasma membrane depolarization. The JC-1 probe does not appear to label the nuclear membrane.

Measurement of Receptor Density and Distribution Using Confocal Microscopy

Using 3D confocal microscopy and a Ca^{2+} channel fluorescent probe, as well as an Ang II fluorescent probe, showed that R-type Ca^{2+} channels as well as Ang II receptors are present on sarcolemmal as well as nuclear membranes of several cell types.[1–3]

We also developed FITC- and Bodipy-conjugated ET-1 as well as Ang II fluorescent probes. Using the ET-1 fluorescent probe, ET-1 receptors were found to be localized at the plasma and nuclear membranes in nonhomogeneous cluster-like patterns.

An Ang II fluorescent probe was also developed in our laboratory. This probe was found to be more specific and more effective than the one available commercially. Also, the newly developed PYY fluorescent probe

[13] S. T. Smiley, M. Reers, C. Mottola-Hartshorn, M. Lin, A. Chen, T. W. Smith, G. D. Steele, and L. B. Chen, *Proc. Natl. Acad. Sci. U.S.A.* **88,** 3671 (1991).

was found to be more effective than those available commercially and Fig. 1E shows an example. The optical settings for these probes are given in Table I.

For the time being, many commercially available receptors labeling dye tested did not show a satisfactory labeling of receptors. In order to develop an agonist or an antagonist-coupled fluorescent probe, the following procedures are recommended: (1) choose a ligand structure that permits attachment of a fluorescent probe without affecting the known active site of the structure; (2) choose a fluorescent probe that is highly stable and does not bind to any structure of the cells; (3) the fluorescent probe alone should be completely washable; (4) no single probe can be coupled with all different ligand types; (5) once the ligand is coupled to the fluorescent probe, the complex must be highly purified in order to isolate only the complex free of dye or ligand; (6) the ligand–dye complex should be tested in isolated cells or preferentially in tissue tension experiments in order to ensure that the ligand is still active and is as specific and powerful as the ligand alone; (7) ensure that the effect of the complex is washable and is antagonized by a well-known antagonist as well as being displaced by the "cold" ligand alone but not with the dye alone; and (8) the stability of the complex should be determined in solution as a function of time as well as a function of storage. It is highly recommended to use the ligand–fluorescent probe complex as soon as it is purified (i.e., within 2 days) and avoid storage if possible. The development of ligand fluorescent probes, although sometimes difficult, is a much worthwhile venue.

Conclusion

Confocal microscopy imaging studies in single cells and tissue sections confirm the importance of this noninvasive technique in the study of cell structure and function as well as the modulation of working living cells by various constituents of cell membranes, organelles, and cytosol.

The use of nonratiometric dyes does not allow adequate expression of intracellular Ca^{2+} concentrations in absolute values. Thus, care must also be taken when expressing results as either intensity or even ratio values. One approach in addressing these difficulties is by using different probes for the same ion in order to ensure that the results do not contain an artificial component due to loading artifacts. A complementary approach is by selecting cells that exhibit fluorescence intensities within fixed values and by determining the R_{min} and R_{max} values of these probes. In addition, in instances where chromophores are conjugated to a particular hormone or drug, steps including washout, use of cold or unlabeled hormones/drugs, and testing of free dye should be performed. Our results also reveal that confocal microscopy can be used successfully in coculture studies.

Photobleaching of the fluorophore can become a major obstacle in certain instances where the sample must be exposed to higher excitation light because of weak signal emission. In the case of mounted material, agents such as propyl gallate,[14] Vectashield (Vector Laboratories, CA), and SlowFade (Molecular Probes) have been used to reduce light-induced photodamage. In our laboratory, 0.1% phenylenediamine (Sigma, St. Louis, MO) in glycerol–PBS (9.1) and Vectashield are used routinely for mounted slide work. Protecting live material from photodynamic damage, however, is more difficult. Additives such as Oxyrase, ascorbic acid, and high doses of vitamin E have been described as having some antifading properties.[15]

With respect to differences in results that may be found between one laboratory and another, they may be due to (1) cell type and origin, (2) experimental conditions, (3) use of 3D measurements, and (4) precise determination of nuclear volume and localization.

Thus, confocal microscopy is extremely powerful in studying contraction, voltage, and chemical coupling between various cell types. One of the limitations, however, is the inability to record serial sections during quick cell responses such as those occurring during cell contraction. Rapid scans can only be performed on single vertical planes, thus limiting access to possible spatial differences in cellular response patterns. However, present results using fluorescent ion and organelle probes illustrate that the three-dimensional study of cell response and function is not only feasible but important in evaluating cell responses to external stimuli. Moreover, confocal microscopy is a highly useful tool in the assessment of cell pathology, whether it be in single cells or tissue sections. Site-selection probes such as receptor, protein, and second messenger probes, organelle probes, and nuclear stains all provide important indicators not only for the determination of actual structure and location of cell components, but also for the study of subcellular distribution and movement of various ions and molecules in working living cells.

For the time being, fluorescence confocal microscopy cannot localize and quantify bound ionic elements of the cell. It is hoped that new technology will soon permit us to determine not only free ions but also bound and/or compartmentalized ions.

Finally, with the help of three-dimensional imaging and volume-rendering capabilities, confocal microscopy constitutes a powerful state-of-the-

[14] H. Chang, *J. Immunol. Methods* **176,** 235 (1994).
[15] A. V. Mikhailov and G. G. Gundersen, *Cell Motil. Cytoskel.* **32,** 173 (1995).

art technique in the continuing investigation of cell structure and function in normal and pathological conditions.

Acknowledgments

This work was supported by an MRCC grant to Dr. Ghassan Bkaily. The authors thank Ms. Susann Topping for secretarial assistance.

[9] Combining Laser Scanning Confocal Microscopy and Electron Microscopy

By XUEJUN SUN

Introduction

Laser scanning confocal microscopy (confocal microscopy) provides a exciting new tool for biological research. For the first time, it is possible to obtain clear optical sections of relatively thick specimens without physically sectioning the tissue. Additionally, because of the high Z-axis resolution of preregistered images it provides, confocal microscopy enables rapid three-dimensional (3D) reconstruction of specimens. The combination of confocal microscopy with powerful computer image-processing techniques gives unprecedented possibilities for visualizing and measuring stained cells in three dimension.[1] The technology has gained wide acceptance in biological research in just one decade of its commercialization.

Although it is true that confocal microscopy provides higher resolution than conventional light microscopy, it is still fundamentally light microscopy, limited in resolution by the wavelength of the light source and the numerical aperture of the lens. This is at best about 150 nm with current optics. In many fields of biological research, one needs subcellular resolution better than 150 nm. For example, the establishment of an association of a gene product with a type of organelle leads to important insights into the role of the product in the cell. For subcellular study, electron microscopy (EM) is still an indispensable tool for the unambiguous identification of organelles. EM provides a resolution that cannot be matched by any other imaging tool currently available. EM, however, suffers from a lack of 3D

---

[1] H. Chen, J. R. Swedlow, M. Grote, J. W. Sedat, and D. A. Agard, in "Handbook of Biological Confocal Microscopy" (J. B. Pawley, ed.), 2nd ed., p. 197. Plenum Press, New York, 1995.

information. Because of the poor penetrating power of electrons, observation with EM is often limited to a fraction (nanometer range) of the cellular structure. High-voltage EM provides an alternative that yields both high resolution and some degree of 3D information. It is, however, not readily accessible to most biologists and even with high-voltage EM, specimens no thicker than 1 μm can be observed. Three-dimensional reconstruction is still required to have a 3D view of cells.

Neuroanatomy provides an extreme situation in which 3D information is particularly important. Neurons often are large and have complicated 3D structure. The intrinsic 3D morphology reflects the connections of neurons within a neuronal network. Yet, ultimately, the discrete synaptic connections, and the neurotransmitter(s) used by them, rule the exact roles a particular neuron plays within the network. Study of stained neurons at the level of light microscopy often pinpoints particular loci of interest. Whether there are synaptic connections in the pinpointed areas provides necessary information for understanding the neuron and the circuit. Assessment of the three-dimensional morphology of neurons requires light microscopic study whereas synaptic information is available only through ultrastructural study. Therefore, many investigators have attempted to combine EM with LM.[2–5] All currently available methods require a certain degree of 3D reconstruction, which, even with advanced computer imaging technology, is still labor-intensive and sometimes impossible for large neurons.

The combination of confocal microscopy and EM is a promising avenue for biologists attempting to solve this problem of 2D versus 3D. Confocal microscopy can provide a valuable overview and 3D information. EM provides ultrastructural detail. A combination of these two technologies is, however, not straightforward because the two technologies require very different tissue processing procedures and staining. First, for confocal microscopy, a specimen must be stained with a fluorescent marker, which is not necessarily electron dense and therefore not suitable for EM study. Second, for confocal observations, the tissue must be translucent, whereas for electron microscopy, heavy metals are used that are opaque to light. Third, for EM observation, tissue is usually embedded in plastic. This often introduces a large amount of tissue shrinkage and distortion, which render the correlation of light microscopic information with ultrastructural data difficult. A combination of confocal microscopy and EM has been attempted

[2] K. Hama, T. Arii, and T. Kosaka, *Microsc. Res. Tech.* **29,** 357 (1994).
[3] J. DeFelipe and A. Fairén, *J. Histochem. Cytochem.* **41,** 769 (1993).
[4] J. K. Stevens, T. L. Davis, N. Friedman, and P. Sterling, *Brain Res.* **2,** 265 (1980).
[5] R. W. Ware and V. LoPresti, *Intl. Rev. Cytol.* **40,** 325 (1975).

using the reflection mode of imaging with the confocal microscope.[6] With this method, a cell labeled with an electron-dense marker can be imaged in the confocal microscope by bouncing light off of it. This method, however, is not widely applied due to the generally poor quality of reflection in confocal images.

We have been studying the neuronal connections within insect central nervous systems by combining standard confocal microscopy with EM.[7–10] The two methods described in this article rely on a very specific interaction between biotin and avidin. Although the methods were developed primarily for neuroanatomy with biotin-stained neurons, it can be adapted to other tissues to localize various substances within cell using biotinylated antibodies or DNA/RNA probes, as shown in this article with an antibody against serotonin, a neuromodulator in insects.[11,12]

Experimental Procedures

Overview of Methods

The methods require the introduction of biotin or biotin derivatives into cells. The biotin can subsequently be detected using a fluorescent tag for confocal microscopy and an electron-dense tag for EM observation. Confocal microscopy can provide a guiding tool for thin sectioning in order to facilitate the correlation of confocal information with EM results. Both methods involve these basic steps of introduction of biotin, introduction of appropriate tags for the biotin, confocal observation, and EM observation.

Both methods have been applied on insect central nervous tissue. Adult central nervous systems of two insect species were used: the moth *Manduca sexta* (Lepidoptera, Sphingidae) and the white-eyed mutant housefly *Musca domestica* (Diptera, Muscidae). The detailed rearing procedures have been described previously (for moth, see Bell and Joachim[13]; for fly, see Fröhlich and Meinertzhagen[14]).

[6] J. S. Deitch, K. L. Smith, J. W. Swann, and J. N. Turner, *J. Microsc.* **160,** 265 (1990).

[7] X. J. Sun, L. P. Tolbert, and J. G. Hildebrand, *J. Histochem. Cytochem.* **43,** 329 (1995).

[8] X. J. Sun, L. P. Tolbert, and J. G. Hildebrand, *J. Comp. Neurol.* **379,** 2 (1997).

[9] X. J. Sun, L. P. Tolbert, J. G. Hildebrand, and I. A. Meinertzhagen, *J. Histochem. Cytochem.* **46,** 263 (1998).

[10] L. P. Tolbert, X. J. Sun, and J. G. Hildebrand, *J. Neurosci. Methods* **69,** 25 (1996).

[11] A. R. Mercer, J. H. Hayashi, and J. G. Hildebrand, *J. Exp. Biol.* **198,** 613 (1995).

[12] A. R. Mercer, P. Kloppenburg, and J. G. Hildebrand, *J. Comp. Physiol. A* **178,** 21 (1996).

[13] R. A. Bell and F. A. Joachim, *Ann. Entomol. Soc. Am.* **69,** 365 (1976).

[14] A. Fröhlich and I. A. Meinertzhagen, *J. Neurocytol.* **11,** 159 (1982).

Method 1: Immunogold Fluorescence Method

This method takes advantage of the finding that immunogold reagents do not bind to all the available biotin in biotin-stained cells.[7] Therefore, the biotin-stained neurons can be incubated with immunogold-conjugated avidin first and then the remaining biotin molecules can be tagged with fluorescent avidin.

Cell Staining with Biotin

Biotin was introduced into cells by one of three methods.

Intracellular Labeling of Neurons. Biotin derivatives, Neurobiotin (Vector Labs., Inc., Burlingame, CA) or biocytin (Molecular Probes, Inc., Eugene, OR), are injected intracellularly into neurons by electrophoresis through a micropipette. For this, an isolated head preparation is used. To gain access to the brain, the mouth parts, frontal cuticle, and muscle systems of the head are removed surgically, and the head is cut off the body and secured with insect pins in a recording chamber where it is superfused continuously with saline solution [149.9 mM NaCl, 3 mM KCl, 3 mM CaCl$_2$, 25 mM sucrose, and 10 mM TES (N-tris[hydroxymethyl]-methyl-2-aminoethanesulfonic acid), pH 6.9]. A standard electrophysiological recording setup is used to inject the dye as described previously.[15] The tip of a sharp glass microelectrode is filled with either 4% Neurobiotin in 1 M KCl or 4% biocytin in a solution of 0.5 M KCl in 0.05 M Tris buffer (pH 7.4). The shaft of the electrode is filled with 2 M KCl solution. After a stable impalement of a neuron or a period of intracellular recording, a depolarizing (in the case of Neurobiotin, usually 2–10 nA) or hyperpolarizing (in the case of biocytin, 2–10 nA) current is passed through the electrode for a period of 5–30 min to inject label into the cell.

Back-Filling of Neurons. Biotin derivatives can also be introduced by back-filling,[7,9] a staining technique widely applied in other animals (reviewed by McDonald[16]). Back-filling is accomplished by cutting surgically accessible axons of the neurons to be labeled. The cut end of the nerve is then exposed to distilled water for approximately 1 min in order to open the damaged neurites. A drop of biotin solution (a few crystals of Neurobiotin in distilled water, estimated concentration <0.2%) is placed over the cut area. The whole animal is then placed in a humidified chamber for 5–20 min for the label to be picked up by the damaged cells. The incubation time depends on the distance that the biotin has to travel. Ten minutes is sufficient for about 2 mm of diffusion. Because biotin travels rather quickly within

[15] S. G. Matsumoto and J. G. Hildebrand, *Proc. R. Soc. Lond. B* **213,** 249 (1981).
[16] A. J. McDonald, *Neuroreport* **3,** 821 (1992).

dendrites, there is no need for additional diffusion time. The area of the cut is flushed with saline after staining to wash out the excess biotin. The brain is then dissected out and processed as described later.

Immunocytochemical Staining by Biotinylated Antibody. Biotin can also be introduced as common biotin-conjugated probes, such as biotinylated secondary antibodies. We tested this method with a biotinylated secondary antibody. The detail procedure for this type of application is described later in a separate section.

Tissue Processing

Fixation. The fixation requirements for confocal microscopy and EM are very different. Confocal applications do not tolerate glutaraldehyde fixation because the high autofluorescence introduced by glutaraldehyde will potentially interfere with the fluorescent label. For EM study, a combination of paraformaldehyde and glutaraldehyde is the most common chemical fixation procedure. We find that lowering the concentration of glutaraldehyde to 0.15–0.40% is compatible with most confocal applications while maintaining adequate tissue preservation for EM. Additionally, using long-emission wavelength fluorophores such as Cy3 reduces the problem associated with autofluorescence.

For EM, the fixative solution must be fresh. Immediately after injection of biotin, the brain is dissected out of the head and fixed at 4° overnight in freshly made fixative solution comprising 4% paraformaldehyde, 0.32% glutaraldehyde, and 0.2% saturated picric acid in 0.1 M sodium phosphate buffer (pH 7.4). Picric acid is optional. It has been suggested that picric acid stabilizes membranes and precipitates proteins for EM observation.[17] It introduces, however, some additional degree of autofluorescence. We do not find a noticeable difference in preservation of ultrastructure in preparations processed with or without picric acid.

Fixation time can vary, depending on tissue size. We routinely fix moth brains overnight at 4° for convenience (approximate size of the brain is 2×3 mm). However, fixation times as short as 30 min can be applied at room temperature.

Tissue Sectioning. It is necessary to cut the tissue into thin slabs because (1) the immunogold reagent has a limited penetration power and (2) the lenses used for confocal microscopy have a short working distance. After fixation, the brain is washed briefly with sodium phosphate-buffered saline (PBS, 0.1 M, pH 7.4) and embedded in 7% agarose [low melting point agarose; gelling temperature of 2% (w/v) solution: 26–30°; Sigma, St. Louis,

---

[17] M. G. Zomboni and C. de Martino, *J. Cell Biol.* **35**, 148A (1967).

MO]. The brains are then sectioned at 50–80 μm with a Vibratome (series 1000; Technical Products International, Inc., St. Louis, MO). The floating sections are collected in PBS.

Permeabilization and Incubation with Immunogold-Conjugated Streptavidin. A common problem in immunogold staining before embedment is the lack of penetration of the gold-conjugated reagent into the specimen because of the relatively large size of the gold particles. To enhance the penetration of the immunogold-conjugated streptavidin, we treat the Vibratome sections with either 0.1% Triton X-100 (J. T. Baker, Inc., Phillipsburg, NJ) in PBS for a period of 10 min or a series of ethanol solutions (30, 50, 70, 50, and 30% ethanol) for 5–10 min at each step. Optimal penetration is achieved with such permeabilization and using streptavidin conjugated to small (1.4 nm) gold particles. We have attempted to use smaller gold particles (0.8 nm, Nanoprobes Inc., Stony Brook, NY) to enhance the penetration of the reagent, but the 0.8-nm gold particle-conjugated streptavidin penetrated even less than the 1.4-nm gold.

After permeabilization, the sections are incubated with a solution of 1.4-nm gold-conjugated streptavidin (Nanoprobes Inc.) diluted 1:50–1:100 in PBS for 2–4 days on a shaker in an 8° cold room. Then 1:200 Cy3-conjugated streptavidin (Jackson ImmunoResearch Labs., Inc., West Grove, PA) is added to the same solution, and the sections are incubated for an additional 8–12 hr in the cold room on the shaker. Following this procedure, the biotin-injected neuron is labeled reliably by gold-conjugated and fluorescently tagged streptavidin up to 40 μm into the tissue and could be detected both by confocal microscopy (fluorescence) and by EM (gold particles) (Fig. 1).

Confocal Observation

Stained slices of brain are mounted on slides in saline or glycerol-based mounting medium (Vectashield, Vector Labs., Inc.). Care must be taken to avoid drying the tissue for potential ultrastructural damage. We normally use nail polish to seal the slide.

We use a Bio-Rad MRC 600 confocal microscopy (Bio-Rad, Cambridge, MA) mounted on a Nikon Optiphot-2 microscope and a Zeiss LSM 410 mounted on a Zeiss Axiovert M100 microscope, both equipped with a krypton/argon laser light source for confocal observations. The tissue is usually imaged with a 20× (NA 0.75) Nikon objective lens (Bio-Rad) or with a 25× (NA 0.8) multi-immersion Plan Neofluar objective lens (Zeiss). A multi-immersion lens is highly suitable for such a study in order to minimize the effects of refractive index mismatch. Serial optical sections, usually at intervals of 1–2 μm, are imaged through the depth of the labeled

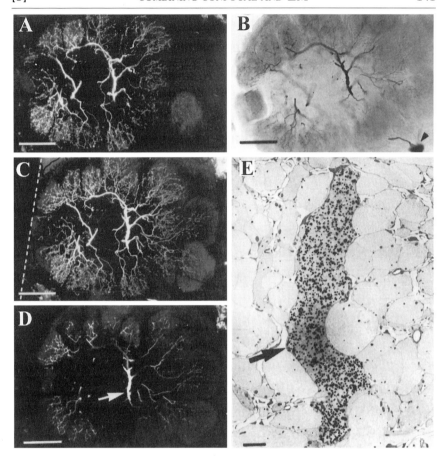

FIG. 1. Example of examination of a Neurobiotin-injected neuron using the immunogold fluorescent method to combine confocal and EM. (A) Intracellular staining of a local interneuron in the antennal lobe of the moth *Manduca sexta*. The neuron was imaged in saline, before plastic embedding, at 3-μm intervals for a total of 18 optical sections. The neuron ramifies extensively throughout the neuropile (antennal lobe). The soma is out of the field of view. (B) Bright-field image of the same neuron after silver enhancement and plastic embedding shows enhanced immunogold staining. The cell body is indicated with an arrowhead. (C) Confocal image of the Epon block demonstrates that the fluorescence is retained. Dashed line indicates the edge of the block. Comparison with confocal image obtained before plastic embedding (A) shows that most of the small processes remain detectable. (D) Single optical section from the 3D image stack at the level of the ultrathin section taken in E. Arrow indicates the corresponding position of the process. (E) Thin section of the same neuron at EM level showing the ultrastructure of the labeled process, which contains numerous gold particles. Bars: (A–D) 100 μm, (E) 2 μm. Adapted, with permission, from X. J. Sun, L. P. Tolbert, and J. G. Hildebrand, *J. Histochem. Cytochem.* **43,** 329 (1995).

neurons and saved as three-dimensional image stacks. Projecting this image stack onto a single two-dimensional plane generates a 2D image of the reconstructed labeled neuron. The image stack is also studied with appropriate software to visualize the neuron in three dimensions. Areas of interest are defined within the stack.

There is no evidence of ultrastructural damage by laser radiation on fixed tissue. However, to avoid potential ultrastructural damage by strong laser radiation with the confocal microscope, we try to minimize both the scanning time and the laser intensity by choosing an acceptable fast scan speed and a low laser power setting. This usually involves a tradeoff of image quality, imaging duration, and laser intensity.

Silver Enhancement of Gold Particles

The 1.4-nm gold particles, too small to allow easy observation under the EM, are enlarged with a silver enhancement solution (Biocell Res. Labs., from Ted Pella, Inc., Redding, CA) in order to facilitate detection in EM. The silver enhancement procedure is modified from the protocol suggested by the manufacture of the kit. In brief, after confocal observation, the selected sections are washed three times (10 minutes each) with PBS to remove the mounting medium. They are then washed in deionized water (two times at 3 min each) and incubated in the silver enhancement solution on ice in the dark for 8 min to allow diffusion of the solution into the sections. The container containing the sections is subsequently transferred to a water bath at room temperature for another 8 min of reaction in the dark. Timing of the reaction is important and care must be taken to avoid a self-nucleation reaction. The silver enhancement reaction is stopped by several washes with deionized water (three times at 3 min each). The sections are then transferred to sodium phosphate buffer (0.1 M, pH 7.4) for further processing for EM study.

Postfixation for EM Study

For EM, it is necessary to stabilize cell membranes by postfixation and then to embed in plastic. A common postfixative for EM is osmium tetroxide. As it was suggested that osmium tetroxide oxidizes silver precipitates of the silver-enhanced gold particles,[18] tissues are lightly osmicated at room temperature (0.1% osmium tetroxide in phosphate buffer for 5–10 min). The postfixation process is stopped by replacing the solution with phosphate buffer. This procedure stabilizes the cell membranes enough for plastic embedding while retaining enough silver precipitate on the gold particles

---

[18] A. I. Basbaum, *J. Histochem. Cytochem.* **37,** 1811 (1989).

for EM observation. Osmium tetroxide and its vapor are highly toxic. This postfixation step must be carried out under a well-ventilated fume hood and the waste solution disposed of properly.

Plastic Embedding

The postfixed brain slices are dehydrated through a graded series of increasing concentrations of ethanol (50, 70, 95, and twice 100% for 10 min at each step), followed by two changes of 10 min each in propylene oxide (Electron Microscopy Sciences, Fort Washington, PA). The sections are then placed in a 1 : 1 mixture of Epon/Araldite and propylene oxide for 30 min. The tissue is infiltrated with pure fresh Epon/Araldite overnight on a rotatory shaker. To flat embed the sections, sections are sandwiched between two Aclar sheets (Ted Pella, Inc., Redding, CA) in Epon/Araldite, and this assembly is further sandwiched between two microscope slides with two paper clips to hold the whole assembly together. The whole assembly is then put into a 60° oven overnight for polymerization. One Aclar sheet can be peeled off easily and the sections can be examined with bright-field microscopy (Fig. 1B) or with the confocal microscope. Selected sections are cut out with a razor blade and glued to blank Epon/Araldite blocks with fast cyanoacrylate adhesive (Electron Microscopy Sci., Inc., Ft. Washington, PA) for microtomy.

Confocal Microscopy as Guide for Microtomy

We found that a certain degree of Cy3 fluorescence is retained after processing tissue as described earlier. Therefore, it is possible to use a confocal microscope as a guide for microtomy. For viewing the block on the confocal microscope, the block is trimmed with a glass knife in order to obtain a smooth surface for confocal observation. It is then held in a specially fabricated brass slide with a hole in the middle. In this way, the block can be imaged in either an inverted or an upright position depending on the configuration of the microscope. These arrangements allow, moreover, a slight tilt of the surface of the block, if the surface of the block is not totally parallel to the image plane, which is a critical prerequisite for determination of the depth of the area of interest. The confocal microscope stage and the z-axis motor are calibrated and the reproducibility of both checked carefully so as to provide accurate depth information. The surface of the block is defined easily by its strong reflecting signal, giving the starting depth of 0 μm. The area of interest can be relocalized with confocal microscopy by reference to the initial set of confocal images. Through the readout of the z-axis motor, the actual depth of the area can be obtained for guiding microtomy. To ease the effort of relocalizing the fiber of interest under EM, the block is trimmed as small as possible.

Electron Microscopic Study

After examination with the confocal microscope, the Epon blocks containing labeled neurons are thin sectioned either at regular intervals by alternating thin sectioning and semithin sectioning or at preselected depths of interest in the tissue defined by the confocal study. Sections are cut with a diamond knife on a MT6000-XL microtome (RMC Inc., Tucson, AZ) or on a Reichert-Jung Ultracut microtome (Leica, Deerfield, IL). To facilitate localization of the chosen depth, thin sections are collected so that the ribbon on each grid contains a defined thickness of tissue. Because of the refractive index mismatch and inaccuracies of the microscopic stage, the depth information obtained using the confocal microscope may not be precisely accurate. Additionally, estimates of section thickness on the microtomes will incorporate some error. Therefore, additional thin sections are collected before and after the supposed depth of interest in order to ensure that the area of interest is contained in the thin sections collected. Thin sections are picked up on Formvar-coated slot grids or on thin-bar copper grids and poststained with saturated aqueous uranyl acetate solution and with lead citrate for 10 min each.

The thin sections are examined with a JEOL JEM-1200EX transmission EM or with a Philips 201C EM. The labeled processes are selectively photographed, usually at magnifications between ×5000 and ×15,000 for the identification of synapses.

Correlation of Confocal Microscopy and EM

In the majority of cases, depth information and the shape of the labeled process in the sections enable easy correlation of the confocal overview with EM detail. However, this correlation must be made with care, as a confocal optical section at the resolution of the lenses used in these experiments may contain the same amount of tissue as is included in as many as 40 thin sections. Moreover, localizing small, labeled profiles in a relative large area under EM is not an easy task. This can be eased by trimming the bock as small as possible to limit the area of view for EM.

One example using the immunogold fluorescence method to combine confocal microscopy and EM is presented in Fig. 1. A neuron in the antennal lobe of the moth brain is stained intracellularly with Neurobiotin and viewed with confocal microscopy before plastic embedding (Fig. 1A) and again in the plastic block (Fig. 1C). The transmitted bright-field microscopy (Fig. 1B) shows the immunogold labeling at the light microscopic level. An electron micrograph (Fig. 1D) of the neuron is shown. Use of this method to investigate the properties of insect neurons has been published in separate papers.[8]

Method 2: Avidin–Biotin Complex Method

This method uses a commercially available kit [avidin–biotin-peroxidase complex (ABC) from Vector Labs. Inc.] to convert some of the fluorescence signal in a fluorescently stained cell into an electron-dense form using 3,3'-diaminobenzadine (DAB) as substrate. It is based on the finding that avidin conjugated to a fluorescent tag can be bridged again with the ABC complex.[9] Then the peroxidase in the ABC complex can be used to convert DAB to an electron-dense reaction product for EM viewing.

Introduction of Biotin

Introduction of biotin to cell is accomplished exactly as the immunogold fluorescence method described earlier.

Fixation and Tissue Sectioning

Fixation is carried out as described earlier for the immunogold fluorescence method.

Because the ABC complex can penetrate much more deeply into the tissue than the immunogold reagent, it is possible to investigate much thicker tissue (up to 150 μm) with this method than with the immunogold fluorescence method. Nevertheless, the maximum thickness of the slices is also limited by the working distance of the objective used for confocal imaging. The 25× multi-immersion lenses used in these experiments has a working distance of 130 μm. Therefore, tissue is usually sectioned on the Vibratome at 100 μm as described earlier and the sections are collected in PBS.

Detection of Labeled Neurons with Fluorescent Marker

Labeled cells are rendered fluorescent by incubating the floating sections in a fluorescently conjugated avidin (or streptavidin). A variety of different fluorophores conjugated to avidin or streptavidin are available commercially (e.g., Molecular Probes, Inc.). We mostly use Cy3 for its photostability and brightness and because it is well suited for imaging with a krypton/argon laser. Because streptavidin or avidin penetrates into tissue much more readily than immunogold particles, the incubation time is much shorter (1–3 hr) than in the immunogold fluorescence method. Moreover, permeabilization is not necessary in this case. We mostly incubate the sections in a Cy3-conjugated streptavidin (Jackson ImmunoResearch Labs., Inc., West Grove, PA) solution (0.5 µg/ml of PBS) for 3 hr to overnight (for convenience). The sections are then washed with PBS and mounted without dehydration in Vectashield under a cover glass sealed with nail polish. Again, for good tissue preservation, care must be taken not to let the tissue dry in this process.

Confocal Microscopy

Confocal microscopy is carried out as described in the immunogold method described previously. However, because the fluorescence signal is no longer detectable after ABC conversion, all confocal examinations must be carried out before the ABC conversion, unlike the immunogold method where some fluorescence is retained in the plastic block.

Avidin–Biotin–Peroxidase Complex Conversion

After confocal microscopic observation, the sections are transferred to PBS and incubated in an avidin–biotin–peroxidase complex (ABC Elite, Vector Labs., Inc.) solution for 2 hr to overnight in a cold room (8°) on a shaker. The concentration used is suggested by the kit supplier. The preparations are then washed with PBS, incubated with DAB (0.25% in PBS) for 15 min, and finally reacted with DAB in the presence of H_2O_2 (0.01% in DAB solution) for 5–10 min. Because the reaction is somewhat light sensitive, it is preferable to carry it out under darkness. It is, however, necessary to check the reaction periodically under a dissecting microscope and to stop the reaction before the background stain rises. The reaction is stopped with several washes of PBS. After such a reaction, the labeled neurons become dark and the fluorescent label is no longer visible.

Postfixation and Plastic Embedding

Postfixation with osmium tetroxide would enhance DAB staining. However, one of the common problems with DAB staining for EM is masking of the internal structure of the labeled cell. We found that a good balance between overall tissue preservation and maintenance of visibility of internal structure is achieved by osmicating tissue slices in 0.5% osmium tetroxide in sodium phosphate buffer (0.1 M, pH 7.4) for 15 min. The sections are then dehydrated and embedded in plastic as described earlier for the immunogold method. The plastic sheet containing the labeled neurons is initially examined at the light microscopic level. Even though the image quality is somewhat impaired by the thickness of the slices and the density of osmication, the filled cells can be identified clearly under bright-field microscopy. Selected plastic-embedded Vibratome sections containing the labeled neuron are cut out with a razor blade and glued to a blank block as described earlier for thin sectioning.

Defining Depth of Areas of Interest with Confocal Microscopy

Because the fluorescence signal is no longer visible after plastic embedding with this method, locating the area of interest in plastic poses a particular challenge. This is overcome by using either the reflection mode of the confocal microscope or the transmitted light detector of the confocal microscope if it is so equipped. Although neither of the methods gives images of high enough quality to allow another 3D reconstruction of the stained cell, both methods allow easy comparison with the original confocal images for identification of the area of interest. Then the depth of that area can be defined from the readout of the z-axis motor of confocal and microtomy can be conducted as described in the immunogold fluorescence method.

EM Study

Once the depth of a particular region of interest is defined, the plastic block is thin sectioned and examined by EM as described earlier for the immunogold fluorescence method. An example of a Neurobiotin-stained neuron visualized using this method is presented in Fig. 2. Preservation of ultrastructure is good judging by the appearance of cell membranes and well-preserved synaptic vesicles (Figs. 2D and 2E).

Application of the Method to Immunocytochemistry

The biotin–avidin interaction is well exploited in many fields of biological research because of the specific and strong interaction between biotin and avidin. The methods described earlier, in theory, should be applicable with any biotinylated probe.

As an example, we use an antibody against serotonin, a neurotransmitter/modulator, to demonstrate the use of the method for immunocytochemistry. A requirement of the method is that the antigen must survive use of the small amount of glutaraldehyde that is required for adequate tissue preservation for EM study. This is the case for this specific antibody against serotonin.[19]

---

[19] X. J. Sun, L. P. Tolbert, and J. G. Hildebrand, *J. Comp. Neurol.* **338,** 1 (1993).

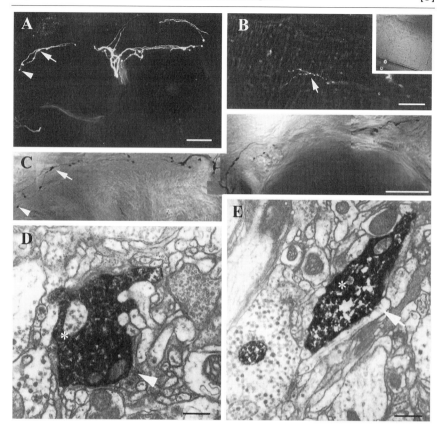

FIG. 2. Example of the use of the ABC conversion method on a Neurobiotin-labeled neuron. (A) Reconstruction of a Neurobiotin-labeled neuron from the fly *Musca domestica*. Neuron was imaged in a 100-μm Vibratome slice. Optical sections were obtained every 2 μm over a depth of 80 μm. Selected varicosities (arrow and arrowhead) indicate areas of interest (varicosities) for further EM investigation. (B) Using the reflecting mode of the confocal microscope, it is possible to locate areas of interest (arrow) in the plastic block to determine the depth of the region (24.5 μm). (Inset) The strong reflecting image of the surface of the block (0 μm). (C) Partial reconstruction from transmitted detector images of confocal indicates that the stain is restricted to the labeled neuron only. Arrow and arrowhead indicate the same regions of the neurons in (A). (D and E) Electron micrographs of the same neurons in the arrowhead (D) and arrow (E) regions in (A–C). Asterisks indicate locations of possible synaptic connections. The quality of tissue preservation is good as indicated by well-preserved cell membrane and synaptic vesicles. Bars: (A) 50 μm, (B) 20 μm, (C) 50 μm, (D, E) 0.5 μm. Adapted, with permission, from X. J. Sun, L. P. Tolbert, J. G. Hildebrand, and I. A. Meinertzhagen, *J. Histochem. Cytochem.* **46,** 263 (1998).

Serotonin immunocytochemical staining is carried out as described previously[19] except that a biotinylated secondary antibody is used to introduce biotin to the labeled neurons. In brief, tissue is fixed and sectioned on a Vibratome as described previously. Then, the tissue is incubated in primary antibody (rabbit antiserotonin, Incstar Corp., Stillwater, MN) in PBS overnight. It is subsequently washed in PBS and incubated with biotinylated goat antirabbit antibody [used at the concentration suggested by the supplier (1:200), Vector Labs, Inc.] for 3 hr. The staining is detected with Cy3-conjugated streptavidin for confocal microscopy. Then, the DAB reaction is carried out as described earlier for the ABC complex. An example of such staining is shown in Fig. 3.

Discussion and Conclusion

The two methods described in this article are complementary. The immunogold method provides better detection than the ABC method because the internal structure is not masked by the stain. Additionally, immunogold particles do not diffuse like the DAB reaction product does. Moreover, the immunogold method provides the possibility to observe the tissue with confocal microscopy after embedding in plastic. Therefore, it avoids the problem of discrepancies caused by tissue shrinkage after plastic embedding when correlating 3D information with ultrastructural detail. The immunogold fluorescence method, however, is slow due to the long incubation time required for the immunogold reagent. Additionally, the immunogold method limits section thickness to about 80 μm, which creates additional problems for large cells. The ABC complex method, however, is rapid. It can be carried out in a single day if necessary. Moreover, the ABC method allows much thicker sections to be studied. In summary, the ABC method is suitable for routine applications, whereas the immunogold fluorescence method should be used when better tissue preservation and detection are required.

Limitations of Methods

The combination of confocal microscopy and EM allows the investigator to correlate 3D information with ultrastructural characteristics with relative ease. However, identifying the same structure under EM and confocal still requires a high degree of skill. This problem is compounded by the fact that confocal optical sections are much thicker than a single thin section. The problem may be minimized by performing a small 3D reconstruction from serial thin sections.

Both confocal–EM methods require the introduction of biotin. This may not be applicable in tissues with a high level of endogenous biotin.

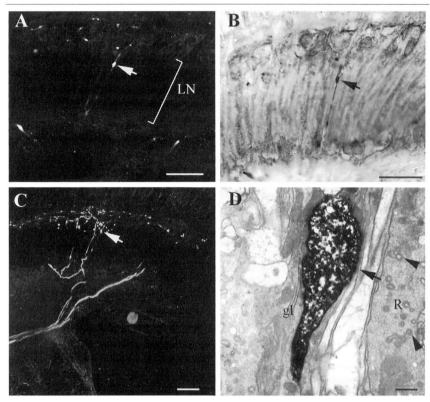

Fig. 3. Extension of the methods to immunocytochemical labeling using labeling with antiserotonin antibody as an example. (A) Single optical section of a serotonin-immunoreactive fiber (which belongs to neuron LBO5HT[21]) in the optical path (lamina, LN) of the fly brain. Arrow indicates a varicosity of interest. (B) Transmitted light image of the same fiber after DAB reaction and plastic embedding. (C) Reconstruction of the entire optical stack (40 optical sections at 1.2-μm intervals) shows the overall view of the neurons in the Vibratome slice. (D) Electron micrograph of the varicosity in B shows that it contains numerous darkly labeled vesicles and is surrounded by glia (gl). A neighboring photoreceptor terminal (R), characterized by capitate projections (arrowheads), shows good tissue preservation. Bars: (A–C) 25 μm, (D) 0.5 μm. Adapted, with permission, from X. J. Sun, L. P. Tolbert, J. G. Hildebrand, and I. A. Meinertzhagen, *J. Histochem. Cytochem.* **46,** 263 (1998).

We used Cy3 almost exclusively as the fluorophore of choice for confocal observation because of its brightness and photostability. Additionally, Cy3 retains a fair degree of fluorescence with the immunogold fluorescence method, allowing confocal microscopic study of the labeled cell in plastic. However, studying plastic sections with confocal microscopy poses other challenges. Opacity of the tissue after osmication made it particularly difficult to image deep into the sections. Moreover, mismatches in refractive indexes between plastic and tissue introduce additional inaccuracies for depth determination.

Adaptation to Other Systems

Major obstacles of combining confocal microscopy and EM are the incompatibilities of the markers used for confocal microscopy and for EM. This article describes methods of simultaneously or sequentially labeling insect neurons with fluorescence for confocal microscopy and with electron-dense labels for EM. The methods should be adaptable to other systems if certain facts are kept in mind. The requirements of tissue fixation for confocal microscopy are much less demanding than those for EM. For EM, however, there is no universal, ideal fixative for every tissue type. Often, a trial-and-error test is needed to find the best fixation procedure for a specific tissue. Finding the best fixative procedure for EM is a good starting point for developing a combined confocal–EM method for the tissue. By systematically modifying the concentration and/or the time of fixation as we did, one should be able to achieve optimum balance between ultrastructural preservation and good signal-to-noise ratios for fluorescent and electron-dense labels.

Similarly, different types of tissues and different antigens (for immunocytochemical labeling) will require different methods and amounts of permeabilization for the access of gold particles. For example, some antigens will not survive dehydration, so alcohol cannot be used as the permeabilization agent. Other methods such as freeze-thawing[20] to permeabilize the tissue may be better.

In summary, we have reviewed two methods we developed for combining the advantages of confocal and electron microscopy. Although other confocal–EM methods have been described,[6] the methods described here are simpler and less tedious. We have proven them to be practical and useful for the combination of confocal microscopy and EM for neurobiological research.

Acknowledgments

The work was carried out in the Laboratories of Drs. Leslie P. Tolbert and John G. Hildebrand at ARL Division of Neurobiology, the University of Arizona, and in the laboratory of Dr. Ian A. Meinertzhagen at the Department of Psychology, Dalhousie University, Canada. The author thanks them for their support. I also thank Dr. L. P. Tolbert for critical reading

[20] A. N. van de Pol, *J. Neurosci.* **6,** 877 (1986).
[21] D. R. Nässel, *Progr. Neurobiol.* **30,** 1 (1987).

of the manuscript and suggestions. Additionally, the author thanks Patty Jasma, A. A. Osman, Lesley Varney, and Maggie Klonowska for technical assistance. The work was supported by an award from the University of Arizona, Center for Insect Science to X.J.S., grants to I.A.M. (NIH EY-03592, NSERC OPG 0000065), to L.P.T. and J.G.H. (NIH NS-28495), and to J.G.H. (NIH AI-23253).

[10] Construction of Line-Scan Confocal Microscope for Physiological Recording

By NICK CALLAMARAS and IAN PARKER

Introduction

The use of confocal laser scanning microscopes (CLSM) together with fluorescent Ca^{2+} indicator dyes allows detailed imaging studies of the spatio-temporal aspects of intracellular Ca^{2+} signaling. For example, the high spatial and temporal resolution of this technique has revealed that release of Ca^{2+} from intracellular organelles into the cytosol occurs as transient, elementary signals, arising at specific subcellular release sites. These elementary events include Ca^{2+} "puffs" mediated through inositol 1,4,5-tris-phosphate ($InsP_3$) receptors in *Xenopus* oocytes and HeLa cells[1–4] and Ca^{2+} "sparks" mediated through ryanodine receptors in cardiac muscle.[5] This article describes the construction and performance of a CLSM, combined with a UV illumination system for the photolysis of caged compounds, presently in use in the authors' laboratory for studies of $InsP_3$-mediated intracellular Ca^{2+} release in *Xenopus* oocytes. This system can be adapted readily to virtually any inverted microscope and offers greater versatility and equal or better performance than comparable commercial instruments for a fraction of the cost.

Commercial confocal microscopes fall broadly into two classes: those designed for morphological studies of fixed or slowly changing specimens and video-rate (30 Hz) full-field imaging systems for monitoring dynamic cellular signals. In either case, the application of these instruments for physiological studies often presents several difficulties. First, design constraints in commercial instruments tend to complicate the optical paths,

[1] I. Parker and Y. Yao, *Proc. R. Soc.* (*Lond.*) *B. Biol. Sci.* **246,** 269 (1991).
[2] Y. Yao, J. Choi, and I. Parker, *J. Physiol.* (*Lond.*) **482,** 533 (1995).
[3] M. Bootman, E. Niggli, M. Berridge, and P. Lipp, *J. Physiol.* (*Lond.*) **499,** 307 (1997).
[4] X.-P. Sun, N. Callamaras, and I. Parker, *J. Physiol.* (*Lond.*) **509,** 67 (1998).
[5] H. Cheng, W. J. Lederer, and M. B. Cannell, *Science* **262,** 740 (1993).

0076-6879/99 $30.00

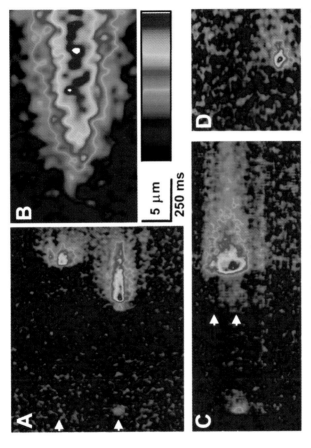

Fig. 8. Images illustrating the use of the CLSM in various modes. (A) Line-scan images of elementary events in the *Xenopus* oocyte. Events were evoked by the photorelease of InsP$_3$ by a photolysis flash delivered before the start of the image and recorded at 8 msec per scan line. Small events (see lower arrow) may represent Ca^{2+} flux through a single channel. Color bar indicates F/F_0 from 0.9 to 2.2. (B) High-resolution line-scan image of the rising phase of a calcium release event (puff). Image shows an average of five sharply focal puffs recorded at a resolution of 1.5 msec per line and 0.2 μm per pixel and formed by aligning, in time and space, the first detectable rise in fluorescence during individual events. Color bar indicates F/F_0 from 0.9 to 2.2. (C and D) Radial distribution of calcium puffs evoked by photorelease of InsP$_3$ before each record. Images were obtained by fast axial scanning and show distance into the oocyte running downward and time from left to right. Color bar indicates F/F_0 from 0.9 to 2.9.

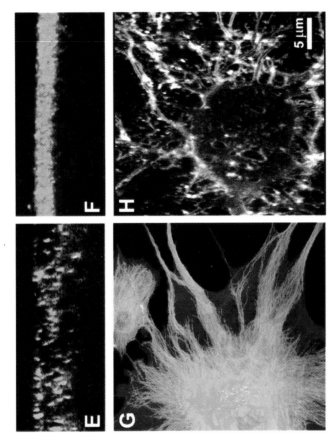

FIG. 8. (*continued*). (E and F) Radial distribution of InsP$_3$ receptors in the oocyte. The pseudo-colored image in (E) is an x–z scan into the vegetal hemisphere of an intact oocyte showing a punctate distribution of InsP$_3$Rs. Image in (F) shows the more dense distribution of InsP$_3$ receptors in the animal hemisphere obtained by x–y scanning of a physically sectioned oocyte. InsP$_3$Rs were visualized with a polyclonal antibody to the InsP$_3$R$_1$ and Cy3-conjugated secondary antibody. The plasma membrane of the cell is aligned near the upper edge of the image; the vertical axis represents depth into the cell (z axis) and the horizontal axis lateral (y) position. (G) Pseudo-colored image of endothelial cells stained for tubulin (green) and F-actin (red). (H) Gray-scale images of cultured cortical neurons stained with Di8-ANNEPS to visualize surface membranes, showing cell soma and processes. The scale bar below (B) applies to (A–D). The distance bar in (H) applies to (E–H). All images were obtained with an Olympus 40× fluor objective; NA. 1.35.

and the detectors and associated electronics are often relatively inefficient.[6] These factors lead to a low overall efficiency in the detection of emitted photons, requiring the use of a high laser intensity and/or a large confocal aperture to obtain images with an adequate signal-to-noise ratio. Second, the "black box" approach of proprietary commercial systems discourages, if not prohibits, the modification of hardware and software often necessary to optimize the device for a specific application. For example, swapping or adding lasers and filters to optimize the fluorescence of a specific dye or customizing software routines for data acquisition and analysis may be constrained severely. Moreover, the high cost of systems (typically >$125,000) makes them difficult to justify by individual investigators.

To overcome these limitations, we constructed a combined CLSM/UV photolysis system with the following advantages: (1) a simple optical path that, together with an avalanche photodiode detector with high quantum efficiency, provides a high efficiency of detection; (2) true photon-counting circuitry and associated software that allows complete flexibility for data acquisition and analysis; (3) access to and control of hardware and software components, allowing great flexibility in the choice of wavelengths, scan rate and length, and data acquisition and processing; and (4) affordability. The total cost to add confocal scanning capability to an existing inverted microscope is <$20,000.

In practice, this system has proven to be a powerful and flexible tool for the study of InsP$_3$-mediated Ca^{2+} release in *Xenopus* oocytes and cardiac myocytes.[7,8] Although this system was designed primarily for rapid (up to 1 kHz) line-scan imaging of Ca^{2+} puffs and sparks, it also works well for many other applications, including high-resolution morphological studies.

Design and Construction of CLSM

Overview

The optical system described in this article consists of two independent units: a laser scanner and confocal detector comprising the CLSM and a UV photolysis system. These are mounted on a 4 × 2.5-ft optical breadboard (Melles Griot, Irvine, CA) and are interfaced to an Olympus IX70 inverted microscope through the video and epifluorescence ports of the microscope,

[6] J. B. Pawley, *in* "Handbook of Biological and Confocal Microscopy" (J. B. Pawley, ed.). Plenum Press, New York, 1995.
[7] N. Callamaras, J. Marchant, X.-P. Sun, and I. Parker, *J. Physiol. (Lond.)* **509.1,** 81 (1998).
[8] I. Parker and W. G. Wier, *J. Physiol. (Lond.)* **505,** 337 (1997).

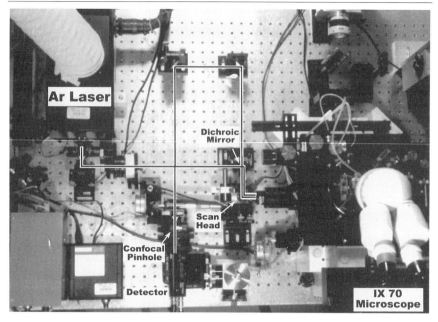

FIG. 1. Overhead photograph of the CLSM after removal of the cover.

respectively (Fig. 1). To maximize light throughput, all mirror and lens surfaces are broadband dielectric coated, and covers over the optical path (especially between the confocal aperture and the detector) are employed to reduce stray light.

At the heart of the CLSM is a photon-counting module (SPCM-AQ-121; EG&G Optoelectronics Inc., Vaudreuil, Canada) that comes complete with an avalanche photodiode, pulse discriminator, cooling circuitry, high-voltage power supply, and an external 5-V power supply. This unit provides a greater photon detection probability (>40% between 500 and 700 nm) than achieved by prismatic-face photomultiplier tubes,[9] while possessing similar dark count rates. Output from the module is TTL pulses (5 nsec duration), with the output rate proportional linearly (<10% deviation) to light intensity for count rates up to 2 MHz and saturating above 15 MHz. In addition, the small active area of the detector (200 μm) rejects stray light greatly, making it possible to operate the microscope under normal room lights.

The following sections describe the construction and use of this system. For simplicity, the instrument is described as set up for use with fluorophores

[9] A. MacGregor, in "Photonics Design and Applications Handbook" H 111, 1993.

with spectra similar to fluorescein. For details pertaining to the electronics and drive circuitry, as well as tips on aligning optical paths, see Parker *et al.*[10] Copies of circuit diagrams and software routines are available from the authors on request.

Fluorescence Excitation Path

Figure 2A shows the schematic layout of the CLSM optics. Fluorescence excitation is derived from a 100-mW multiline argon ion laser (Omnichrome, Chino, CA), bolted to the optical breadboard using insulating plastic brackets to raise the beam height to the axis of the video port on the microscope. Because the output from the laser far exceeds that required (typically <100 μW at the specimen), the laser is run in "standby" mode to prolong the life of the laser tube. After passing through an electronically controlled shutter (S1), allowing blocking of the beam between recordings, the beam is steered by mirror (M1) through a narrow-band interference filter (F1) selecting the 488-nm emission line. A rotating polarizer (P) allows a continually variable attenuation of the polarized laser light over a >100-fold range, and additional neutral density filters are placed in the light path as needed. The laser beam is then expanded by lens L1 ($f = -10$ cm) to overfill the back aperture of the microscope objective and is reflected by a dichroic mirror (DM; λ 500 nm, Omega Optical, Brattleboro, VT, mounted in a Nikon filter cube) onto a galvanometer-driven scan mirror in the confocal scan head. This mirror is mounted at a conjugate telecentric plane, formed by a scan lens (L2), consisting of a wide-field 10× eyepiece (Zeiss) mounted to the bayonet adapter of the video port. The beam diameter at the scan mirror is about 2 mm, allowing use of a small mirror (3 mm square), with low inertia, permitting fast scan rates. Rotations of the scan mirror result in deflections in position of the diffraction-limited laser spot focused in the specimen by the microscope objective. The maximum deflection in our system is limited by the aperture of L2 and corresponds to about two-thirds of the field viewed through the microscope oculars or about 200 μm with a 40× objective. Recordings were made with a 40× oil-immersion fluorescence objective (1.35 NA) with good UV transmission to maximize confocal sectioning ability and allow simultaneous photolysis of caged compounds.

Fluorescence Emission Path

Fluorescence emission is collected by the objective and passed back through the video port. Because the ability to directly view fluorescence

[10] I. Parker, N. Callamaras, and W. Wier, *Cell Calcium* **21**, 441 (1997).

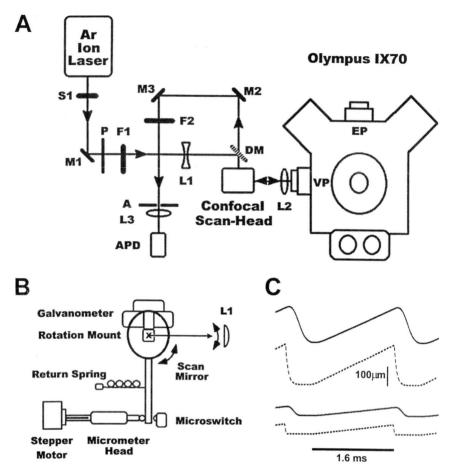

Fig. 2. Optical layout of the CLSM and drive waveforms. (A) The CLSM is interfaced through the video port (VP) on the left of the Olympus IX70 microscope. S1, shutter; M1–3, fully reflecting front surface mirrors; P, rotating polarizer (variable attenuator); F1, 488-nm narrow-band interference filter; L1, diverging lens, focal length = −10 cm; DM, dichroic mirror, λ = 500 nm; L2, scan lens (Zeiss 10× wide-field eyepiece); F2, barrier filter (515-nm long-pass color glass: Schott OG515); A, confocal aperture; L3, converging lens, f = 5 cm; APD, avalanche photodiode module. (B) Diagram of the scan head looking from the front of the microscope. The x-scan galvanometer is mounted through a metal block (serving also as a heat sink) to a stage allowing rotation around the axis marked by the cross. A lever attached to the stage is driven by a motorized micrometer to rotate the scan mirror in the y axis (vertical), while the galvanometer allows simultaneous fast scanning in the x axis (horizontal). Alignment of repeated scans is achieved by a microswitch, contacted at a preset position, that gates data acquisition. (C) Fidelity of mirror scan. Traces show ramp drive waveforms (dashed curves) and actual mirror position (solid curves) for small (lower) and large (upper) scan displacements, both at a scan rate of 2 msec per line. With a 40× objective, the galvanometer deflections correspond to 50- and 200-μm scan lengths, respectively. The horizontal bar indicates the linear ramp time during which image data are acquired. The flyback interval was set to 400 μsec.

excited by the laser scan line during experiments is of great assistance, the microscope is equipped with an 80/20 beam splitter directing light to the video port and oculars, respectively. The loss of 20% of the excitation light can easily be compensated for by increasing laser power. Some manufacturers (e.g., Nikon Inc., Melville, NY) obviate this minor difficulty by equipping their inverted microscopes with both 80/20 and 100/0 beam splitters. For safety while viewing, a 510-nm long-pass filter is installed before the microscope oculars to block 488-nm laser light. Fluorescence light passing through L2 is descanned by the galvanometer mirror so that a stationary beam passes through the dichroic mirror and is deflected by mirrors M2 and M3 onto the detector system (Fig. 2A). The purpose of M2 and M3 is to provide a longer optical path so that a highly magnified (about ×1000) image of the fluorescent spot is formed at the plane occupied by the confocal aperture A, and the z-axis resolution of the system can be optimized by varying the size of this pinhole. Because of the high magnification, the required apertures are relatively large (0.5–2 mm), and a range of sizes were made using a broken glass microelectrode to punch holes in pieces of aluminum foil.

F2 is a sharp cutoff 510-nm long-pass filter, which blocks the 488-nm laser light while transmitting fluorescence at $\lambda > 510$ nm. Use of a long-pass filter maximizes detection efficiency, but for some purposes, a broad band-pass filter provides better rejection of stray light and autofluorescence at wavelengths away from the emission maximum of the fluorophore. Finally, the light passing through the aperture is focused by lens L3 ($f = 5$ cm) to a small spot centered within the active area of the avalanche photodiode detector, which, to assist in alignment, is mounted on a low-profile micropositioner (New Focus Inc., Santa Clara, CA).

Scan Head

Fast x-scanning of the confocal spot is achieved by a galvanometer-based mirror positioning system (Model 6800/CB6588; Cambridge Technology, Inc., Watertown, MA), which is mounted in the scan head assembly to scan the laser beam in the horizontal (x) plane (Fig. 2B). For correct operation, it is critical that the axis of the mirror is positioned at the conjugate telecentric plane imaged by the scan lens L2 and that the galvanometer be oriented so light is reflected by the front surface of the mirror. Correct alignment can be determined by placing a piece of frosted glass over an open position on the microscope nosepiece. When adjusted correctly, the glass should be illuminated uniformly and the laser light stable as the mirror is scanned. A practical point is to mount the galvanometer in a holder that acts as a heat sink, as the drive coil generates appreciable heat at high scan rates.

To permit two-dimensional (x–y) scans, the galvanometer is mounted on a rotation stage (prism mount; New Focus Inc.) allowing simultaneous rotation around a horizontal axis orthogonal to and passing through the center of the scan mirror (Fig. 2B). An arm attached to the stage is driven by a micrometer and stepper motor, allowing vertical deflection of the laser beam in the y axis at a precise rate determined by the pulse rate of the stepper motor controller. To obtain square pixels at various "zoom" settings, the motor speed settings are calibrated for a range of commonly used line lengths and scan rates. Use of a single mirror scanned around two orthogonal axes, rather than a dual-mirror system, simplifies the optical path, minimizing light loss and problems of alignment, although it does limit the maximum acquisition rate of x–y images to about 1 frame every 2 sec.

The galvanometer scanner is driven by a servo amplifier, such that the ramp drive signal derived from the electronics unit described later produces a linear change in scan angle and hence linear displacement of the confocal spot. During experiments, it is convenient to monitor the analog signal from the detector using an oscilloscope to provide immediate visual feedback, e.g., to locate sites of Ca^{2+} release as the specimen is moved using the mechanical stage of the microscope. Figure 2C illustrates ramp waveforms (at a scan rate of 2 msec per line) and the corresponding mirror positions reported by the optical position sensor on the galvanometer.

Electronics

A custom electronics unit was designed to both process output from the detector and provide a ramp drive signal for the galvanometer driving the x-scan mirror (Fig. 3). This is driven by a pixel clock, generating pulses at switch-selected intervals between 1 μsec and 1 msec. During each pixel interval, photon counts from the detector are summed by an 8-bit counter. Following the next clock pulse, this sum, representing the pixel intensity, is made available on both digital or analog interfaces for sampling by a computer, and the counter is reset to zero to begin counting again. An LED provides visual warning if the pixel counts exceed the maximum count of 255. The pixel clock also drives the ramp generator used to drive the scan mirror. Pixel pulses are summed by a 10-bit counter, the output of which is fed continuously through an analog-to-digital converter to produce a linear stepwise ramp. A thumbwheel switch is used to set the desired number of pixels per scan line (1–999), and when this count is reached, the counter is reset to zero, a TTL pulse is sent to a "line sync" output

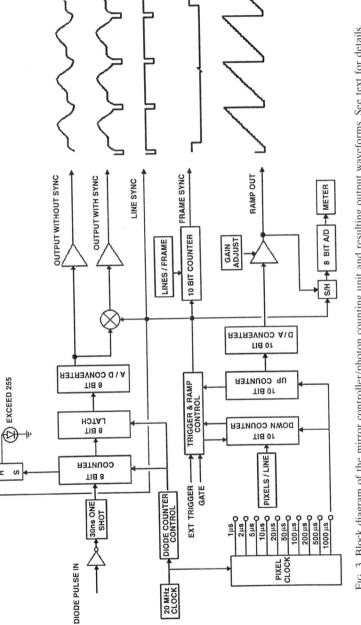

FIG. 3. Block diagram of the mirror controller/photon counting unit and resulting output waveforms. See text for details.

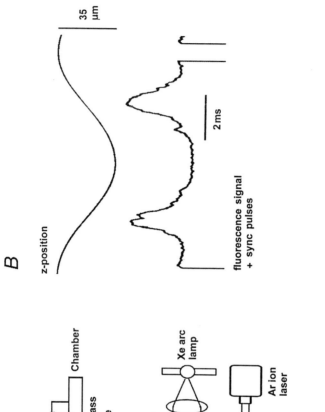

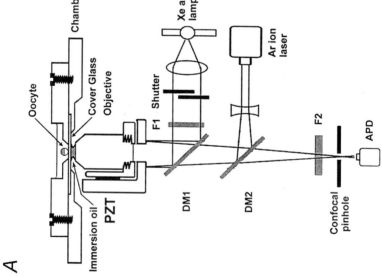

socket, and the mirror is returned to the starting position following a flyback waveform, which is low-pass filtered by an RC circuit to minimize overshoot and thus decrease the settling time. The length of the scan line in the specimen is adjusted by varying the gain of an amplifier placed after the analog-to-digital (AD) converter and, because the line length also depends on the selected number of pixels per line, a sample-and-hold circuit and digital meter are used to display the peak voltage at the maximum excursion of each line. These readings can then be calibrated readily in terms of length of scan line in micrometers for use with each objective lens. Output waveforms from the ramp generator, sync outputs, and photon counter are illustrated in Fig. 3.

Piezo-Objective System

To study the radial distribution of Ca^{2+} release sites within the large (1 mm diameter) and turbid oocyte, we developed a system allowing rapid (100 Hz) line-scan imaging along a line extending axially (z dimension) into the cell, rather than parallel to the cell membrane. For this application the scan mirror is stopped, providing a stationary confocal spot in the x and y directions, and the microscope objective is focused rapidly up and down to image along an axial into the oocyte. Figure 4A shows a schematic of the axial-scanning confocal system with the microscope objective mounted on a piezoelectric focusing drive (Model P-721.10; Polytec PI, Inc., Costa Mesa, CA) incorporating a position sensor and driven by a servo-feedback amplifier module (Model E-810.10; Polytec PI, Inc.).

The piezo drive is unable to follow linear ramp command signals at rates >10 Hz, so a sine wave command is used and images are collected during the relatively linear portion of the scan. The sinusoidal output from the position sensor is connected to a threshold comparator to produce

---

FIG. 4. System for rapid confocal scanning along the z axis of the microscope. (A) Schematic diagram of the piezoelectric translator (PZT) system to focus rapidly through a range of up to 35 μm at rates up to 200 Hz. The oocyte was viewed through a cover glass cemented in a small (5-mm diameter) aperture in a rigid Plexiglas chamber bolted firmly to the microscope stage. DM1, dichroic mirror reflecting $\lambda < 400$ nm; DM2, dichroic mirror reflecting $\lambda < 500$ nm; F1, bandpass filter λ 340–400 nm; F2, barrier filter, $\lambda > 510$ nm; Ar ion laser, attenuated 488-nm beam from a 100-mW argon ion laser; APD, avalanche photodiode photon counting module. (B) Traces showing position of the piezo drive during one cycle of the sine-wave scan (upper) and corresponding output from the fluorescence detector after the addition of sync pulses (lower). The fluorescence signal was obtained while scanning into an oocyte, and the extreme downward deflection of the scan (through of sine wave) corresponds to the position of the cover glass. The fluorescence profiles on the "down" and "up" scans are not perfect mirror images because of lateral hysteresis in the translator.

synchronization pulses, which are then added to the continuous stream of fluorescence data from the confocal detector (Fig. 4B). Images are formed at a rate of 100 Hz, with integration times of 10 or 20 μsec per pixel and a peak-to-peak scan amplitude of about 32 μm, providing nominal pixel sizes of about 0.05 or 0.1 μm throughout the linear range of the scan.

A practical difficulty is that the oil-immersion objective transmits vibration to the specimen through the oil. To minimize this, the oocyte is imaged through a cover glass cemented to a small (5-mm diameter) aperture in the base of a rigid Plexiglas recording chamber that is bolted firmly onto the microscope stage (Fig. 4A). This rigid recording chamber allows recordings at rates of 100 Hz to be made without difficulty. Another concern is that lateral hysteresis of the piezo drive unit results in the upward scan imaging of a slightly different (<1 μm) region of the cell; therefore only data collected during the downward scan of each cycle are used to construct images.

Computer Software

The photon count output and line sync signal are combined, digitized as a gap-free continuous record using a Digidata 1200 interface and Axoscope software package (Axon Instruments, Foster City, CA), and stored directly on hard disk for durations limited only by the available disk space. Custom routines written using the IDL programming language (Research Systems Inc., Boulder, CO) are then used to construct and analyze images formed from the raw data files. In brief, a routine identifies the positions of line sync pulses in the data file and forms an image array in which successive columns represent successive line scans aligned by the rising edges of preceding sync pulses.

Photolysis System

To allow simultaneous confocal fluorescence imaging and photolysis of caged intracellular compounds, UV (<400 nm) light sources are interfaced through the epifluorescence port of the microscope in conjunction with a UV filter cube. Figure 5 shows the layout of the optical components and the photolysis light paths in our system. UV light is provided by a continuous arc lamp source and a pulsed UV laser, with light from both being combined by a 50% beam-splitting mirror. The choice of light source depends on the experimental requirements. For photorelease of caged compounds over wide areas (10–100 μm), UV light is derived from a 75-W xenon arc lamp mounted in a Zeiss housing and operated from a stabilized constant-current power supply. An electronic shutter (Uniblitz; Vincent Associates, NY)

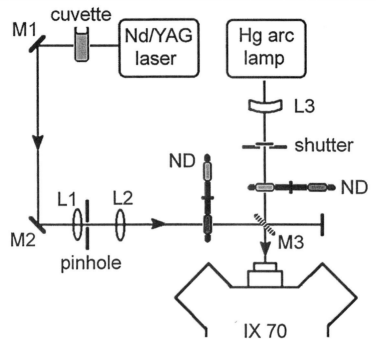

Fig. 5. Optical schematic of the systems for wide-field and point photolysis of caged InsP$_3$. See text for details.

controls exposure duration while a set of neutral density wheels (3.0 OD in steps of 0.1 OD, New Focus Inc.) allow control of light intensity.

For "point" photorelease, a laser system is employed, which consists of a Mini-Lite frequency-tripled (355 nm) Nd-YAG laser (Continuum, Santa Clara, CA) mounted on an optical table with insulating stand-offs to avoid electrical noise. The laser can be operated in single-shot mode (5-nsec pulse, triggered by a push switch or TTL input) or pulsed repeatedly at up to 10 Hz. A quartz cuvette containing a solution of FeSO$_4$ is used to initially attenuate the beam (which otherwise is sufficiently powerful to destroy mirror coatings) about 10-fold, and neutral density filter wheels provide further graded attenuation. Lenses L1 and L2 form a beam expander and spatial filter, so that the laser beam fills the back aperture of the objective lens, and can be focused to a near diffraction limited spot in the specimen. A beam dump is placed after M3 to prevent the beam passing into the room. Appropriate safety precautions, including laser safety goggles, a UV-blocking filter in the Olympus filter cube, and an additional long-pass filter ($\lambda > 510$ nm) inserted permanently in the microscope binocular head, are all mandatory to permit safe viewing while the laser is in use, particularly during alignment of the UV laser. For further details on the implementation and alignment of the system see Callamaras and Parker.[11]

Choice of Indicators

We currently employ the Oregon Green family of Ca^{2+} indicator dyes (Molecular Probes, Eugene, OR), which are optimized for excitation by the 488-nm line of the argon ion laser and fluoresce in the green (emission 520 nm). The excitation wavelengths of these dyes are sufficiently separated from the photolysis wavelengths of caged compounds so that no appreciable photolysis results from the 488-nm fluorescence excitation light, and photolysis flashes ($\lambda < 400$ nm) cause little or no artifacts in fluorescence recordings. In addition, dyes with different Ca^{2+} affinities are available (e.g., Oregon Green 488 BAPTA-1, K_D 150 nM; and Oregon Green BAPTA-5N, K_D 10 μM). A limitation of all currently available long-wavelength indicators is that, unlike the UV-excited dyes fura-2 and indo-1, none are ratiometric. A "pseudo ratio" signal, however, may be obtained by expressing Ca^{2+}-dependent fluorescence signals relative to the resting fluorescence before stimulation.[4]

Performance of CLSM System

Spatial Resolution

Various approaches may be used to characterize the lateral and axial (z axis) resolution of a confocal microscope. A simple test of z-axis resolution is to measure reflected light from a stationary laser spot while the microscope is focused through a front-surface mirror.[6] This technique yields the full-width at half-maximum intensity (FWHM), a number that can be compared with other systems and aids in aligning the system. Measurement of the normalized reflected intensity as the microscope was focused through a mirrored slide, using various confocal apertures, showed that the FWHM decreases with decreasing size of the aperture and reduces to about 500 nm with an aperture of 0.75 mm diameter.[10] A more physiologically relevant test was made by imaging fluorescent latex beads (green fluorescent latex microspheres: Molecular Probes Inc.) in aqueous medium. Figure 6A shows intensity profiles from axial (x–z) and lateral (x–y) scans through the center of subresolution (0.1 μm) beads, providing measures of the point-spread function of the CLSM. The FWHM of the axial and lateral point-spread

[11] N. Callamaras and I. Parker, *Methods Enzymol.* **291**, 497 (1998).

A   **B**

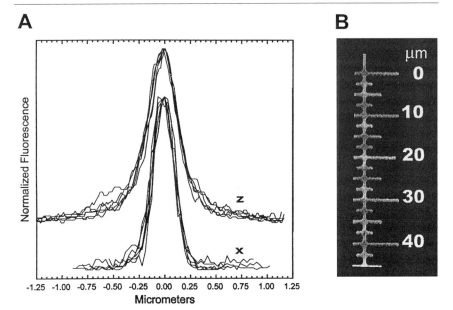

FIG. 6. Measurements of axial and lateral resolution. (A) Intensity profiles from images of six beads scanning laterally (x) and axially (z) through 0.1-μm beads. (B) x–y scan of a mirrored micrometer calibration slide. All images were obtained with an Olympus 40× fluor objective; NA, 1.35.

functions are about 300 and 400 nm, respectively, with no correction applied for the finite size of the beads. Finally, the linearity of the overall scanning system (including both the galvanometer and optics) was tested by imaging (in reflectance mode) a mirrored micrometer calibration slide (Fig. 6B). Any deviations from linearity in the image of this graticule were <500 nm over a 50-μm scan line (i.e., <1%). A similar performance was also achieved for scans in the y axis. Therefore, when aligned properly, the spatial resolution of the CLSM is close to the diffraction-limited performance expected for a confocal microscope,[12] and the user has complete control over the desired pixel resolution (e.g., up to 1000 × 2000 pixels in x–y scan mode).

To calibrate the distance scale and linearity of axial-scan images, reflected laser light was monitored from a mirrored slide after removing the barrier filter (F2; Fig. 2A), resulting in a sharp peak of reflected laser light as the microscope focus knob was advanced manually in 2-μm increments

[12] S. Inoue, in "Handbook of Biological and Confocal Microscopy" (J. B. Pawley, ed.). Plenum Press, New York, 1995.

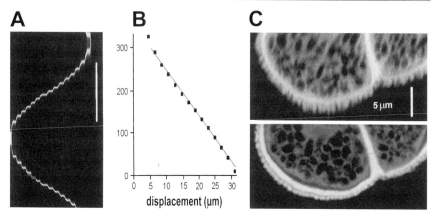

Fig. 7. Axial resolution of the fast piezo z-scanning system. (A) Axial line-scan image showing reflectance from a mirrored slide as the microscope focus was advanced manually in 2-μm increments at roughly 1-sec intervals. Each vertical line is one scan cycle, and time runs from left to right. (B) Plot of pixel position versus focus displacement for the selected region (indicated by the bar in A) of the image in (A). The central portion of the scan was approximately linear over a range of 22 μm. (C) Images of fluorescent pollen grains (mixed pollen grain slide; Carolina Biological Supply). The upper frame shows a z–y image obtained using the piezo drive to scan rapidly (100 Hz) in the z axis (depicted vertically), while the laser spot was scanned slowly in the y axis by a rotating mirror. For comparison, the lower frame shows a conventional lateral x–y scan of a matching region of a different pollen grain. All images were obtained with an Olympus 40× fluor objective; NA, 1.35.

to move the mirror through the entire scan range of the objective (Fig. 7A). Steps in the image thus correspond to 2-μm displacements of the objective lens, and the distance scale of the image could be calibrated by plotting the pixel position of peak reflectance against displacement of the microscope nosepiece. Figure 7B shows a plot of pixel position of peak reflectance against displacement of the objective over the region marked by the bar in Fig. 7C. With the system set to give a peak-to-peak displacement of 32 μm, a linear range of >20 μm was obtained around the center of the scan. The image in the upper trace shows a z–y image obtained with the piezo scanned rapidly (100 Hz) in the z axis while the laser spot was scanned in the y axis by rotating the scan head, whereas the lower trace shows an x–y image of a similar pollen grain.

A consideration in axial imaging is the refractive index mismatch among water, the cover glass, and immersion oil resulting in a change in focal distance in the specimen that is slightly less than the corresponding movement of the objective. To correct for this effect, measurements based on the physical position of the objective were scaled by dividing by an experimentally derived factor of 1.17, arrived at by focusing on the advancing tip of a microelectrode in water using a calibrated manipulator, and confirmed by comparing the axial and lateral diameters of spherically symmetrical 15-μm fluorescent beads.[13] A comparison of z–y and x–y scans of fluorescent pollen grains is shown in Fig. 7C.

Temporal Resolution

Other aspects of the fidelity of dynamic imaging concern the maximum possible temporal resolution; factors that are determined primarily by the mechanical properties of the galvanometer unit used to rotate the scan mirror (in x-scan mode) or by the piezo-objective scanner (in z-scan mode). The speed of the galvanometer was tested by monitoring the output from the position encoder while driving the servo board with rapid (2 msec per line) ramp signals (Fig. 2C). For small deflections, corresponding to a 50-μm scan with a 40× objective, the flyback and settling time after each line is only about 200 μsec (lower traces), but increases to about 400 μsec with large (200 μm) deflections. As a compromise, the scanner is operated with a dead time of 400 μsec following flyback so that image data are collected for 1.6 msec of each 2-msec waveform during the linear part of the ramp (bar in Fig. 2C). If the length of the scan line is restricted, however, it is possible to increase the scan speed to 1 msec per line and decrease the dead time to 200 μsec to collect data for 800 μsec each scan. In addition, line-scan data can be recorded continuously, unlike some commercial instruments that impose gaps after 512 or 1024 lines. This allows the activity of individual Ca^{2+} release sites to be monitored for prolonged periods and has revealed that individual sites may undergo long-term changes in "mode" or frequency of response.[8] When acquiring x–y images, a 1000 × 1000 pixel image (250 μm with 40× objective) takes about 21 sec to acquire using a pixel dwell time of 20 μsec. Images with lower resolution can be acquired with correspondingly shorter scan times. When acquiring axial line-scan (z–t) images the piezo drive could follow a sine-wave command signal at rates up to 200 Hz with a maximum excursion of about 35 μm. This response time is sufficiently rapid that it should be possible to use the piezo drive in conjunction with commercially available video-rate confocal microscopes so as to obtain x–z images at frame rates as fast as 60 Hz.

Applications

Imaging of Local Ca²⁺ Transients

The confocal scanner was designed primarily for the line-scan confocal recording of localized transient intracellular Ca^{2+} signals: InsP$_3$-mediated

---

[13] N. Callamaras and I. Parker, *J. Gen. Physiol.* **113**, 199 (1999).

puffs in *Xenopus* oocytes[14] and sparks in cardiac cells.[5,15] These events have rapid (tens or a few hundred millisecond) time courses and involve Ca^{2+} signals localized to within a few micrometers. Thus, imaging systems with high spatial and temporal resolution are required, and the use of confocal microscopy provides a particular advantage by avoiding blurring and out-of-focus fluorescence. Figures 8A and 8B (see color insert) illustrate Ca^{2+} puffs evoked in an oocyte by photoreleasing $InsP_3$ throughout a roughly 100-μm spot around the scan line using a flash of UV light provided by the arc lamp system shown in Fig. 5. Two puffs arising independently at sites about 10 μm apart are shown in Fig. 8A (see arrows). In addition, yet smaller "blips" can be observed occurring about 500 ms before and just preceding the lower puff in Fig. 8A. Such events were too small to resolve previously in video-rate confocal images obtained using a Noran Odyssey system[14] and probably represent Ca^{2+} flux through individual ion channels. For this application, a scan rate of 8 msec per line provided sufficient resolution of the time course of puffs; four times faster than the frame rate of a conventional video system. The image in Fig. 8B shows the average of five sharply focal puffs recorded at a faster temporal resolution (1.5 msec per line). The beginning of this image shows that the release of Ca^{2+} arises from a virtual point source of approximately 0.25 μm (at or below the limit of resolution of the system).

$InsP_3$-sensitive release sites in the oocyte are distributed irregularly in the x–y axes (i.e., parallel to the surface membrane) at a mean spacing of a few micrometers,[7,14] but little is known about their radial distribution. The piezoelectric objective scanning system described here provided a means to obtain time-resolved confocal images of transient Ca^{2+} release events along an axis running into the oocyte. Figures 8C and 8D illustrate Ca^{2+} puffs at varying depths in the oocyte evoked by brief photorelease of $InsP_3$. Figure 8C (see color insert) shows differentially activated, yet closely adjacent, sites whereas Figure 8D (see color insert) shows activation of a relatively deep site. Studies using this technique have revealed that puffs arise largely from release sites distributed in a thin (6 μm), superficial band a few micrometers below the cell surface.[13]

Using CLSM for Morphological Studies

Although the confocal scanner was built primarily for fast line-scan imaging of Ca^{2+} transients, it also works well for morphological studies using x–y and x–z scanning at slow frame rates, as shown in Figs. 8E–8H (see color inserts). The optical sectioning ability of the CLSM was used to

[14] I. Parker, J. Choi, and Y. Yao, *Cell Calcium* **20**, 105 (1996).
[15] J. R. Lopez-Lopez, P. S. Shacklock, C. W. Balke, and W. G. Wier, *Science* **268**, 1042 (1995).

localize InsP$_3$R antibody staining in unsectioned intact oocytes (Fig. 8E, see color insert) to correlate with the distribution of functional release sites in a thin subplasmalemmal band. Receptor staining is visualized in an x–z image of the vegetal hemisphere, made by slowly advancing the focus knob of the microscope with a synchronous motor. In the vegetal hemisphere, a sparse punctate distribution of InsP$_3$R was observed in a thin superficial layer similar to the distribution of functional release sites. For comparison, Fig. 8F (see color insert) shows InsP$_3$Rs localized using conventional x–y scanning in fixed sections of the animal hemisphere of the oocyte. The more intense staining observed is likely due to a higher density of InsP$_3$Rs in the animal hemisphere.[16]

A commercially available slide of multiply labeled endothelial cells (Fluo Cells, Molecular Probes) is shown in Fig. 8G (see color insert) as a readily available standard specimen allowing comparison with other CLSM systems. The image shows a pseudo-colored overlaid image of bovine artery endothelial cells imaged at 488ex./515em. to visualize fluorescein-labeled tubulin and 488ex./560em. to visualize Texas Red-labeled F-actin. A further example of resolution is shown in Fig. 8H: an x–y scan (488ex./515em) of a cultured primary cortical neuron stained with the lipid-soluble membrane dye Di8-ANNEPS (Molecular Probes) to visualize the surface membrane. Thin processes (<1 μm diameter) are resolved readily, and the apparent thickness of sections through the plasma membrane is 200–300 nm.

Conclusions

The objective of this article is to demonstrate the feasibility of constructing a custom confocal microscope that offers superior performance and versatility for dynamic Ca^{2+} imaging than commercial instruments at a fraction of the cost. No new principles are involved in our instrument and it is subject to the same theoretical limits of any confocal microscope; however, in practice, we have found its sensitivity and spatial resolution to exceed that of some commercial designs. The improvement probably lies in the simple optical design, the highly efficient detector, and that once having built the microscope the experimenter is well qualified to optimize and align the system. In our hands, this system has proved to be a versatile tool for the optophysiological study of many cellular processes, and construction of a similar instrument is well within the grasp of many investigators wishing to construct their own microscope "customized" to particular questions.

[16] N. Callamaras and I. Parker, *Cell Calcium* **15,** 66 (1994).

Section III

Measurement of Subcellular Relations and Volume Determinations

[11] Three-Dimensional Visualization of Cytoskeleton by Confocal Laser Scanning Microscopy

By WERNER BASCHONG, MARKUS DUERRENBERGER,
ANNA MANDINOVA, and ROSEMARIE SUETTERLIN

Overview

Functional studies of the cytoskeleton require confocal laser scanning microscopy for the three-dimensional (3D) documentation of the cytoskeletal constituents at optimal resolution. To facilitate this, specific techniques for cell cultivation, such as three-dimensional collagen cultures, can be used to provide an environment reminiscent to that in tissue. Practical suggestions for specific cell cultivation techniques and tissue excision will be provided. Key problems with permeabilization and fixation of cells and tissue will be addressed. Fixation, especially cross-linking by aldehydes, may lead to substantial background signals, either by the binding of fluorescent labels to free aldehyde groups or by the formation of fluorescent aldehyde polymers. Moreover, tissue constituents, especially the extracellular matrix, may display substantial autofluorescence. The removal of autofluorescence, therefore, is critical for achieving adequate resolution of the cytoskeleton in multilayer cultures or in tissue biopsies. Proper treatment with $NaBH_4$ can minimize aldehyde-related background labeling and autofluorescence. Immunolabeling techniques and sample mounting are mentioned briefly, as these techniques have been discussed extensively elsewhere. Accordingly, hardware requirements, data acquisition, and image processing will relate to analyzing the 3D nature of the cytoskeleton.

Cytoskeletal Functions and Consequences for Sample Preparation

Imaging of the three-dimensional architecture in the cytoskeleton of cells in culture and in tissue by confocal laser scanning microscopy depends on two features: proper sample preparation and adequate resolution.

The cytoskeleton consists of three sets of filaments: microfilaments containing actin, microtubules containing tubulin, and intermediate filaments containing keratin-like proteins. Microfilaments measure 6 nm in diameter,[1,2] microtubules 25 nm,[3] and intermediate filaments range between 8 and 10 nm.[4] Actins and tubulins are ubiquitous, whereas intermediate filament proteins are cell specific. Keratins constitute the intermediate filaments in keratinocytes, vimentin in fibroblasts, desmin and vimentin in muscle, neurofilament proteins in neurons, and so on. Free microfilaments, microfilaments bundled as stress fibers, intermediate filaments, and microtubules span the cytoplasm as cytoskeletal networks extending in a cell body over 10 to 100 μm. Equilibria between cytoskeletal structures and their monomers are regulated by associated proteins, by intracellular calcium, and by nucleotides such as ATP or GTP.

The cytoskeleton participates in many processes and functions in the cell. Intracellular transport, changes of the cellular shape, and cell motility are connected to changes in the 3D architecture, including the association and dissociation of cytoskeletal protein monomers.[5] An important consequence is that alterations induced by specimen preparation, such as changes of the substrate texture or the mechanical load, of temperature, pH, osmotic pressure, or ischemia and energy depletion, may influence the dynamics and architecture of the cytoskeleton artificially.

It has been known for a long time that actin and tubulin are present in nearly all eukaryotic cells. However, the visualization of such structures was originally restricted to regular arrays of actin or tubulin in muscle or cilia. The combination of aldehyde cross-linking techniques with indirect immune fluorescence revealed that actin, tubulin, and the intermediate filament proteins constituted three-dimensional intracellular cytoskeletons in many kinds of cells.

Electron microscopy of individual thin sections of embedded cells permits resolution of single cytoskeletal filaments or microtubules. However, visualization of the spatial arrangements of the cytoskeletal networks from electron micrographs requires extensive serial sectioning, digitization, and computer reconstruction, an expensive and time-consuming task.

In a more economical way, cytoskeletal filaments may be visualized by conventional fluorescence microscopy if they are arranged in the same focal plane of a very thin specimen, where light scattered from out-of-focus planes is minimal and background fluorescence is negligible. Indeed, the existence of cytoskeletal networks in whole cells had first been documented

---

[1] A. Bremer, R. C. Millonig, R. Sütterlin, A. Engel, T. D. Pollard, and U. Aebi, *J. Cell Biol.* **115,** 689 (1991).
[2] A. Bremer, C. Henn, K. N. Goldie, A. Engel, P. R. Smith, and U. Aebi, *J. Mol. Biol.* **242,** 683 (1994).
[3] A. V. Grimstone and A. Klug, *J. Cell Sci.* **1,** 351 (1966).
[4] U. Aebi, M. Häner, J. Troncoso, R. Eichner, and A. Engel, *Protoplasma* **145,** 73 (1988).
[5] D. E. Ingber, *Cell* **75,** 1249 (1993).

in monolayer cell cultures, where cells are flat, i.e., extend only a few nano-meters in the z axis, and comparable to thin sections of embedded tissue.

In the cytoplasm of such cells, the group of Weber revealed, using immunofluorescence microscopy, stress fibers or microfilament bundles that spanned cells parallel to the substrate,[6] microtubules,[7] and intermediate filaments.[8] Improvement of fixation and labeling techniques eventually per-mitted the concomitant visualization of the three cytoskeletal networks by three different sets of fluorescence labels.[9]

As an example of conventional fluorescence microscopy, Fig. 1 (see color insert) shows triple-labeled human fibroblasts grown as a monolayer on a glass coverslip. The microfilaments have been visualized by fluorescent phalloidin (green), the vimentin network (blue), and the microtubuli (red) by the corresponding primary and secondary antibodies (for details, see legend for Fig. 1). The photographical overlay of the fluorescent images presents the three cytoskeletal networks in a two-dimensional projection.

Confocal laser scanning microscopy offers the possibility to image cells in much thicker specimens. To demonstrate that this type of microscopy is comparable or better than conventional techniques, Fig. 2 illustrates the distribution of the three cytoskeletal elements—actin, vimentin, and tubulin—as visualized by confocal laser scanning microscopy. MG-63 osteo-blasts have been triple labeled as described for Fig. 1 (for details, see legend to Fig. 1 and later). Cells have been scanned along the z axis at steps of 0.23 μm and to a depth of 2.75 μm. The latter represents the thickness of the cell monolayer. The overlay of stepwise recorded optical sections by computer-aided image processing has been used to gain a three-dimensional image of the spatial organization of the cytoskeleton. The spatial distribu-tion of the microfilaments (Fig. 2, top left, see color insert), of the vimentin intermediate filaments (Fig. 2, bottom left), and of the microtubuli (Fig. 2, top right) are documented as composites of the respective fluorescence images. The architecture of all three networks is presented as the projection of the three sets of images recorded (Fig. 2, bottom right).

The quality of the projection image acquired by conventional fluorescent microscopy of the triple-labeled cytoskeleton (Fig. 1) apparently compares well with that obtained by confocal laser scanning microscopy (Fig. 2, bottom right). Both techniques reveal the cytoskeleton as three complete interwoven networks. F-actin containing microfilaments and actin stress fibers are located specifically at the periphery of the cells and parallel to

---

[6] E. Lazarides and K. Weber, *Proc. Natl. Acad. Sci. U.S.A.* **71,** 2268 (1974).
[7] K. Weber, R. Pollack, and T. Bibring, *Proc. Natl. Acad. Sci. U.S.A.* **72,** 459 (1975).
[8] M. Osborn, W. W. Franke, and K. Weber, *Proc. Natl. Acad. Sci. U.S.A.* **74,** 2490 (1977).
[9] J. M. Small, S. Zobeley, G. Rinnerthaler, and H. Faulstich, *J. Cell Sci.* **89,** 21 (1988).

the underlying glass support, and microtubules and intermediate filaments arrange preferentially around the nonstained nucleus. Both figures represent cells cultured on a plane support, where interference from other focal planes is minimal. However, as discussed earlier, the cytoskeletal architecture of cells growing on a flat substrate may be biased by the fact that the cytoskeleton itself can adapt its shape to the texture of the underlying substrate.[10] Cytoskeletal imaging has therefore been extended to three-dimensional specimens such as tissue and cells grown in 3D cultures reminiscent of the natural three-dimensional environment. In this situation, confocal laser scanning microscopy has proven to be a unique tool for visualizing functional changes of the cytoskeleton in space and time.

Sample Preparation for Fluorescence Labeling

Cultivation of Cells: Preparation of Cells and Tissue

Monolayer Cultures. Monolayer cultures have been extremely useful for studying the influence of physiological and physical stimuli on cytoskeleton architecture. We routinely grow cells on glass coverslips placed in 24-well cell culture plates (Figs. 1 and 2). It is important to use glass coverslips that are plane and 0.17 ± 0.005 mm thick. Aberrations will influence the precision of imaging, especially along the z axis (see Fluorescence Labeling and Mounting). If plastic coverslips have to be used they should show minimal autofluorescence (check supplier's specification). When using fibroblast cultures for optimizing cell permeabilization and fixation or immunolabeling, we seed 10,000 cells onto each 12-mm-diameter glass coverslip (10,000 cells/well in 0.5 ml growth medium). This concentration corresponded to about a half-confluent cell layer. We fix and label the cells for visualization after overnight attachment. Otherwise, we plate correspondingly fewer cells to reach half-confluency at the time preestablished for fixing.

When we analyze the cytoskeleton in cells growing on metal surfaces, reminiscent of the tissue/metal interface of orthopedic implants, we seed 10,000–20,000 cells onto a titanium disk (1–2 mm thick and 1 cm in diameter) that fits into a well of a 24-well culture plate. The cells are visualized on the metal surface by confocal laser scanning microscopy in the fluorescence mode. Alternatively, the combination of the fluorescence image with the reflection mode image displays also the underlying structure of the metal surface clearly as by scanning electron microscopy. The overlay of the composite image of TRITC–phalloidin-labeled MG-63 osteoblasts and

---

[10] C. Oakely and D. M. Brunette, *J. Cell Sci.* **106,** 343 (1993).

the reflection image of the machine-polished titanium surface reveals a well-spread cell with strong stress fibers growing on the flat, occasionally scratched, metal surface (Fig. 3, see color insert).

Multilayered Cultures. Cells forming multiple layers can be grown on glass slips in 24-well culture plates. Ultra-long-term cultures, especially bone marrow cultures that are grown without antibiotics, require absolute sterility, which is difficult to maintain in culture dishes. We use 25-cm^2 culture bottles (Falcon, Becton Dickinson Labware, Lincoln Park, NJ), where the risk of contamination is much lower. Because glass coverslips cannot be placed in such bottles and the isolation of such cells from the plastic surface by scratching or trypsination would induce cytoskeletal changes, we permeabilize, fix, and NaBH$_4$ reduce the cells within the bottle (see later). After removal of the storage buffer, we use a soldering device equipped with a thin hot wire and cut out one or several slices (about 15 × 15 mm) from the bottom of the flask for immunolabeling and visualization. Figure 4 (see color insert) shows a hematopoietically active long-term culture (about 50 μm thick) of bone marrow cells double-labeled for F-actin (red) and tubulin (green).

Three-Dimensional Cell Cultures. Ideally, the cytoskeleton architecture of cells growing in a natural three-dimensional environment should be identical to that encountered *in vivo.*

Three-dimensional cell culture systems had been known for some decades.[11] They became especially popular when Bell *et al.*[12] reported a "dermal equivalent," a fibroblast populated collagen lattice able to contract or reduce its volume over time. They related this reduction to wound contraction and used it as a corresponding *in vitro* model. We grow fibroblasts in collagen matrices, visualizing them between days 2 and 4 after seeding. If they are populated with fibroblasts with a long lag phase and a long doubling time such as KD cells,[13] their diameter remains unaltered for more than a week after seeding.[14] Populating such matrices with more rapidly dividing cells leads to faster matrix contraction.

Figure 5 shows KD cells and KD cell-derived HuT fibroblasts[15] growing within attached, low contracting collagen matrices. The matrix (Fig. 5, left) has been scanned at steps of 0.5 μm to a depth of 30 μm. A continuous tubulin network extending within the whole cell body is revealed in the top most HuT cells. KD cells (Fig. 5, right) labeled for vimentin have been scanned at steps of 0.5 μm to a depth of 75 μm. Vimentin covers the whole

[11] R. L. Ehrmann and G. O. Gey, *J. Natl. Cancer Inst.* **16,** 1375 (1956).
[12] E. Bell, B. Ivarsson, and Ch. Merrill, *Proc. Natl. Acad. Sci. U.S.A.* **76,** 1274 (1979).
[13] T. Kakunaga, *Proc. Natl. Acad. Sci. U.S.A.* **75,** 1334 (1978).
[14] W. Baschong, R. Sütterlin, and U. Aebi, *Eur. J. Cell Biol.* **72,** 189 (1997).
[15] J. Leavitt and T. Kakunaga, *J. Biol. Chem.* **255,** 1650 (1980).

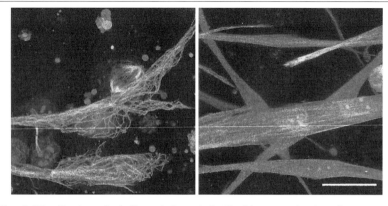

FIG. 5. Visualization of tubulin and vimentin in fibroblasts growing in collagen matrices. Human KD fibroblasts (right) and Hut-12 fibroblasts [J. Leavitt and T. Kakunaga, *J. Biol. Chem.* **255**, 1650 (1980)] (left) cultured in attached, low-contracting collagen matrices. [W. Baschong, R. Sütterlin, and U. Aebi, *Eur. J. Cell Biol.* **72**, 189 (1997).] Fixed as described in Fig. 1, but prefixed for 10 min and postfixed for 30 min. Removal of autofluorescence: $2\times$ 15 min by $NaBH_4$. (*Labeling*) Tubulin (left): anti-β-tubulin (monoclonal mouse) and FITC-antimouse (sheep, Amersham). Vimentin (left): antivimentin (monoclonal mouse, Amersham) and Cy3-antimouse (donkey, Jackson Imm.Res.). (*Visualization*) (Left) z scan at 0.5-μm steps, 61 focal planes. (Right) z scan at 0.5-μm steps, 151 focal planes. Composite of optical sections of tubulin fluorescence (left) and vimentin image (right). Clearly visible tubulin networks extend all over the cells. Vimentin filaments are discernible in the top cells, but due to reduced resolution by beam attenuation, not in the lower cells. Bar: 25 μm.

cell as a fine network. Fibers in the upper cells are clearly discernible from each other, whereas those in the lower cells are not fully resolved (see Limits for Optical Sectioning).

We prepare attached dermal equivalents in 24-well cell culture plates as described by Baschong *et al.*[14] We mix (at 20°) 1.5 ml of $10\times$ Dulbecco's modified Eagle's medium (DMEM, GIBCO, Life Technologies, Gaithersburg, MD), 2.8 ml of 0.1 N NaOH, 150 μl of 100 units/ml penicillin/100 μg/ml streptomycin (Pen/Strep, $100\times$ stock solution; GIBCO), 150 μl of 0.2 M glutamine (GIBCO), 150 μl of 0.1 M sodium pyruvate (GIBCO), 120 μl of 7.5% $NaHCO_3$, 375 μl of 1 M HEPES, and 1.5 ml of fetal calf serum (FCS), sterilize by filtration, and combine it with 8.25 ml of sterile Vitrogen 100 solution (type I collagen; Collagen Corporation, Palo Alto, CA) and with 2×10^5 cells/ml suspended in 1/50 of the final volume of DMEM (supplemented with 10% (v/v) FCS and 1/100 Pen/Strep stock solution). Portions (0.5 ml) of the final collagen matrix solution are cast into each well of a 24-well culture plate. Best results are obtained when leaving the culture plates for gelation in a warm room (37°) for 1 hr before transferring them to the 37° incubator (95% air/5% CO_2, v/v). The gelation is strongly pH dependent, and to prevent immediate precipitation of the collagen, the solutions have to be kept acidic. The final mixture is at pH 6.9, as measured by pH indicator paper strips and usually gels within 15 to 30 min. The amount of NaOH required to compensate for the acidic Vitrogen is determined empirically in mixtures not containing the cells.

The collagen matrices attach to the wall of the culture well, yet they usually do not stick to the bottom of the well. This is probably due to the presence of fetal calf serum already in the collagen mixture used for casting the gels. Lee et al.[16] reported that collagen matrices would stick properly to the bottom only in the absence of fetal calf serum.

Sample Preparations from Biopsies. Tissue extracted by biopsy undergoes significant mechanical stress and damage imparted by surgical instruments, by muscle contraction, and by an unphysiological environment. As mentioned previously, the cytoskeletal architecture is susceptible to changes of the mechanical load, temperature, osmotic pressure, pH, and tissue oxygenation or nucleotides, respectively. Microfilaments, intermediate filaments, and microtubules react differently to the various physiological changes.

Mechanical shock seems to alter the microfilament architecture especially and cell shape concomitantly.[14,17] Figure 6 illustrates the influence of mechanical damage to KD cells grown in a collagen matrix. At the edge of a 1-mm stencil wound (Fig. 6, left), cells have rounded up immediately after wounding. They show no actin fibers. The strong fluorescence in the cytoplasm indicates the presence of small F-actin elements. If a similar hole is excised by a CO_2 laser (Fig. 6, right), the heat-denatured cells at the wound edge retain their stellate form and display distinct stress fibers.

For cells grown in contracting, attached collagen matrices, it has been reported that detachment induces remarkable changes in microfilament architecture,[18,19] perhaps similar to that encountered when tissue retracts after surgery. However, mere detachment from the sidewalls of the culture well of low-contracting matrices (which do not stick to the bottom) and handling during fixation and labeling did not seem to influence the cytoskeletal organization.[14] Accordingly, attached low-contracting collagen matrices fixed directly in the culture well and those detached and transferred to further wells for washing and fixation display the same cytoskeletal organization.[14,18]

[16] T.-L. Lee, Y.-Ch. Lin, K. Mochitate, and F. Grinnell, *J. Cell Sci.* **105,** 167 (1993).

[17] P. Weiss and B. Garber, *Proc. Natl. Acad. Sci. U.S.A.* **38,** 264 (1952).

[18] K. Mochitate, P. Pawelek, and F. Grinnell, *Exp. Cell. Res.* **193,** 198 (1991).

[19] J. J. Tomasek, C. J. Haaksma, R. J. Eddy, and M. B. Vaughan, *Anat. Rec.* **232,** 359 (1992).

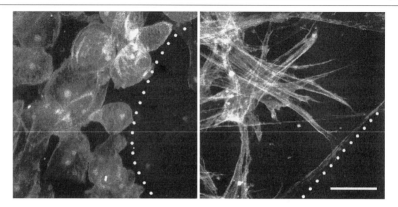

FIG. 6. Influence of mechanical shock on actin cytoskeleton. A 1-mm punch wound was excised from collagen matrices populated with KD fibroblasts as described by Baschong *et al.* [W. Baschong, R. Sütterlin, and U. Aebi, *Eur. J. Cell Biol.* **72**, 189 (1997)] (left). For comparison, a 1-mm hole was evaporated by a CO_2 laser. (*Fixation*) less than 5 min after wounding as described in Fig. 5. (*Labeling*) Actin: TRITC–phalloidin (Sigma). (*Visualization*) (Left) *z* scan at 0.5-μm steps, 110 focal planes. (Right) *z* scan at 0.3-μm steps, 35 focal planes. Composite of optical sections displayed. Cells at the wound edge (dotted line) in mechanical wounds (left) have rounded up and show no actin fibers, but label strongly for F-actin, indicating the presence of small actin fragments. Cells at the edge of the laser wound have retained their stellate form and show distinct, slightly distorted actin stress fibers. Bar: 25 μm.

Temperature shifts seem to be critical for microtubule stability.[20] Moreover, other parameters, such as osmotic pressure, pH, energy depletion, and nucleotide concentration, which all influence association and dissociation of cytoskeletal proteins, have to be considered.

For reliable analysis of the cytoskeletal architecture, we therefore either freeze tissues immediately in liquid nitrogen or keep them at physiological conditions until fixation. Ideally, tissue biopsies should be fixed at excision.

Permeabilization and Fixation

The goal of fixation and permeabilization is to preserve the three-dimensional structure of the cell and cytoskeleton, while at the same time ensuring access to labeling agents. Cytoskeleton structures have been reported to be stabilized either by chilled organic solvents such as alcohol or acetone in the cold, i.e., at $-20°$ below freezing, or by aldehyde fixation either at room temperature (20–25°) or at 37°. The principal steps during both types of cytoskeleton fixation are permeabilization of the membrane

[20] C. Archangeletti, R. Sütterlin, U. Aebi, F. De Conto, S. Missorini, C. Chezzi, and K. Scherrer, *J. Struct. Biol.* **119**, 35 (1997).

concomitant with lipid extraction to provide adequate penetration to the cell interior, stabilization of the intracellular cytoskeletal network, and extraction of the cytosolic components, including monomers of the cytoskeletal proteins.

Fixation by Solvents. Fixation by solvents chilled to $-20°$ is common in immunohistochemistry. For cytoskeletal labeling, it has been used especially when processing plant or yeast material for conventional fluorescence or confocal laser scanning microscopy.[21] These organic solvents disrupt the membrane concomitantly with precipitating the proteins in the cytosol. Protein precipitation onto cytoskeletal elements might limit the access for labeling agents and induce thin filaments such as microfilaments or intermediate filaments to coalesce.

Fixation by Aldehydes. Various aldehyde fixation methods have been used for stabilizing the cytoskeleton during processing for fluorescent labeling and immunoelectron microscopy. These methods originate with the use of aldehyde fixation[22] for the stabilization of cytoskeletal elements[23,24] and with detergent extraction introduced subsequently[25,26] to improve the visualization of cytoskeletal proteins.

The effect of detergent extraction in the presence of a low concentration of glutaraldehyde followed by glutaraldehyde fixation is documented in Fig. 7. Labeling of detergent-permeabilized cells displays the vimentin network (Fig. 7, top left) and the actin filaments and stress fibers (Fig. 7, bottom left). In nonpermeabilized cells, vimentin fibers are not discernible at all (Fig. 7, top right) and actin stress fibers are barely visible (Fig. 7, bottom right). Poor membrane permeabilization and poor extraction of cytosolic components may still allow some penetration for the small phalloidin (~1250 Da), yet not to the larger fluorescent antivimentin IgG (~150,000 Da).

The preferred method in our laboratory is essentially based on the procedure of Small *et al.*[27] We retained the use of modified Hanks' buffer [MHB: Ca^{2+}-free, containing 2 mM EGTA and 5 mM MES (2-morpholinoethanesulfonic acid), pH 6.2–6.4] and the postfixation with glutaraldehyde, but replaced Triton X-100 with the nonionic detergent octyl-POE (*n*-octylpolyethoxyethylene). The latter was introduced by Garavito and Rosen-

[21] J. P. Pringle, A. E. M. Adams, D. G. Drubin, and B. K. Haarer, *Methods Enzymol.* **194**, 565 (1991).
[22] D. D. Sabbatini, K. Bensch, and R. J. Barnett, *J. Cell Biol.* **17**, 19 (1963).
[23] D. Slauterbach, *J. Cell Biol.* **18**, 367 (1963).
[24] M. Ledbetter and K. Porter, *J. Cell Biol.* **19**, 239 (1963)
[25] S. Brown, W. Levinson, and J. Spudich, *J. Supramol. Struct.* **5**, 119 (1976).
[26] J. V. Small and J. E. Celis, *Cytobiologie* **16**, 308 (1977).
[27] J. V. Small, D. O. Fürst, and J. De Mey, *J. Cell Biol.* **102**, 210 (1986).

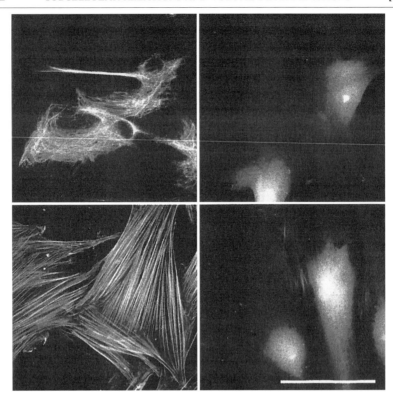

FIG. 7. Cell permeabilization influences cytoskeletal labeling. Human periodontal ligament cells grown on glass coverslips isolated from extracted teeth and cultivated as described by Carnes *et al.* [D. L. Carnes, Ch. L. Maeder, and D. T. Grave, *J. Periodontol.* **68,** 701 (1997)]. (*Fixation*) With (left) and without (right) permeabilization/prefixation as described in Fig 1. (*Labeling*) Vimentin (top): Cy3-antivimentin (monoclonal mouse); F-actin (bottom): TRITC–phalloidin. (*Visualization*) z scan at 0.2-μm steps, 10–14 focal planes. Projections of optical sections. Cytoskeletal labeling is evident only in permeabilized cells. In nonpermeabilized cells, actin is partially accessible by the small phalloidin label (1250 Da) but no vimentin structures are visualized by larger IgG label ($\sim$150,000 Da). Bar: 25 μm.

busch[28] for membrane extraction and is now available commercially (Alexis Corporation, San Diego, CA).

We commonly use 24-well cell culture plates for fixation and process our samples at 20–25° (room temperature). For each of the steps, a well is filled with the respective solution (0.5–1 ml/well). We first transfer the glass coverslip, metal disk, collagen matrix (which has been detached using

[28] M. Garavito and J. P. Rosenbusch, *Methods Enzymol.* **125,** 309 (1986).

a short bevel gauge needle), or biopsy to DMEM supplemented with 10% FCS, then to the next well containing MHB, and then quickly to the well with permeabilization buffer, i.e., to MHB additionally containing 0.125% glutaraldehyde and 0.5–2% octyl-POE. Subsequently, samples are fixed by 1% glutaraldehyde in MHB and then placed into MHB. Before removing autofluorescence and reducing the free aldehyde groups by $NaBH_4$ treatment (see later), samples are washed three to four times in MHB.

Permeabilization and fixation conditions have to be adapted to specific requirements of cells and tissue. We tested the robustness of our standard method and varied the conditions for permeabilization and fixation. We subsequently labeled microfilaments, vimentin, and microtubules and judged the structure preservation by conventional fluorescence microscopy. The three networks were complete and well-discernible against the background, whether we permeabilized fibroblasts with 0.5–2% octyl-POE (2 min, 20°) followed by 1% glutaraldehyde (20 min), or with 1–2% Triton X-100. The concentration of glutaraldehyde during prefixation could be varied from 0.125 to 0.25% and the pH of the fixation buffer varied between pH 5.5 and pH 9.5. The 1% glutaraldehyde used for postfixation could be replaced by 1% paraformaldehyde. In contrast, the omission of prefixation or postfixation, the use of 1% paraformaldehyde for prefixation or of 1% paraformaldehyde for postfixation, or pH values below pH 5.5 or above pH 9.5 led to artifacts or incomplete labeling. The processing temperature during labeling appeared to be less critical, with comparable labeling at 20–25° (room temperature) or at 37°.

In the standard procedure, the time is increased for thicker specimens during prefixation with octyl-POE/0.125% glutaraldehyde and postfixation by 1% glutaraldehyde. For three-dimensional collagen matrix cultures and for long-term organotypic cultures of bone marrow, we also extend the time for prefixation and fixation accordingly. Approximate times for prefixation, postfixation, and removal of background fluorescence are listed in Table I.

We would like to emphasize that there is no such thing as a fixation method generally applicable to all situations. The visualization of specific cytoskeletal constituents may require different fixation procedures. For example, MG63 osteoblast monolayers on glass slips have to be permeabilized by 0.5% octyl-POE, whereas 1 or 2% detergent detaches or even lyses these cells. When labeling microfilaments in the slime mold *Dictyostelium* with actin-specific antibodies, the octyl-POE/glutaraldehyde method performed poorly, which is in contrast to fixation with paraformaldehyde in the presence of picric acid and without detergent treatment.[29] Humbel *et*

---

[29] B. M. Humbel and E. A. Biegelmann, *Scan. Microsc.* **6,** 817 (1992).

TABLE I

TIME REQUIRED FOR PREFIXATION, POSTFIXATION, AND REMOVAL OF
BACKGROUND FLUORESCENCE[a]

Tissue	Prefixation	Postfixation	NaBH$_4$ treatment
Monolayer	2–5 min	20 min	2× 10 min, 0.5 mg/ml
Multilayered cultures	5–10 min	20–30 min	2×
Collagen matrix cultures	10–15 min		15–20 min, 0.5–5 mg/ml
Bone marrow biopsies	20–30 min	30–60 min	2× 30 min
Cartilage biopsies	20–40 min		5 mg/ml

[a] Time increases with thickness of the specimen. Prefixation with 0.5–2% octyl-POE in the presence of 0.125% glutaraldehyde in MHB and fixation with 1% glutaraldehyde at 20–25° (room temperature). Removal of background fluorescence at 0°.

al.[30] reported that permeabilization of U$_2$SO osteosarcoma cells by 0.2% Brij 58 followed by 4% formaldehyde fixation led to comparable brilliant labeling and good preservation of the tubulin network as did 2% formaldehyde, followed by 2% formaldehyde containing 0.02% glutaraldehyde and subsequent permeabilization with Triton X-100 or Brij 58. When investigating the possible attachment of proteasomes to the cytoskeleton, 10% (v/v) Triton in the presence of 1% formaldehyde at 37° proved to be the most efficient.[31] In contrast, the same procedure with only 5% Triton X-100 proved detrimental. However, 1% Triton in combination with 0.125% glutaraldehyde followed by 1% glutaraldehyde as postfixation again provided labeling that was satisfactory, but not as precise and brilliant. In this case, the rapid extraction of cytosolic components together with filament stabilization has been proposed as the mechanism for "instantaneous" fixation.[31] Protein precipitation due to the elevated concentration of organic compounds may contribute to this effect.

Removal of Background Fluorescence

The inherent fluorescence of aldehyde polymers, especially those of glutaraldehyde polymers,[32] the unspecific location of fluorescent labels attaching to nonblocked aldehyde groups, and the autofluorescence of the tissue add to a background fluorescence that obscures true cytoskeletal

[30] B. M. Humbel, D. M. M. de Jong, W. H. Müller, and A. Verkleij, *Microsc. Res. Techn.* **42,** 43 (1998).
[31] C. Archangeletti, R. Sütterlin, U. Aebi, F. De Conto, S. Missorini, C. Chezzi, and K. Scherrer, *J. Struct. Biol.* **119,** 35 (1997).
[32] T. J. A. Johnson, *Eur. J. Cell Biol.* **45,** 160 (1987).

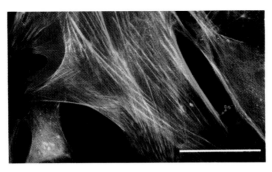

Fig. 1. Localization of F-actin, vimentin, and tubulin in fibroblasts (monolayer) by conventional fluorescence microscopy and photographical overlay. Rat 2sm6 fibroblasts [J. Leavitt and T. Kakunaga, *J. Biol. Chem.* **255,** 1650 (1980)], monolayer on glass coverslips. (Fixation) Prefixation: 2% octyl-POE (*n*-octylpolyethoxyethylene)/0.125% glutaraldehyde, 2 min in MHB [Hanks' buffer, Ca^{2+}-free, containing 2 mM EGTA and 5 mM MES (2-morpholinoethanesulfonic acid), pH 6.2–6.4]. Postfixation: 1% glutaraldehyde in MHB (20 min) at ~20°. Removal of background fluorescence with 0.5 mg/ml $NaBH_4$ (2× 10 min at 0° [cf. W. Baschong, R. Sütterlin, and U. Aebi, *Eur. J. Cell Biol.* **72,** 189 (1997)]. (Labeling) F-actin (green): FITC–phalloidin (Sigma, St. Louis, MO); vimentin (blue): antivimentin (rabbit, Eurodiagnostics, Dardilly, F) and AMCA-antirabbit (goat, Jackson Imm.Res. Lab., Westgrove, PA), tubulin (red): anti-β-tubulin (monoclonal, mouse, Boehringer, Mannheim, D) and Cy3-antimouse (donkey, Jackson Imm. Res.). (Visualization) Excitation using fluorescence filters (Zeiss, Oberkochen). Composite image by triple exposure using 50 ASA film. Bar: 25 μm.

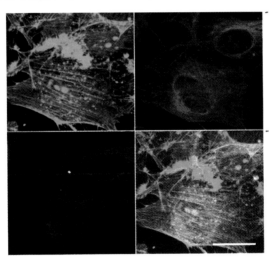

Fig. 2. Localization of F-actin, vimentin, and tubulin in MG-63 osteoblasts (monolayer) by confocal laser scanning microscopy. Human MG-63 osteoblasts (CRL-1427, ATCC, Rockville, MD), monolayer on glass coverslips. Fixed as described in Fig. 1 but prefixed with 0.5% Octyl-POE. (Labeling) F-actin (green): FITC–phalloidin; vimentin (blue, pseudo color): Cy3-antivimentin (monoclonal mouse, Sigma), tubulin (red, pseudo color): anti-β-tubulin (monoclonal mouse) and Cy5-antimouse (donkey, Jackson Imm. Res.). (Visualization) z scan at 0.23-μm steps, 13 focal planes. Excitation at 488, 568, and 647 nm. Image registration and channel separation using AOTF (acousto-optic tunable filter, Leica). Projections of optical sections registered as single fluorescence images: actin (top left); vimentin (bottom left); tubulin (top right); and as overlay of the three fluorescence images (bottom right). Bar: 25 μm.

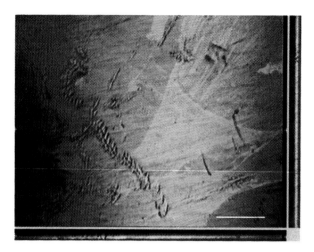

Fig. 3. Concomitant visualization of actin stress fibers and the substrate structure of fibroblasts growing on a metal disk. Human KD fibroblasts [T. Kakanuga, PNAS 75, 1334 (1987)] grown on a polished titanium surface. Fixed as described in Fig. 1. (Labeling) F-actin, TRITC–phalloidin (Sigma). (Visualization) z scan at 0.3-μm steps, 22 focal planes in the fluorescence and the reflection mode. Fluorescence (red): excitation at 568 nm. Reflection registered at 488 nm (gray scale). x–y overlay of fluorescence and reflection images, x–y projection (bottom), and y–z projection (right). Bar: 20 μm.

Fig. 4. Localization of F-actin and tubulin in multilayered, bone marrow long-term cultures. Human bone marrow cultured for 6 weeks from bone marrow biopsies (after informed consent and ethical approval) in 25-cm^2 culture bottles [from W. Baschong, R. Sütterlin, Ch. Racine, and S. Hauser, *Bone* **22**(Suppl.), C141 (1998)]. (Fixation) As in culture bottle and as described in Fig. 1 but prefixation for 5 min. The bottom of the culture bottle was cut out and mounted. (Labeling) F-actin (red): TRITC–phalloidin (Sigma); tubulin (green): anti-β-tubulin (monoclonal mouse) and Cy2-antimouse (donkey, Amersham, Amersham, UK). (Visualization) z scan at 0.29-μm steps, 20 focal planes. Excitation at 488 and 568 nm. Both channels have been registered separately. (Right) Fibroblast-like blanket cells (bottom) display discrete actin stress fibers. Emerging macrophage-like cell (top) with strong actin labeling. (Left) Tubulin network in blanket cells parallel to stress fibers. Macrophage-like cell demonstrates dense tubular scaffold. Bar: 25 μm.

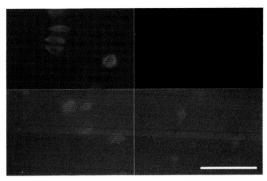

FIG. 8. Influence of background fluorescence and removal by NaBH$_4$. Bovine articular cartilage (100-μm-thick sections). Fixation as described in Fig. 1, but prefixation is for 40 min and postfixation for 60 min. Treatment by chondroitinase (200 mU/ml) and keratinase (400 mU/ml) for 14–16 hr in Tris–HCl, pH 7.2, at 37°. From E. Langelier, R. Sütterlin, U. Aebi, and M. Buschmann, 4th Canadian Connective Tissue Conference, London, Ontario, 1998. Removal of background fluorescence by 5 mg/ml NaBH$_4$ 2× 30 min (top: left and right) and without NaBH$_4$ treatment (bottom: left and right). (Labeling) (Left) Cy3-antivimentin; right: not labeled. Visualization by conventional fluorescence microscopy. (Top) Exposure time 2 sec. (Bottom) Exposure time 0.2 sec.

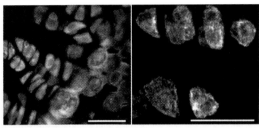

FIG. 9. Conventional background fluorescence and immunofluorescent labeling after the removal of background fluorescence visualized by confocal laser scanning microscopy. Bovine articular cartilage (100-μm-thick sections). Prepared conventionally by glutaraldehyde, ruthenium hexamine trichloride fixation according to Hunziker *et al.* [E. B. Hunziker, R. E. Schenk, and L. M. Cruiz-Orive, *J. Bone Joint Surg.* **69A,** 162 (1987)] (left) and NaBH$_4$ reduced as described in Fig. 8 (right). (Labeling) Tubulin: anti-β-tubulin (monoclonal mouse) and Cy2-antimouse (donkey, Amersham, Amersham, UK). (Visualization) z scan, step size: left: 0.3 μm, 66 focal planes; right: 0.5 μm, 57 focal planes. Excitation at 488 nm. In conventionally processed chondrocytes, microtubules are obscured by cellular background fluorescence (left), whereas NaBH$_4$-treated cells reveal microtubular elements. Bar: 25 μm.

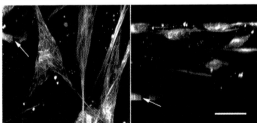

FIG. 10. Depth for optical sectioning is limited by the working distance of the objective. Human KD fibroblasts cultured and fixed as described in Fig. 5. (Labeling) Tubulin: anti-β-tubulin (monoclonal mouse) and Cy3-antimouse (donkey, Jackson Immun.Res.). (Visualization) (Left) z scan at 0.5-μm steps, 188 focal planes. The cell in the bottom layer (arrow) in the projection along the z axes plane (left) and in the $x–z$ projection is blurred. The blurred image in the $x–z$ projection indicates the mechanical block of the objective lens at the limit of its working distance. Bar: 25 μm.

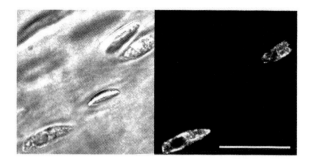

FIG. 11. Discrimination between tubulin-labeled and unlabeled cells within bovine articular cartilage tissue by differential interference contrast (DIC). Prepared and fixed as described in Fig. 8. (Labeling) Tubulin: anti-β-tubulin (monoclonal mouse) and Cy2-antimouse (donkey, Amersham, Amersham, UK). (Visualization) z scan, step size: left: 0.5 μm, 71 focal planes. Excitation at 488 nm. The focal plane 20 has been imaged by DIC in gray scale (left) and by tubulin fluorescence (right). Of the chondrocytes visualized in the focal plane 20 by DIC, only two cells label positive for tubulin. Bar: 25 μm.

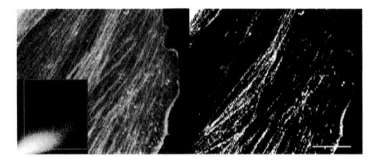

FIG. 12. Colocalization of a microfilament-associated protein. HVSMC (CRL-1999, ATCC, Rockville, MD) cells from human arterial smooth muscle tissue grown on glass coverslips and fixed as described in Fig. 1. Protein S100 A1 was localized as described by Mandinova *et al.*[37] (Labeling) F-actin (green): FITC–phalloidin; protein S100 A1 (orange): anti-S100 A1 antibody (goat) and Cy3-antigoat (rabbit, Sigma). (Visualization) z scan at 0.12-μm steps, 20 focal planes. The mixed false color image (left) represents the overlapping of the fluorescence signals registered in the 488- and 568-channels. The probability for colocalization calculated by the Imaris (Bitplane, Zurich, Switzerland) program is presented as a scatter plot (inset, left). Voxels that colocalize distribute in a cloud along the diagonal. Voxels registered in only one channel would distribute along the axes. The differential image (right) represents fluorescence signals registered in both channels and indicates the extent of colocalization of S100 A1 protein with microfilaments. Bar: 20 μm.

labeling. In our laboratory, we reduce the overall background fluorescence by treatment with $NaBH_4$ to an acceptable level. This is documented by the conventional fluorescent micrograph of bovine cartilage, which has been labeled by Cy3-antivimentin (Fig. 8, see color insert). $NaBH_4$ treatment of permeabilized and aldehyde-fixed cartilage reduces the background fluorescence of labeled (Fig. 8, top left) and nonlabeled (Fig. 8, top right) cartilage to a negligent level. However, micrographs of untreated, labeled (Fig. 8, bottom left), and unlabeled tissue (Fig. 8, bottom right) appear blurred by the high background, e.g., chondrocyte contours are barely discernible from the background.

Blurring does not show in confocal laser scanning microscopy. However, as documented in Fig. 9 (see color insert), the fluorescence background may give way to false-positive interpretations. In nonreduced cartilage, the background fluorescence of tubulin-labeled chondrocytes (Fig. 9, left) obscures the tubulin distribution, revealed by indirect immunofluorescence after treatment with $NaBH_4$ (Fig. 9, right).

For monolayer cultures, we routinely use 0.5 mg/ml of $NaHB_4$ for the removal of background fluorescence, for three-dimensional cell cultures, and tissues up to 5 mg/ml. After fixation samples are washed by changing MHB three to four times. Cells or tissue are then treated by two cycles of $NaHB_4$ in MHB on ice (see Table I). The $NaBH_4$ solution should be made immediately prior to use at $0°$ (ice water), using chilled water to limit borohydride hydrolysis. Sodium borohydride is best stored under a dry atmosphere in the cold. Our experience is that background fluorescence in aldehyde-fixed biopsy material is diminished to an acceptable level by reducing immediately after fixation. Removal of background fluorescence with $NaHB_4$ in samples exposed to aldehydes for longer periods is generally unsuccessful.

Storage

The preparation time from harvesting cells to the removal of background fluorescence often requires 4 to 6 hr. We usually store such samples at $4°$ in MHB supplemented with 0.005% sodium azide within closed 24-well plates in the dark. These samples may be shipped without problems. We found that subsequent labeling was unchanged even after 1–2 months of storage.

Fluorescence Labeling and Mounting

Because general immunofluorescent techniques have been dealt with extensively elsewhere in this volume, we will comment only on problems specific to the visualizaton of cytoskeletal proteins.

Fluorescent Phalloidin

Phalloidin with fluorescent markers such as TRITC or FITC have proven to be remarkably useful for the visualization of microfilaments.[9,33,34] It is worth noting that phalloidin may also associate with F-actin oligomers or short sheared stretches of F-actin. Such stablized F-actin fragments increase in length by monomer association and may eventually lead to a high unspecific background.[35]

At low magnification, microfilaments are not resolved, but form a bright background in the cytosol. We therefore use TRITC–phalloidin or FITC–phalloidin labeling also for visualizing whole cells in three-dimensional matrices or on metal surfaces.

Immunofluorescent Labeling

In our experience, the reaction of each antibody has special characteristics that must be determined empirically. We routinely optimize labeling conditions using monolayer cultures of the respective cells or of similar cells. We incubate antibodies for 30 min and wash three to four times for 10 min with MHB. We then evaluate the procedure by conventional fluorescence microscopy.

When we label cytoskeletal constituents in three-dimensional cell cultures or in tissue, we incubate samples for 3–4 hr at ~20° (room temperature) with appropriate concentrations of TRITC–phalloidin or antibodies. We then wash four to five times for 10–20 min with MHB. For secondary antibody labeling, we further incubate with fluorescent-labeled secondary antibodies for another 3–4 hr at ambient temperature. Samples are then washed as described for primary antibodies. We do not use specific blocking reagents.

Multiple Fluorescent Labeling. We now preferably use cyanine-labeled antibodies (Cy series from Amersham, Amersham, UK; Sigma, St. Louis, MO; Jackson Imm.Res. Labs., Westgrove, PA) as they appear to be significantly more stable than conventional FITC or TRITC labels. Cy labels also have a narrower window for fluorescence excitation and emission, which makes them ideal for multiple labeling.

The sequence of labeling may be crucial in limiting cross reactions and weak labeling due to steric hindrance. As a rule of thumb, indirect labeling using fluorescent secondary antibodies should precede direct labeling with

---

[33] Th. Wieland, A. Deboben, and H. Faulstich, *Just. Lieb. Ann. Chem.* **1980,** 416 (1980).
[34] H. Faulstich, H. Trischmann, and D. Mayer, *Exp. Cell Res.* **144,** 73 (1983).
[35] M. Steinmetz, K. Goldie, and U. Aebi, *J. Cell Biol.* **138,** 559 (1997).

fluorescent antibodies or fluorescent phalloidin, and polyclonal antibodies should be used prior to monoclonals.

Specimen Mounting

The quality of the cover glass implies a major influence on the z-direction precision of confocal stacks. A cover glass, such as a lens, is an optical element in the light path ideally with two parallel and plane surfaces. Objectives used for confocal laser scanning microscopy are adjusted conventionally to a 0.17 ± 0.005-mm-thick coverslip. Thinner or thicker coverslips interfere by altering the optical path. We specifically select for coverslips within these specifications. Alternatively, we use a Zeiss C-APO water objective, where the correction for the cover glass can be set manually from 0 to 2 mm.

Various permanent water-soluble and water-insoluble resins containing an antifading compound are in use for specimen mounting.[36] We routinely employ Mowiol-1188 (Hoechst, Frankfurt BRD) with n-propyl gallate (Sigma) as an antifading reagent acting as a radical scavenger after sample illumination. We usually prepare a 100-ml batch of mounting solution. For this purpose, we suspend 20 g Mowiol in 80 ml MHB by stirring for at least 8 hr at room temperature. To this mixture we add 40 ml of glycerol and stir for another hour. We then centrifuge the suspension for 15 min at 10,000g. We subsequently dissolve 0.75% of n-propyl gallate in the upper 100 ml of the supernatant by stirring for 4–5 hr in the dark. We divide the mounting solution into smaller batches and store them at 4° in the dark.

Cells grown on glass slips and collagen matrices are mounted bottom up, i.e., we invert the glass or the collagen matrix prior to mounting. We first remove the buffer of the glass slips by placing the edge of the glass on blotting paper. We then place the glass on a drop of Mowiol. When collagen matrices are mounted, we hold the matrix with the forceps and drain excess buffer with the tip of a strip of blotting paper. We then place the matrix on a drop of Mowiol on the microscopy slide and place a second drop on top of the matrix and cover it with the glass coverslip. The immiscible aqueous phase drains out of the matrix and is removed with the tip of a piece of blotting paper.

Metal disks and cutouts from culture bottles are glued with cyanoacrylate glue onto the glass slides with the cell side up and then mounted accordingly. Mounted slides are dried for 24 hr at room temperature in the dark and then are stored at 4° in the dark until viewed.

---

[36] A. Longin, C. Soucher, M. Ffrench, and P.-A. Bryon, *J. Histochem. Cytochem.* **41,** 1833 (1993).

Hardware- and Software-Based Limits and Possibilities for
  Visualization of the Cytoskeleton

Limits for Optical Sectioning

Optical sectioning by confocal techniques is limited by the working distance of the objective lens.

Cytoskeletal elements are visualized using 60 to 100× objectives with oil immersion having a lateral resolution of approximately 0.3 μm. A 40× objective will still resolve thick actin stress fibers and isolated microtubules. Intermediate filaments such as vimentin or microtubules in the microtubular organizing center are not resolved. Conventional oil-immersion objective lenses have a numerical aperture of 1.4, which limits their working distance to 170 μm. In contrast, water-immersion objectives with a numerical aperture of 1.2 still resolve cytoskeletal elements and offer a working distance of up to 500 μm because of the slightly smaller collection angle achieved by the lower refraction of water. Optical sectioning is limited further by the transparency of the object. In planes above the point of focus, the beam intensity is attenuated by absorption by the specimen and the resolving power is reduced by beam scattering and fluorescence in the upper regions of the object. Attenuation may be the cause of the lower resolution of single vimentin filaments in the cells located deeper in the matrix (Fig. 5, right).

These limitations may become really crucial when visualizing cytoskeletal elements in tissue. However, in three-dimensional collagen matrices, scattering and fluorescent attenuation are much lower. Figure 10 (see color insert) represents KD fibroblasts grown in three-dimensional collagen matrices that have been labeled for tubulin. The projection of the optical sections acquired along the *z* axis (Fig. 10, left) reveals the microtubular network spanning the KD fibroblasts. The 90° projection in *x*, *z* direction of the image stack (Fig. 10, right) reveals the low-beam attenuation in matrices. The working distance of the objective lens is limited by mechnical contact with the cover glass. This is manifested by the blurred projection of the cell at the bottom of the scanned matrix section (Fig. 10, bottom, corner left).

Software-Aided Localization and Colocalization of Fluorescent
  Cytoskeletal Elements

Software for image recording and image processing usually offer the calculation of projections, three-dimensional image reconstruction, and colocalization of fluorescent objects. This can be applied by three-dimensional imaging of the cytoskeleton, as long as the confocal step size is close to the 0.3-μm lateral resolution and the cover glass applies to specifications (flat, $\sim$0.17 μm thick).

Localization of Fluorescent Elements within Cells and Tissue. Confocal laser scanning microscopes can also visualize an object with normal light by differential interference contrast (DIC). Most image processing software are able to display the digitalized DIC image in gray scale and to overlay the fluorescent information in color.

As illustrated in Fig. 11 (see color insert), comparing the DIC image and the confocal fluorescence image of the same optical section can identify specifically labeled cells in tissue. Whereas DIC provides a differential contrast image of a cartilage preparation showing various chondrocytes emerging out of the extracellular matrix (Fig. 11, left), confocal laser scanning microscopy identifies only two of these chondrocytes, which label positively for tubulin (Fig. 11, right).

Colocalization. The study of cytoskeleton functions requires the identification of associating proteins. This is documented in Fig. 12 (see color insert) by the example of S100 proteins that regulate the contraction of heart muscle. To decide whether one of these proteins, S100 A1, is associated with actin filaments of primary heart muscle cells, they were double-labeled by antibodies against S100 A1 and by TRITC–phalloidin labeling for F-actin.[37] The program Imaris (Bitplane, Zurich, Switzerland) was used on a Silicon Graphics platform to calculate the probability of the colocalization of S100 A1 and F-actin. A scatter plot is displayed showing the correspondence between the fluorescent channels for each voxel. Voxels that colocalize distribute in a cloud along the diagonal (Fig. 12, left inset), whereas voxels registered only in one channel will distribute along the axes. The mixed color image in Fig. 10 (left) represents the overlapping of the fluorescence signals registered in the two separate channels. The differential image representing the fluorescence of signals registered in both channels indicates the extent of colocalization of S100 A1 protein with microfilaments (Fig. 12, right).

Acknowledgments

We thank Drs. H. Epstein, B. Heymann, and U. Aebi for critical discussions and R. Häring, M. Imholz, and M. Schwager for technical assistance. We are obliged to the late Dr. S. Hauser (1954–1998) for bone marrow cultures and to Dr. M. Buschmann for cartilage preparations. This work was sponsored by the M. E. Müller Foundation, by the Canton of Basel Stadt, and by the Swiss Cancer League.

[37] A. Mandinova, D. Atar, B. W. Schäfer, M. Spiess, U. Aebi, and C. W. Heizmann, *J. Cell Sci.* **111,** 2043 (1998).

[12] Chromosome Spread for Confocal Microscopy

By Nadir M. Maraldi, Silvano Capitani, Caterina Cinti,
Luca M. Neri, Spartaco Santi, Stefano Squarzoni,
Liborio Stuppia, and Francesco A. Manzoli

Introduction

Despite the limits in resolution with respect to electron microscopy, light microscopy is still the method of choice for studying the structural organization and composition of chromosomes. Classical cytogenetic methods based on the use of a variety of banding techniques have been improved greatly by molecular biology techniques. Indeed, *in situ* hybridization, nick translation, specific nuclease digestion by restriction enzymes, polymerase chain reaction, and primed *in situ* DNA synthesis offer the chance of exploring chromosome structure and function in greater detail.[1–3]

Techniques that identify *in situ*-specific DNA sequences along fixed chromosomes can utilize fluorescent probes instead of staining or immunoenzyme labeling, thus improving not only the sensitivity and specificity of the reactions, but also resolution. In order to take complete advantage of fluorescence labeling techniques, the use of confocal laser scanning microscopy (CLSM) appears to be mandatory. In fact, CLSM provides improved lateral and axial resolution; moreover, the series of optical sections can be recombined digitally to render three-dimensional reconstruction. Finally, CLSM can combine fluorescence with reflectance signals, allowing the combination of fluorochromes and colloidal gold labeling, thus providing resolution performances very close to the theoretical limit of the system.[4]

This article describes the use of CLSM in the detection of specific DNA sequences in human chromosomes. Particular attention will be devoted to the methods of chromosome spreading that allow the maintenance of a suitable chromosome structure and to the labeling procedures utilized for fluorescence and reflectance CLSM identification of specific chromosome probes.

[1] A. T. Sumner, M. H. Taggert, R. Mezzanotte, and L. Ferrucci, *Histochem J.* **22,** 639 (1990).
[2] M. A. Ferguson-Smith, *in* "Chromosome Today" (A. T. Sumner and A. C. Chandley, eds.), Vol. II, p. 3. Chapman and Hall, London, 1993.
[3] A. T. Sumner, J. De La Torre, and L. Stuppia, *J. Mol. Evol.* **37,** 117 (1993).
[4] D. M. Shotton, *J. Cell Sci.* **94,** 175 (1989).

General Considerations

Methods of preparing chromosomes for microscopic examination have generally involved fixation. Fixation appears to be necessary both for obtaining good quality metaphase preparation and for allowing chromosome banding.[5] Classic fixation methods for metaphase chromosomes involve the use of a mixture of acetic acid with either alcohol or water. A drop of the fixed suspension is then dropped onto a slide and chromosome spread is obtained either by "squashing" under a coverslip or, more frequently, by "air drying."[6,7]

Acetic acid has little fixative action on proteins and cytomembranes, but is a strong precipitant of nucleic acids. Therefore, chromosomes are well preserved, whereas the other cell components are either lost or dispersed by the spreading and do not contaminate chromosome preparations.

The removal of histones has been estimated to vary from 10 to 90%, according to the starting material[8]; however, the distribution of histones remaining in fixed chromosomes tends to be inhomogeneous and mimicks banding patterns.[9] Also, nonhistone chromosome proteins are removed selectively from fixed chromosome.[8] However, the loss of DNA is very reduced, although acid treatment induces both denaturation and depurination, as well as nicks; this can be partly prevented by using alcohol–acetic acid instead of aqueous acetic acid fixation.[10]

Nevertheless, after standard spreading treatment the chromosome structure evaluated by electron microscopy is maintained, being the 20- to 30-nm chromatin fiber organization preserved.[11]

It has been demonstrated by CLSM measurements during the spreading procedure that the primary factor for better maintenance of chromosome integrity is the correct evaporation of the fixative, which ensures increasing concentration of acetic acid on the sample.[12]

The following section describes the principles of the chromosome spreading technique and a protocol that can be utilized for preparing chromosomes for CLSM observation.

[5] A. T. Sumner, *Cancer Gen. Cytogen.* **6,** 59 (1982).

[6] P. S. Moorhead, P. C. Nowell, W. J. Mellman, D. M. Battips, and D. A. Hugenford, *Expt. Cell Res.* **20,** 613 (1960).

[7] J. R. Baker, "Principles of Biological Microtechnique." Methuen, London, 1958.

[8] J. M. Hancock and A. T. Sumner, *Cytobios* **35,** 37 (1982).

[9] B. M. Turner, *Chromosoma* **87,** 345 (1982).

[10] R. Mezzanotte, D. Peretti, M. G. Ennas, R. Vanni, and A. T. Sumner, *Cytogenet. Cell Genet.* **50,** 54 (1989).

[11] D. A. Welter and L. D. Hodge, *Scan. Electr. Microsc.* **2,** 879 (1985).

[12] R. Hliscs, P. Muhling, and U. Clausen, *Cytogenet. Cell Genet.* **76,** 176 (1997).

Chromosome Spread Preparation

Proper techniques for harvesting dividing cells and preparing slides for chromosome analysis are critical for attaining the quality needed for the correct labeling of DNA sequences. At the beginning of the 1960s, Moorhead et al.[6] described a technique for air drying peripheral blood chromosomes using methanol–acetic acid fixative because of its property of quick evaporation. Quick loss of fluid from the cell flattens it out and forces the chromosomes to spread without additional mechanical force. Interestingly, many decades later, we still have nothing superior to this method.

Large quantities of metaphases can be obtained by colchicine treatment. Colchicine disrupts the spindle apparatus, which causes chromosomes to string together and bend at the centromere. After treatment with hypotonic solutions, the cells swell and resemble a water balloon with chromosomes suspended inside. When fixative is added, the cells are preserved in their swollen shape. Next, the drop of fixative with the cell is put on a slide. The fixative then becomes a thin layer unable to support the cell and it evaporates quickly from under and around the cell, and the cell flattens completely, forcing the chromosomes to spread out. Throughout the whole process, the cell membrane is present, but if it breaks open at any stage of the harvest or slide-making procedures, some or all of the chromosomes will spill out, causing random loss of chromosomes.

The quality of chromosome spread and condensation can be affected greatly by the way the chromosomes adhere to the glass slide during the evaporation of the fixative. There is no unique, foolproof method. Optimal techniques vary significantly from laboratory to laboratory and from geographic location to location, depending on many technical and environmental conditions. The many variables affecting slide quality can finally be shown to affect the drying of cells and chromosomes by modifying the evaporation of the fixative. However, variables and their usual effects on chromosome morphology can be used to come up with a method that works for any given set of conditions and that can satisfactorily ensure its reproducibility. Once the slide-making technique has been mastered for a set of conditions (on a given day), it is important to be consistent with the technique so that the batch of slides will give uniform responses for whatever staining method is required. The knowledge of the following variables is useful in achieving good chromosome preparations.

Quality and Condition of Glass Slides

Clean the slides before use to remove any residual oils or dust. A useful slide should fit the following triad: cold, clean, and wet. It should be remembered that some laboratories prefer to use dry slides and that a cold slide instead of a room temperature one appears not to be so critical. The film of water should be as thin as possible so that drying is not impeded. When the fixative hits the film of water, the cell is first spread out on the water surface briefly and then is deposited on the slide. There is no mixing of water and fixative unless water is allowed to move back onto the slide from the edges, where it has retreated by fixation. The method used to ensure the complete removal of water from the slide and uniform drying across the slide is to flood it with drops of fixative after the cell suspension has been applied. Heavier applications of fixative will increase spreading.

Fixative

The reference fixative suggested is methanol : acetic acid (3 : 1). However, many laboratories have substituted methanol with ethanol, which is less toxic and expensive, with comparable results. Increasing the methanol in the 3 : 1 fixative to up to 6 : 1 will hasten evaporation in high humidity and harden cell membranes to help control chromosome scattering. Increasing the acetic acid will soften cell membranes, increase spreading, and slow down evaporation. Variations in fixative types and ratios (methanol, ethanol, 45% aqueous acetic acid solutions) may be useful when experiencing difficulties, but a conservative approach of keeping the standard 3 : 1 fixative is suggested unless all else fails.

Humidity, Temperature, and Drying

Neither fast nor slow evaporation is ideal: slides that dry too slowly will result in low–contrast (grayish) chromosomes and chromosome scattering, whereas a very rapid drying rate will increase the contamination by the cytoplasm and induce very dark and glassy (hyaline) chromosomes. Speeding up or slowing down the evaporation process will induce more or less spreading.

Drying time should be roughly 30 to 45 sec, but in dry climates it is necessary to slow drying time and in humid climates to speed it up. In very humid climates it is more likely to use a technique that improves drying, such as the use of 6 : 1 ethanol acetic acid fixative. Many methods are also available to control humidity and evaporation rates: dehumidifiers or humidifier devices and steaming slides.

The following temperatures may affect results: temperature of air in the room, temperature of the slides and the water they are coated with, and temperature at which the slide is allowed to dry. It is important, if possible, to standardize the temperature of the laboratory. Because laboratories are usually adjusted for what feels comfortable and therefore too cool in summer and too hot in winter, protocols must be adjusted to compensate for cold temperatures by warming slides and hot temperatures by slowing drying.

The slide can be dried at room temperature, on a slide warmer or on a hot plate (for a few seconds up to overnight), over an alcohol flame, a humidifier, or hot steam.

Cell drying speed can be increased or decreased by changing the air flow over the slide. Be aware of air movements in the slide-making area; air-conditioning vents may be a source of drying problems.

The most critical and variable step is air drying the slides. In general, poorly spread chromosomes with halos are drying too fast. To slow the drying, chill the fixative or place the slides on a wet paper towel. Pale, gray, or overspread chromosomes are drying too slowly. In this case, lower the drying angle or warm the slide briefly.

Protocol for CLSM Chromosome Spread

1. Wash microscope slides first with a harsh cleaning solution (Alconox, Chemipol). Be sure to remove all the cleaning solution by rinsing several times in distilled water. This step is crucial.
2. Store slides in a Coplin jar with absolute methanol (or ethanol).
3. Remove a slide from methanol (ethanol) and clean it with a folded Whatman (Clifton, NJ) paper for microscope objective lenses. Dip the slide back into the methanol (ethanol) and plunge it in a beaker of distilled water until the methanol (ethanol) is dissolved and a uniform layer of water covers the slide.
4. Holding the ground glass end of the slide, drain the excess water on paper towels, long edge down. While draining the slide, keep the edge in contact with the paper towel.
5. While draining the slide, fill a Pasteur pipette with a well-mixed cell suspension. Make sure that there are no bubbles while withdrawing cells.
6. Lower the slide until it forms about a 30° angle with the bench top with the water film facing up. Angle the slide slightly with the paper towel, but not so much that the cells will collect at the lower edge.
7. Slowly drop 3 or 4 evenly spaced, controlled drops of cell suspension onto the slide, progressing toward the frosted end. Drop cells from 5 to 8 cm above the slide. The droplets should strike the slide closer to the upper edge of the slide and should spread evenly as they strike the slide. Nevertheless, the cells should still appear in round circular patterns on the water layer.
8. Drain excess water on the paper towel by tilting the slide to a vertical position for a moment. Return to the previous low angle

and flood the slide with 4 to 5 drops of fresh fixative, starting from the upper corner and moving toward the ground glass end. This removes most of the water and gives more control over the drying environment and timing.

9. Dry the slide flat on a paper towel, or upright, according to the temperature and humidity in the slide-making area, checking that the time is between 25 and 45 sec. A convenient approach more independent of environment humidity could be the use of a water bath set at 56°: slides can be held about 10 cm above the water with metaphases facing down for a time of roughly 20–25 sec. At the end of the procedure check if the slide is dry (when the slide has dried, from certain angles rainbow colors appear on the surface). Adjust the drying time on the water bath to standardize the protocol.

10. Take into account the humidity of the environment; if the humidity is above 45%, it will usually be necessary to speed up drying to prevent chromosome scattering. Possible solutions include warming the slide with the back of the hand or thigh, increasing all flow by waving the slide in the air or increasing the air flow over the slide, flaming the slide once or twice, or keeping the slide in a oven for a brief period. If the humidity drops to 35% or less, too fast drying will produce dark cells surrounded by cytoplasm. Possible modifications include drying slides on wet paper towels, protecting slides from air currents, drying the slides flat instead of upright, increasing the amount of acetic acid in the floating fixative (e.g., 5:2), or holding the slides over a source of steam or over a humidifier. The last part of the drying seems to be in general the critical part.

11. Check on the phase microscope. If cells are gray or the chromosomes are scattered, warm longer. If cells are very dark and show visible cytoplasm, reduce or eliminate warming.

Evaluation of Chromosome Spread Quality and
  High-Resolution Morphology

Several chromosome preparation methods have been reported that allow comparison between light and electron microscope observation; however, they generally require nonstandard fixation and spreading techniques.[13–15] In addition, none of the methods described made it possible

[13] G. D. Burkholder, *Nature* **247**, 292 (1974).
[14] E. J. Du Praw and G. F. Bahr, *Acta Cytol.* **13**, 188 (1969).
[15] J. Gosalvez, A. T. Sumner, C. Lopez-Fernandez, R. Rossino, V. Goyanes, and R. Mezzanotte, *Chromosoma* **99**, 36 (1990).

to use all types of microscopes for observing the same chromosome preparation.

This section describes a method of mounting chromosome spreads obtained by standard cytogenetic methods (methanol–acetic acid fixation, air drying spreading) that allowed us to observe the very same chromosomes with light and electron microscopes in transmission (TEM), scanning (SEM), or scanning–transmission (STEM) mode, as well as with CLSM.[6] This method makes it possible to carry out more complete morphological studies on chromosomes by allowing direct correlation of the results obtained from different techniques using the same sample, such as banding pattern analysis, fluorescent probe localization, *in situ* hybridization, structural surface or whole mount studies, and fine localization of colloidal gold markers.

An example of chromosome spread is reported in Fig. 1. The partial decondensation of the chromosomes after digestion with DNase I and staining with propidium iodide (PI) is visible by CLSM (Fig. 1A). The same field, observed by TEM (Fig. 1B), STEM (Fig. 1C), and SEM (Fig. 1D), shows some details of the chromatin fiber organization, even at low magnification.

Preparation of Spreads on Carbon Film and Transfer to Grids

Normal glass slides are cleaned by rubbing with a few drops of undiluted commercial liquid hand soap and wiping with lint-free paper until they appear completely clean. It is important that a very thin detergent film remains on the glass surface in order to facilitate detachment of the carbon layer from the slide. A thick (dark gray) carbon film is evaporated on the slides with an Edwards E 306 apparatus (Edwards High Vacuum, Crawley, UK); sufficiently thick films are obtained by repeated evaporations. These can be generated, without changing the carbon electrodes, by rapidly increasing the current values to the maximum as soon as the vacuum is sufficient. Two drops of methanol–acetic acid-fixed, metaphase-blocked cells are spread on the carbon-coated slide. After air drying the glass edges are scored with a finely pointed metal tool (e.g., electron microscopy tweezers). With the aid of an inverted microscope (at 10× magnification) 100-mesh nickel grids are placed on the carbon film. Try to localize at least two to three metaphase plates at the center of the grid holes in each grid. It is convenient to position the grids with the apparently flat face in contact with the carbon film; this procedure requires a firm hand and is facilitated by touching the carbon film with one side of the grid first and then gently dropping the grid from the tweezers (some position adjustments are generally possible). Finally the grid-covered carbon film is detached from the

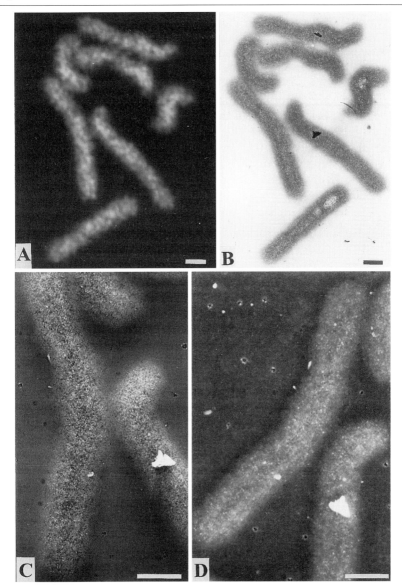

Fig. 1. Fixed human chromosome spread preparations for CLSM and electron microscopy observation. (A) After digestion with 1 U/ml DNase I in phosphate buffer, pH 6.0, for 30 min at room temperature, and staining with 1 μg/ml propidium iodide, the sample was mounted on a glass slide with 60% glycerol and a coverslip and was observed by a Sarastro CLSM. (B) The same field observed by TEM. The correspondence with CLSM is evident, but the resolution is highly improved. The same chromosomes show some surface details when observed by STEM (C) and SEM at 5 kV (D). Bars: 1 μm.

slide by immersing it at a low angle in distilled water. The floating carbon film is collected from above with a clean Parafilm sheet (American National Can, Greenwich, CT) and air dried in a dust-free container. In some cases not all the grids can be collected in the first attempt, but after a number of tries this step is always successful; the addition of 5–10 μl 10% Triton X-100 to the water, after adhesion of the grid-covered film to the Parafilm, may be helpful in lowering the water surface tension. After complete drying, the grids are picked up with fine tweezers and stored on a filter paper in a petri dish. If pieces of the carbon film fold up on the grids, they are removed later by slowly dipping the (perfectly) dry grids vertically in distilled water. At the end of this procedure the carbon film is transferred to the bottom of the grids, with chromosomes (and some cells) protruding upward in the grid holes. During all of the following steps mechanical damage by direct exposure of the grids to the turbulence of liquids must be avoided.

Preparation of Grids for Different Purposes

Samples are rehydrated for 15 min in distilled water, cleaned with 0.0125% trypsin (Gibco Ltd., Paisley, UK) in 0.15 M phosphate buffer, pH 6, for 60 sec, washed in distilled water, maintained in phosphate buffer, pH 6, for 1 hr at 37°, and kept in distilled water overnight at 4°. After this initial step, it is possible to carry out different treatments (e.g., DNase digestion or staining with Giemsa dye).

Processing of Grids for Electron Microscopy

The samples are fixed with 1% glutaraldehyde–0.1 M phosphate buffer, pH 7.2, for 30 min, washed in distilled water (if desired they can be postfixed in 1% aqueous OsO_4 for 30 min and washed in distilled water), and dehydrated to absolute ethanol and dried at the critical point. Shadowing can be performed in order to obtain electron microscopy images of the chromosome surface. Rotary shadowing gives optimal results: this is best carried out with 2-nm platinum evaporated at 45° with an electron gun at the lowest pressure (e.g., 10^{-6} mbar) (a film thickness-measuring device is useful).

Light Microscopy (LM) Observation

The grids can be observed in a LM simply held on a glass slide (Giemsa staining can be made after the trypsin-cleaning step). For CLSM the processed grids are mounted gently on a glass slide with 60% glycerol and coverslip and observed; after observation the immersion oil can be removed with filter paper and the slide soaked in distilled water to take the coverslip away and wash the grids, which can be processed further for electron microscopy.

Scanning Electron Microscopy Observation

For SEM observation the grids are mounted usefully on a specially made carbon support in order to improve the quality of the image and to reduce the back scattered electron (BSE) contribution and X-ray spurious signal.[16] For easier observation of the grids in the SEM, it is better to examine the grids at 5 kV first and to increase the accelerating voltage later to obtain high magnification details on the identified chromosome. Nevertheless, it is possible to obtain good quality images even at 5 kV at short (5 mm) working distances.

Transmission and Scanning Transmission Electron Microscopy

Transmission and scanning transmission micrographs can be taken in a TEM equipped with a scanning attachment at sufficiently high voltage (100 kV and more). Stereo pairs can be recorded generally with 6° tilt difference. STEM micrographs taken in dark-field/bright-field ratio mode (Z contrast) offer the great advantage of representing a true surface image, obtained by collecting selectively the transmitted electrons scattered by high molecular weight elements, i.e., the platinum layered on the chromosome surface. This is obtained easily by setting the camera length at the proper value for collecting on the annular (dark field) detector the brightest diffraction band generated by the platinum coating in a planar zone of the grid, devoid of chromosomes.

Identification of DNA Probes by CSLM

The following sections describe some applications of the chromosome spreading technique for the identification of specific DNA probes utilizing CLSM.

Restriction Enzyme–Nick Translation (RE/NT) Procedure

In addition to the fluorescence *in situ* hybridization (FISH) technique, other methods involving nuclease digestion have been found useful for detecting DNA sequence distribution in chromosomes. The restriction enzyme–nick translation procedure, which involves the synthesis of labeled

---

[16] S. Squarzoni, C. Cinti, S. Santi, A, Valmori, and N. M. Maraldi, *Chromosoma* **103,** 381 (1994).

DNA starting from the nicks induced by restriction enzymes, provided valuable information about the distribution of genes on chromosomes.[3,17]

Instead of utilizing immunoenzyme labeling methods (biotin–streptavidin–peroxidase–DAB; biotin–streptavidin–alkaline phosphatase), which have been used mainly for detection purposes in the RE/NT reaction, we reported the use of different fluorescent labels by using CLSM.[18]

A comparison was made among biotin–dUTP–FITC–avidin, DIG–dUTP–anti-DIG–FITC, fluorescein-12–dUTP, and resorufin-15–dUTP. Results indicated that the DIG–dUTP method utilizing FITC-labeled anti-DIG presented the greatest sensitivity and a minimal background, whereas the use of fluorescein–dUTP without amplification resulted in the most rapid procedure. Moreover, CLSM allowed us to determine the precise correlation between DNA distribution and the RE/NT pattern because of its ability to detect different signals simultaneously.[18]

Some examples of the RE/NT labeling are reported in Fig. 2. Fixed human chromosomes digested with DraI for different times are viewed either with the red light detector for PI staining (Figs. 2A, 2C, and 2E) or with the green light detector for FITC (Figs. 2B, 2D, and 2F). Chromosomes digested for 1 hr and labeled with biotin–dUTP/FITC–avidin show a staining pattern corresponding to C-negative bands as well as a less detailed G-banding (Fig. 2B). After the same digestion procedure, labeling with DIG–dUTP/anti-DIG–FITC is much brighter (Fig. 2D). A C-banding pattern is revealed by DraI digestion for 12 hr, as observed in a higher magnification of a single chromosome (Fig. 2F).

Methods for RE/NT Procedure

Slides are aged for a few hours to 1 week before being used in the experiments. Restriction enzyme (RE) digestion is carried out using 30 U DraI endonuclease (target site TTT ↓ AAA) dissolved in 100 μl of the buffer supplied by the manufacturer. Digestion is carried out at 37° in a moist chamber for 1 to 12 hr and is stopped by washing the slides briefly in tap water. The nick translation (NT) reaction is performed as reported by De La Torre *et al.*[19] and is modified to allow the use of fluorescent-labeled reagents. In detail, the slides are incubated in 50 μl of a mixture containing 5 U DNA polymerase I, 10 nM each of dATP, dCTP, and dGTP, and 10 nM of a differently labeled dUTP for each experiment, namely

[17] S. Adolph and H. Hameister, *Hum. Genet.* **69,** 117 (1985).
[18] L. Stuppia, C. Cinti, S. Santi, R. Peila, G. Calabrese, G. Palka, N. M. Maraldi, and A. T. Sumner, *Genome* **38,** 1032 (1995).
[19] J. De La Torre, A. T. Sumner, J. Gosalvez, and L. Stuppia, *Genome* **35,** 891 (1992).

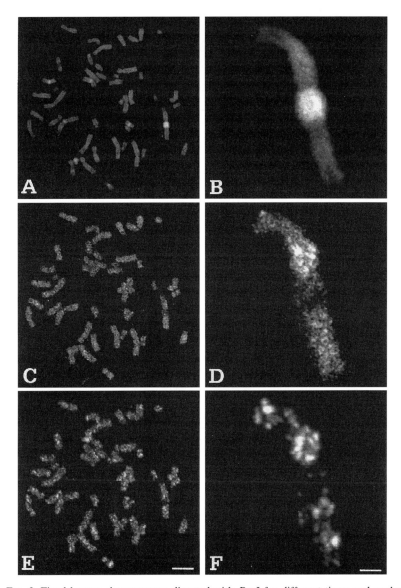

FIG. 2. Fixed human chromosomes digested with *Dra*I for different times and analyzed by the RE/NT technique with different labeling procedures by CLSM. (A, C, E) Detection of PI staining. (B, D, F) Detection of FITC signal. (A, B) *Dra*I digestion for 1 hr; biotin–dUTP/FITC–avidin labeling. The staining pattern consists of C-negative bands plus a pattern broadly similar to G-bands. (C, D) *Dra*I digestion for 1 hr; DIG–dUTP/anti-DIG–FITC labeling. The image shows a similar staining pattern, but the brightness is enhanced with respect to biotin labeling. (E, F) *Dra*I digestion for 12 hr; DIG–dUTP/anti-DIG–FITC labeling. The staining pattern in a single chromosome appears localized in the C-band, whereas PI counterstaining shows the differential extraction of DNA in the centromeric region. Bars: (A–D) 5 μm; (E, F) 1 μm.

biotin-16–dUTP, digoxygenin (DIG)-11–dUTP, fluorescein-12–dUTP, or resofurin-15–dUTP, for 30 min at room temperature. After a blocking step with 2% bovine serum albumin, slides labeled with biotin-16–dUTP or DIG-11–dUTP are incubated with fluorescein isothiocyanate (FITC)-conjugated avidin or with FITC-conjugated anti-DIG antibodies. The samples are counterstained with propidium iodide (PI). Slides are rinsed briefly in phosphate-buffered saline (PBS) and mounted for CLSM observation.

Localization of Single Copy Genes by PRINS

The resolution of FISH allows the detection of DNA sequences separated a few megabases apart in chromosomes.[20] The oligonucleotide-primed *in situ* DNA synthesis (PRINS) technique represents an alternative to FISH, whose utilization has been limited to the detection of repeated sequences.[21]

We reported a method that demonstrated the possibility of identifying single copy genes by using very short sequences obtained from synthesized oligonucleotides in the X human chromosome by a modification of the PRINS technique that utilizes the high efficiency and resolution of CLSM.[22] In fact, the signal-to-noise ratio and the fading did not allow the detection of the signals from single copy gene by conventional fluorescence microscopy.

Methods for Single Copy Gene Localization by PRINS

Two oligonucleotides for the polymorphic region of the factor IX gene are used (X1 = 5' ACCTTATGGGACCACTGTCG 3'; X2 = 5' ATATTTCTCCTTCCCTCCCTC 3'; Biotech, Italy). The fixed human chromosome spreads are allowed to dry for at least 10 days. The samples are dehydrated in ethanol (70, 90, and 100%) and then denatured in 70% formamide, 2× SSC, pH 7.0, for 10 min at 72°. The samples are dehydrated further in ethanol at 4° and air dried. The slides are incubated in a reaction mixture containing (in 50 μl) 200 pmol of each oligonucleotide, 5 μl of 10× polymerase chain reaction (PCR) buffer II (AmpliTaq, Perkin-Elmer, Norwalk, CT), 2 μl of DIG DNA labeling mixture (1 mM dATP, 1 mM dCTP, 1 mM dGTP, 0.65 mM dTTP, 0.35 mM DIG–dUTP; Boehringer, Mannheim, Germany), and 2 U of *Taq* I DNA polymerase (AmpliTaq, Perkin-Elmer) for 10 min at 58° (the annealing temperature for the oligonucleotide used) and for 30 min at 72°. The slides are then washed twice in 2× SSC, pH 7.0, and in 4× SSC, pH 7.0, for 5 min at room temperature.

[20] B. J. Trask, D. Pinkel, and G. van den Eng, *Genomics* **5,** 710 (1989).
[21] J. Gosden, D. Hauratty, J. Starling, J. Fantes, A. Mitchell, and D. Porteous, *Cytogenet. Cell Genet.* **57,** 100 (1991).
[22] C. Cinti, S. Santi, and N. M. Maraldi, *Nucleic Acid Res.* **21,** 5799 (1993).

The slides are then placed in 4× SSC, 0.5% (w/v) bovine serum albumin (BSA), pH 7.0, for 5 min and incubated for 45 min at room temperature with anti-DIG–FITC (Boehringer) diluted 1:100 in 4× SSC, 0.5% (w/v) BSA, pH 7.0. The slides are finally washed in 4× SSC, 0.05% Triton X-100 and counterstained with 1 μg/ml PI.

Crucial steps with respect to the previously reported procedures[14,21] are (i) drying of the samples onto the glass for at least 10 days, (ii) avoiding preheating the glass and the incubation mixture and, (iii) omitting treatment with 50 mM NaCl, 50 mM EDTA at 70°.

Multiple Fluoresce and Reflectance Detection of Restriction Enzyme-Digested DNA Sequences

The RE/NT procedure utilizing the *Hae*III restriction enzyme specific for GC/CC-rich regions appears to be selective for the identification of cytosine and guanine-rich sequences (CpG islands) corresponding to a R-band pattern on human chromosomes, characteristic of potentially active genes.[3,17,19]

To overcome localization problems due to the use of DAB or alkaline phosphatase labeling, whose staining reactions cover wide areas of the sample, we utilized a combination of fluorescent and colloidal gold-labeled probes that can be detected simultaneously by CLSM.[23] The overlaying of three different signals allowed the simultaneous observation of distinct chromosome structures: total DNA by propidium iodide, fluorescein and gold-labeled sequences after RE/NT, or total DNA and centromeric sequences of different chromosomes after *in situ* hybridization. The identification of *Hae*III RE-sensitive regions by fluorescent signals revealed large areas of labeling, whereas a more defined signal, often organized in spot pairs that resembled an R-like banding, was detected when gold probes were revealed by the CLSM reflected signal.[23]

The lateral resolution of CLSM in detecting gold-labeled *Hae*III digestion sites utilizing the 488-nm laser line and the reflected light detector has been evaluated to reach 150–160 nm. GC-rich regions organized in pairs of spots have been identified that conceivably correspond to symmetrical single pairs of alleles coupled on the arms of the same chromosome.[24]

Some examples of multiple fluorescence and reflectance detection of DNA sequences by the RE/NT procedure after digestion with *Hae*III at

[23] L. M. Neri, S. Santi, C. Cinti, P. Sabatelli, A. Valmori, C. Capanni, S. Capitani, L. Stuppia, and N. M. Maraldi, *Eur. J. Cell Biol.* **71,** 120 (1996).
[24] L. M. Neri, C. Cinti, S. Santi, M. Marchisio, S. Capitani, and N. M. Maraldi, *Histochem. Cell Biol.* **107,** 97 (1997).

the CLSM are reported in Fig. 3. The centromeric DNA, undigested by *Hae*III, is visualized by PI staining in a chromosome plate (Fig. 3A) and in chromosome 1 (Fig. 3B). DNA sequences identified by the RE/NT procedure are identified by FITC (Figs. 3B and 3C) and, with a higher definition, by gold reflectance (Figs. 3E and 3F).

Methods for RE/NT

Digestion is performed with 30 U of *Hae*III RE (specific for GG/CC-rich regions) in 100 μl of the appropriate buffer. The reaction is carried out for 90 min at 37° in a moist chamber and is stopped by several washings in distilled water. The *in situ* NT reaction is performed as reported in the previous paragraph, utilizing a modified version of the method described by De La Torre *et al.*[19]

Methods for Multiple Labeling after in Situ Nick Translation

For the detection of DIG-11–dUTP-labeled sequences, a FITC–anti-DIG sheep antibody (Boehringer) is employed diluted 1 : 100 in 1 *M* Tris–HCl, 1 *M* NaCl, 20 m*M* MgCl$_2$, pH 7.5, supplemented with 2% BSA for 45 min at room temperature. Slides are counterstained with PI. For gold labeling the samples are washed in 20 m*M* Tris–HCl, pH 8.2, 225 m*M* NaCl and incubated with 1-nm gold-labeled sheep anti-DIG antibody (Biocell, Cardiff, UK) diluted 1 : 100 for 2 hr at room temperature. Slides are then treated with an IntenSE silver enhancement kit (Amersham, Bucks, UK) for 5 min at room temperature and counterstained with PI. The samples are mounted in Tris–HCl/glycerol containing 1,4-diazobicyclo[2.2.2]octane.

Methods for Fluorescent and Reflected Signal Detection by CLSM

Specimens are imaged by a PHOIBOS 1000-SARASTRO (Molecular Dynamics, Sunnyvale, CA) confocal laser scanning microscope mounted on a Nikon Optiphot microscope (Nikon, Tokyo, Japan) equipped with a 25-mW multiline argon ion laser that produces a line at 488 nm, selected with a band-pass filter and used to reveal all the three signals detected. The laser power is tuned at 10 mW, and the laser beam is attenuated to 30% of transmission.

In the illumination path a dichroic excitation filter of 50/50 is used as a beam splitter reflecting light, and a polarizing lens of one-fourth wave turn is inserted in the light path at the level of the filter cube of the microscope. To avoid image shift, all three signals are acquired with the reflectance mode microscope configuration: no image alteration is observed. Images obtained before and after this insertion do not show any modification either in terms of intensity or of specificity of the labeling.

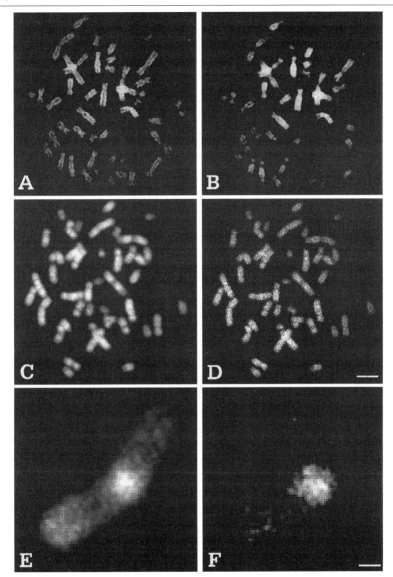

FIG. 3. Fixed human chromosomes digested with *Hae*III and revealed by the RE/NT technique utilizing multiple detection of PI (A, B), FITC (C, D), and gold (E, F) CLSM signals. (A, C, E) Chromosome plate. (B, D, F) High-resolution image acquisition of chromosome 1. The reflected signal shows a labeling organized in discrete dots (F). Bars: (A, C, E) 5 μm; (B, D, F) 1 μm.

Samples are observed with a 100×, 1.4 numerical aperture (NA) plana-pochromat objective lens. To obtain the highest resolution (R), we employ the lowest laser light wavelength (λ) suitable and the highest NA objective $(R = 0.46\lambda/\text{NA})$. The refractive index of the immersion oil is 1.518 (Nikon); the oil is dissolved up to a concentration of 20% in the mounting medium to minimize distortion of the confocal spot during the laser beam penetration inside the specimen.

The back pinhole aperture with a diameter of 50 μm is in front of the detectors, two photomultiplier tubes (PMTs). To block any reflected contribution signal when observing fluorescein, a 515 OG long-pass filter has been inserted in front of the dichroic emission filter. In front of the two PMTs, a dichroic emission filter of 565 nm and barrier filters are used. The barrier filter in the "green" channel has a pass band of 515 to 545 nm (530 DF 15), and in the "red" channel, a long-pass filter of 590 nm is used (OG 590). Thus the green detector is sensitive to light in the region of 515 to 545 nm, whereas the red detector is sensitive to wavelengths over 590 nm. The green and red PMTs are set at 950 and 850 mV, respectively. When the green PMT is used to detect the reflected signal, it is set at 515 mV and the barrier filter band pass 530 nm removed and substituted with a polarizing filter in front of the PMT. The reflected image never shows undesirable reflections such as bright diffraction rings or reflection images of the microscope slide. Using a polarizing filter with the low setting of the PMT never allows one to observe any fluorescent signal, thus separating in practice FITC and gold signals. As a control, a 500-nm dichroic emission filter is used instead of that of 565 nm to observe gold and PI and to skip fluorescein. The image obtained is identical to that obtained with the instrument configuration described earlier.

Digitized optical sections are stored on the graphics workstation Indigo Iris XS24 (Silicon Graphics, Mountain View) with a scanning mode format of 512 × 512 pixels and 256 gray levels. Optical series of horizontal x/y images are obtained, through the entire samples, at increments of 0.3 μm along the z axis. A vertical scan mode is used to obtain a y–z plane. Image processing and volume rendering are performed using the ImageSpace software (Molecular Dynamics). Each kind of signal (FITC, gold, or PI) is elaborated independently to optimize the contrast, the brightness, and the intensity of the images and is then superimposed to be analyzed as a multiple signal.

Conclusions

CLSM represents a powerful tool for the detection of specific DNA sequences in metaphase chromosomes. The only limitations are the cost and the complexity of the acquisition–elaboration procedure, which do not permit the use of CLSM in routine cytogenetics. However, its use improves the limits of the different chromosome mapping procedures greatly.

The main prerequisites for successful utilization of CLSM in chromosome analysis are (1) preparation of high-quality chromosome spreads in which contaminant proteins are removed completely and chromosome ultrastructural organization is maintained and (2) setting of the CLSM when different signals are collected in order to maintain lateral and axial resolution without signal displacement.

In fact, provided that chromosome spreading is obtained carefully, the maintenance of chromosome organization can ensure a very precise localization of DNA probes. The three-dimensional reconstruction capability of CLSM, associated with high lateral and axial resolution, is particularly suitable to detect DNA probes that are localized very close to each other (e.g., in the two arms of a chromosome). Finally, the possibility of detecting colloidal gold-labeled probes in the reflection mode and the possibility of a simultaneous detection of multiple signals represent a formidable tool in studying the *in situ* genome organization.

Acknowledgments

The authors thank Mr. A. Valmori for photographic help and Ms. M. Bolognini for typing the manuscript. Research was supported by grants from the University of Bologna (Funds for Selected Research Topics), the Italian Research Council (PF Biotecnologie), the Ministero della Ricerca Scientifica e Università (60%), the Associazione Italiana per la Ricerca sul Cancro (AIRC), the Biomedical Research Arcispedale S. Anna, Ferrara, and the Institute "Rizzoli" Bologna.

[13] Visualization of Nuclear Pore Complex and Its Distribution by Confocal Laser Scanning Microscopy

By ULRICH KUBITSCHECK, THORSTEN KUES, and REINER PETERS

Introduction

The nuclear pore complex (NPC) is a cylindrical structure approximately 120 nm in diameter and 70 nm in height spanning the nuclear envelope (NE). The NPC is of great importance to cell function. On the one hand it mediates the exchange of matter and information between the genetic material inside the nucleus and the protein-synthesizing apparatus in the cytoplasm. On the other hand it plays a key role in the spatial organization of nucleus and genome. Thus, the NPC is one example of an important cellular structure that is quite large, yet still substantially smaller than the resolution limit of the light microscope (approximately 225 nm). This article describes how the NPC, despite its submicroscopic dimensions, can be visualized by confocal light microscopy and how it, although structurally unresolved, can be localized in space with nanometer accuracy. The high localization accuracy together with a potential for *in vivo* studies provides a whole set of novel analytical opportunities.

Nuclear Pore Complex

Structure

The nuclear pore complex is an assembly of globular and fibrillar structures composed of probably more than 100 different proteins and having a total mass of approximately 125 MDa.[1] The total number of NPCs per nucleus as well as its area density varies among different cell types within wide limits.[2] Thus, mammalian cells typically harbor about 2000–4000 NPCs per nucleus at a density of 5–15 NPCs per μm^2 whereas amphibian oocytes feature several million NPCs per nucleus at 40–50 NPCs per μm^2.

Transport Properties

The NPC mediates both the import into and the export from the nucleus.[3–7] Molecules up to about 50 kDa can pass the NPC by passive diffusion via a channel approximately 10 nm in diameter and 50 nm in length. However, the transport of molecules and particles having a diameter of 10–25 nm is highly specific, requiring molecular signals, several soluble cofactors, and metabolic energy.

Several nuclear localization sequences (NLS), i.e., short stretches of amino acid residues contained in the transport substrate, have been identified with the one of the SV40 (simian virus 40) large T antigen being characterized best.[8] In addition to an NLS, at least four soluble cofactors are involved: importin α and β (also designated as karyopherin α and β),

[1] R. Reichelt, A. Holzenburg, E. L. Buhle, M. Jarnik, A. Engel, and U. Aebi, *J. Cell Biol.* **110,** 883 (1990).
[2] G. G. Maul, *Int. Rev. Cytol. Suppl.* **6,** 76 (1977).
[3] N. Panté and U. Aebi, *Curr. Opin. Cell Biol.* **8,** 397 (1996).
[4] F. Melchior and L. Gerace, *Curr. Opin Cell Biol.* **7,** 310 (1995).
[5] D. Görlich and I. W. Mattaj, *Science* **271,** 1513 (1996).
[6] D. Görlich, *Curr. Opin. Cell Biol.* **9,** 421 (1997).
[7] E. A. Nigg, *Nature* **386,** 779 (1997).
[8] D. Kalderon, B. L Roberts, W. D. Richardson, and A. E. Smith, *Cell* **39,** 499 (1984).

nuclear transport factor 2 (also known as p10, pp15), and Ran/TC4. The latter is a small G protein undergoing a GTP/GDP cycle. According to current knowledge, NLS-dependent import involves at least five steps: In the cytoplasm a molecule containing a NLS binds to the importin α subunit of an importin $\alpha\beta$ heterodimer. The ternary complex associates with cytoplasmic filaments of the NPC via its importin β subunit and then migrates on or is transported by the cytoplasmic filaments to the center of the NPC. The ternary complex is translocated through the NPC, a step mediated by Ran/TC4. At the karyoplasmic face of the NPC the ternary complex dissociates into its components and thus releases the transport substrate. Among these steps the association of the ternary complex with cytoplasmic filaments is inhibited by the lectin wheat germ agglutinin (WGA) whereas translocation through the NPC is inhibited by energy depletion or temperature reduction.

A similar scheme, involving different signal sequences and cofactors, applies to nuclear export. Nuclear export signals (NES) have been identified and export cofactors are currently being studied intensively. The functional coupling between import and export is still speculative.

Role in Nuclear Morphogenesis

In addition to nucleocytoplasmic transport the NPC apparently has another, equally important but less obvious function: the induction and maintenance of the three-dimensional architecture of the genome. Chromosomes occupy defined territories in the cell nucleus, particularly during interphase.[9] Chromosomal territories are probably separated by interchromatin channels serving as pathways for the transport of mRNA transcription factors between genetic loci and nuclear periphery.[10] Possibly, the genome as a whole has a three-dimensional (3D) structure characteristic and specific for the cell type as well as the cell cycle phase and developmental stage.[11] Among structural elements, which may affect the 3D architecture of the genome, the nuclear matrix and the NE have to be considered in the first place.[12] While the debate of the molecular composition and physiological relevance of the nuclear matrix, despite considerable progress, continues,[13] the structure, molecular composition, and function of the NE have been

---

[9] D. A. Agard and J. W. Sedat, *Nature* **302,** 676 (1983).

[10] T. Cremer, A. Kurz, R. Zirbel, S. Dietzel, B. Rinke, E. Schrock, M. R. Speicher, U. Mathieu, A. Jauch, P. Emmerich, H. Scherthan, T. Ried, C. Cremer, and P. Lichter, *Cold Spring Harb. Sym. Quant. Biol.* **LVIII,** 777 (1993).

[11] G. Blobel, *Proc. Natl. Acad. Sci. U.S.A.* **82,** 8527 (1985).

[12] S. D. Georgatos, *J. Cell. Biochem.* **55,** 69 (1994).

[13] R. Berezney and K. W. Jeon, eds., "Structural and Functional Organization of the Nuclear Matrix," *Int. Rev. Cyt.* **162b,** 1995.

clarified to a certain extent. For instance, at least in vertebrate cells, the chromosomes are attached to the NE by means of a considerable number of specific sites.[14] On the ultrastructural level, a large fraction of chromatin appears to be coaligned with the NE,[15] which raises the possibility that chromosomes are specifically attached to the inner surface of the NE. In fact, Marshall et al.[16] reported that there are approximately 15 NE–chromosome contacts per chromosome arm in *Drosophila* embryos. These contacts are thought to organize the genome such that each genetic locus occupies a highly determined position within the nucleus, thereby the interaction between chromosomes and NE is thought to be a major determinant of the intranuclear architecture. The nature of NE–chromosome contacts is still unresolved. One possibility is that the nuclear lamina provides contacts via lamin B, the lamin B receptor, and lamina associated proteins.[17–19] Another possibility is that the NPC itself harbors contact sites. Filaments radiating from the internal face of the NPC into the karyoplasm and connecting in the fashion of a basket or fish trap[20,21] seem to suit that purpose. The NPC itself contains proteins having zinc finger domains that might serve as DNA-binding sites. In accordance with this contention, the gene-gating hypothesis[11] assumes that the NPC is the major factor determining the 3D architecture of the genome. Each NPC is thought to be associated to a specific genetic locus and to channel its transcripts to the cytoplasm. If this hypothesis applies, the distribution of NPCs in the NE would directly reflect the 3D organization of the genome and its changes during cell cycle and ontogenesis.

Lateral Mobility

Surprisingly, it has been observed that the NPC is very mobile in yeast nuclei.[22,23] This observation was made by an elegant application of green fluorescent protein (GFP) technology. GFP is a small protein having a very

---

[14] M. W. Goldberg and T. D. Allen, *Curr. Opin. Cell Biol.* **7**, 301 (1995).
[15] M. R Paddy, A. S. Belmont, H. Saumweber, D. A. Agard, and J. W. Sedat, *Cell* **62**, 89 (1990).
[16] W. F. Marshall, A. F. Dernburg, B. Harmon, D. A. Agard, and J. W. Sedat, *Mol. Biol. Cell* **7**, 825 (1996).
[17] J. H. Worman, J. Yuan, G. Blobel, and S. D. Georgatos, *Proc. Natl. Acad. Sci. U.S.A.* **85**, 8531 (1988).
[18] G. Simos and S. D. Georgatos, *EMBO J.* **11**, 4027 (1992).
[19] C. A. Glass, J. R. Glass, H. Taniura, K. W. Hasel, J. M. Blevitt, and L. Gerace, *EMBO J.* **12**, 4413 (1993).
[20] H. Ris, *EMSA Bull.* **21**, 54 (1991).
[21] M. Jarnik and U. Aebi, *J. Struct. Biol.* **107**, 291 (1991).
[22] N. Belgareh and V. Doye, *J. Cell Biol.* **136**, 747 (1997).
[23] M. Bucci and S. R. Wente, *J. Cell Biol.* **136**, 1185 (1997).

strong endogenous fluorescence.[24] By construction and expression of genes in which cellular proteins are fused to GFP it has become possible to fluorescently label cellular proteins *in vivo*. When nuclei containing nucleoporin–GFP are fused with normal nonfluorescent nuclei the fluorescence spreads over the entire new nucleus within 20–30 min. The large lateral mobility of the NPC raises many questions. As discussed earlier, at least in vertebrate cells, the chromosomes are attached to the NE by means of a considerable number of specific sites, a situation in which a large mobility of the NPC in the NE appears to be quite improbable. A lateral mobility of the NPC might also interfere with transport through the NPC because the protein p210 that extends through the nuclear membrane (gp210) is assumed to anchor the NPC in the nuclear membrane and to regulate passive transport through the NPC. It is therefore an open and important question whether the lateral mobility of the NPC that is observed in yeast also holds for vertebrate and mammalian cells.

Theoretical Methods

Basic Concepts

Independent of an internal architecture single subresolution objects such as fluorescently labeled NPCs are imaged by a light microscope as "spots", i.e., small 3D intensity distributions. The basic idea of single particle localization is to determine, as accurately as possible, the center of the experimentally observed 3D intensity distribution. Two methods are used for that purpose most frequently. The centroid algorithm[25] simply calculates the center of mass of the intensity distribution above a certain background threshold. The second method involves fitting of an appropriate theoretical intensity distribution, the point-spread function (PSF) or a valid approximation of the PSF, to the experimental intensity distribution using a nonlinear χ^2-minimization algorithm, e.g., the Levenberg–Marquardt method.[26] The following sections focus on the latter method because it is more accurate and, most importantly, permits to explicitly determine the accuracy of localization.

Approximation of Point-Spread Function

The mathematical function describing the intensity distribution of an image of a subresolution object is referred to as PSF. The PSF is a complex

[24] M. Chalfie, Y. Tu, G. Euskirchen, W. W. Ward, and D. C. Prasher, *Science* **263,** 802 (1994).
[25] A. Patwardhan, *J. Microsc.* **186,** 246 (1997).
[26] W. Press, S. A. Teukolsky, W. V. Vetterling, and B. P. Flannery, "Numerical Recipies in C." Cambridge Univ. Press, Cambridge, 1992.

function and can be computed directly for idealized systems only. For nonideal but largely simplified systems the computation requires numerical simulation. Therefore we employ an approximation of the PSF that can be handled easily and yet is sufficiently precise. The approximate PSF is determined by a semiempirical procedure as illustrated in Fig. 1: Immobilized subresolution fluorescent beads, for instance, 100 nm in diameter are imaged in the z-scan mode. The radial and axial profiles of the images are determined and fitted by a 3D Gaussian of the following form:

$$f(x, y, z) = A \exp\left[-\frac{(x - x_c)^2 + (y - y_c)^2}{2\sigma_{xy}^2} - \frac{(z - z_c)^2}{2\sigma_z^2} \right] + I_{bg} \qquad (1)$$

where x_c, y_c, and z_c designate the center coordinates of the Gaussian and σ_{xy}^2 and σ_z^2 designate the radial and axial variance. A is the amplitude, and I_{bg} is an offset intensity. The limit at which two particles can be recognized as separate entities is approximately equal to the full-width at half-maximum (FWHM) of the approximate PSF. For the confocal microscope employed in this study (Leica, TCS) the FWHM amounted to 0.27 μm in the lateral and 0.49 μm in the axial direction using an objective of 1.3 numerical aperture and optimal imaging conditions.

Accuracy of Single-Particle Localization

The image of a subresolution object contains both signal and noise, as exemplified in the simulation of Fig. 2. The characteristic parameter here is the signal-to-noise ratio (SNR), which may be defined as

$$\text{SNR} = \frac{I_0}{\sqrt{\sigma_{bg}^2 + \sigma_I^2}} \qquad (2)$$

where I_0 designates the maximum signal intensity, σ_{bg}^2 the variance of the background intensity, and σ_I^2 the variance of the maximum signal intensity. σ_I^2, in particular, depends strongly on image acquisition parameters such as laser power, gain of the photomultiplier tube, size of the confocal aperture, pixels and lines per frame, and the number of frames averaged.

Because the image of a subresolution particle has a certain, usually rather small SNR, the method for determining the object center described earlier yields only an estimate. As a characteristic parameter for the accuracy of localization we use certain distances Δx, Δy, and Δz for which the probability of including both the estimated and the true object center amounts to 0.68, a value corresponding to the standard deviation of a normal distribution. Δx and Δy are equal in theory whereas Δz differs from

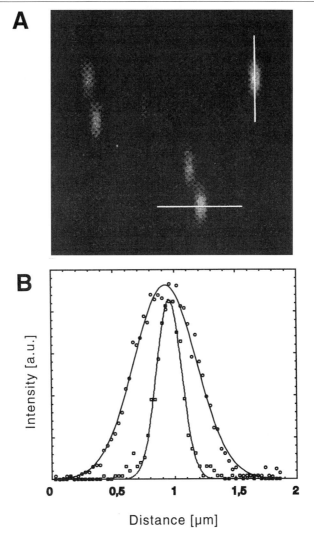

FIG. 1. Experimental determination of the parameters of a 3D Gaussian in an approximation to the confocal point-spread function. (A) Perpendicular confocal section (xz image) of a sample of immobile fluorescent microbeads with a 100-nm diameter. The white lines have a length of 1.9 μm. (B) Intensity profile along the white bars as determined in (A) in the focal plane (squares) and along the optical axis (circles). The full and dotted lines represent Gaussians determined by a fit to the respective data. The thus determined FWHM values are 0.24 and 0.61 μm lateral and axial, respectively.

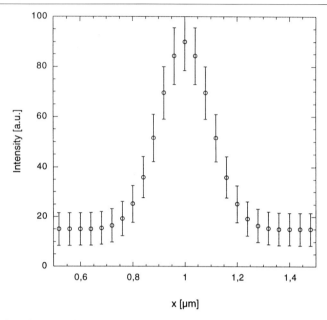

FIG. 2. Gaussian intensity profile with the associated theoretical local standard deviations in signal intensity $\sigma_1(x)$. This graph was constructed according to the amplitudes and noise parameters found in a typical NPC localization (see the section focusing on the accuracy of single NPC localization). Pixel size was 40 nm, background intensity was 15 with a $\sigma_{bg} = 6.5$, maximum signal amplitude above background was 72, and $\sigma_I(x_i, y_j, z_k) = [1.25I\,(x_i, y_j, z_k)]^{1/2}$. Under these conditions, the center of the corresponding Gaussian can be found with an accuracy of <5 and <13 nm in radial and axial direction, respectively (see Fig. 3).

Δx and Δy because the PSF has an axially symmetrical but elongated shape. For the lateral direction the localization accuracy is derived to be[27,28]:

$$\Delta x = \sqrt{4.72 \left[\sum_{i=-N}^{N} \sum_{j=-M}^{M} \sum_{k=-L}^{L} \frac{\dfrac{I_0^2 x_i^2}{\sigma_{xy}^4} \exp\left[-\dfrac{x_i^2 + y_j^2}{2\sigma_{xy}^2} - \dfrac{z_k^2}{2\sigma_z^2} \right]}{\sigma_{bg}^2 + \sigma_I^2(x_i, y_j, z_k)} \right]^{-1}} \qquad (3)$$

Here x_i, y_j, and z_k designate the three discretized spatial coordinates, whereas $2N + 1$, $2M + 1$, and $2L + 1$ designate the number of pixels in the spatial directions used for the localization. The absolute pixel size enters

[27] N. Bobroff, *Rev. Sci Instrum.* **57,** 1152 (1986).
[28] T. Kues, "Zur Genauigkeit der Lokalisierung Immobiler und Mobiler Submikroskopischer Partikel durch Konfokale Laser-Scanning-Mikroskopie und Bildanalyse." Diploma thesis, Phys. Department, University of Münster, 1997.

this expression in only an indirect way, namely in the ratio between radial and axial variances and the discretization intervals $x_{i+1} - x_i$, $y_{j+1} - y_j$, and $z_{k+1} - z_k$.

$\sigma_I^2(x_i, y_j, z_k)$ is the local variance of the signal intensity above background (cf. Fig. 2). It depends on the maximum signal intensity and the number of frames or lines averaged. If the signal could be measured in units of detected photons, this variance would be equal to the photon number because the detection of light corresponds statistically to a Poisson process. In general, however, the photon detector introduces an arbitrary amplification factor. Hence, it is necessary to measure the variance as a function of signal intensity for the respective instrument parameters. This can be done easily by imaging an object repetitively with intensities covering the complete dynamic range. From this image series, the standard deviations can be determined as a function of the respective average image intensities. For computing the axial localization accuracy Δz, Eq. (3) has to be modified slightly, replacing $\dfrac{I_0^2 x_i^2}{\sigma_{xy}^4}$ by $\dfrac{I_0^2 z_k^2}{\sigma_z^4}$. Δz is about twofold larger than Δx.

Figure 3 shows, for a specific set of instrumental parameters, how both theoretical and experimentally determined (see later) localization accuracies depend on the SNR. At a SNR of 5, for instance, both theoretical and experimental localization accuracies amount to ± 3 nm in radial and ± 8 nm in axial direction.

Distribution Analysis

The methods described so far yield the locations of single submicroscopic objects and an estimate about the associated localization accuracy. After determining the positions of an ensemble of single objects in this way, an important question is whether the particles are distributed in an organized or in a random pattern. In the following paragraph three procedures are presented that allow the detection of deviations from randomness in object distributions. It can be determined whether the objects are distributed in either a more aggregated or dispersed manner than would be expected for a random distribution. Statistical distribution functions, such as the nearest neighbor distribution function (NNDF) and the pair correlation function, yield information about the type of object distribution, and additional characterization is conveyed by a cluster analysis. In general it is necessary to combine several statistical procedures in order to obtain reliable results.

The general procedure is as follows. The measured coordinates of the objects are used to calculate their statistical distribution functions (or "estimators" of these, as discussed later). In a first step these may be compared

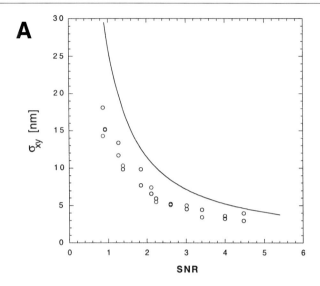

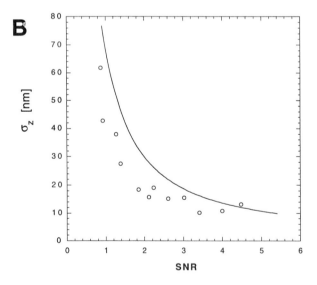

to the distribution functions of randomly distributed point objects of the same number density. In case of significant deviations between experimentally determined estimators and point object functions a new assumption is made according to which interaction potential the objects in question are distributed. Employing Monte Carlo methods estimators of the distribution functions are computed. Good correspondence between these and experimentally determined functions then argues for a distribution of the observed objects according to the assumed interaction potential. In case of a bad correspondence, the hypothetical interaction potential must be modified until a good agreement is achieved.

Nearest Neighbor Distribution Function

The NNDF $d(r)$ quantifies the probability of finding the nearest neighbor of an arbitrarily chosen object (or *mean* object) within a distance r:

$$d(r) = P(\text{distance from a mean object to the nearest object is at most } r) \quad (4)$$

where P is the accumulated probability.

However, each experimentally observed object pattern is always a random sample of the underlying real distribution. Hence, from experimental data the real distribution cannot be determined but only estimated. The estimated distribution is designated as the *estimator*. Because experimentally determined object patterns are always limited to a certain spatial region, difficulties arise due to the region borders. Such problems are designated as edge effects, which may be circumvented by a respective definition of the estimator.[29–31] This results in the following expression for

[29] P. J. Diggle, "Statistical Analysis of Spatial Point Patterns." Academic Press, London, 1983.
[30] D. König, S. Carvajal-Gonzalez, A. M. Downs, J. Vassy, and J. P. Rigaut, *J. Microsc.* **161,** 405 (1990).
[31] L. M. Karlsson and A. Liljeborg, *J. Microsc.* **175,** 186 (1994).

FIG. 3. Theoretical and experimental radial and axial localization accuracies in dependence on SNR. (A) Localization accuracy in the focal plane, and (B) localization accuracy along the optical axis. Graphs of the theoretical localization accuracies were calculated according to Eqs. (3) and (4) for a specific set of instrumental parameters. Results of an experimental determination of the localization accuracy are indicated by symbols. A specimen of immobile fluorescent subresolution microbeads was imaged repetitively with different laser powers and correspondingly different maximum signal intensities and hence SNRs. From the resulting image stacks the center positions of the objects were determined by χ^2 fits according to Eq. (1), and the distances between the found centers were evaluated. The standard deviation of the repetitively measured distances divided by $\sqrt{2}$ represents the experimental localization accuracy. This approach yields localization accuracies that are independent of mechanical drifts of the object stage during the relatively long data acquisition.

the edge corrected estimator $\hat{d}(r)$ for the NNDF for an object ensemble Φ_W that is confined to a region W:

$$\hat{d}(r) = \frac{\text{Number of objects of } \Phi_W \text{ with } d < R \text{ and } d < r}{\text{Number of objects of } \Phi_W \text{ with } d < R} \tag{5}$$

Here, d designates the distance from an arbitrarily chosen object to its next neighbor, and R is the distance to the nearest point on the border of region W. The edge effect is taken into account by evaluating only objects whose next neighbor distance is smaller than the minimum distance to the region edge.

Pair Correlation Function (PCF)

The PCF $g(r)$ is defined by the mean number of objects $n(r)\,dr$ found in a shell of radii r and $r + dr$ around a mean object normalized by the shell area times the mean density, ρ[32]:

$$g(r) = \frac{n(r)\,dr}{\rho 2\pi r\,dr} \tag{6}$$

Therefore, the PCF quantifies the deviation of the local density from the mean density. The correction of $g(r)$ for edge effects can be attained using the K function, $K(r)$, which is defined as the expectation value of the number of objects that are positioned within a distance r from a mean object (that itself is not counted) divided by the object density. The edge-corrected estimator of $K(r)$ in the case of a circular region W of radius a, $\hat{K}_{W_a}(r)$, can be determined as

$$\hat{K}_{W_a}(r) = \frac{\pi a^2}{n^2} \sum_{i=1}^{n} \sum_{j=1}^{n} \frac{I_r(d_{ij})}{\left[1 - \dfrac{a}{\pi} \cos\left(\dfrac{a^2 - r^2 - d_{ij}}{2rd_{ij}}\right)\right]} \tag{7}$$

where n designates the number of objects in W, and d_{ij} is the distance between objects i and j. The function $I_r(d_{ij})$ is defined as

$$I_r(d_{ij}) = \begin{cases} 1 & \text{for} \quad d_{ij} \leq r \\ 0 & \text{for} \quad d_{ij} > r \end{cases} \tag{8}$$

After determination of $\hat{K}_{W_a}(r)$ for discrete distances $m\,\Delta r$, $m = 1, 2\ldots$, the edge-corrected estimator $\hat{g}_{W_a}(r)$ for the PCF can be computed as

$$\hat{g}_{W_a}(m\,\Delta r) = \frac{\hat{K}_{W_a}[(m + 1)\,\Delta r] - \hat{K}_{W_a}[(m - 1)\,\Delta r]}{4\pi r\,\Delta r} \tag{9}$$

[32] D. A. McQuarrie, "Statistical Mechanics." Harper Collins, New York, 1976.

Numerical Computation of Estimators

Usually it is not straightforward to obtain edge-corrected reference functions for the NNDF and PCF, even in the comparatively simple case of point objects. To render the statistical analysis more realistic and thus applicable to the distribution of the NPC, one would have to consider hard, extended objects—discs—instead of points. The distribution of discs is determined by their hard core potential, meaning that the center-to-center object distance is at least equal to the particle diameter. More complicated interaction potentials occur in the case of attraction or repulsion between objects. Although in principle the NNDF and PCF for objects with arbitrary interaction potentials may be calculated using methods of statistical mechanics,[32] it is usually difficult to yield explicit numerical solutions. More often, numerical means are used to determine the distribution functions of objects, where the distribution is determined by a nontrivial interaction potential. Monte Carlo methods can be used to generate object coordinates at random while taking the object interaction potential into account. For example, in the case of hard core objects the procedure is as follows. An ensemble of object coordinates is produced in a sequential manner. Whenever new object coordinates are generated using a random number generator,[26] it is examined whether the center-to-center distance between the new object and all existing objects is greater than the object diameter. If that is not the case the new coordinates are rejected and another set of coordinates is drawn until the hard core condition is met. After this a new set of coordinates may be generated until the desired mean object density is obtained. As in the case of the experimental determination of the PCF and NNDF, each numerical simulation resembles only one random realization of the theoretical distribution. Numerically, however, it is simple to repeat the simulation 1000–10,000 times. Subsequently the estimator functions of the respective distributions are calculated and finally averaged. These reference curves with corresponding standard deviations are finally compared with the experimentally determined estimator.

NNDF and PCF for Randomly Distributed Point Objects

Distribution functions for a random distribution of point objects with an average surface density ρ in a plane are especially simple and can be computed analytically. Such random distributions of point objects are generated by a two-dimensional stationary Poisson point process. For a Poisson process, the NNDF, $d_{\mathrm{Poi}}(r)$, is given by

$$d_{\mathrm{Poi}}(r) = 1 - e^{-\pi r^2 \rho} \tag{10}$$

By comparing the NNDF for this random distribution with the NNDF calculated for the object pattern in question, it can be decided whether the objects are aggregated, which would result in a relative excess of small nearest neighbor distances compared to the random distributed ones, or repelling each other, which would result in a relative deficiency of nearest neighbor distances for small distances. Likewise, the PCF can be calculated explicitly for objects that are distributed according to a Poisson process because there are, on average, no deviations between local and global density, which results in

$$g_{\text{Poi}}(r) = 1 \qquad \text{for } r > 0 \tag{11}$$

As in the case of the NNDF, a comparison between the PCF for a Poisson process with a PCF, which is determined from an experimentally observed pattern, reveals attractive or repulsive interaction potentials of the objects. Values of the PCF for small radii below unity indicate a repulsion, whereas values above one indicate an attraction between the objects in the observed sample.

Cluster Analysis

In a cluster analysis, the mean nearest neighbor distance and their variance for a given observed ensemble of objects is compared with the corresponding values of objects, which are distributed according to a hypothetical interaction potential. First, the nearest neighbor distances are determined for all observed objects and the mean value, $\langle d_{nn} \rangle$ and its variance, σ^2, are calculated. Then, these values are normalized by their expectation values, $E(\langle d_{nn} \rangle)$ and $E(\langle \sigma^2 \rangle)$, yielding the ratios $Q = \langle d_{nn} \rangle / E(\langle d_{nn} \rangle)$ and $R = \sigma^2 / E(\langle \sigma^2 \rangle)$. For objects with a complex interaction potential, $E(\langle d_{nn} \rangle)$ and $E(\langle \sigma^2 \rangle)$ must be computed by Monte Carlo methods as described earlier. For a simple Poisson point process, these are given as

$$E_{\text{Poi}}(\langle d_{nm} \rangle) = \frac{1}{2\sqrt{\rho}} \tag{12}$$

and

$$E_{\text{Poi}}(\sigma^2) = \frac{4 - \pi}{4\pi\rho} \tag{13}$$

If the objects of the sample data set are distributed according to the interaction potential, which was assumed for calculation of the expectation values, the Q and R values are unity. This would verify an object distribution according to the respective hypothetical interaction potential. Deviation of Q from unity indicates clustering ($Q < 1$) or repulsion ($Q \geq 1$) beyond the forces that are described by the potential. The R value contains additional information about fluctuations of the actual in comparison to an ideal distribution.[33]

Experimental Methods

Determination of Localization Accuracy

Localization accuracy varies among different microscopes and, furthermore, for a particular microscope, depends on several parameters, such as objective lens, size of confocal aperture, and PMT voltage. Therefore, the localization accuracy has to be determined for specific imaging conditions. For that purpose a specimen is prepared that contains small immobile fluorescent particles. We frequently use fluorescent microbeads with a diameter of $\leq$100 nm. Immobilization is achieved by dispersing the beads in a concentrated polyacrylamide gel. The specimen is imaged repetitively, obtaining five image stacks of the same location. Each stack consists of 16 confocal scans in which the specimen is shifted through the focus between the scans in 200-nm steps. Each image stack is used to determine the center of all particles by fitting the 3D Gaussian of Eq. (1) to the intensity distributions of the particles. For each stack the distances between the centers of all particles are determined. Then, the standard deviation of the repetitively measured distances between any two particles divided by $\sqrt{2}$ is equal to the localization accuracy. It is noteworthy that the method yields localization accuracies that are independent of possible drifts during data acquisition.

For the confocal microscope employed in this study (Leica, TCS), we found the localization accuracy, depending on the SNR, to be equal or larger than $\pm$4 nm in radial and $\pm$12 nm in axial directions (Fig. 3, symbols). Figure 3 also reveals that at small SNR the localization accuracy is even better than expected theoretically. This may be due to the fact that the theory contains the assumption that signal statistics follow Gaussian statistics, while in reality it follows a Poisson statistic. Despite this the theory provides a good estimate for the upper limit of the achievable localization accuracy.

Fluorescent Labeling of NPC

The NPC can be labeled very well by immunofluorescence methods.[34] An established cell line, for instance, is cultured at standard conditions and

[33] H. Schwarz and H. E. Exner, *J. Microsc.* **129**, 155 (1983).
[34] L. Davis and G. Blobel, *Cell* **45**, 699 (1986).

seeded on coverslips. One to 3 days after seeding, the cells are permeabilized *in situ* by treatment with digitonin.[35] A stock solution is prepared by dissolving digitonin (Calbiochem, Bad Soden, Germany) in dimethyl sulfoxide (DMSO) at a concentration of 20 mg/ml. The stock solution is partitioned into small aliquots and stored at $-20°$. The stock solution of digitonin is diluted in transport buffer [TB; 50 mM HEPES/KOH, pH 7.3, 110 mM potassium acetate, 5 mM sodium acetate, 2 mM magnesium acetate, 1 mM EGTA, 2 mM dithiothreitol (DTT)] to yield a final concentration of 50 μg digitonin/ml. A coverslip carrying a cell monolayer is washed thoroughly with ice-cold transport buffer and is then incubated with 1 ml of the digitonin solution for 5 min on ice. After this, cells are washed carefully. An anti-NPC antibody is diluted appropriately in TB supplemented with 1% bovine serum albumin (BSA). The coverslip is deposited in a wet chamber and covered with 40 μl of antibody solution. After incubation for 45 min, cells are washed, and 40 μl of a solution of a fluorescent secondary antibody at appropriate dilution is added to the cell monolayer. After an incubation time of 45 min, cells are washed. Cells are finally mounted in threefold concentrated TB–BSA (33.3%), glycerol (66.6%) containing an antifading agent. All incubations are performed at $4°$.

There are several other methods for labeling the NPC. Thus, wheat germ agglutinin inhibits the transport of NLS-containing proteins efficiently through the NPC by binding with high affinity to N-acetylglucosamine residues of NPC proteins. Fluorescent WGA can therefore be used to label the NPC. However, because N-acetylglucosamine residues not only occur in NPC proteins but also in other components of the NE, the labeling is not very specific. In the transport of NLS-containing proteins, metabolic energy is only required for translocation but not for the initial binding. Therefore, the NPC can be labeled with fluorescent NLS-containing proteins if cells are depleted carefully of energy or kept at reduced temperature. Last but not least, the NPC can be labeled fluorescently *in vivo* by means of green fluorescent protein (GFP) technology.[22,23] In particular, the NPC proteins Nup49p and Nup133p were fused to GFP and the constructs expressed in yeast.

Optimization of Imaging Conditions

The goal is to obtain a complete stack of optical sections of single nuclei at the highest possible resolution and SNR. It is therefore necessary to optimize the imaging conditions carefully. In general, photobleaching will inevitably occur to a certain extent and has to be monitored and corrected

---

[35] S. Adam, R. Sterne-Marr, and L. Gerace, *J. Cell Biol.* **111**, 807 (1990).

for as good as possible. How the imaging conditions of a confocal laser scanning microscope are optimized has been discussed in great detail in the literature[36] and therefore will only be outlined here briefly. The objective lens should be designed for fluorescence microscopy and have a large numerical aperture (NA 1.3 or 1.4). The size of the confocal pinhole should be adjusted to a value between two and four optical units for an optimal SNR and spatial resolution. Excitation laser power should be as low as possible and the field size should be chosen such that the imaged object is imaged with a good sampling (ca. 40–80 nm/pixel). Line or frame averaging should be chosen such that the SNR is high and bleaching still acceptable (four- to eightfold). The axial step size should be 0.2–0.3 μm for a NA of 1.3–1.4. PMT saturation must be avoided, and contrast settings should exploit the dynamic range of the instrument optimally. After data acquisition, the extent of photobleaching per image should be determined and then corrected in the original data. For a particular confocal microscope, the model TCS of Leica, optimum imaging conditions are achieved by employing the following components and instrument settings: objective lens 63×, NA 1.4 or 100×, NA 1.3 oil-immersion objectives, pinhole size of 80 corresponding to ca. two optical units, pixel size approximately 40 nm/pixel radially and 200 nm axially, eightfold line averaging, and PMT voltage of 640 V. After acquisition the image stacks are smoothened with a 3 × 3 × 3 Gaussian filter in order to reduce the inherent photon noise. The localization algorithm works more reliably with low noise images. The width of the smoothing kernel should be small compared to the width of the point-spread function. Otherwise, the intensity peaks due to single objects will be artificially smeared out, increasing the problem of differentiating closely neighbored objects.

NPC Localization

Imaging of a particular nucleus is usually restricted to its upper spherical shell; 11–16 *xy* images separated in the *z* direction by 0.2 μm are acquired, thus covering a total volume of 20 × 20 × 2–3 μm. The stacks of confocal images are used to determine the 3D coordinates of the center of each imaged NPC (Fig. 4). For this purpose, we have developed an interactive image-processing program by which small 3D cubes can be selected within the complete image stacks. The cubes defined a region of interest (ROI) and thus reduce computation time requirements. Object center localization is straightforward in the case of isolated NPCs but is more complex for

[36] J. B. Pawley, ed., "Handbook of Biological Confocal Microscopy," 2nd ed. Plenum Press, New York, 1995.

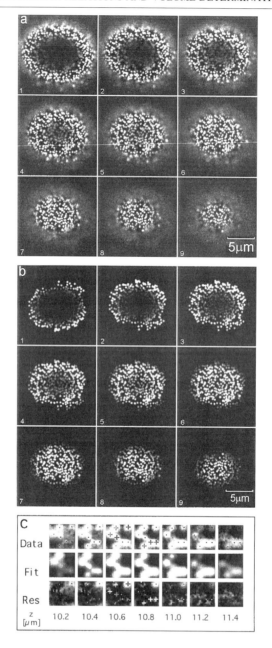

NPC aggregates. Therefore the localization procedure is performed in two steps. First, all ROIs containing single spots are selected, and the centers and amplitudes of the respective spots are determined automatically by a fit to the 3D Gaussian of Eq. (1) utilizing an appropriate offset level. This results not only in a coordinate list of the spot centers, but also in minimum and maximum amplitudes for single NPC intensities. These values are used as restrictions in further analysis. Second, intensity distributions presumably representing two or more NPCs are evaluated individually. The analysis ROI is selected, and the putative centers of NPCs are selected interactively by clicking with the mouse onto respective ROI voxels. The coordinates and intensity values of these voxels serve as starting values for the fitting procedure that simultaneously uses up to six 3D Gaussians to reproduce the ROI intensity distribution optimally, thereby locating the centers of NPCs.

After convergence of the fitting routine, the analysis program inverts the voxel values nearest to the centers such that three-dimensional crosses indicating the detected bead centers are created in the image stack. In this way the fitting results can be displayed without losing the original data. After each fitting process, inspection of an image stack comprising the original data, the fitted profiles, and the absolute residuals allows one to judge the fitting results (Fig. 4c). In case of an unsatisfying result, the voxel values of the marker crosses are inverted a second time to restore the original intensity data and the process is started again after the definition of new start values. The analysis finally results in a list containing the x, y, and z coordinates of 300–500 individual fluorescent spots representing

---

FIG. 4. Confocal imaging and 3D localization of single NPCs[37] (a) 3T3 cells were permeabilized with digitonin, and NPCs were labeled with the antinucleoporin antibody MAb 414 and a FITC-labeled secondary antibody. Nine confocal sections acquired with an axial step size of 0.2 μm from a single 3T3 cell nucleus are shown. The images were corrected for photobleaching and smoothed by a $3 \times 3 \times 3$ Gaussian kernel. Many single NPCs and clusters of NPCs are discernible. (b) Result of the fitting process utilizing the determined NPC centers and intensities. The 3D intensity profiles of the NPCs shown in (a) were fitted by 3D Gaussians. A single 3D Gaussian was used in the case of isolated NPCs and up to six 3D Gaussians simultaneously in the case of NPC clusters. (c) Data analysis cube, fitting result, and absolute difference image. The outcome of the fitting process was assessed for each analysis cube by the visual correspondence between image data (first row) and results of the fitting process (second row) and by an image showing the absolute residuals (third row). In the first image row the 3D crosses mark the voxels that correspond most closely to the positions of the found NPC centers. The respective planes where the NPC centers were located were marked with 2D crosses. The image planes above and below the center positions were indicated by a dot. Because the center of the NPC in the upper right-hand corner of the data trace was determined beforehand, it was not simulated during the shown fitting process.

single NPCs. These coordinates represent a warped surface, namely that of the NE. This curved surface may be approximated by a sphere surface. Thus, the x, y, z coordinates of the NPCs must be converted to spherical coordinates and can then be projected to the best-fitting sphere surface. The arc length between the projected points on the spherical surface finally gives the unambigous distance between a pair of coordinates.

Accuracy of Single NPC Localization

The just introduced theoretical procedure to estimate the NPC localization accuracy results in the following. For an eightfold line average and a PMT voltage of 640 V, we found $\sigma_I^2(x_i, y_j, z_k) = 1.25I(x_i, y_j, z_k)$. The averaged amplitude I_0 of all NPCs from several image stacks was $I_0 = 72$, whereas the pixel size amounted to 40 nm. The standard deviation of the background signal intensity corresponded to $\sigma_{bg} = 6.5$. With these values the SNR corresponded to ca. 6. The theoretical localization accuracy for this SNR in radial and axial directions can be determined from Fig. 3 as <5 and <13 nm, respectively. The accuracy of determining distances between single NPCs is $\sqrt{2}$ times these values. Hence, the lateral distances between single, isolated NPCs in three dimensions may be determined with an accuracy as small as ≈10 nm. The localization accuracy is certainly worse for NPCs that were determined in the second step of the procedure because for very closely neighbored NPCs parameter correlations unavoidably occur when fitting the 3D Gaussians to the respective intensity profiles.

Representation of Localization Results

Primarily the described methods yield a list in which the x, y, and z coordinates of many NPCs are contained. This spatial distribution of the NPC can be best visualized by a 3D computer representation, which can be rotated deliberately. An example of an alternative representation, pertaining to studies of 3T3 cell[37] is shown in Fig. 5. Herein the NPCs are represented by spheres 200 nm in diameter, which is actually 10 times larger than the actual localization accuracy.

Distribution Analysis

Figure 6 (left column) displays several object patterns, which were distributed according to distinct interaction potentials. The object patterns were computed by Monte Carlo calculations together with their NNDF

---

[37] U. Kubitscheck, P. Wedekind, O. Zeidler, M. Grote, and R. Peters, *Biophys. J.* **70,** 2067 (1996).

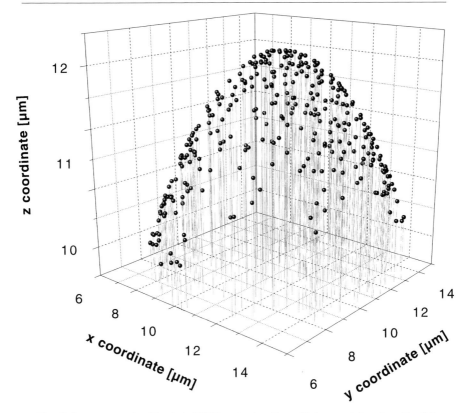

Fig. 5. A representative 3D plot of NPC center locations. Data were taken from the fitting results to the image data shown in Fig. 4. Diameters of the spherical symbols marking the NPC centers correspond to ca. 200 nm and are therefore greater than the 10-fold 3D localization accuracy.

and PCF (middle and right column, respectively) to give a first intuitive understanding of the approach. In Fig. 6E the analysis of the NPC distribution from a single cell nucleus by the NNDF and PCF estimators is illustrated. It is obvious that the qualitative appearance of the NNDF and PCF corresponds very well to those of the random hard core particle distribution (Fig. 6D), but do not agree with the distribution functions for point particles, regularly distributed hard core objects, and aggregated hard core objects as shown in Fig. 6A, 6B, and 6C, respectively. Furthermore, in Fig. 6E the estimators for randomly distributed hard core disks with a diameter of 145 nm are shown (dotted lines with standard deviations). These curves agree very well with data obtained from the NPC distribution (symbols). Extrapolation of the downward slope of the curves for the NPCs indicates the

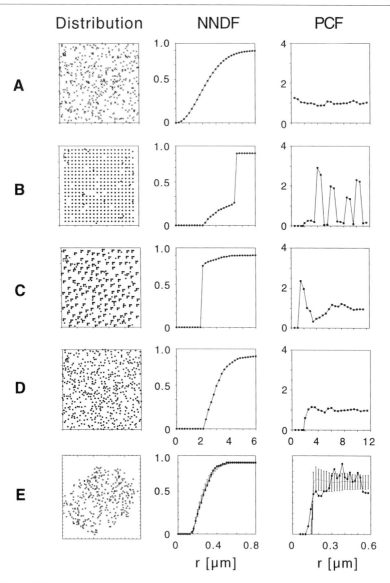

FIG. 6. Object distributions and corresponding estimators for the nearest neighbor distribution and pair correlation functions $d(r)$ and $g(r)$. (A) Random points, (B) predominantly regular hard disks, (C) clustered hard disks, (D) random hard disks, and (E) a measured set of NPCs. The diameter of the hard disks in (B–D) was defined as $d = 2$. The middle and right columns show the corresponding NNDF and PCF estimators, respectively. Functions of the random points were calculated according to Eqs. (10) and (11), and the estimators for the distributions shown in (B–E) were calculated using Eqs. (5) and (9). Each of the different

NPC diameter because it represents the lower limit for nearest neighbor distances, d_{nn}. The extrapolation for eight different cell nuclei yields a value of $d_{nn,ave} = 138 \pm 17$ nm, which is in good agreement with the results of electron microscopic and *in vivo* atomic force microscopic measurements of minimum inter-NPC distances.[38,39] Altogether, the distribution analysis shows that the NPC is distributed at random in the NE of 3T3 cells.[37] In particular, the distribution of the NPC in the NE is like that of randomly distributed, impenetrable, rotationally symmetrical objects 138 ± 17 nm in diameter.

Conclusion and Prospects

Considering the current knowledge and the hypotheses about the structural and functional properties of the NPC and its role in cell function, the possibility of visualizing the NPC by light microscopy and determining its relative position in the cell at an accuracy of a few nanometers stimulates many questions. Starting with simple questions, one can ask for the smallest distance any two NPCs can approach each other. Such information would provide new clues to the functional radius of the NPC, which may be different from the structural radius as determined by electron or atomic force microscopy and furthermore depend on cellular conditions. A more demanding type of question concerns the nature of the distribution. Is it random or does it follow an order? A higher organization of any kind would indicate that interactions among NPCs or among NPCs and other components of the NE or nucleus exist. Very important questions, for instance, concern topographic and functional relationships between specific NPCs and genes. Such interactions could be studied by labeling both the NPC and its partner. Still other types of questions concern the dynamics of the distribution. Is the NPC mobile in the NE? If it exists, is mobility based on lateral diffusion or active transport? Here an important aspect also relates to the biosynthesis of the NPC and its dispersion in the NE.

[38] C. W. Akey, *J. Cell Biol.* **109**, 955 (1989).
[39] H. Oberleithner, E. Brinckmann, A. Schwab, and G. Krohne, *Proc. Natl. Acad. Sci. U.S.A.* **91**, 9784 (1995).

object distributions shows qualitatively distinct distribution functions. (E) *xy* coordinates of NPCs of a single 3T3 cell nucleus and estimators for the NNDF and PCF. Data are shown together with the averaged estimators of randomly distributed hard core objects with a diameter of $d = 145$ nm (dashed line) with the corresponding SD. For hard core objects, NNDF and PCF estimators were calculated for 1000 individually simulated random distributions of objects with a number density corresponding to that of the NPC data set.

From a technical point of view the prospect of extending, by means of GFP technology, visualization and localization from permeabilized to living cells is most exciting. This would permit correlating the NPC distribution and its dynamics with (patho)physiological parameters, such as cell cycle phase and apoptosis. Finally, it may be noted that the described methods are not restricted to the NPC but can be applied, with little modification, to many other cellular structures.

Acknowledgment

Support from the Deutsche Forschungsgemeinschaft (Grants Pe138/14, Pe138/15, and Ku975/3-1) is gratefully acknowledged.

[14] Measurement of Tissue Thickness Using Confocal Microscopy

By James V. Jester, W. Matthew Petroll, and H. Dwight Cavanagh

Introduction

Measurement of tissue thickness, density, and depth of tissue structures provides important quantitative data that can be used to characterize physiological function, evaluate tissue organization, and establish or compare tissue responses during injury and repair. For the cornea of the eye, thickness measurements also provide uniquely valuable information about the functional status and transparency of the tissue, as well as information necessary for many surgical procedures. While various methods including light biomicroscopy and ultrasound pachymetry have been used to make these measurements, none have the spatial resolution to detect cells and other important internal structures.

Confocal microscopy, with dramatically improved lateral and axial resolution, has the capability of optically sectioning thick tissue specimens at different depths and can determine the axial displacement between different structures. Additionally, reflected light confocal microscopy also has the ability to detect cells, vessels, nerves, and other light-reflecting structures important in characterizing tissue organization.[1-3] While thickness and den-

---

[1] H. D. Cavanagh, W. M. Petroll, and J. V. Jester, *Neuro. Biobehav. Rev.* **17,** 483 (1993).

sity measurements have been made previously under *in vitro* conditions, problems associated with movement have hindered the confocal measurement of tissue thickness in the intact, living biologic organism.

Over the past decade, improvements in confocal microscope configuration, digital image acquisition, and objective lens design have made it possible to reconstruct living tissue and quantitatively measure changes in thickness and cell density in the same tissue noninvasively over time. These measurements provide a more dynamic, "four-dimensional" view of the tissue at a cellular level, which have led to important insights into cellular responses and interactions involved in ocular irritation,[4,5] contact lens wear,[6] wound healing,[7,8] and kidney function.[9] This article provides a review of the major technical details concerning the *in vivo* measurement of tissue thickness using reflected light confocal microscopy.

Reflected Light Confocal Microscope

Critical to the *in vivo* measurement of tissue thickness using confocal microscopy is the imaging of tissue in real time. This is necessary to avoid unwanted irregular movement of the tissue due to pulse and respiration and to capture fast temporal events, such as blood flow or acute cellular responses, that can blur images at high magnification. While there are several laser scanning confocal microscopes that provide "real time" imaging, these instruments have not been routinely used clinically because of very real concerns of tissue toxicity due to the intensity and wavelength of the illuminating coherent light. Therefore, the vast majority of the research evaluating *in vivo* tissue responses has used non coherent (white light) confocal microscopes. At present, two basic confocal microscope designs, a Nipkow disk and a scanning slit, are capable of collecting images at video frame rates using noncoherent illumination.

[2] J. V. Jester, P. M. Andrews, W. M. Petroll, M. A. Lemp, and H. D. Cavanagh, *J. Elect. Microsc. Tech.* **18**, 50 (1991).

[3] W. M. Petroll, J. V. Jester, and H. D. Cavanagh, *Int. Rev. Exp. Pathol.* **36**, 93 (1996).

[4] J. V. Jester, J. K. Maurer, R. D. Parker, J. Bean, W. M. Petroll, and H. D. Cavanagh, *Invest. Ophthalmol. Vis. Sci.* **39**, 2610 (1998).

[5] J. V. Jester, H.-F. Li, W. M. Petroll, R. D. Parker, H. D. Cavanagh, G. J. Carr, B. Smith, and J. K. Maurer, *Invest. Ophthalmol. Vis. Sci.* **39**, 922 (1998).

[6] M. Imayasu, W. M. Petroll, J. V. Jester, S. K. Patel, J. Ohashi, and H. D. Cavanagh, *Ophthalmology* **101**, 371 (1994).

[7] T. Moller-Pedersen, H.-F. Li, W. M. Petroll, H. D. Cavanagh, and J. V. Jester, *Invest. Ophthalmol. Vis. Sci.* **39** (1998).

[8] T. Moller-Pedersen, W. M. Petroll, H. D. Cavanagh, and J. V. Jester, *Curr. Eye Res.* **17**, 736 (1998).

[9] P. M. Andrews, W. M. Petroll, H. D. Cavanagh, and J. V. Jester, *Am. J. Anat.* **191**, 95 (1991).

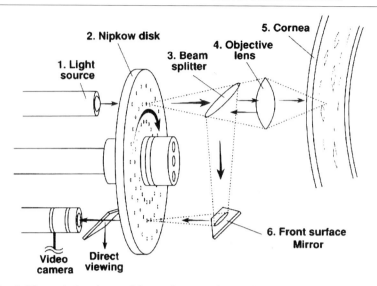

FIG. 1. The optical pathway of the tandem scanning confocal microscope. Light from the illuminating source (1) passes through pinholes on one side of a Nipkow disk (2) and on through a beam splitter (3) and objective lens (4) and then to the tissue (5). The light reflected or backscattered is diverted by the beam splitter (3) and front surface mirror (6) to the conjugate pinholes on the Nipkow disk (2). Reproduced, with permission, from H. D. Cavanagh, W. M. Petroll, H. Alizadeh, Y.-G. He, J. P. McCulley, and J. V. Jester, *Ophthalmology* **100**, 1444 (1993).

The Nipkow disk-based confocal microscope was designed originally by Petran *et al.*[10] in 1968 and used a modified Nipkow disk containing multiple, optically conjugate (source : detector) pinhole apertures arranged in Archimedian spirals. In this design (Fig. 1), light from the illuminating source passes through the pinhole apertures on the source side of the disk and is focused within the specimen by means of an objective lens. Light reflected back from the specimen is then focused back through the objective lens and is diverted to the detector side of the disk by a beam splitter and front surface mirror or prism. Light focused at the detector apertures passes through to the detector while out-of-focus light is blocked. Rapid rotation of the disk then provides even scanning of the tissue, which can be captured using a low-light level video camera or CCD. Because illumination and detection using this microscope occur in tandem, this microscope has been referred to as a tandem scanning confocal microscope (TSCM).

[10] M. Petran, M. Hadravsky, M. D. Egger, and R. Galambos, *J. Opt. Soc. Am.* **58**, 661 (1968).

There are several advantages to the TSCM for use in evaluating living tissue, first of which is the real time imaging capabilities. The white light used is also not toxic to tissues and produces minimal discomfort. Additionally, the use of diffraction-limited pinholes provides optimal lateral and axial resolution when using objectives of the same numerical aperture. Disadvantages of this system include the very low light budget obtained using Nipkow disks, which can only transmit from 0.25 to 1% of the total source light onto the specimen. While transmittance can be increased by increasing pinhole size and packing density, this leads to decreased lateral and axial resolution and the detection of out-of-focus light at the detector, which reduces the signal-to-noise ratio.

As an alternative to the tandem scanning design, Xiao and Kino[11] have developed a one-sided Nipkow disk system where illumination and detection occur simultaneously through the same pinhole. While a single light path microscope does not require the complicated alignment of conjugate pinholes, the disk surface, which reflects light back toward the detector, tends to produce higher background noise levels compared to the TSCM design. This increased background noise greatly reduces the sensitivity of the instrument to detect low contrast, poorly reflecting structures that are common to biologic tissues.

In the second type of "real time" confocal microscope, the "tandem pinholes" are replaced with two variable width slits located in conjugate optical planes.[12,13] The advantages of using a variable slit design is that the intensity of light focused within the tissue volume can be adjusted for depth of focus, thus maximizing the signal-to-noise ratio and producing higher contrast images. However, disadvantages include decreased axial resolution and potential geometric distortion of the detected image caused by tissue movement when nonreal time video imaging is employed.

The microscope that was used for the measurement of tissue thickness reported here is a modification of the original Nipkow disk-based microscope and was produced by Tandem Scanning Corporation (Reston, VA). The microscope is horizontally configured on a standard slit-lamp table, such that the microscope objective extends out from the microscope housing and a patient can sit comfortably in front of the microscope with their eye directly facing the microscope objective. The Nipkow disk is a metallic-coated glass disk and has 64,000 photo-etched pinholes, 20 or 30 μm in diameter, with 0.25 or 0.33% transmittance, respectively. Pinholes are arranged in an interlacing pattern that provides "streak-free" single video

[11] G. Q. Xiao and G. S. Kino, *Proc. SPIE* **809,** 107 (1987).
[12] C. J. Koester, *Appl. Opt.* **19,** 1749 (1980).
[13] B. R. Masters and A. A. Thaer, *Appl. Opt* **33,** 695 (1994).

images at a disk rotation speed of 900 rpm. The microscope uses a 100-W mercury arc lamp and the light path is a combination of front surface mirrors and prisms. Images are detected in real time using a 30-Hz, DAGE/ MTI VE-1000 video camera (Michigan City, IN) with adjustable gain, kV, and black level, and are recorded on Super VHS videotape.

Digital Image Acquisition

While viewing living tissue the field of view is often moving due to pulse and respiration resulting in blurring of single images and poor resolution of cellular detail. Additionally, the use of an intensified camera system may also introduce random noise and shading artifacts that further degrade the image quality. Digital image acquisition can be used to select single images that show the least blurring and highest resolution. Furthermore, digital processing techniques, such as image averaging, filtering, and background subtraction, can be used to reduce random noise, adjust shading, and improve image quality. Empirically, there is often 0.1 to 0.13 sec of videotape or three to four sequential frames that show no movement and can be used for single image acquisition and frame averaging. Frame averaging reduces noise by the square root of N, where N is the number of frames with no positional changes that are averaged.[14] Therefore, the maximum number of images needs to be selected for averaging in order to obtain the best resolution of cellular detail.

In our studies, a Silicon Graphics INDY system with a VINO video board and 192 Mbytes of system RAM (Silicon Graphics, Mountain View, CA) is used to acquire digital images. A specialized program has been developed that allows images to be digitized either directly from the DAGE/MTI camera or from the Super VHS videotape. Up to 600 consecutive images (640×480 pixels) can be digitized and stored sequentially in memory in real time (30 frames/second). Once digitized, the images are shown as a movie in a window on the computer screen as they are scrolled through interactively using a computer mouse. Images from the sequence can then be saved individually or averaged with other images in a sequence. A similar program for the personal computer has been developed in Microsoft Visual C++ (version 1.0) and used for digital image acquisition and processing.[15]

---

[14] R. C. Gonzalez and P. Wintz, "Digital Image Processing." Addison-Wesley, Reading, MA, 1987.

[15] W. M. Petroll, J. V. Jester, and H. D. Cavanagh, *Scanning* **18**, 45 (1996).

Objective Lens Design and Control

The measurement of tissue thickness under living conditions places unique constraints on system design that are not encountered when viewing cells or tissues in culture. In particular, tissue movement due to pulse and respiration leads to rhythmic changes in the axial distance between the front element of the objective lens and the tissue surface. For standard microscope objectives, these movements result in structures within the tissue appearing to come in and out of focus as the optical section moves through the tissue. While use of viscous immersion fluids may help couple the objective to the tissue and dampen the tissue movements, it is not possible to calibrate the z axis, which is necessary for measuring tissue thickness accurately. This problem has been addressed previously for ophthalmic specular microscopy using a surface contact or applanating objective that touches the tissue during evaluation, reducing axial movements.[16,17]

Based on this approach, a special 24×, 0.6 numerical aperture objective lens has been designed by Tandem Scanning Corporation that has a concave (7.5 mm radius of curvature similar to that of the cornea) applanating tip and a 1.5-mm working distance.[18] The design is similar to other ophthalmic specular microscope objectives except that the internal lenses of the objective can be moved within the objective casing in order to vary the distance between the focal plane of the objective and the applanating tip (Fig. 2). A mechanical drive within the microscope housing, which moves the internal lens elements, is controlled by an Oriel 18011 Encoder Mike Controller (Stratford, CT) with digital readout. The axial coordinate or focal depth can then be determined by relating lens position to the focal plane position in micrometers using a look-up table or solving a third order polynomial equation that is provided by the manufacturer. The axial response of this lens has been measured previously using the 20-μm, pinhole-diameter disk. The full-width at half-maximum (FWHM), an indicator of the optical slice thickness, was determined to be 9.02 and 9.27 μm at focal plane positions 400 and 800 μm from the tip of the objective, respectively (Fig. 3).[18] Using this objective with the current microscope and camera system, the field of view observed on the video monitor or digital image measures 475 × 350 μm.

[16] R. A. Laing, M. M. Sandstrom, and H. M. Leibowitz, *Arch. Ophthalmol.* **93,** 143 (1975).
[17] D. M. Maurice, *Invest. Ophthalmol. Vis. Sci.* **13,** 1033 (1974).
[18] W. M. Petroll, H. D. Cavanagh, and J. V. Jester, *J. Microsc.* **170,** 213 (1993).

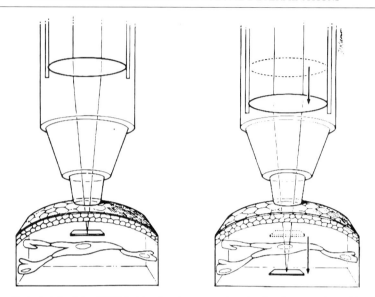

FIG. 2. A sketch of the TSCM objective demonstrating the depth or z-axis position of the focal plane and how the depth can be changed by moving the internal lens elements. Reproduced from W. M. Petroll, H. D. Cavanagh, and J. V. Jester, *J. Microsc.* **170,** 213 (1993), with permission from the Royal Microscopical Society, Oxford, UK.

Measurement of Corneal Thickness

The measurement of tissue thickness also requires a depth encoding system (DES), the details of which have been published previously (Fig. 4).[15] Briefly, the DES software, which runs on a personal computer, was developed using Microsoft QuickBasic (version 4.5) and the PDQComm supplemental communications library and operates under Microsoft DOS version 5.0 or higher. The major purpose of the DES program is to read, translate, record, and display the focal depth for easy determination of the z-axis position within the tissue being evaluated. Essential elements of the system include a serial interface between the Oriel encoder and the personal computer, which relays the positional information of the internal lens elements. The computer then calculates the focal plane position using the equation provided by the manufacturer. The depth information is then transferred through a second serial interface to a time code generator (TCG, Fast Forward Video Model F30, Irvine, CA). The depth information is then encoded on the audio track of the videotape and/or displayed as a numeric overlay on the video monitor displaying the confocal images.

Using the DES, the axial separation between two structures can be

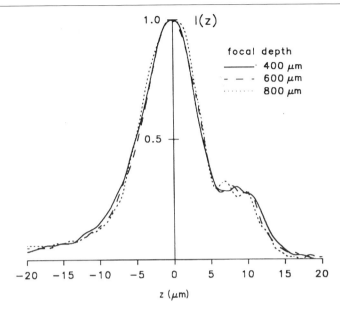

F ${}_{\text{IG}}$. 3. Experimentally measured axial response curves generated by focusing through a perfect planar reflector. Curves were generated with the reflector surface at a focal depth of 400, 600, and 800 μm. $I(z)$ is the measured intensity with the focal plane positioned z μm above (negative) or below (positive) the reflector surface. I is normalized to 1.0 at z 0 μm. Reproduced from W. M. Petroll, H. D. Cavanagh, and J. V. Jester, *J. Microsc.* **170**, 213 (1993), with permission from the Royal Microscopical Society, Oxford, UK.

calculated by determining the difference between the displayed depths. As an example, while focusing through the cornea, unique structural landmarks can be detected that define the anterior and posterior margins of the cornea. These are shown in Fig. 5 and include (A) the superficial corneal epithelium, (B) the epithelial basal lamina, (C) the nuclei of the corneal stromal fibroblasts or keratocytes, and (D) the corneal endothelium. The axial depth displayed at each of these locations can then be used to measure epithelial thickness, stromal thickness, and the entire corneal thickness, which for the eye shown in Fig. 5 measured 43.2, 328.9, and 372.1 μm, respectively. Using this approach, the accuracy of confocal microscopic measurements was compared to tissue thickness measurements using ultrasound, a standard method of measuring corneal thickness.[15] In a series of enucleated eyes from different species showing a range of corneal thickness from 300 to 700 μm, there was a significant correlation between the confocal and ultrasound measurements ($p < 0.01$, $R = 0.995$, $n = 15$), with an average percentage difference of only $0.50 \pm 2.58\%$. In a series of seven patients, the

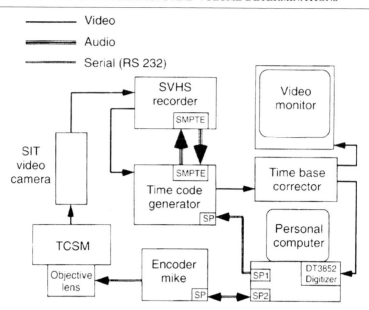

FIG. 4. Schematic of the hardware interfaces used for depth encoding and image digitization. Copyrighted and reprinted with the permission of *Scanning* and/or the Foundation for Advances of Medicine and Science (FAMS), Box 832, Mahwah, New Jersey, from W. M. Petroll, J. V. Jester, and H. D. Cavanagh, *Scanning* **18,** 45 (1996).

average percentage difference between the *in vivo* confocal and ultrasound measurements was only 1.67 ± 1.38%. Overall, these results confirm the accuracy of the axial calibration methods and the DES program for making depth measurements in living tissue.

Thickness Measurements Using Rapid Through Focusing Approach

Although accurate measurement of the tissue thickness is possible with the DES approach, there are several technical problems. First, DES requires an observer to focus on each structural landmark to record the individual depths. Because the axial resolution of the microscope is ~9.0 μm, thin planar structures such as the superficial epithelium and basal lamina may appear to be in focus over a range of depths, leading to errors in the recorded versus the actual depth of the structure. For thick tissue measurements, these errors will be small and not likely to effect the overall measurement as shown earlier. However, for thinner structures such as the corneal epithelium, a 3- to 4-μm error in the recorded focal depth of the surface epithelium and basal lamina may lead to a substantial % error in the

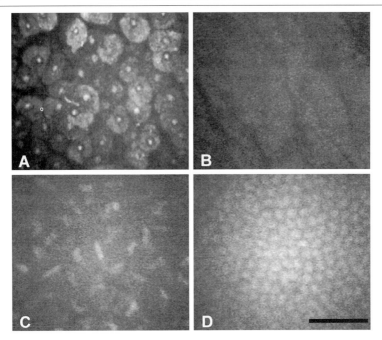

Fig. 5. TSCM images digitized from a living rabbit cornea showing the (A) superficial corneal epithelium, (B) the basal lamina, (C) the corneal stroma, and (D) the corneal endothelium. Bar: 100 μm.

thickness measurement. Second, as the distance between structures increases, the time required to focus on each structure also increases, increasing the likelihood of drift or lateral movement of the tissue, causing loss of axial calibration. Finally, considerable time is required to focus between thick structures, making repeat measurements time-consuming and difficult to perform in a clinical setting.

In order to address these problems, we have developed a confocal microscopy through focusing (CMTF) approach that provides a more accurate, rapid, and easily repeatable means of measuring tissue thickness. This approach takes advantage of the mechanical advance mechanism for the internal lens elements, which can be set by the Oriel encoder to a specific speed to facilitate fast or slow focusing through the tissue. While advancing the lens elements and simultaneously video recording the images, the axial distance between consecutive video frames can be calculated based on the lens advance speed. For the current system, a lens advance speed of 80 μm/second translates to an axial distance of ~1.06 μm between consecutive images. Using this approach, approximately 400 to 500 images or 15 sec

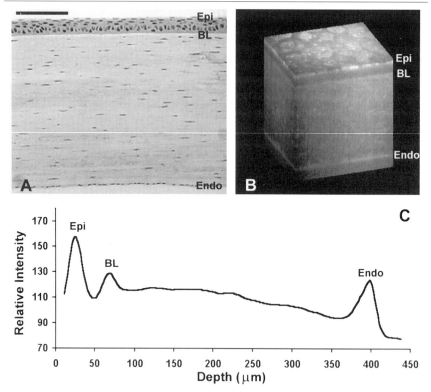

FIG. 6. (A) Histologic photomicrograph of a normal rabbit cornea showing the corneal epithelium (Epi), basal lamina (BL), and corneal endothelial layer (Endo). (B) Three-dimensional reconstruction of a living rabbit cornea taken from a through focus data set. (C) Plot of image pixel intensity versus focal plane depth of the three-dimensional reconstruction shown in (B). Bar: 100 μm.

are required to focus through and capture the entire corneal thickness. This speed can be increased to 160 μm/second without loss of axial resolution due to the interlacing of video lines from the DAGE/MTI camera, albeit at the expense of horizontal resolution. At this higher speed, recording of the entire cornea only takes approximately 8 sec, whereas the corneal epithelium can be captured in less than 1 sec, substantially reducing the risk of axial drift interfering with thickness measurements. Furthermore, multiple scans through the cornea, or just the corneal epithelium, can be performed rapidly and easily for a more accurate measurement of tissue thickness.

Digitizing the through focus scan and reconstructing the cornea three dimensionally using a volume/surface projection generated from the ANALYZE software program (Mayo Medical Ventures, Rochester, MN) produces a cross-sectional $(x–z)$ view of the living cornea that appears similar to that of histologic sections (Figs. 6A and 6B). Furthermore, the relative thickness of the corneal epithelial sheet and stroma as defined by the respective separation between the superficial epithelium (Epi) and basal lamina (BL), and the basal lamina and endothelium (Endo), appears equivalent to that observed histologically. While thickness measurements from the surface projection are possible, an axial, light-scattering profile can be calculated by measuring the average pixel intensity of each image comprising the through focus scan and plotting as a function of axial depth (Fig. 6C). This representation identifies the major tissue structures that backscatter or reflect light as peaks in the depth–intensity profile. Interestingly, tissue interfaces between cells and matrix (epithelium and stroma)

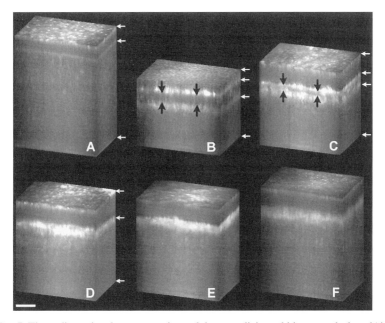

FIG. 7. Three-dimensional reconstructions of the same living rabbit cornea before (A) and 1 (B), 2 (C), 3 (D), 7 (E), and 17 (F) weeks after excimer laser photoablation of the central cornea. In the preoperative cornea (A), three reflective layers were detected corresponding to the superficial epithelium, the epithelial basal lamina, and the endothelium (white arrows). One (B) and 2 (C) weeks after injury showed two new reflective layers corresponding anteriorly to the photoablated stromal surface and posteriorly to the migrating stromal fibroblasts (black arrows). Note the dark band between the photoablated surface and migrating cells, which corresponded to the acellular zone in the anterior stroma. Bar: 100 μm in the x axis and 70 μm in the z axis. Reproduced from T. Moller-Pedersen, H.-F. Li, W. M. Petroll, H. D. Cavanagh, and J. V. Jester, *Invest. Ophthalmol. Vis. Sci.* **39,** 487 (1998), with permission from the Association for Research in Vision and Ophthalmology.

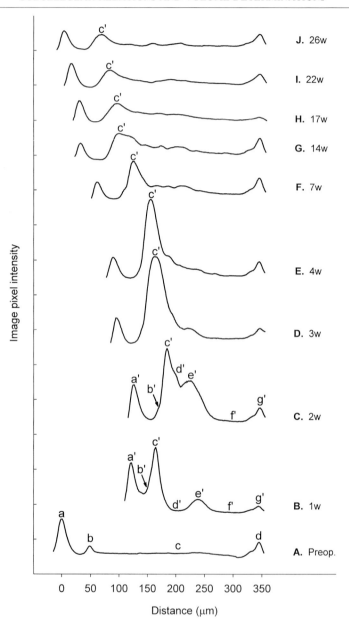

or tissues and fluid (tears or anterior chamber aqueous), where there are changes in the refractive index, appear to be the major sources of backscatter.

Using the two-dimensional depth–intensity profile, tissue thickness can be measured objectively, without operator bias, by measuring the distance between the intensity peaks. For the cornea, the corneal thickness (CT), epithelial thickness (ET), and stromal thickness (ST) can be calculated using the following equations:

$$CT = Z_{ENDO} - Z_{EPI} \qquad (1)$$
$$ET = Z_{BL} - Z_{EPI} \qquad (2)$$
$$ST = Z_{ENDO} - Z_{BL} \qquad (3)$$

where Z_{EPI}, Z_{BL}, and Z_{ENDO} are the axial depths of the intensity peaks for the superficial epithelium, basal lamina, and endothelium, respectively. For the rabbit, these thicknesses have been determined to be 381.6 ± 27.3 μm for the corneal thickness and 47.7 ± 2.2 μm for the corneal epithelial thickness in a series of 10 rabbits, which is in agreement with the known thickness of these structures using other techniques.[19] For the corneal epithelium, the standard deviation of repeat scans ranged from 1.3 to 3.1 μm with a coefficient of variation of 2.5%, suggesting that this method is reproducible and accurate. For human patients, similar accuracy and repeatability have been reported and intracorneal structures as thin as 16 μm (Bowman's membrane) have been measured reproducibly.

In addition to measuring tissue thickness, CMTF analysis also provides a simple and quantitative means of studying tissue responses in four dimensions. This has been demonstrated in studies of corneal wound healing following excimer laser keratorefractive surgery, where a portion of the

[19] H.-F. Li, W. M. Petroll, T. Moller-Pedersen, J. K. Maurer, H. D. Cavanagh, and J. V. Jester, *Curr. Eye Res.* **16**, 214 (1997).

FIG. 8. CMTF profiles of the same rabbit cornea before (A) and at 1 (B), 2 (C), 3 (D), 4 (E), 7 (F), 14 (G), 17 (H), 22 (I), and 26 (J) weeks after excimer laser photoablation. The preoperative scan shows three CMTF peaks corresponding to the superficial epithelium (a), the epithelial basal lamina (b), and the corneal endothelium (d). At 1 week (B), four peaks corresponding to the superficial epithelium (a'), the photoablated stromal surface (c'), the layer of migrating fibroblasts (e'), and the endothelium (g') were detected. No peaks were detected in the acellular zone (d') or the posterior stroma (f'). Postoperative scans are arbitrarily offset in the y axis and all curves are aligned with the corneal endothelial peak to illustrate changes in the corneal thickness. Reproduced from T. Moller-Pedersen, H.-F. Li, W. M. Petroll, H. D. Cavanagh, and J. V. Jester, *Invest. Ophthalmol. Vis. Sci.* **39**, 487 (1998), with permission from the Association for Research in Vision and Ophthalmology.

anterior cornea is photoablated, changing the central corneal curvature.[7] Using CMTF image data sets, the same cornea can be reconstructed three dimensionally at various times before and after surgery (Fig. 7). By comparing the images from one time point to another, the immediate loss of corneal tissue can be recognized and the effects of laser ablation and later tissue responses can be detected. These three-dimensional images can then be represented as a pixel intensity–depth profile (Fig. 8) and measurements of tissue loss and regrowth determined, as well as other corneal responses quantified.

In particular, as shown in three-dimensional surface projections, there is a substantial decrease in overall corneal thickness 1 week after surgery compared to the same cornea before surgery (compare Figs. 7A and 7B). This change in thickness can be quantitated by measuring the change in peak locations of the depth–intensity profile (Figs. 8A and 8B). Using this approach, the actual amount of tissue photoablated per pulse of laser light was measured and showed considerable variation between individual eyes, ranging from 112 to 130 μm of tissue for the same 9.0 diopter refractive correction. In addition, evaluation of the tissue responses showed two brightly reflecting layers (Fig. 7B; black arrows); an anterior layer immediately below the epithelium and a posterior layer some distance into the stroma. Evaluation of the individual images comprising the three-dimensional stack indicated that the anterior layer represented the residual photoablated damage to the stromal extracellular matrix whereas the more posterior layer represented migrating stromal fibroblasts. Between these two regions was an acellular zone where stromal cells had apparently been killed following the injury.

Dynamic changes in cell migration and tissue growth are also suggested by the changes in the three-dimensional appearance of the cornea at different times (Fig. 7). Initially, the posterior layer noted in Fig. 7B appears more anterior at 2 weeks (Fig. 7C) and merges with the anterior layer by 3 weeks (Fig. 7D) after surgery. This is also shown in the depth–intensity profile (Fig. 8, lines B and C and D), suggesting migration of the stromal cells into the wound region. A rate of cell migration can then be calculated based on the movement of peak e' (Fig. 8). The cornea also appears to increase in thickness after surgery as suggested by in the three-dimensional reconstruction. A rate of growth of the stroma can then be calculated by measuring the distance between the basal lamina peak (Fig. 8, b') and the endothelial peak (Fig. 8, g'). Actual measurements showed a range in the rate of growth from 21 ± 9 μm/week at 3 weeks to 6 ± 5 μm at 7 weeks after surgery.

Finally, CMTF profiles can also be used to estimate the amount of backscattering of light from the tissue. If the camera gain, kV, and black level are set to constant levels, then the amount of light detected by the camera is principally related to changes in the reflectivity or backscattering of light by the tissue. Light backscattering can then be estimated by integrating the area under each peak, or the entire curve, using the equation

$$\text{Area} = \Sigma\{(I_i + I_{i+1})/2 - I_{\text{AC}}\}\Sigma z_i$$
$$\Sigma z_i = z_i - z_{i-1} \tag{6}$$

where I_i is the intensity value of image i; I_{AC} is the background intensity of the anterior chamber; and z_i and z_{i-1} are the focal plane positions of image i and $i - 1$. The unit of measurement (U) is defined as μm $\times$ pixel intensity. Using this method, a close correlation between the loss of corneal transparency, as measured by the clinical examination of the cornea, and the CMTF estimate of light backscattering has been shown.[20] Furthermore, integration of the CMTF intensity peak has also been used to evaluate ocular irritation and shown to provide an objective, quantitative measure of the corneal response to injury, including inflammation, fibrosis, and repair.[4] Although the pixel intensity values do not by themselves indicate any specific response, values can be correlated directly to actual images taken during scanning to identify unique pathologic processes underlying the elevated intensities. Integration of intensity values over a region of the cornea, i.e., stroma, then allows objective, quantitative measure of the area and depth of response occurring within the same tissue.

Acknowledgments

This work was supported, in part, by grants from the National Eye Institute (EY07348), Bethesda, Maryland, the Procter & Gamble Company, Cincinnati, Ohio, and Research to Prevent Blindness, Inc., New York, New York.

[20] T. Moller-Pedersen, M. Vogel, H.-F. Li, W. M. Petroll, H. D. Cavanagh, and J. V. Jester, *Ophthalmology* **104,** 360 (1997).

[15] Measurements of Vascular Remodeling by Confocal Microscopy

By SILVIA M. ARRIBAS, CRAIG J. DALY, and IAN C. MCGRATH

Vascular Remodeling

Over a short time scale the morphology of the vascular system is viewed as being very stable. However, the vasculature is a dynamic structure capable of sensing changes within its milieù and changing its structure in response to long-term modifications in hemodynamic or humoral conditions. These changes in vessel morphology have been termed by some authors as "vascular remodeling." However, this term is still defined poorly and has different meanings to different authors. Therefore, in this methodological article, we must clarify our definition of vascular remodeling.

Blood pressure and flow, along with humoral factors, are constantly changing in response to demands imposed by different activities and environmental stimuli, and there is a continuous vascular adaptation to these changes from development to aging. Vascular remodeling can be considered a physiological process in this sense. Angiogenesis (elaboration of vessels from preexisting ones), vasculogenesis (*de novo* formation of vessels) in the embryo, and adaptations to flow are good examples of vascular remodeling during development. Mature arteries also exhibit substantial remodeling in response to flow changes, not only in diameter but also in major wall constituents.[1] Other examples of physiological vascular remodeling are the changes that occur during pregnancy[2] or the vascular adaptations to exercise or disuse.[3,4]

Pathological disorders can also chronically alter vascular structure as an adaptation to the demands imposed by the disease state. These adaptive morphological responses may also affect the progression of the disease. Vascular remodeling can therefore contribute to the pathophysiology of circulatory disorders. A clear example is the association between hypertension and structural changes in the resistance vasculature, which contribute to the increase in peripheral resistance and to the maintenance of high blood pressure.

[1] B. L. Langille, *J. Cardiovasc. Pharmacol.* **21**(Suppl. 1), S11 (1993).
[2] M. V. Hart, M. J. Morton, J. D. Hosenpud, and J. Metcalfe, *Am. J. Obstet. Gynecol.* **154,** 887 (1986).
[3] H. B. Chew and S. S. Segal, *FASEB J.* **4,** A722 (1990).
[4] D. H. Wang and R. L. Prewitt, *Am. J. Physiol,* **260,** H1966 (1991).

The study of vascular remodeling in hypertension has gained a great deal of attention in recent years and is still a subject of much debate.[5-7] First, there is no consensus on the definition of vascular remodeling. Part of the discussion comes from the work of Baumbach and Heistad,[8] who introduced the term vascular remodeling to describe a structural alteration in cerebral arterioles from stroke-prone spontaneously hypertensive rats. These authors referred to a reduction in the outer and inner diameter without medial hypertrophy of the vessel wall. Thereafter, "vascular remodeling" and "vascular hypertrophy" were regarded as different vascular structural alterations. However, it is now clear that this classification of remodeling/hypertrophy is not completely accurate. The mechanisms underlying the restructuring of the vessel wall during remodeling are still not known; they might also include cell proliferation and growth as well as cell death and migration. This has created more difficulties in finding an accurate nomenclature and fueled the debate on vascular remodeling.[6,7]

As a general methodological article, we consider it more appropriate to use a broad definition of vascular remodeling that covers all aspects of vascular morphological changes from physiology to pathology. In this article, vascular remodeling is used to describe any change in vascular morphology from one situation to another. This definition could include modifications of the vascular structure of a given vessel over time, which are physiological, pathological, or due to intervention (i.e., after treatment).

In our view, vascular remodeling is related not only to gross structural differences such as an increase or reduction in lumen and wall dimensions, but also modifications at the cellular level, such as cell number, distribution, phenotype, and orientation. These questions have been addressed using two approaches: on the one hand, classical histological methods that allow for detailed cellular studies but require tissue processing and on the other hand, physiological techniques for live blood vessels that give information only on gross structural changes. The technique described in this article combines both approaches and permits the study of vascular remodeling in intact blood vessels with the detail of histological methods.

Methods Used to Study Vascular Remodeling

Advances in the study of blood vessel structure have been paralleled with the development of microscopy and imaging techniques. From the

[5] M. J. Mulvany, G. L. Baumbach, C. Aalkjaer, and A. M. Heagerty, *Hypertension* **28,** 505 (1996).
[6] R. M. K. W. Lee, M. A. Adams, P. Friberg, and P. Hamet, *J. Hypertens.* **15,** 333 (1996).
[7] P. I. Korner and J. A. Angus, *Hypertension* **29,** 1065 (1997).
[8] G. L. Baumbach and D. D. Heistad, *Hypertension* **13,** 968 (1989).

early work of Anitschkow on vascular structure in the last part of the 19th century, who identified different cell types in rabbit vessels with bright-field microscopy, a substantial number of techniques have been developed to assess vascular structure.

The study of vascular remodeling can be performed noninvasively. Laser Doppler imaging of blood flow, capillaroscopy, and fluorescein angiography provide information on capillaries and small vessel structure in the eye and skin. Echo-tracking techniques, B-mode ultrasound, nuclear magnetic resonance imaging, and intravascular ultrasound can be used to assess vascular remodeling in large arteries. Noninvasive techniques are very useful for the study of human blood vessel structure (for review, see Schiffrin and Hayoz[9]), but still do not provide us with a detailed description of the actual changes to the structure.

The detailed analysis of vascular remodeling needs *in vitro* methods. A wide range of histological techniques have been used to analyze the vascular structure of muscular and elastic arteries under fixed conditions. The process of vascular remodeling can also be studied under nonfixed conditions. Physiological approaches offer the advantage of providing information on gross structure in live vessels and allow the correlation of changes in vessel structure and function. The importance of resistance vasculature in the development and/or progression of vascular diseases, such as hypertension, has stimulated the development of techniques to determine structure and function in small vessels. Two methods have been widely used: wire and perfusion myography.

Wire Myography

Wire myography was introduced by Bevan and Osher[10] in the early 1970s to study functional responses in resistance arteries. This technique involves threading small arteries of approximately 2 mm on two 40-μm wires. The two wires are held in position between two heads: one is attached to a micrometer and the other to a force transducer. Once the vessel is set between the wires, it is normalized in order to determine the resting tension–internal circumference relationship. The normalization procedure was first proposed in 1977 by Mulvany and Halpern.[11] This protocol calculates the internal circumference that the vessels would have when relaxed and under a transmural pressure of 100 mm Hg (L100). The vessel is subsequently stretched to an arbitrary degree of tension that has been

[9] E. L. Schiffrin and D. Hayoz, *J. Hypertens.* **15,** 571 (1997).
[10] J. A. Bevan and J. V. Osher, *Agents Actions* **2,** 257 (1972).
[11] M. J. Mulvany and W. Halpern, *Circ. Res.* **41,** 19 (1977).

shown to result in the optimal magnitude of active tension development in response to agonist stimulation; most investigators use 0.9_{L100}.

The gross dimensions of an artery can be determined by placing the wire myograph onto the stage of a light microscope fitted with a $\times 40$ water-immersion objective and a filar eyepiece. Morphological measurements are determined (before normalization) on the vessel held under just enough tension to hold the walls separated. Under these conditions the media width and lumen dimensions are measured at different sites along the vessel and averaged. The vessel is then normalized and set to the optimum tension and wall thickness, internal diameter, and cross-sectional area (CSA) are calculated for this tension. Finally, function can also be assessed at the same level of tension.

Wire myography enables the measurement of gross morphology and functional responses. Among the disadvantages of this method are the artificial conditions used: (i) the geometry of the artery is distorted: the vessel wall is not distended radially with the pressure distributed evenly, as occurs *in vivo*; (ii) part of the endothelial surface is under the pressure exerted by the wires and possibly damaged: and (iii) it is not possible to distinguish clearly the adventitia/media and the media/intima boundaries so wall measurements might be inaccurate.

Pressure Myography

A more physiological approach is pressure myography. Pressure myography was developed as a tool for studying the function of small artery preparations from normotensive and hypertensive rats.[12] This technique involves securing a small vessel between two fine glass cannula with strands of silk. One cannula is closed (or it can be opened for flow studies) and the other one is connected to a system containing physiological salt solution (PSS), which in turn is linked to a pressure Servo unit. This system allows the pressure (or flow) in the system to be precisely controlled and altered at will. Once the vessel is set between the cannulas, the myograph is placed onto the stage of a microscope fitted with a video camera. The signal from the camera is shown on a monitor to visualize the vessel throughout the experiment and it is also fed into a video dimension analyzer that allows continuous measurements of the lumen and wall thickness dimensions. Function can also be assessed based on changes in lumen diameter upon agonist stimulation.

This method, like wire myography, allows the correlation of functional responses with vessel morphology. Pressure myography has some advan-

---

[12] W. Halpern, M. J. Mulvany, and D. M. Warshaw, *J. Physiol.* **275,** 85 (1978).

tages over wire myography: (i) vessel shape is maintained; (ii) endothelium is intact; and (iii) wall thickness and lumen dimensions can be studied at a wide range of pressures and therefore it is possible to assess the mechanical characteristics of the vessel. The main disadvantage is the lack of accurate measurements of the different wall layers: adventitia, media, and intima, as the adventitia/media and intima/media boundaries are not clear enough.

Histology

Histology allows for the study of vascular structure in great detail (even at the subcellular level), which is not possible to achieve in live vessels with wire or perfusion myography.

Histology requires a great amount of vessel processing. First, the vessel has to be fixed, embedded in paraffin wax or resin, and sometimes dehydrated. Then it has to be cut into thin sections and stained. Each section is then visualized with a microscope and photographed. Finally the images taken are analyzed for quantification. Morphometric and histometric methods have been used for many years to quantify vascular morphology. They utilize either planimetric or point and intersect counting methods to measure the vessel wall components. In the morphometric method,[13] a point-counting system is usually used to determine the cross-sectional areas of the lumen, intima, media, and adventitia in a relaxed vessel. Depending on the size of the vessels and magnification used, appropriate transparent morphometric test grids are used. Histometric methods[14] are used for vessels that are either collapsed or contracted on sampling and reconstructs the hypothetical relaxed state of the arteries. It is based on two assumptions: that the length of the internal elastic lamina (IEL) and that the CSA of the vessels remain unchanged during contraction. A number of researchers have demonstrated that CSA of a vessel is in fact a reliable constant of the vessel wall under most physiological conditions, but IEL length is changed during contraction in some types of arteries. The consequence is that lumen size will be underestimated in some cases by this method (for review, see Lee[15]). The dissector technique, a method developed more recently, is based on a similar principle. Successive specimen sections are placed under two specially equipped microscopes, projected side by side onto a table top, and the number of nuclei counted with the aid of a grid.[16]

[13] E. R. Weibel and R. P. Bolender, *in* "Hayat: Principles and Techniques of Electron Microscopy: Biological Applications," p 237. Van Nostrand Reinhold, New York, 1973.
[14] T. A. Cook and P. O. Yates, *J. Pathol.* **108,** 119 (1972).
[15] R. M. K. W. Lee, *Scan. Microsc.* **1,** 128 (1987).
[16] M. J. Mulvany, U. Baandrup, and H. J. G. Gundersen, *Circ. Res.* **57,** 794 (1985).

The main disadvantages of classical histological methods are (i) the need of sample processing: fixation, embedding, and cutting, which implies, apart from the loss of tissue viability, some degree of distortion, and (ii) the methods of analysis, which are very time-consuming and not always bias free.

Cell Culture Studies

In recent years the study of vascular remodeling has also been taken to the molecular level and biochemical and molecular techniques are now being used to address the problem of the structural changes in the vasculature. Most of these studies have been performed in cells in culture, mainly smooth muscle cells (SMC) and also endothelial cells. These studies provide information on the possible biochemical pathways involved in the process of vascular remodeling and also enable the determination of cellular function. Cell culture is also valuable in determining the mediators controlling cell death, growth, migration, and proliferation, which are crucial in understanding the process of vascular remodeling.[17] The main disadvantage of cell culture studies is the lack of the natural interaction between different types of cells and extracellular matrix components that exist in intact blood vessels.

Measurements of Vascular Remodeling by Confocal Microscopy

All methods used to assess vascular remodeling have their pitfalls. In general, physiological methods such as wire or perfusion myography allow the measurement of vascular structure in intact live vessels and can provide information on vascular mechanics, but not on cellular or subcellular vascular components. However, methods such as histology and cell culture permit a detailed analysis of vascular components but are based either on nonviable tissue or on isolated cells.

For a number of years we have been interested in the development of a method that allows for the simultaneous study of vascular structure and function, a technique that could give us information on what cellular events take place during vessel vasomotion. Using conventional fluorescent microscopy, we developed novel applications for the study of blood vessels to identify cell type, location, and viability within living myograph-mounted arteries.[18] However, with conventional fluorescence microscopy, the image is distorted by out-of-focus glare from structures above and below the plane

[17] G. H. Gibbons and V. J. Dzau, *N. Engl. J. Med.* **330,** 1431 (1994).
[18] C. J. Daly, J. F. Gordon, and J. C. McGrath, *J. Vasc. Res.* **29,** 41 (1992).

of focus and limits imaging within thick tissues, such as blood vessels. An alternative approach is to use confocal microscopy. Confocal microscopes produce optical sections throughout translucent tissue, commonly to a maximum depth of 200–400 μm, without the need for cutting thin slices. They eliminate blur and flare from out-of focus planes in an object and axial resolution is improved greatly.

Our recent objective has been to develop methods based on laser scanning confocal microscopy (LSCM), which allow the study of vascular remodeling and function in intact small blood vessels at the cellular or subcellular level. This technique lies in an area between traditional work for the study of vascular structure and function at the supracellular level and methods for single or cultured cells. This entails modifying conventional means of maintaining the preparations and techniques and developing new protocols first on cell cultures before application to vessels.

This article describes some of the uses of LSCM to study vascular remodeling in intact small blood vessels. We first describe some of the fluorophores we have used, their advantages and pitfalls, and examples of the images of blood vessels that can be obtained with a LSCM system. Next, we explain in detail a method we have developed to analyze small vessel structure based on the combination of LSCM and pressure myography. We have used this method to assess vascular remodeling in fixed tissues from various pathological animal models. We are now extending the use of this technique to study vascular remodeling in live myograph-mounted vessels. This aspect will also be discussed.

Fluorophores Used for Study of Vascular Remodeling

Specimens to be imaged with fluorescent LSCM must be labeled with a fluorescent probe. Several factors must be considered when selecting a fluorophore. The major considerations are the excitation/emission wavelengths of the probe to be used, the laser lines available, and the filters used. Many laser sources are available for LSCM, ranging from UV wavelengths to the infrared. Some of the important considerations for selecting a laser for LSCM are cost, wavelength of emission, output power at each wavelength, efficiency, and stability. When performing multilabeling experiments, it is important to choose fluorophores whose excitation/emission wavelengths overlap minimally.

Fluorescent dyes are available for a wide range of biological preparations, permitting the use of LSCM to image from cellular or extracellular vascular components to subcellular structures, including receptors or enzymes. The list of fluorescent molecules useful for the study of vascular remodeling is growing continuously. This section describes the use of some of these dyes, focused mainly on the use of nuclear stains to study vascular structure.

Nuclear Dyes. Nuclear dyes have several advantages over other fluorophores in the study of vascular remodeling: (i) they produce a bright stain of the nuclei of a wide range of vascular beds, (ii) are easy to use, (iii) allow identification of different cell types located in the vessel wall by their nuclear shape and orientation, and (iv) produce simple models for image analysis.

Nuclear dyes allow us to distinguish the vascular tunics: adventitia, media, and intima. The tunica adventitia is the outermost layer, consisting of fibroelastic tissue, fibroblasts, macrophages, vasa vasorum, and nerves.[19] On staining with a nuclear dye, adventitial cells show irregular-shaped nuclei (apparently) located randomly (Fig. 1a). The tunica media is made up of SMC, which can be arranged circularly, longitudinally, or running diagonally, depending on vessel type, a varied number of elastic laminae, and a network of collagenous and elastic fibrils.[19] SMC stained with a nuclear dye show spindle-shape nuclei, oriented circularly in the majority of muscular arteries (Fig. 1b). The intima is the innermost layer of all blood vessels. It is composed of a single layer of endothelial cells lining the vascular wall, a thin basal lamina, and, in some blood vessels, a subendothelial layer.[19] Endothelial cells have oval-shaped nuclei oriented in the direction of flow (Fig. 1c).

The list of nuclear dyes is vast and it is beyond the scope of this article to describe all of them. We will only describe those which, in our experience, have proved to be more useful for the study of vascular structure.

Hoechst 33342 (excitation 345 nm, emission 460 nm) is a lipophilic dye that crosses the plasma membrane and intercalates with DNA. Hoechst 33342 is very stable; arteries stained with this dye and fixed with 10% formaldehyde–saline solution remain stained in good conditions for prolonged periods of time, as long as 1 year. Hoechst 33342 is a vital dye and can be used *in vivo.* We have demonstrated previously that a bolus injection of this dye in rats (10 mg/kg) had no effect on blood pressure or heart rate. Injected in three boluses of 10 mg/kg, the dye is able to stain all vascular cells (except those of the cerebral vasculature, probably due to the blood–brain barrier). The staining is time and concentration dependent.[18] Hoechst 33342 is visualized with a UV line. This allows for multilabeling experiments combining Hoechst with dyes fluorescent in the fluorescein or rhodamine bands to locate simultaneously cell nuclei and other cellular or

[19] J. A. G. Rhodin, *in* "Handbook of Physiology: The Cardiovascular System: Vascular Smooth Muscle" (D. F. Bohr, A. P. Somlyo, H. V. Sparks, Jr., and S. R. Geiger eds.), p 1. American Physiological Society, Bethesda, MD, 1980.

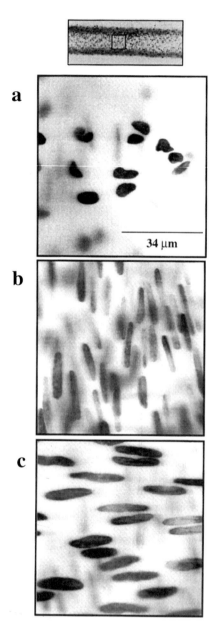

FIG. 1. Extended focus model of nuclei of cells located in a region of the adventitia (a), media (b), and intima (c) of a rabbit cerebral artery. The vessel was fixed under pressure on a perfusion myograph, stained with the nuclear dye propidium iodide, and visualized mounted on a slide with the 488/515-nm line of a laser scanning confocal microscope using a ×10 air objective (top image) and a ×40 water-immersion objective. MetaMorph software was used to acquire 1-μm-thick serial optical sections and for three-dimensional reconstructions.

extracellular vascular components. Alternative nuclear dyes for live tissue are dihydroethidium (excitation 510 nm, emission 595 nm) or the more recently developed Syto series (fluorescent in the fluorescein band). Dihydroethidium is a lipophilic dye that is cleaved in the cytoplasm by an intracellular dehydrogenase into ethidium bromide, a nonpermeable dye that is retained inside the cell.

Propidium iodide (PI) is a hydrophilic dye that intercalates with DNA of cells having a permeabilized membrane. PI can be detected with both fluorescein or rhodamine bands as it stains nuclei of all vascular cells very brightly. The best results are obtained at low concentrations and several washouts, or prolonged washout periods. We have observed that concentrations as low as 1 nM are able to stain most of the nuclei in small arteries. The main problem with PI is the instability of the dye; after 1–2 days the dye starts leaking out of the nuclei into the cytoplasm. We have observed this phenomena in several vascular beds and have used this property to stain the cell cytoplasm. An example of this is shown in Fig. 2, a simulated fluorescence projection reconstruction from a rat cerebral arteriole stained with PI. In small vessels, as in the one shown in Fig. 2, the cytoplasm of all vascular cells can be observed. However, in thicker arteries, such as rat tail or carotid arteries, only the cytoplasm of SMC, but not of adventitia or endothelial cells, is stained with PI. Alternative hydrophilic dyes are ethidium bromide (excitation 510 nm, emission 595 nm) and ethidium homodimer.

Fading and bleaching of a labeled specimen are major problems in LSCM. High power and a focused beam of LSCM cause enhanced fading of a specimen as compared to a conventional epifluorescence microscope, which essentially bathes the entire specimen in low-power, wide-beam excitatory light. In some cases the use of an antifade reagent is needed. Both Hoechst 33342 and PI are quite resistant to bleaching by the laser, whereas other nuclear dyes such as Syto bleach relatively quickly under laser excitation; again this is variable and depends on the type of blood vessel.

Another factor in data acquisition using LSCM is the signal/noise ratio in the optical sections of the specimen. In this respect the adjustment of the detector pinhole is of critical importance. If the ratio signal/noise is too low, the diameter of the pinhole can be increased, thereby allowing increased detection of the signal at the expense of resolution. However, if the pinhole is too large, the confocality is compromised. Nuclear dyes such as Hoechst 33342 and PI give a very good signal/noise ratio and can be used with a 10- to 15-μm slit, allowing for good confocality.

What are the applications of nuclear dyes in the study of vascular remodeling? First, nuclear dyes can be used to screen cell viability. Nuclear dyes have been used for many years by cell biologists and can also be

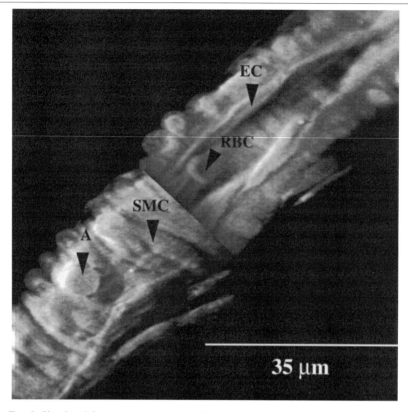

Fɪɢ. 2. Simulated fluorescence projection (sfp) reconstruction of a rat cerebral arteriole. The vessel was stained with the nuclear dye propidium iodide, which was released into the cytoplasm. The vessel was visualized mounted on a slide with the 488/515-nm line of a laser scanning confocal microscope using a ×60 oil-immersion objective, and 1-μm-thick serial optical sections were then captured with MetaMorph image analysis software. A Silicon Graphics workstation was used for sfp three-dimensional reconstruction and AdobePhotoshop for image montage showing the inner and the outer part of the vessel. A, Adventitial cell; SMC, smooth muscle cell; RBC, red blood cell; and EC, endothelial cell.

applied to vascular tissues to determine the rate of dead/live cells. We have used a combination of permeable/nonpermeable dyes fluorescent at different wavelengths (Hoechst/PI, for example) to identify dead cells in blood vessels based on dye exclusion.[18] In intact nonfixed vessels, all cells will be stained with Hoechst 33342, whereas only dead cells with a permeabilized membrane will capture PI. Nuclear dyes also allow identification of apoptotic nuclei based on nuclei shape. This has been used mainly to measure the rate of apoptosis in cultured cells.[20] In combination with LSCM, which allows a large amount of vascular tissue to be studied in detail, it is possible to identify this phenomena in intact arteries.

Second, nuclear dyes can also be used to assess gross morphological parameters such as lumen size, wall, and layer thickness and to determine cell number in different layers of the vessel wall and cell distribution. An important advantage of nuclear dyes is that they show the cells as individual objects in the image. This facilitates image thresholding with image analysis software, also allowing for the semiautomated counting of different cell types and for the quantification of several nuclei characteristics, such as nuclei shape, dimensions, and orientation. The detailed protocol is described in a later section.

Finally, the use of vital nuclear dyes allows for the tracking of nuclei positions during vasomotion and can also be used to assess vascular function.[18,21]

Cytoplasmic Dyes. Nuclear dyes do not offer a complete description of cell dimensions. However, a complete description is possible with fluorescent probes that stain the cytoplasm. An example is BCECF-AM [2',7'-bis(carboxyethyl)-5-(and -6-)carboxyfluorescein, acetoxymethyl ester; excitation 439–505 nm/emission 530–540 nm]. This lipophilic dye is an AM-ester. Once in the cytoplasm the dye is cleaved by an intracellular esterase and the resulting compound BCECF is therefore retained inside the cells. This dye is pH sensitive and has been used for pH studies in vascular tissues and appears to have no toxic effects.[22] It has been shown in live vascular tissues that 5 μM BCECF-AM produces good staining of the SMC, some staining in the adventitia, but no staining in the endothelium.[23] The best results are obtained with two incubation periods of 30 min in a gassed physiological solution at 37° and the samples should also be protected from light. An example of the stain obtained with BCECF-AM in SMC from mesenteric and cerebral arteries is shown in Fig. 3.

In theory, cytoplasmic dyes provide an excellent two-dimensional illustration of cell shape, dimensions, and arrangement. However, images obtained with cytoplasmic dyes are much more complex to analyze than those obtained with nuclear dyes. The cells are not always separated clearly from each other, as occurs with the nuclear dyes, even when visualized with high-

[20] M. J. Pollman, T. Yamada, M. Horiuchi, and G. H. Gibbons, *Circ. Res.* **79,** 748 (1996).
[21] M. McAuley, S. M. Arribas, and J. C. McGrath, *Neurogen. Neuron. Dev. Plast.* (abs. 2-31) (1995).
[22] J. G. Matthews, J. E. Graves, and L. Poston, *J. Vasc. Res.* **29,** 330 (1992).
[23] S. M. Arribas, C. J. Daly, J. F. Gordon, and J. C. McGrath, *in* "The Resistance Arteries: Integration of Regulatory Pathways" (W. Halpern, J. A. Bevan, J. Braiden, H. Dustan, M. Nelson, and G. Osol, eds.), p. 259. Humana Press, New Jersey, 1994.

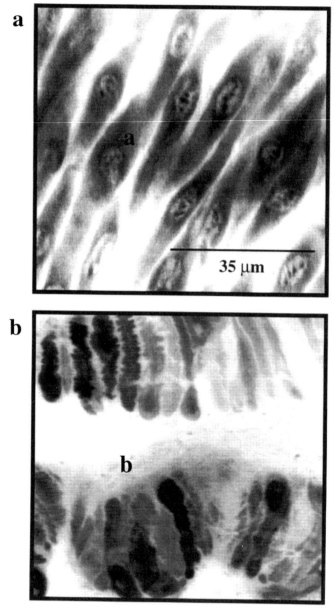

FIG. 3. Images of smooth muscle cells stained with the cytoplasmic dye BCECF-AM. The vessel was visualized mounted on a slide with the 488/515-nm line of a laser scanning confocal microscope using a ×60 oil-immersion objective. Images were captured with MetaMorph image analysis software. (a) Rat mesenteric resistance vessel and (b) two rat cerebral arterioles.

power LSCM. This causes difficulties in thresholding and semiautomated measuring with conventional image analysis packages.

Extracellular Dyes. Extracellular stains also provide, as do cytoplasmic dyes, definition of the cell outlines and can be used to determine cell shape, dimensions, and orientation. Fluorescein isothiocyanate bovine albumin (FITC–albumin; excitation 528 nm, emission 553 nm) at a concentration of 1 mg/ml can be used to stain connective tissue. However, this dye does not provide a good resolution of cell shape.[23] This can be achieved with 5(6)-carboxyfluorescein (5,6-CF; excitation 488 nm, emission 515 nm). This dye provides very good resolution of the cell outlines in cerebral arteries and arterioles (Fig. 4). We have used this characteristic of 5,6-CF to analyze SMC orientation and to detect areas of cellular disorganization in intact arteries from hypertensive rats.[24] The vessel is incubated overnight at 4° in PSS containing 1 mg/ml 5,6-CF. The arteries are then mounted on a slide and viewed in the presence of the dye with a LSCM. Optical sections corresponding to various SMC layers located at different depths within the media are captured and image analysis software is used to measure the angle of orientation of the SMC with respect to the longitudinal axis of the vessel. Vessel orientation is determined in the images by the orientation of the IEL. The folds of the IEL lie parallel to the longitudinal axis of the vessel and can be detected with the same filters as 5,6-CF due to the autofluorescent properties of elastin, the main component.

5,6-CF does not bind to any particular vascular structure; it surrounds the cells and fills space between them. This provides the advantage of being easier to use than BCECF-AM. However, it also limits its use to slide-mounted unfixed tissue and makes it impractical to use on myograph-mounted vessels; the dye has to be present while the tissue is being visualized, thereby producing an unacceptable level of fluorescence in the column of fluid between the tissues and the lens.

Measurement of Vascular Remodeling in Pressurized Vessels

The combination of fluorescent dyes and LSCM provides a useful tool in studying vascular morphology in great detail and clearly facilitates histological studies. With confocal microscopy there is no need for embedding, dehydrating, and sectioning the specimen; even fixation can be avoided. Image analysis software also provide a means for taking semiautomated measurements and reconstruction of the images. However, LSCM is not just an improved histological method to determine vascular structure; it can also be used to address the problem of vascular remodeling from a more

---

[24] S. M. Arribas, J. F. Gordon, C. J. Daly, A. F. Dominiczak, and J. C. McGrath, *Stroke* **27**, 1118 (1996).

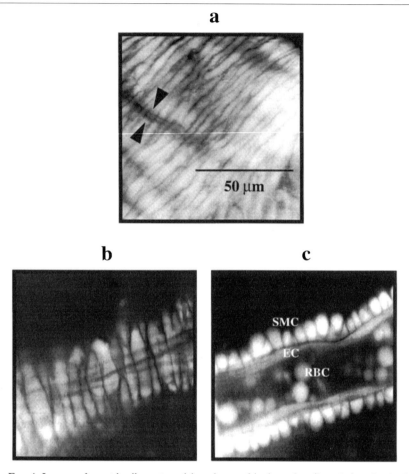

FIG. 4. Images of a rat basilar artery (a) and one of its branches (b and c) stained with the extracellular dye 5(6)-carboxyfluorescein. The vessel was visualized mounted on a slide in the presence of the dye with the 488/515-nm line of a laser scanning confocal microscope using a ×40 oil-immersion objective. Images were captured with MetaMorph image analysis software. The dye is shown in black in all the images.(a) Smooth muscle cells (SMC) in the media of the main artery; arrowheads show one of the grooves of the internal elastic lamina (IEL). (b) SMC and IEL grooves in a branch. (c) Image of the inner part of the branch showing red blood cells (RBC), endothelial cells (EC), and the profile of SMC.

physiological point of view. This can be achieved by combining existing myograph techniques with LSCM. The next section describes the method that has been developed to analyze vascular structure in intact vessels based on the combination of pressure myography and LSCM. Pressure myography was chosen because it provides a more physiological method in which the endothelium remains intact, vessel shape is maintained, and vessel structure can be studied at a wide range of pressures.

Pressurization and Fixation Protocol. The first part of the protocol involves pressurization of the vessels as described previously for pressure myography, staining, and pressure fixation. It is also possible to maintain an intact nonfixed preparation and to visualize the live vessel mounted on the pressure myograph. We will describe in detail the protocol for pressure-fixed-vessels. The method involving live vessels, under development at present, will be discussed later.

Several questions need to be addressed in order to obtain the most accurate measurements of vessel morphology. (1) What is the most appropriate pressure at which vessels should be set? (2) What are the physiological conditions for the vessels? (3) What fixative should be used?

What pressure to set vessels at is an important issue, especially for comparative studies of blood vessels, which are distended at different vascular pressures *in vivo* (i.e., vessels from normotensive and hypertensive animals or humans). *In vivo* operating pressures are not known for most vascular beds. Arteries of 125–175 μm in diameter have been estimated to operate at close to half the systolic pressure of the animal.[25] We have used this approach previously to pressurize rat cerebral vessels in the previous mentioned range of lumen diameters to study vascular remodeling in hypertension.[26] A different approach was used by Dickhout and Lee[27] to study small mesenteric resistance arteries from normotensive and hypertensive rats. To address the problem of which pressure to mount vessels at, the length of the arteries was measured and marked *in situ* before sampling in anesthetized rats. After dissection the artery was set on the myograph and pressurized to a pressure that restored the *in vivo* length of the artery. The authors found that the required pressure was 70 mm Hg.[27] In our opinion, the best approach would be to obtain measurements from a live myograph-mounted vessel at different distending pressures. This can be achieved with the classical method (video-dimension analyzer) or, in more detail, with the LSCM.

The physiological state of the blood vessel has also to be taken into account as the degree of vessel contraction will influence measurements of lumen and wall thickness. Most authors take these measurements under conditions where the vessels are maximally relaxed. This can be achieved in several ways. We have used 3×10^{-5} *M* sodium nitroprusside to achieve

[25] W. Halpern and M. Kelly, *Blood Vessels* **28**, 245 (1991).
[26] S. M. Arribas, C. González, D. Graham, A. F. Dominiczak, and J. C. McGrath, *J. Hypertens.* **15**, 1685 (1997).
[27] J. G. Dickhout and R. M. K. W. Lee, *Hypertension* **29**, 781 (1997).

a maximal relaxation. Other alternatives are Ca^{2+} free medium in the presence of EGTA (5 mM), 10^{-5} M isoproterenol, or 10^{-5} M papaverine.

What fixative should be used? The ideal fixative should penetrate tissues quickly, act rapidly, and preserve the cellular structure before the cell can react to produce structural artifacts. Unfortunately, there is no ideal fixative and the use of a particular agent is determined by the experimental requirements. Most researchers who are involved in quantitative morphological studies recognize that fixation has important effects on tissues; some procedures cause swelling and others shrinkage. Glutaraldehyde (2.5–5%, v/v) in buffer is used commonly as a fixative for vascular morphological studies. However, it gives high levels of autofluorescence and therefore is not ideal for fluorescent LSCM. A combination of 3.5% (v/v) formaldehyde, 0.75% (v/v) glutaraldehyde in 0.05 M phosphate buffer has been shown to preserve the artery structure with little nonspecific autofluorescence and has been used for LSCM studies of vascular structure.[27] Formaldehyde, another commonly used cross-linking fixative, is a good alternative, as it produces less autofluorescence than glutaraldehyde. We have demonstrated that 10% formaldehyde in saline solution does not induce significant changes in the adventitia, media, or intima CSA in small rat mesenteric arteries. In order to minimize vascular changes that could create artifacts, it is important to note that the fixative should be added at 37°. If it is added at low temperature, the artery tends to contract.[28] For small size arteries mounted on a pressure myograph, 30 min of fixation with 10% formaldehyde–saline solution at 37° is enough to preserve arterial structure. Lower concentrations of formaldehyde (4%) in saline buffer, preferred for immunohistochemical studies, can also be used.

If a functional study is not required, an alternative to pressure myograph fixation is *in vivo* fixation. Some authors are of the opinion that in order to preserve the *in vivo* dimensions of blood vessels, it is desirable to fix the vessels at physiological pressures. However, high perfusion pressures can cause perivascular edema, and other investigators using low pressures (approximately 16–24 mm Hg) and low flow rates (1 ml/min/100 g) have shown that the luminal diameter of the arteries is not affected. Clearance of the blood with an oxygenated physiological solution containing a vasodilator is advised before perfusion of the fixative to avoid contraction (for review, see Lee[15]).

Staining Protocol. We have used the nuclear dyes Hoechst 33342 (for both live or fixed tissue; 0.01 mg/ml) or PI (only for fixed tissue; 10^{-6} M)

---

[28] S. M. Arribas, C. Hillier, C. González, S. McGrory, A. F. Dominiczak, and J. C. McGrath, *Hypertension* **30**, 1455 (1997).

with very good results. As mentioned previously, Hoechst 33342 is more stable than PI. Both dyes act very quickly and a 15-min incubation period is enough to stain all vascular cells *in vitro*. Adventitial nuclei always appear more stained than SMC, probably due to differences in the DNA content. The stain of endothelial cell nuclei is always weaker that the rest of the artery. To enhance the stain of the endothelium, the dye can be perfused intraluminally for 5 min at low flow (0.1 ml/min) to avoid endothelial damage. We have observed that endothelial cells are only stained if the lumen is expanded (i.e., after pressurization). If the vessel is collapsed, the dye is kept between the folds of the IEL and nuclei of the endothelial cells are not visible. If PI is used, the arteries should be washed several times in PSS after staining to avoid background fluorescence. Once stained, the vessels are prepared for imaging on the confocal microscope: they are placed on a glass slide and covered with a coverslip attached to the slide by a thick layer of vacuum grease. This forms a small chamber and ensures that the coverslip does not press against the artery, therefore minimizing any changes that might occur in the wall of the vessel if the coverslip lays directly on top of the artery.

LSCM and Image Capturing. We have used an upright or an inverted Odyssey LSCM (Middleton, WI, Noran Instruments) coupled to a Nikon Optiphot or Diaphot microscope, respectively (Nikon, Kingston on Thames, Surrey, UK). To visualize the lumen, a ×10 air objective (Nikon, NA 0.5) can be used and for images of the wall, a ×40 oil-immersion objective (Nikon, NA 1.3) or a ×40 water-immersion objective (Nikon 1.15) can be used. For detection of Hoechst 33342-stained vessels, the 364-nm line (400-nm barrier filter) of the LSCM was used and for PI the 515-nm line (550-nm LP barrier filter) gives the best detection, although PI can also be detected with the NA 488/515-nm line. Both dyes give a very good signal, and a pinhole aperture of 10 or 15 μm is sufficient.

Images are captured with MetaMorph software (Universal Imaging Corporation, West Chester, PA). Lumen images are taken in the X–Y plane by placing the focus in the middle of the vessel. This allows the visualization of the lumen and cell nuclei in profile (Fig. 5a). A group of 5–10 images of the lumen (depending on vessel length) of different regions are captured along the artery and stored for further analysis.

We have demonstrated previously that X–Y images of the wall taken with a ×40 objective clearly showed adventitial, medial, and intimal layers. However, the boundaries between the tunics were not clear enough to ensure accurate measurements of the adventitia, media, and intima.[28] Wall morphology can then be determined more accurately in the Z axis from stacks of serial optical sections. Z-axis resolution, given by the microscope

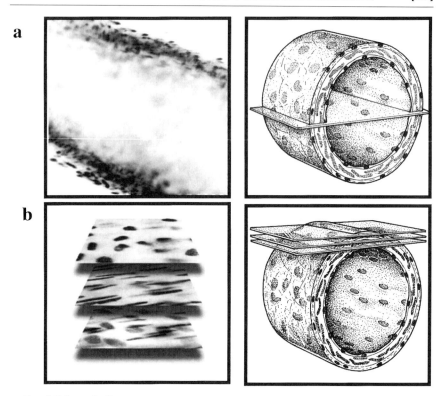

FIG. 5. Schematic diagrams and images of the lumen diameter and cellular wall components obtained with a laser scanning confocal microscope (LSCM). Vessels were pressurized on a perfusion myograph, stained with nuclear dye Hoechst 33342, and visualized with the 346-nm/open filter line of the LSCM. MetaMorph software was used for image acquisition and measurements. (a) Lumen dimensions were measured in the X–Y plane from images obtained by focusing on the middle of the vessel ($\times 10$ air-immersion objective, image dimensions $300 \times 300 \ \mu$m). (b) Wall morphology was determined in the Z axis from stacks of serial images (1 μm thick) taken from the top of the adventitia to the bottom of the intima ($\times 40$ oil-immersion objective). (Left) Representative examples of adventitial (top image), smooth muscle (middle image), and endothelial cell nuclei (bottom image) as observed with a LSCM (inverted contrast, image dimensions $95 \times 95 \ \mu$m). From S. M. Arribas *et al., J. Hypertens.* **15,** 1685 (1997).

manufacturer, was 0.5 μm for $\times 40$ objectives described here. Each stack of images is composed of 1-μm-thick serial optical slices taken from the first adventitial to the last endothelial cell nuclei.

To ensure accurate wall thickness measurements, all stacks of images must include regions with clearly stained cells in all the layers. From each vessel it is recommended to capture two or three of these stacks from different, randomly chosen, regions. This is particularly important in avoiding inaccurate measurements of adventitial cell number; the adventitia is irregular in cellular composition and different regions from the same vessel may have very different number of cells. In arteries from hypertensive rats, the distribution of endothelial cells is not homogeneous along the vessel length. Both mesenteric and cerebral vessels from different models of hypertension show areas of the intima almost depleted of cells, whereas there are other regions in the same segment with normal cell number.[26,28] Therefore, to ensure accurate measurements of endothelial cell number and nuclei characteristics, it is best to capture an additional group of images of the endothelial layer covering the whole length of the vessel.

Compared to conventional histological methods, this LSCM-based protocol reduces the time of processing greatly. On average, image capturing at 64–128 frames, which ensures good quality images, takes approximately 10 min per stack of 100 images. The whole process of image capturing, including lumen images, three stacks from different regions of the vessel wall, and a stack of images from the endothelial layer, would take less than 1 hr for a 100-μm-thick artery and less time for a thinner vessel.

Image Quantification. Image quantification is still a limiting factor for the assessment of vascular structure with the aid of confocal microscopy. Currently available image analysis software packages do not allow for automated quantification of the images and are not developed specifically to analyze blood vessels. For a number of years we have used MetaMorph image analysis software to quantify some vascular parameters useful in describing vessel morphology. In addition to the classical parameters measured with wire or pressure myography (lumen size and wall thickness), MetaMorph software also allows the quantification of adventitia, media, and intima thickness, the number of cells in each layer, cell orientation, and some useful nuclei characteristics (shape, length, breadth, area, perimeter) of the different cells in the wall of a blood vessel.

Lumen measurements are obtained in the image with the aid of a toolbar. They are acquired in pixels and then calculated in micrometers based on the calibration factor of the objective used. Wall thickness is obtained from the number of planes in the stack, i.e., the number of optical sections captured in the Z axis, provided the thickness of each section is known. In addition, it is also possible to measure the thickness of the adventitia, media, and intima, as nuclei of cells located in each layer differ significantly in shape and orientation. This method of measurement has to be reliable and reproducible. From experience we know that the beginning and the end of a stack of images can differ between different observers; even the same observer may take a different number of sections from the same region at different times. We therefore need a standardized method of image capturing and measurement. We have used a method based on computer measurements of fluorescence intensity and differences in nuclei shape. With MetaMorph software it is possible to determine its average (or total) intensity and its shape factor automatically for each object in the image, given as a value in the range 0–1 representing how closely the object resembles a circle. We have established a code based on the following: first nuclei appearing on the stack of images at its maximum intensity locate the first adventitial cell and plane 1 of the adventitial layer (and of the vessel wall). Of course we are considering the situation of an intact vessel where the adventitia is the outermost layer. First nuclei with the shape of a SMC at its maximum intensity locate the first plane of the media and the same is applied for the endothelium. This classification is possible as the shape of the different cells in the vascular wall can be defined clearly. As an example, in rat cerebral vessels, adventitial cell nuclei have an irregular shape, usually round (shape factor = 0.76 ± 0.01), SMC have an elongated shape (shape factor = 0.35 ± 0.01), and endothelial cells have an oval-shaped nuclei (shape factor = 0.49 ± 0.02). With this protocol, adventitia, media, and intima thickness can then be delimited accurately and measured by different observers. We have demonstrated that this is a bias-free method, and statistical analysis showed no significant difference between observers.[28] Based on the same principle of differences in shape between nuclei, it has also been possible to count the cells located in each layer of the vessel wall, to determine nuclei characteristics (orientation, length, breadth, perimeter, area), and to detect and quantify possible abnormalities. We have developed a method that allows for semiautomated quantification of these parameters. The method is described in the next section.

From the measurements of lumen and adventitia, media, and intima thickness and cell number in each layer, several parameters useful in characterizing vessel morphology (CSA, volume, luminal surface area of the vessel and cell density in each layer) can be calculated. For simplicity, these calculations can be performed on the basis of 1-mm segment lengths as

$$\text{Wall CSA (mm}^2) = \text{external CSA} - \text{internal CSA}$$
$$\text{External CSA} = \pi[\text{lumen}/2 + \text{wall thickness}]^2$$
$$\text{Internal CSA} = \pi[\text{lumen}/2]^2$$
$$\text{Wall volume (mm}^3) = \text{CSA} \times 1 \text{ mm}$$

Adventitia, media, and intima CSA and volumes can be calculated in a similar way by substituting the respective layer CSA for wall CSA:

$$\text{Luminal surface area (mm}^2) = [2\pi \times \text{lumen}/2] \times 1 \text{ mm}$$

The total number of SMC or adventitial cells in 1 mm length of vessel can be calculated as

$$\text{Cell number} = \text{No. nuclei counted per stack} \times$$
$$\text{No. of stacks per vessel volume}$$
$$\text{No. nuclei counted per stack} = \text{average of two to three stacks}$$
$$\text{from different regions}$$
$$\text{No. of stacks per vessel volume} = \text{vessel volume/stack volume}$$
$$\text{Stack volume} = \text{image area} \times \text{stack thickness}$$

Endothelial cell (EC) number can be calculated in two ways:

EC/mm^2
Total $EC = [EC/mm^2] \times$ luminal surface area

Semiautomated Quantification of 3D Objects and Volumes. Although reliable, the method of quantification described earlier requires a fairly large amount of time and user intervention. An ideal situation would be to have a totally automated method of measurement where it could be possible to pass a complete data volume (i.e., image stack) to a software routine that could identify discrete objects, count and classify them, and then report the findings to the user. However, any computerized measurement system requires a set of classifiers that first tells the software within which parameters to operate. For a nuclei-stained volume this could consist of a set of arguments that define the limits of individual nuclear dimension (e.g., minimum and maximum length and width). In this respect, the measurement algorithm will still be semiautomated in that it requires the user to set the classifiers at the beginning and then perhaps alter the classification depending on the output (results). If a given tissue and staining protocol yielded uniform staining each and every time then it may be possible to fully automate the process. However, tissue variability makes it almost impossible to predict on any occasion what the classifier values should be, particularly where intensity of staining is a factor. For this reason, and because of the problems described later, we have yet to find an algorithm that can fully automate this process. Furthermore, even our own semiautomated method is not 100% accurate.

Thresholding and segmentation routines are at the heart of many image-based measurements. To avoid confusion and for the purposes of this article we will define "thresholding" as being the process of selecting intensity ranges; "segmentation" is defined as the process of extracting an object from a volume. Nuclei-stained blood vessels present a particular challenge to the thresholding and segmentation routines. Essentially, the vascular wall can be treated as a 3D volume containing several objects (in our case nuclei) that have different sizes, shapes, orientations, and intensities (see

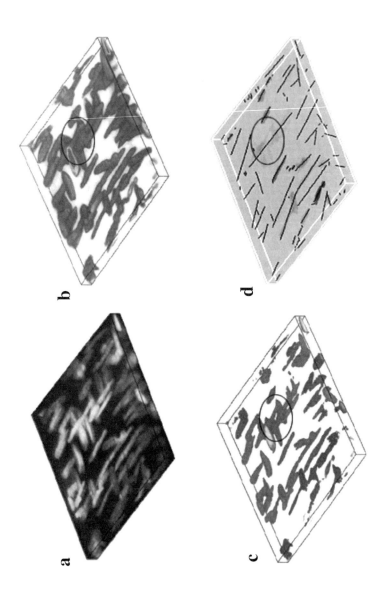

Fig. 6). The challenge is to segment each nucleus accurately from the volume with the minimum input from the user. Several segmentation methods have been suggested. In biomedical image processing, Ong et al.[29] gives a review of four categories for the segmentation of tissue section images: thresholding, region growing, edge detection, and pattern matching. Most methods deal only with 2D images and although some can be extended to 3D, it is far more complicated. Specific routines for confocal-derived data are particularly difficult to find. The need for such routines has prompted us to develop our own methods designed to handle volumetric data from vascular segments.

Problems Associated with LSCM. LSCM data sets of thick biological specimens present a unique set of problems for segmentation. For example, (1) objects (i.e., nuclei) can be as little as a few micrometers apart and may appear to fuse together in the segmented volume. (2) Neighboring objects can have different intensities and individual objects can themselves have a wide range of intensity values. (3) Intensity will be lower deep within the specimen due to diffraction of light. (4) Even in LSCM data sets, some objects will be out of focus. The problems just described are general and will be relevant to most biological applications employing thick "live" tissue. We shall now consider each problem with respect to the confocal study of blood vessels.

OBJECT FUSION. With the exception of cells undergoing mitosis, there will be no cases where two nuclei occupy the same cell and therefore be touching. In general, cell nuclei must be separated by at least the thickness of two cell membranes. In this extreme case it will probably not be possible

---

[29] S. H. Ong, X. C. Jin, Jayasooriah, and R. Sinniah, *Comput. Biol. Med.* **26,** 269 (1996).

FIG. 6. Three-dimensional reconstructions of raw and processed data from a segment of myograph-mounted (pressurized) rabbit cutaneous resistance artery showing identical views of the same data set after different rendering and analysis processes. (a) Extended focus view of nuclei within the wall of the vessel. Note that some nuclei are brighter than others (object intensity heterogeneity). (b) The volume has been rendered using a back-to-front (BTF) algorithm, which allows for the control of opacity of individual voxels. In this case the background (i.e., darkest) voxels have been made transparent. This method is effective for volume examination and identification of potential problems. One particular group of nuclei was found to be touching (object fusion) and is circled in the figure. (c) The data volume was then processed using the IMTS routine (without object classifiers) which segmented the objects. Some individual nuclei were segmented clearly whereas others were not. Some degree of object fracturing and fusion was observed. (d) The segmented volume was then passed for automated analysis. Only the longest cord of each object is shown. These cords are 3D vectors, which describe the length and orientation of the object. It is clear that the software has failed to identify certain "individual" objects (circled).

to resolve the distance between nuclei, particularly in the optical (Z) axis. Fortunately, the architecture of the vessel wall is such that cells are generally offset and are not stacked along the optical axis. Therefore, it is a reasonable assumption that where two nuclei appear to fuse it is probably the result of some optical aberration that may be correctable. If the objects cannot be separated, then the user must ensure that classifier parameters are stringent enough to eliminate objects whose volume is greater than the average by a factor of two (i.e., is it a double nuclei?).

OBJECT INTENSITY HETEROGENEITY. This creates the greatest problem for efficient thresholding. Vital nuclear stains report the concentration of DNA. Therefore, if a nucleus has regions of high or low DNA concentration (i.e., an apoptotic nuclei), the resulting image of that nucleus will have a range of intensities. Selecting a fixed range of intensities for thresholding will undoubtedly result in fracturing of the object where only the brightest (or dimmest) regions will be segmented. One solution would be to broaden the thresholding range. However, this will increase the incidence of object fusion. In addition, two nuclei occupying the same optical plane can have completely different intensity ranges. This means that efficient segmentation of one nuclei in a plane does not guarantee segmentation of all other nuclei in that plane.

INTENSITY ATTENUATION WITH DEPTH. This is a more general problem that has a significant effect on our ability to image the complete vascular wall. As the focal plane is increased along the Z-axis, the intensity of the fluorescent signal drops. This is mainly due to the diffraction of light and the efficiency of penetration of the laser. Practically, it means that nuclei deep within the wall (near the lumen) are more difficult to resolve cleanly. Moreover, it makes visualization of the endothelial cells particularly difficult. One solution is to simply increase the PMT gain or laser power slightly with each increase in Z-axis position during stack acquisition. This solution is only valid if intensity data per se is not meaningful or required.

DIFFRACTION OF LIGHT. Even in the best confocal system there will be some out-of-focus glare from fluorescent structures above and below the plane of focus. In vascular samples this is not a major problem because the relative size of the objects is greater than the actual diffraction from the point light source. Much has been written about the use of deconvolution methods to correct for this.[30,31] We have tested several methods of deconvolution on vascular samples and have found the *Iterative Constrained*

[30] H. T. M. Van der Voort and K. C. Strasters, *J. Microsc.* **178,** 165 (1995).
[31] P. J. Shaw, *in* "Handbook of Biological Confocal Microscopy" (J. Pawley, ed.), p 373. Plenum Press, New York/London, 1995.

Tikhonov Miller and *Maximum Likelihood* methods to produce the best results for vascular tissues.

Multilevel Thresholding and Segmentation (IMTS). To overcome the problems described earlier we have developed a specialized semiautomated algorithm for thresholding and segmenting objects from confocal volumes of vascular structure. The algorithm was developed in collaboration with Dr. Daisheng Luo at the University of Glasgow. An evaluation copy of the software is currently available on our web site (www.cardiovascular.org).

The iterative multilevel thresholding and splitting (IMTS) method segments 3D images into volumes iteratively by increasing the threshold value and splitting larger volumes into smaller volumes. The volumes are extracted by the slice merging method. The object segmentation is controlled by intensity homogeneity and volume size criterion.

Unlike the automatic multilevel thresholding,[32] where threshold values are found from the global histogram and different objects are segmented in different threshold bands, IMTS segments different objects at different threshold values that match the object themselves. It is beyond the scope of this article to describe the details of the method but we have found it to be particularly effective for segmenting nuclei in vascular segments. Operation of IMTS is simple. The routines have been built into an existing 3D-rendering package (Microvision; Fairfield Imaging). The software runs on Silicon Graphics hardware and has a friendly windows-like interface. Once the volume has been loaded the user selects the thresholding range and the incremental step size. The software then segments the volume interactively and displays the segmented volume in the render window. The user can then examine the volume visually and can view the object data text file. If the results are not satisfactory, the thresholding step size and range are altered and the process is repeated.

Object Classification and Measurement. As stated earlier, and shown in the figures, cell type can be identified by nuclei position, shape, and orientation. Therefore, segmented objects can be used to report the number, position, and orientation of the different cell types. Once a data volume has been analyzed by the IMTS segmenter, object data are output to a simple text file. Objects can then be ordered by size, shape factor orientation, and as on. If only smooth muscle cells are to be examined then classifiers that are unique to these cells can be defined (i.e., ratio of length to width, shape factor, orientation with respect to a fixed axis).

This section described briefly the problems associated with the automated analysis of 3D data volumes derived from biological confocal microscopy. Until now our analysis has relied on the fully interactive model where

[32] F. J. Chang, J. C. Yen, and S. Chang, *IEEE Trans. Image Process.* **4**, 370 (1995).

the user identifies each nucleus by hand (using a mouse). We have now developed a method that can accurately segment individual objects in 3D. However, the method will need further development to enable it to recognize touching objects and automatically separate them.

Measurements in Live Myograph-Mounted Vessels. We have described the protocol used to characterize small vessel morphology in pressure-fixed vessels. This method, based on LSCM, represents important advantages over conventional histological techniques: it reduces vessel processing and allows for semiquantitative analysis. Disadvantages of the method are (i) the use of fixatives and therefore a certain degree of tissue manipulation and (ii) the assumption that we are working at the right physiological pressure. The ideal situation would be to develop a similar technique in live vessels, which would constitute an important step forward: the possibility of analyzing vessel structure at the physiological level. In principle this is possible. We have shown that live myograph-mounted arteries stained with nuclear or cytoplasmic dyes can be visualized using a conventional fluorescent microscope or LSCM.[18,21,23] However, this is still an area under development that requires first improvement of some methodological aspects.

First, vital dyes have to be used. We have demonstrated previously that the nuclear dye Hoechst 33342 does not affect function and is useful in monitoring cell location and also cell movement during vasomotion.[18,21] The use of Hoescht 33342 is restricted to LSCM with a UV line, not so common as LSCMs with primary laser lines above 400 nm. Syto series, detected in the fluorescein band, are a possible alternative to study structure and function in live blood vessels *in vitro* (provided they have no toxic effects on vascular cells).

The second requirement in assessing vascular structure and function in myograph-mounted vessels is the need for adequate water-immersion objectives. They should fulfill two conditions: long working distances to avoid touching the preparation and high numerical apertures for good imaging.

Finally, an important methodological consideration is the need of special chambers for vessel pressurization with the following characteristics: (i) be able to fit the objective lenses between the cannulae, (ii) ensure a perfect alignment of the cannulas, and (iii) have an inner heating system to maintain physiological conditions.

These methodological aspects are being improved at present. We are confident that this system will be fully developed in the near future and will allow a more precise assessment of vascular structure and function.

Summary and Conclusions

The introduction of myographs was crucial for the study of structure and function of resistance arteries. The ability to support and maintain small blood vessels paved the way for true microscopic studies of the vascular cells. However, even after decades of study, we still do not know very much about the "normal" arrangement of smooth muscle cells in the vascular wall and how their distribution affects function. It was clearly time for the next technological step forward. We have shown here how the combination of myography and confocal microscopy creates a platform for the study of vascular structure at the cellular level and in 3D. In addition, the possibility of using live myograph-mounted vessels in combination with LSCM opens a new field of research to assess vascular remodeling from a physiological point of view and to study vascular function at a level not achieved by any other method at present.

Now that the hardware is in place it is time to concentrate on the software and improve the methods of analysis. We have used 2D analysis of 3D data sets to describe differences in vascular structure and, at the same time, developed methods to semiautomate the process. The success of the 3D methods will ultimately depend on the reliability and accuracy of the analysis routines. There are still problems to overcome en route to finding a complete solution. However, we believe that the search for a robust fully (or semi-) automated method of 3D analysis will be more than worthwhile.

We have defined vascular remodeling to include any changes in cellular arrangement or morphology. However, on a more subtle level, changes in receptors, enzymes, and proteins leading to altered functionality could also be regarded as remodeling. In that respect it may be interesting to map the distribution of the many receptors, channels, and proteins that regulate vascular growth, death, and function. Currently, there is a growing list of fluorescent ligands and antibodies that can be used in conjunction with confocal microscopy. It is possible that multiple stains could be used and imaged at different wavelengths with a view to constructing full 3D models of various structures and their colocalization.

It is our belief that the confocal approach will prove to be a major tool in unraveling the complexities of cell–cell interactions and arrangements and will allow a better understanding of the process of vascular remodeling and function.

[16] Cell Volume Measurements by Fluorescence Confocal Microscopy: Theoretical and Practical Aspects

By LOTHAR A. BLATTER

I. Introduction: Significance of Cell Volume Measurements

Many questions in cell biology remain unanswered because of the lack of detailed knowledge about the three-dimensional organization of complex cellular structures that maintain cellular functions. These questions relate to how intracellular organelles are organized in space to efficiently perform their tasks such as nuclear transcription, protein synthesis, and release and storage of ions and second messengers; how intracellular as well as surface membrane receptors are organized to perform cell signaling tasks; how ion channels are localized in membranes to provide sources of ion fluxes; and what are the exact structures through which cells interact with each other. Furthermore, information is still sparse about how cells maintain or adapt their shape and three-dimensional (3D) morphology. Many aspects of cell volume regulation itself have remained a mystery. Virtually every cell type has developed well-regulated mechanisms to maintain a constant volume or to adjust its volume as it is required by cellular functions or changes in the intracellular or extracellular environments.[1] Maintenance and regulation of cell volume is a vital function of all cells. Any net changes in transmembrane solute fluxes that result from ion fluxes through membrane channels or changes in the Donnan forces arising from changes in concentrations of intracellular impermeant charged macromolecules affect the cellular osmotic equilibrium. As a result, water moves across the membrane and ultimately changes cellular volume.

Over the past decade the widespread use of confocal microscopy and the development of many useful fluorescent probes has led to significant progress in volume investigation and 3D studies of cell morphology and function.[2] In particular, new insights have been gained from methods that involve complex image processing techniques that extract three-dimensional information on structure and function of cells and cellular organ-

---

[1] W. E. Crowe, J. Altamirano, L. Huerto, and F. J. Alvarez-Leefmans, *Neuroscience* **69**, 283 (1995).

[2] J. K. Stevens, *in* "Three Dimensional Confocal Microscopy: Volume Investigation of Biological Systems" (J. K. Stevens, L. R. Mills, and J. E. Trogadis, eds.), p. 3. Academic Press, San Diego, 1994.

elles.[3-7] Laser scanning confocal microscopy (LSCM) has allowed the detection of fluorescence signals with high spatial and temporal resolution from uniformly thin optical sections at variable cell depths.[8] Optical cell sections obtained with LSCM have been used for three-dimensional reconstructions of subcellular structures[2] and for the quantitative analysis of cell volume of various cell types. In particular, neurons,[9-14] with their complex structures of cell body, axon, dendrites, and dendritic spines,[15] which form a small anatomical compartment that is critical for synaptic transmission, have been subjected to 3D volume studies and reconstructions. Quantitative cell volume studies using LSCM have not been performed exclusively on neuronal cells. Other tissues studied include epidermal Langerhans cells,[16] renal collecting duct cells,[17] rat lung Clara cells,[18] chondrocytes,[19,20] pancreatic islets of Langerhans,[21] and cardiac myocytes.[22-24] Three-dimensional

[3] G. J. Brakenhoff, H. T. M. van der Voort, E. A. van Spronsen, and N. Nanninga, *Scann. Microsc.* **2**, 33 (1988).
[4] G. J. Brakenhoff, H. T. M. van der Voort, M. W. Baarslag, B. Mans, J. L. Oud, R. Zwart, and R. van Driel, *Scann. Microsc.* **2**, 1831 (1988).
[5] H. T. M. van der Voort, G. J. Brakenhoff, and M. W. Baarslag, *J. Microsc.* **153**, 123 (1989).
[6] D. König, S. Carvajal-Gonzalez, A. M. Downs, J. Vassy, and J. P. Rigaut, *J. Microsc.* **161**, 405 (1991).
[7] L. Lucas, N. Gilbert, D. Ploton, and N. Bonnet, *J. Microsc.* **181**, 238 (1996).
[8] S. J. Wright, V. E. Centonze, S. A. Stricker, P. J. De Vries, S. W. Paddock, and G. Schatten, *in* "Cell Biological Applications of Confocal Microscopy" (B. Matsumoto, ed.), p. 1. Academic Press, San Diego, 1993.
[9] P. Wallen, K. Carlsson, A. Liljeborg, and S. Grillner, *J. Neurosci. Methods* **24**, 91 (1988).
[10] J. N. Turner, D. H. Szarowski, K. L. Smith, M. Marko, A. Leith, and J. W. Swann, *J. Electr. Microsc. Tech.* **18**, 11 (1991).
[11] A. Kriete and H.-J. Wagner, *J. Microsc.* **169**, 27 (1993).
[12] Y. S. Prakash, K. G. Smithson, and G. C. Sieck, *Neuroimage* **1**, 95 (1993).
[13] A. R. Cohen, B. Roysam, and J. N. Turner, *J. Microsc.* **173**, 103 (1994).
[14] M. Hanani, L. G. Ermilov, P. F. Schmalz, V. Louzon, S. M. Miller, and J. H. Szurszewski, *J. Auton. Nerv. Syst.* **71**, 1 (1998).
[15] T. Hosokawa, T. V. P. Bliss, and A. Fine, *Microsc. Res. Tech.* **29**, 290 (1994).
[16] A. Emilson and A. Scheynius, *J. Histochem. Cytochem.* **43**, 993 (1995).
[17] H. Tinel, F. Wehner, and H. Sauer, *Am. J. Physiol.* **267**, F130 (1994).
[18] D. E. Dodge, C. G. Plopper, and R. B. Rucker, *Am. J. Respir. Cell Mol. Biol.* **10**, 259 (1994).
[19] R. J. Errington, M. D. Fricker, J. L. Wood, A. C. Hall, and N. S. White, *Am. J. Physiol.* **272**, C1041 (1997).
[20] F. Guilak, *J. Microsc.*, **173**, 245 (1994).
[21] F. A. Merchant, S. J. Aggarwal, K. R. Diller, K. A. Bartels, and A. C. Bovik, *Biomed. Sci. Instrum.* **29**, 111 (1993).
[22] E. Chacon, J. M. Reese, J. Nieminen, G. Zahrebelski, B. Herman, and J. J. Lemasters, *Biophys. J.* **66**, 942 (1994).
[23] H. Satoh, L. M. D. Delbridge, L. A. Blatter, and D. M. Bers, *Biophys. J.* **70**, 1494 (1996).
[24] L. M. D. Delbridge, H. Satoh, W. Yuan, J. W. M. Bassani, M. Qi, K. S. Ginsburg, A. M. Samarel, and D. M. Bers, *Am. J. Physiol.* **41**, H2425 (1997).

visualization techniques have also been applied to subcellular structures such as cell nuclei,[25] mitochondria,[26] chromosomes,[27] microtubules,[28] and receptors.[29] In all these studies confocal microscopy, in conjunction with appropriate fluorescent probes and image processing techniques, has proven to be an invaluable tool for the three-dimensional analysis of cellular structures and functions.

The purpose of this article is to review briefly the theory of cell volume investigation and to discuss the fundamental steps involved in this process as well as the potential pitfalls. The steps of quantitative volume investigation using confocal microscopy are illustrated with a specific example of a volume study[23] that was performed previously in our laboratory on isolated cardiac myocytes.

II. Principles of Quantitative Volume Studies: A Brief Overview

In biological and life sciences the recent technical advancements in confocal microscopy, the availability of a vast number of fluorescent probes, and the decreasing costs of high-powered computer hardware have led to a widespread rekindled interest in quantitative volume and surface studies of cellular and subcellular elements. A novel technology, termed volume investigation,[2] has emerged that allows unprecedented visualization and quantitative analysis of complex three-dimensional objects and structures.[7,30–32] In its basic approach,[2] volume investigation creates a mathematical model of an object based on the analysis of a stack of a series of

[25] Q. Zhu, P. Tekola, J. P. A. Baak, and J. A. M. Belien, *Anal. Quant. Cytol. Histol.* **16**, 145 (1994).

[26] H. Fujii, S. H. Cody, U. Seydel, J. M. Papadimitriou, D. J. Wood, and M. H. Zheng, *Histochem. J.* **29**, 571 (1997).

[27] R. Eils, S. Dietzel, E. Bertin, E. Schrock, M. R. Speicher, T. Ried, M. Robert-Nicoud, C. Cremer, and T. Cremer, *J. Cell Biol.* **135**, 1427 (1996).

[28] J. E. Trogadis and J. K. Stevens, *in* "Three Dimensional Confocal Microscopy: Volume Investigation of Biological Systems" (J. K. Stevens, L. R. Mills, and J. E. Trogadis, eds.), p. 301. Academic Press, San Diego, 1994.

[29] R. L. Zastawny, G. Y. K. Ng, J. E. Trogadis, S. R. George, J. K. Stevens, and B. F. O'Dowd, *in* "Three Dimensional Confocal Microscopy: Volume Investigation of Biological Systems" (J. K. Stevens, L. R. Mills, and J. E. Trogadis, eds.), p. 233. Academic Press, San Diego, 1994.

[30] V. Interrante, W. Oliver, S. Pizer, and H. Fuchs, *in* "Three Dimensional Confocal Microscopy: Volume Investigation of Biological Systems" (J. K. Stevens, L. R. Mills, and J. E. Trogadis, eds.), p. 131. Academic Press, San Diego, 1994.

[31] R. Bacallao and A. Garfinkel, *in* "Three Dimensional Confocal Microscopy: Volume Investigation of Biological Systems" (J. K. Stevens, L. R. Mills, and J. E. Trogadis, eds.), p. 169. Academic Press, San Diego, 1994.

[32] N. S. White, *in* "Handbook of Biological Confocal Microscopy" (J. B. Pawley, ed.), p. 211. Plenum Press, New York, 1995.

cross-section images. The mathematical model represents three-dimensional objects through small unitary volume elements, termed "voxels," which are extracted from the original stack of cross-sectional images. The basic purpose of the model is to extract quantitative information from a set of two-dimensional gray scale images (i.e., two-dimensional arrays of image elements or "pixels") to create the three-dimensional representation of pixels, i.e., the voxel. Arrays of voxels are used to quantitatively represent, in three spatial dimensions, the original biological object. Voxel models then allow the visualization and quantitative exploration of complex biological structures and objects in terms of various morphometric parameters such as distances, surfaces, areas, volumes, and geometric relationships among various objects and structures. Quantitative volume studies typically encompass four basic steps[2] that will be discussed briefly first in a more general sense. The four steps are (1) data acquisition, (2) two-dimensional segmentation, (3) three-dimensional visualization, and (4) quantitative three-dimensional analysis. After a brief theoretical discussion, we will illustrate these four basic steps with a specific example of volume and surface exploration of single cardiac myocytes using laser scanning confocal microscopy. Although we will illustrate cell volume investigation using LSCM, these principles are not restricted to this technique and can be applied to a wide variety of imaging methodologies that are capable of producing stacks of cross-sectional images. Such techniques include electron microscopy, magnetic resonance imaging, conventional light microscopy, and computerized axial tomography to name but a few.

In the following discussion of quantitative volume measurements, we basically follow the terminology forwarded by Stevens in an excellent review article on this topic.[2] The very same article is also an excellent source of references on theory and application of volume investigation techniques.

Step 1. Data Acquisition: The Image Stack

The basic data set for any kind of volume investigation typically consists of a stack of serial cross-sectional images of an object of interest (in the example illustrated in Section III, a single isolated cardiac cell). Stevens[2] lists several criteria that should be adhered to when acquiring the image stack. In practice, however, these criteria are rarely met in their entirety. Criteria for data acquisition relate to the degree of destructiveness of the acquisition procedure itself, the spacing of the z sections, density inhomogeneities and alignment in all three dimensions, and density and spatial calibration of the images.

Spatial calibration of images within the focal plane (x, y dimension) is rather straightforward as magnification and zoom factors are typically known precisely. To a first approximation the spatial calibration in the z dimension is given by the stepping size during z sectioning. Spherical aberration related to mismatch of refractive indices may complicate accurate spatial measurements in the axial dimension, as will be discussed in more detail later. Misalignment, particularly in the z dimension, can occur due to a faulty mechanism that advances the focal plane during the sectioning procedure. However, this can be checked easily by optical sectioning of an object of known geometry such as fluorescent beads. A problem related to alignment of the image stack may arise from motion artifacts of the specimen itself. Changes in the three-dimensional shape of cells during data acquisition, for example, due to changes in cell volume or contraction of muscle cells, can affect the accuracy of volume determination obtained by sectioning methods. Misalignment of the excitation source (laser) in fluorescent studies can lead to the inhomogeneous excitation of fluorophores and result in apparent density inhomogeneities.

Ideally, serial sections should be isotopic, i.e., x and y dimensions of the pixels should be identical to the z dimension, and sections should be acquired adjacently, i.e., with no gaps or overlaps. In practice, however, this criterion is seldom met. This is related directly to the fact that the spatial resolution of confocal microscopes is significantly less in the z dimension as compared to the x and y dimension. The spatial resolution (typically expressed as the point-spread function) of a typical LSCM is in most cases only ≤ 1 μm, or two to three times less than in the x and y dimension. As a consequence of the limited spatial resolution in the z dimension in conventional LSCM, the axial distance between cross-sectional image planes is often chosen to be larger than the voxel size in the x and y dimension.

The most serious practical limitations for the acquisition of an image stack stem from the fact that LSCM is inherently destructive. High light intensities, which are typical for laser illumination, can lead to phototoxic side effects through the generation of phototoxic by-products and free radical formation from fluorescent probes[33,34] and to photobleaching. While problems related to phototoxicity typically are restricted to living specimens, photobleaching applies to fixed specimens as well. Photobleaching results from the fact that during the acquisition of the image stack, virtually the entire specimen is exposed continuously to the excitation light. The light, which is focused into a specific focal plane, also penetrates through

[33] R. Y. Tsien and A. Waggoner, *in* "Handbook of Biological Confocal Microscopy" (J. B. Pawley, ed.), p. 267. Plenum Press, New York, 1995.
[34] J. Hüser, C. E. Rechenmacher, and L. A. Blatter, *Biophys. J.* **74,** 2129 (1998).

adjacent regions above and below the plane and thereby bleaches the fluorescent probe in these regions.

Among the problems listed, photobleaching represents probably the most serious limitation in volume investigation studies based on fluorescence confocal microscopy. In practice the actual protocol for the data acquisition is often the result of a compromise among several factors, including acceptable signal-to-noise ratio; bleaching "tolerance" of the specimen and the fluorophore; sensitivity of the recording system; required spatial resolution and accuracy; and the application of off-line image processing routines to reduce noise further. These parameters can vary widely depending on the biological specimen and the fluorescent probe used. Unfortunately, there is not any clear-cut protocol for the image stack acquisition, and individual protocols have to be established for different specimens. Although, as a rule of thumb, a number of 20 or more sections are typically necessary to reliably reconstruct spherical or ellipsoidal test objects.[20] However, it is rarely necessary to acquire more than 100 serial sections to obtain a meaningful set of data for volume investigation.[2]

Step 2. Two-Dimensional Segmentation

After acquisition of the image stack, the next step consists of defining the object of interest in the acquired images. This process is often referred to as "segmenting" and involves defining the pixel intensities or densities that define the object in the gray scale images of the stack, and/or some properties that link voxels together.[32] Segmentation is the process by which objects are extracted from the rastered voxel image data. The process can consist simply of defining a threshold or an intensity window, although it often involves more advanced image processing steps such as various filtering procedures. Segmenting can be based solely on subjective criteria chosen by the observer or they can involve more objective criteria. Section III discusses an example of an attempt to define objective criteria for segmenting and thresholding. Figure 2C illustrates how different segmenting criteria affect the outcome of the volume reconstruction in the example of cardiac myocytes.

Step 3. Three-Dimensional Visualization of Reconstructed Object

One of the most intriguing benefits of computer-based volume reconstruction is the capability of actually visualizing the reconstructed object. Three-dimensional visualization provides the means to understand the geometric relationships of complex biological objects, such as the three-dimensional organization of cell organelles and cytoskeletal elements, to name just two examples. Despite a wide variety of methods and algorithms used to visualize 3D data (see later), most methods fall into one of two categories referred to as (1) solid body voxel-based images and (2) translucent ray-cast images.[2] With voxel-based solid body reconstruction, surfaces are placed on objects or groups of voxels of the same (ranges of) intensity, forming so-called shells. These shells contain a specific number of voxels of known volume, which then allow for the quantitative determination of the shell volume. Figure 2B illustrates a solid body-type visualization of a cardiac cell and Fig. 2C illustrates how the process of thresholding or segmenting affects the reconstruction of the shells. In contrast to solid body images, translucent ray-cast images are constructed by sending an imaginary ray of light, pixel by pixel, through the stack of gray scale images and projecting the resulting image on a new surface, analogous to a medical X-ray image. (It is beyond the purpose of this article to give a detailed discussion of 3D visualization methods; however, there is broad literature on 3D visualization and many commercial software products for volume investigation are available. Among the many useful articles on this topic, there are two excellent publications worth mentioning that summarize the literature and give many helpful practical hints.[30,32])

Step 4. Quantitative Three-Dimensional Analysis

Although 3D visualization is extremely helpful for the intuitive understanding of the three-dimensional organization of complex biological structures, the usefulness of accurate mathematical models of volume reconstruction goes beyond mere visualization. The true value of these models lies in their potential to allow *quantitative* analysis of geometric relationships of biological structures, i.e., in the quantitative determination of cell volumes, cell organelles, distances between subcellular objects, surface areas of cellular membranes, and much more. Volume investigation based on quantitative analysis of high-resolution image stacks has become an invaluable analytical tool in quantitative biology. This is illustrated in the next section with a quantitative analysis of cell volume and surface of single heart cells and these findings are related to functional aspects of these cells.

III. Practical Illustration of Volume Investigation: Quantitative Volume Measurements in Cardiac Myocytes

The purpose of this section is to illustrate, with a practical example, the four basic steps of volume investigation. The example is taken from a previous study[23] using laser scanning confocal microscopy that was undertaken to comparatively measure cell volumes of cardiac myocytes from several mammalian species and to explore how the developmental stage of animals affected cell surface area and volume.

Step 1. Data Acquisition: The Image Stack

Cell Preparation. Cell volumes are measured in three different mammalian species: New Zealand White rabbits, ferrets, and Sprague–Dawley rats. For the investigation of the relationship between developmental stage and cell volume, 3-month-old rats are compared with 6-month-old animals.

Ventricular myocytes are isolated by an enzymatic procedure (retrograde Langendorff perfusion). During the experiments, cells are superfused with a modified Tyrode solution containing (in mM) 140 NaCl, 6 KCl, 1 MgCl$_2$, 5 N-2-hydroxyethylpiperazine-N'-2-ethanesulfonic acid (HEPES), 10 glucose, and 2 CaCl$_2$ (1 CaCl$_2$ for rat). Adjust to pH 7.4 with NaOH. Experiments are carried out at 23°.

Fluorescent Probe. To label the intracellular space, myocytes are loaded with the fluorescent probe calcein. Calcein is distributed evenly throughout the cell, is largely unaffected by changes in the intracellular environment, is highly fluorescent and not contaminated by cellular autofluorescence, and has been used previously for volume measurements.[20,22,35,36] Cardiac myocytes are loaded with the probe by exposure to the membrane-permeant calcein/acetoxymethyl (AM) ester (5 μM; Molecular Probes, Inc., Eugene, OR) for 30 min at room temperature.

Laser Scanning Confocal Microscopy. Confocal image stacks of cross sections of calcein-loaded single cardiac myocytes are obtained with a LSM 410 laser scanning confocal microscope (Carl Zeiss, Germany) coupled to an inverted microscope (Axiovert 100, Carl Zeiss). The objective lens used in these experiments is a Zeiss 40× oil-immersion Plan-Neofluar lens with a numerical aperture (NA) of 1.3. Calcein fluorescence is excited with the 488-nm line of the argon ion laser. Emitted fluorescence is collected at wavelengths >515 nm.

For the acquisition of a stack of images the objective is moved in the axial direction (z dimension) with a stepping motor under computer control. Depending on the thickness of individual myocytes, 20–35 sections at 1.2-μm steps are obtained. At each optical plane the cell is scanned once to produce a cell fluorescence image (x, y scan) of 512 by 512 pixels. Magnification is set by the objective and the zoom factor of the confocal microscope to give a pixel size of approximately 0.1 μm^2 (~0.3 by ~0.3 μm). This magnification allows imaging of the entire myocyte at a spatial resolution near the theoretical x, y resolution of the system. The analog output of the photomultiplier tube is digitized at 8-bit resolution representing fluorescence intensities with 256 levels of gray.

[35] C. A. Poole, N. H. Brookes, and G. M. Clover, *J. Cell Sci.* **106,** 685 (1993).
[36] J. Farinas, V. Simanek, and A. S. Verkman, *Biophys. J.* **68,** 1613 (1995).

Photobleaching. As outlined in the previous section, photobleaching of fluorescent probes is potentially one of the most critical sources of artifactual volume estimates. In our experiments, care is taken to keep photobleaching to a minimum by choosing the lowest excitation light intensity while maintaining a reasonable signal-to-noise ratio in the acquired images. Photobleaching of intracellular calcein is estimated quantitatively by repetitively scanning cardiac myocytes at the same focal plane. The residual fluorescent intensity after 30 consecutive scans is 86.8 ± 8.1% of the initial value ($n = 10$ cells). These control experiments show that, with appropriated precautions, photobleaching can be kept within a reasonable range. Because of the small effect of photobleaching under our experimental conditions, no special correction for bleaching is applied.

Motion Artifacts. Volume measurements are performed on intact living cardiac cells that maintain their ability to contract. A practical approach is chosen to avoid motion artifacts. First, only quiescent cells are selected, and spontaneously contracting cells are excluded from the study. If noticeable motion starts to occur during the acquisition of the data stack, these data are discarded. Second, the adherence of the cells to the bottom of the experimental chamber is enhanced by pretreating the coverslips with laminin. This procedure contributes significantly to the elimination of motion artifacts.

Step 2. Two-Dimensional Segmentation

Background Subtraction, Thresholding, and Intensity Segmenting. As outlined in Section II, the correct discrimination of cellular fluorescence intensities (thus representing cell volume elements) from background fluorescence, image noise, and out-of-focus fluorescence is pivotal for the quantitatively accurate cell volume investigation. Several steps are undertaken to eliminate noncellular volume elements (voxels) from the volume calculations. In each individual experiment, offset and gain of the system is chosen to represent noncellular background near zero and fluorescence intensities from cellular elements to cover the full range of 256 intensity levels (8-bit digitization) without saturation of the recording system. With this approach the need for additional background subtraction procedures is eliminated. Because the spatial resolution in the axial direction (z dimension) is estimated to be ≤1 μm in our system, out-of-focus contributions from adjacent sections are minimized by choosing a sectioning interval of 1.2 μm. Due to the limits of spatial resolution of the system and the finite size of the sampled volume elements, voxels at the cell border only partially represent cellular volumes. On the level of an individual voxel, this results in a variable decrease of fluorescence intensity depending on the (*a priori* unknown) degree by which an individual voxel is filled by cell volume. Elimination of out-of-focus fluorescence and noncellular contributions from voxels at the cell border are achieved by intensity segmentation (thresholding). In order to achieve this in an objective manner, we perform a quantitative analysis of the distribution of the pixel intensities in the image stack. Figure 1A shows a frequency histogram of fluorescence intensities of a central section of a cardiac myocyte. The frequency histogram shows a clear peak around a fluorescence intensity value of about 150 that can be fitted perfectly by a simple Gaussian distribution (dashed line). It is assumed that the pixel intensities encompassed completely by the Gaussian fit would represent volume elements entirely located within the cell borders. The Gaussian fit is used to estimate the contribution from extracellular pixels and "partial" cellular pixels. The Gaussian fit is subtracted from the measured frequency distribution resulting in the frequency distribution shown in the inset of Fig. 1A. The frequency distribution after subtraction does not reveal any specific peaks, except a small peak near intensity zero representing noncellular background. From the lack of peaks in the remaining distribution of intensities, we conclude that pixels with a random fractional cell volume between 0 and 1 account for most of the residual area shown in the inset of Fig. 1A. For that reason, we assume that on average these pixels contribute 50% to the cell volume and we therefore attribute a weighing factor of 0.5 to them. This translates into an effective threshold level of 50 for the stack of fluorescence images. In summary, for our volume reconstruction, we apply an intensity window to our image stacks that ranges from gray scale levels 50 to 255.

Figure 1B, however, illustrates how dramatically the estimated cell volume depends on the chosen value for thresholding. For example, increasing the threshold level from intensity level 50 to 100 (assuming that all the pixel value histogram shown in the inset of Fig. 1A would be noncellular elements) would reduce the estimated cell volume by approximately 25%. Although we think that this assumption is unrealistic, the example still illustrates unequivocally that accurate thresholding and intensity segmenting are crucial.

Step 3. Three-Dimensional Visualization of Single Cardiac Myocytes

Method and Algorithm for 3D Visualization. For the three-dimensional volume representation the stack of images (Fig. 2A) is processed with a volume-rendering algorithm to produce a voxel-based solid body-type volume reconstruction. Typically, for the volume reconstruction of a single cardiac myocyte, a stack of 20–35 fluorescence images, depending on the thickness of the cell, are obtained at 1.2-μm axial spacing ($\Delta z = 1.2\ \mu$m).

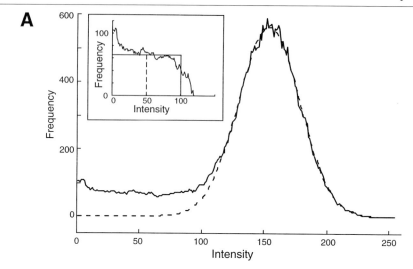

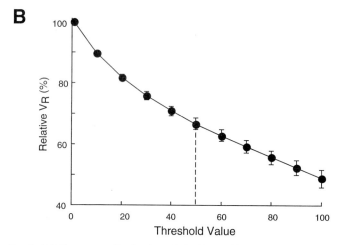

FIG. 1. (A) Frequency distribution of pixel intensities from a central cross section through a rabbit ventricular myocyte. The single myocyte was stained with the fluorescent dye calcein. Fluorescent cellular cross-sectional images were digitized at 8-bit resolution yielding pixel intensities between 0 and 255 (2^8 levels of gray). The dashed line represents a Gaussian fit to the frequency distribution centered around a pixel intensity of 150. (Inset) Difference distribution of pixel intensities obtained by subtraction of the Gaussian fit from the measured frequency histogram. To a small extent, these pixels represent noncellular background and partial cellular regions encountered at the cell border. The difference pixel distribution did not reveal a distinguishable peak. As discussed in the text, from the difference frequency distribution a thresholding value of 50 was derived. (B) Effect of level of thresholding on estimates of rendered cell volume (V_R). Data from eight cells were normalized to the volume value measured before thresholding. The dashed line indicates the threshold intensity level used in this study. Modified from H. Satoh *et al.*, *Biophys. J.* **70**, 1494 (1996), with permission.

Figure 2B illustrates a 3D volume representation of a single cardiac myocyte. The cell is reconstructed by describing its three-dimensional surface contour as a set of polygons. By this method, each volume element (voxel) is visited to find the polygons formed by the intersections of the contour surface and the voxel edges. The method used is based on the method proposed by Klemp et al.[37] and is similar to the "marching cube algorithm" described by Lorensen and Cline.[38] The marching cube algorithm has been used successfully to perform 3D surface reconstructions from stacks of two-dimensional anatomical and medical data. It has been applied to image data obtained from computed tomography and magnetic resonance[38] as well as confocal microscopy.[20] Volume rendering and surface topography reconstruction are performed using the programming language IDL (Interactive Data Language, Research Systems Inc., Boulder, CO).

Effect of Intensity Segmenting on Volume Visualization. As illustrated in Fig. 1B, volume estimation and reconstruction depend critically on thresholding and the chosen window for intensity segmentation. For the reconstruction shown in Fig. 2B, all pixels with intensity levels <50 are eliminated from the reconstruction, i.e., the segmenting intensity window ranges from an intensity value of 50 to 255. This is achieved by setting the numerical value in the algorithm that specifies the constant density surface (also called an isosurface) to be rendered at the threshold value of 50. Figure 2C illustrates the effect of changing the segmenting or thresholding value. The segmenting window for pixel intensities is decreased incrementally by increasing the lower threshold. Computationally this is achieved by adjusting the numerical value in the algorithm for constant-density surface to the values shown in Fig. 2C. As can be expected, the outcome of the 3D reconstruction procedure depends critically on this value and affects the reconstructed volume with regard to three-dimensional appearance and estimated size.

Step 4. Quantitative Analyses of Cell Volume and Surface of Single Mammalian Cardiac Myocytes

Three-dimensional volume and surface reconstructions do not stop at the point of impressive computer-generated displays of biological specimens. Although these visualizations can be extremely useful for the intuitive understanding of complex three-dimensional biological structures, they would be of only limited use if they could not be advanced further to obtain precise quantitative analysis of these three-dimensional relationships. This

---

[37] J. B. Klemp, M. J. McIrvin, and W. S. Boyd, *in* "Sixth International Conference on Interactive Information and Processing Systems," p. 286. American Meterological Society, Boston, 1990.
[38] W. E. Lorenson and H. E. Cline, *Comput. Graph.* **21,** 163 (1987).

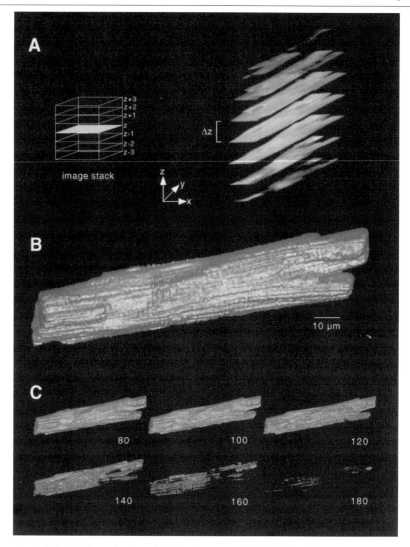

FIG. 2. (A) Serial sectioning of single fluorescently labeled cardiac myocytes by laser scanning confocal microscopy. Confocal images from calcein-loaded rat myocytes were recorded at 1.2-μm axial intervals to generate an evenly spaced stack of cross-sectional images. The pixel size in x and y dimension was 260 nm. z represents the central section across the cell, and sections above and below the central plane are labeled $z + n$ and $z - n$, respectively. (Right) Perspective view of the cellular cross sections. For presentation purposes only, every other cross-sectional image of the stack is shown, i.e., the axial distance between the sections shown was 2.4 μm. (B) Three-dimensional volume representation of a single cardiac myocyte. The solid body volume reconstruction is based on a marching cube algorithm using the programming language IDL. The isosurface value for this reconstruction was 50, corresponding

section illustrates this point with quantitative measurements of cell volumes and surfaces in single cardiac cells and discusses how these quantitative data can provide information regarding the physiology of these cells.

Accuracy of Cell Volume Measurements by Optical Sectioning: Spherical Aberration. We have already discussed and illustrated the importance of correct thresholding or segmenting of the image stack, and we have introduced criteria to choose the correct value that are based on an objective analysis of the distribution of pixel intensities within the image stack. This analysis allows distinction and compensation for noncellular background near the cell border, volume elements completely filled by cellular elements, and image elements that are only partially, but to a various degree, filled with cell elements. The developed criteria are also sensitive to the irregular three-dimensional topography (membrane invaginations, dents, and branches) that are typical for isolated cardiac myocytes. The approach efficiently eliminates volume elements that, in a given section through the cell, particularly near the bottom or the top, would appear within the boundaries of the cell, but in fact represent extracellular space.

A very important issue that can have a profound impact on the accuracy of volume estimates has been left untouched so far. This issue relates to potential artifacts that arise from distortions due to erroneous estimations of axial distances. Mismatch of refractive indices of the specimen, mounting medium (in the case of living specimens the physiological bathing solution), and lens immersion medium cause spherical aberrations and distortions of morphological data.[39,40] This can amount to a serious problem in any quantitative measurement using light microscopy that involves the depth (axial) dimension of the specimen. Mismatch of the refractive index of lens immersion medium (oil) and mounting medium (water) has the consequence that the movement of the focal plane in a specimen does not necessarily follow the movement of the objective when acquiring optical z sections through that specimen. If the refractive index of the immersion medium (in our study, an oil-immersion objective is used; refractive index

---

[39] S. W. Hell and E. H. K. Stelzer, *in* "Handbook of Confocal Microscopy" (J. B. Pawley, ed.), p. 347. Plenum Press, New York, 1995.
[40] H. E. Keller, *in* "Handbook of Confocal Microscopy" (J. B. Pawley, ed.), p. 111. Plenum Press, New York, 1995.

---

to a segmentation window of gray scale levels between 50 and 255. (c) Effect of segmenting on the result of volume reconstruction. The same stack of images was processed by the same algorithm; however, the width of the segmenting window of gray scale levels was decreased incrementally by increasing its lower threshold. The numbers beneath the individual reconstructions represent the cutoff intensity level used.

r_{oil} = 1.48) is larger than the refractive index of the specimen and its mounting medium (in our case the cellular environment is approximated by the refractive index for water: r_{water} = 1.33) then the distance by which the plane of focus is advanced into the specimen in axial direction is actually smaller than the nominal axial movement of the objective.[41] As a consequence, the z extension of the specimen and therefore the reconstructed volume would be overestimated. Even with the significantly higher spatial resolution of confocal microscopy, uncompensated spherical aberration can generate errors up to 50% in quantitative measurements that include the depth dimension.[41] The deeper the focal plane is moved into the specimen, the more severe becomes the distortion.

Distortion due to spherical aberration is illustrated in Fig. 3. We have investigated the degree of spherical aberration by optically sectioning an object of known geometry and dimensions, and with fluorescent properties similar to calcein, i.e., the probe we used for the cell volume measurements. For this purpose we choose fluorescent latex beads with a diameter of 15.5 μm (exact diameter measured by electron microscopy, Molecular Probes). This diameter size is selected because it compares well to the average thickness of a cardiac cell (we[23] measure an average cell depth of cardiac myocytes from different mammalian species of 12–14 μm). The fluorescent properties of these beads are very similar to calcein (excitation 490 nm, emission 515 nm). Beads are placed onto the coverslip of the experimental chamber and, after sufficient time to allow for attachment to the glass surface, are covered by a droplet of water. Thus, the beads mimic the fluorescently labeled cardiac cells with respect to mounting medium, fluorescent properties, and axial dimension. Optical sections are obtained on the same confocal instrument. We compared two different objective lenses to evaluate the effect of immersion medium. The lenses used were a 40× oil-immersion lens with a numerical aperture of 1.3 (Plan-Neofluar, Carl Zeiss) and a 40× water-immersion NA 1.2 objective (C-Apochromat, Carl Zeiss). Figure 3A shows that with both lenses the obtained x, y images (cross sections) are perfectly circular. The x, z reconstructions of the same beads, however, reveal elongations of the object in the axial dimension. The degree of this axial distortion depends on the type of objective lens used. As shown in Fig. 3B, the distortion is much more pronounced with the oil-immersion objectives. With the water-immersion lens the mismatch of refractive indices between immersion and mounting media is minimized and the distortion of the reconstructed object becomes smaller. Unfortunately, at the time the cell volume measurements were performed, the high-

[41] L. Majlof and P.-O. Forsgren, in "Cell Biological Applications of Confocal Microscopy" (B. Matsumoto, ed.), p. 79. Academic Press, San Diego, 1993.

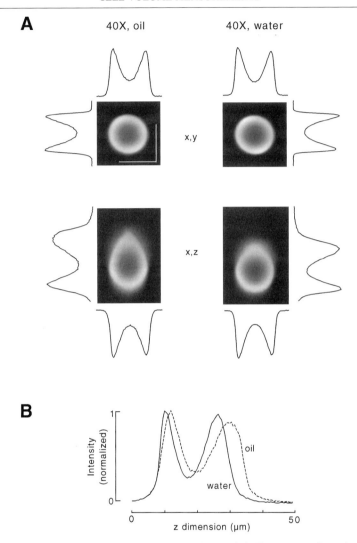

FIG. 3. Confocal imaging of fluorescent microspheres. (A) Fluorescent microspheres with a diameter of 15.5 μm were sectioned optically by recording x, y cross-sectional images at 0.5-μm axial steps. The images were recorded with an oil-immersion 40× (Plan-Neofluar, NA = 1.3; left) and a water-immersion 40× objective (C-Apochromat, NA = 1.2; right) x, y images (top) and x, z reconstructions (bottom) of the microspheres and corresponding normalized fluorescence intensity profiles through the center of the microsphere are shown. The doughnut-shaped appearance of the microsphere is due to the lack of fluorophores in the center of the beads. Fluorescence was excited at 488 nm and the emitted light was recorded at wavelengths >515 nm. Calibration bar: 15 μm (B) Overlay plot of axial intensity profiles through the center of the micropheres recorded with the two objectives showing enhanced distortion with the oil-immersion lens.

quality C-Apochromat water-immersion lens was not available. The study[23] was performed with the 40× Plan-Neofluar oil-immersion lens. We therefore used the stack of the cross-sectional images of the latex beads to derive a correction factor for the cell volume measurements in order to correct for distortion due to spherical aberrations. Intensity profiles are drawn through the center of the microsphere and the full-width at half-maximum diameter (FWHM) is used as a measure of distortion in the three spatial dimensions. FWHM of the sphere is identical in the x and y dimension, indicating lack of distortion within the focal plane. The FWHM in the z dimension is approximately 1.46 times larger than the FWHM in the x and y dimension, respectively. We calculated a volume correction factor based on the 3D reconstruction of beads. The apparent volume of the beads as reconstructed from the optical section is calculated by approximating the geometry of the imaged sphere as an ellipsoid with a diameter of the FWHM in the x dimension and a height corresponding to the FWHM in the z dimension. A correction factor is calculated by dividing the true spherical volume by the apparent ellipsoidal volume. This correction factor $V_{sphere}/V_{ellipsoid} = 0.68$ is applied to the estimated cell volumes of cardiac cells (see later). Although this method has the advantage of applying a correction factor that is obtained empirically by a relatively simple procedure on the same instrument under similar experimental conditions, it has its limitations. For example, cardiac myocytes have a more complex geometry than microspheres. In addition, the microspheres do not fully account for the fact that the degree of spherical aberration becomes larger as the depth into the specimen increases. Nevertheless, the method proves valuable for the correction of spherical aberration related to over- or underestimations of cell volume based on optical sections.

Results of Quantitative Volume Estimates of Single Cardiac Cells from Different Mammalian Species. We applied our volume estimate procedure to single myocytes isolated from three different mammalian species. The cellular cross-sectional area from each individual section of the image stack is determined by multiplying the number of cellular pixels above the chosen threshold value with the size of an individual pixel (given by the magnification factor of the objective and the zoom factor of the LSCM). The raw value for the cell volume (pl or μm^3) is calculated by multiplying the number of all significant pixels from all sections with the z interval of individual sections (1.2 μm). Raw volume values are then scaled by the volume correction factor of 0.68 in order to account for volume distortions due to spherical aberration.

After the correction factor for spherical aberration is applied, the average cell volumes are 30.4, 30.9, and 34.4 pl, respectively, for adult rabbits, ferrets, and rats. These values are not significantly different from each other. There are no differences in average maximal cell length (138–143 μm) and width (31–32 μm) among different species. The cell thickness (z dimension) ranges from 12.2 μm (rabbits) to 13.3 μm (rats) to 14.0 μm (ferrets). Our volume estimates of 30–35 pl compare well with the range reported in the literature. For rat myocytes, volume values between 16 and 45 pl (measured with light scatter flow cytometry and Coulter analysis) have been reported.[42–45] Gerdes *et al.*[46] reported a range of 5–25 pl based on morphometric measurements of stained rat tissue. Values of 36 and 29 pl were reported for ferret[47] and guinea pig hearts,[44] respectively. Poole-Wilson[48] reviewed studies on human myocytes that reported values between 15 and 52 pl. Cell volumes ranging from 35 to 43 pl were calculated from rabbit cell cross-sectional area and cell thickness.[49]

Cardiac Cell Volume Estimates in Relation to Other Morphological Parameters. Although recent years have seen a tremendous increase in the number of laboratories using laser scanning confocal microscopy, volume-rendering procedures based on optical sectioning and computational image processing routines are not always suitable or applicable to experimental studies in which knowledge of the cell volume would be essential. We therefore aimed in a previous study[23] to provide a simple morphological measurement that could serve as a reliable estimate of cell volume in cardiac cells. Figure 4A shows the reconstructed cell volume (V_R), measured by optical sectioning, as a function of the simple product of maximal cell length and width ($L \times W$) that can be measured easily with conventional nonconfocal microscopy. There is clear linear correlation between $L \times W$ and estimated cell volume for all species. The slopes of the regression lines that describe the correlation between $L \times W$ and V_R are very similar among the three species investigated. This result indicates that cell length and width are useful parameters to extrapolate cell volume with conventional light microscopy, provided the scaling factor that relates the two-dimensional parameter $L \times W$ to the three-dimensional volume has been determined accurately.

[42] G. B. Nash, P. E. R. Tatham, T. Powell, V. W. Twist, R. D. Speller, and L. T. Loverock, *Biochim. Biophys. Acta* **587,** 99 (1979).
[43] S. E. Campbell, A. M. Gerdes, and T. D. Smith, *Anat. Rec.* **219,** 53 (1987).
[44] H. W. Vliegen, A. van der Laarse, J. A. N. Huysman, E. C. Wijnvoord, M. Mentar, C. J. Cornelisse, and F. Eulderink, *Cardiovasc. Res.* **21,** 352 (1987).
[45] A. Fraticelli, R. Josephson, R. Danziger, E. Lakatta, and H. Spurgeon, *Am. J. Physiol.* **257,** H259 (1989).
[46] A. M. Gerdes, J. A. Moore, J. M. Hines, P. A. Kirkland, and S. P. Bishop, *Anat. Rec.* **215,** 420 (1986).
[47] S. H. Smith and S. P. Bishop, *J. Mol. Cell Cardiol.* **17,** 1005 (1985).
[48] P. A. Poole-Wilson, *J. Mol. Cell Cardiol.* **27,** 863 (1995).
[49] D. Drewnowska and C. M. Baumgarten, *Am. J. Physiol.* **260** C122 (1991).

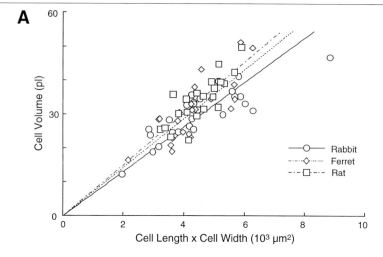

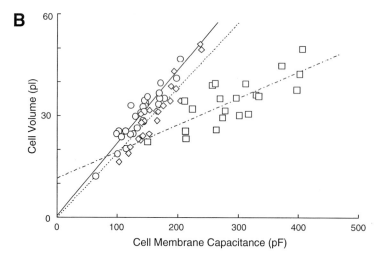

FIG. 4. Illustration of quantitative analysis of cross-sectional image data and cell surface measurements. (A) Correlation between the product of maximal cell length and width ($L \times W$) and reconstructed cell volume (V_R) in three different mammalian species (rat, rabbits, and ferrets). $L \times W$ was measured in the central plane of the image stack (plane z in Fig. 2A). From the regression lines, cell volumes can be predicted reasonably well from $L \times W$ by V_R (in picoliters, pl) = slope $\times 10^{-3}$ pl/μm^2 $\times L \times W$ (μm^2). The slope factors were 6.59, 7.21, and 7.59 for rabbits, ferrets, and rats, respectively. (B) Correlation between V_R and membrane surface. The membrane surface was estimated by electrophysiological cell membrane capacitance (C_m) measurements. Using the patch clamp technique (voltage clamp), the total membrane surface was estimated from whole cell capacitance measurement using a

Cardiac Cell Volume Estimates in Relation to Electrophysiological Parameters. Electrophysiological studies using the voltage clamp technique are typically undertaken to characterize membrane currents carried by the flux of ions through specific ion channels in the cell membrane. The amplitude of these ionic currents is dependent not only on membrane potential and transsarcolemmal ionic gradients but also on cell size. For comparative purposes, membrane currents and membrane conductance in general are often normalized to the cell membrane capacitance, thereby expressing ion movements through channels as current densities (e.g., pA/pF). This approach takes advantage of the fact that the "specific capacitance" of cell membranes is approximately 1 $\mu F/cm^2$ and is fairly constant among different muscle cell types and species.[50] Therefore, cell capacitance measurements by electrophysiological methods are widely used to estimate the surface membrane area of cells and as an indirect index of cell volume. Because it is ultimately the critical change of the concentrations of intracellular ions (e.g., calcium and protons) produced by ionic currents that trigger and control cellular events such as contraction, exocytosis, and cell proliferation, it would be desirable to relate ionic currents *directly* to cell volume. In patch clamp studies membrane capacitance is measured easily and routinely. We therefore explored whether the relatively easy-to-measure membrane capacitance could provide a useful estimator of cell volume. The usefulness of the parameter membrane capacitance (or total electrically accessible membrane surface) for this purpose may not be immediately obvious. The degree of membrane folding and the abundance of membrane invaginations are highly variable among different cell types and animal species[51,52] and in cardiac cells depends on the developmental stage.[53,54]

To answer this question we performed a set of combined measurements of membrane capacitance using the whole cell voltage clamp technique (ruptured patch) and simultaneous optical sectioning of calcein-loaded

---

[50] B. Hille, "Ionic Channels of Excitable Membranes." Sinauer, Sunderland, MA, 1992.

[51] E. Page, *Am. J. Physiol* **235**, C147 (1978).

[52] D. M. Bers, "Excitation-Contraction Coupling and Cardiac Contractile Force." Kluwer Academic Publishers, Norwell, MA, 1991.

[53] E. Page, J. Early, and B. Power, *Circ. Res.* **34** and **35**(Suppl II), II-12 (1974).

[54] J. B. Delcarpio, W. C. Claycomb, and R. L. Moses, *Am. J. Anat.* **186**, 335 (1989).

---

value of 1 $\mu F/cm^2$ for the specific cell membrane capacitance. There was a significant linear correlation between C_m and V_R in all three species. The regression lines were $V_R = 0.215 \times C_m + 0.718$, $r^2 = 0.86$ ($p < 0.01$, $n = 28$) in rabbits, $V_R = 0.192 \times C_m + 0.070$, $r^2 = 0.86$ ($p < 0.01$, $n = 23$) in ferrets, and $V_R = 0.078 \times C_m + 11.755$, $r^2 = 0.58$ ($p < 0.01$, $n = 21$) in rats, respectively. Modified from H. Satoh *et al., Biophys. J.* **70**, 1494 (1996), with permission.

myocytes. For whole cell voltage clamp experiments, patch electrodes contain (in mM) 125 CsCl, 10 MgATP, 20 HEPES, 0.3 GTP, and 10 ethylene glycol-bis(β-aminoethyl ether) N,N,N',N'-tetraacetic acid (EGTA); pH is adjusted to 7.4 with CsOH. The membrane capacitance (C_m) is calculated from 5-mV hyperpolarizing and depolarizing steps (20 msec) applied from a holding potential of -80 mV. Routinely, optical sectioning is performed immediately before the voltage clamp experiments.

Figure 4B illustrates the correlation between C_m and V_R for the three mammalian species. The average membrane capacitance is 138 pF in rabbits, 162 pF in ferrets, and 289 pF in adult rats. C_m correlates linearly with V_R; however, the regression lines are different among animal species, and the variability of data is larger in rats (correlation coefficient $r^2 = 0.58$) than in rabbits ($r^2 = 0.86$) and ferrets ($r^2 = 0.86$). Despite the large variability of cell sizes (range: 12–51 pl), the C_m–V_R relationship is highly constant in rabbits and ferrets and, to a lesser degree, in rats. Even though the C_m–V_R relationship varies substantially among species, it still serves as a reasonable estimate or predictor for cell volume within an individual species. We also compared the developmental dependence of the C_m–V_R relationship by comparing cells from 3-month-old rats with 6-month-old animals (data not shown; see Satoh *et al.*[23]). Both parameters, V_R and C_m, are higher in the older animals. Also the ratio of membrane capacitance to estimated volume (C_m/V_R) increases as the animal matured, most likely due to an increased degree of surface membrane folding and membrane invaginations. In cardiac cells the degree of membrane folding and the fraction of transverse tubular membrane are species dependent and are influenced by developmental factors.[55] In mammalian ventricular muscle, approximately 20–50% of the total surface membrane area forms the t tubules, and there are significant species differences in the surface area-to-cell volume ratio (for summary, see Bers[52]).

IV. Conclusions

The availability of laser scanning confocal microscopes, suitable fluorescent probes, and advanced computational image processing routines has enabled morphometric studies of intact living cells at a significantly higher spatial resolution and with improved fidelity. Furthermore, the combination of these optical methods with other methods, such as patch clamp recordings, has provided a means to study various morphological and functional properties of single cells simultaneously. Nevertheless, these methods

[55] J. S. Frank, G. Mottino, F. Chen, V. Peri, P. Holland, and B. S. Tuana, *Am. J. Physiol.* **267,** C1707 (1994).

can be subject to various sorts of artifacts that can hamper accurate quantitative measurements. As discussed and illustrated in the previous sections, quantitative studies using fluorescence microscopy can encounter many pitfalls. However, many of them can be avoided by careful precautions or can be corrected by appropriate strategies. Inaccuracies in volume measurements using fluorescence confocal microscopy can arise from motion artifacts occurring during data acquisition, inaccuracies of background subtraction and thresholding, and erroneous estimation of axial distances due to optical limitations of spatial resolution due to spherical aberration and other problems related to optical sectioning of thick specimens. Photobleaching or loss of fluorescent probes from the cellular compartment of interest can interfere with the accuracy of the measurements, and the combination of intense illumination and the presence of fluorescent probes can have deleterious effects, particularly on living specimens.[33,34] Nevertheless, with careful experimental protocols, one should be able to navigate around these potential traps and pitfalls. In practice, the appropriate experimental protocol requires compromising sometimes opposing strategies to avoid pitfalls in order to obtain quantitative results that are scientifically meaningful (e.g., signal averaging of multiple scans at the same focal plane to increase the signal-to-noise ratio has to be weighed against the concomitant increase in the degree of photobleaching and potential photodamage). Using the specific example of volume measurements in single cardiac myocytes, we have illustrated how such strategies can be developed in practice.

We have discussed the three-dimensional visualization techniques that provide helpful tools for the intuitive understanding of the three-dimensional organization of complex biological structures and objects. The caveat, however, remains that three-dimensional visualization is not necessarily always a precise science or, as N. S. White[32] puts it in his seminal article on visualization of multidimensional confocal imaging data, "seeing should never be used as the sole criterion for believing." Visualization algorithms need to be benchmarked and validated by objective methods. Validation of visualization techniques on objects of known geometries and dimensions have proved to be of invaluable importance in this context. Without these precautions the scientific value of these advanced imaging methods would remain limited.

Acknowledgments

I thank my collaborators on the volume study in cardiac muscle cells[23] for their permission to use this material in this article, Rachel L. Gulling for secretarial help, and Drs. J. Hüser and S. L. Lipsius for critical comments on the manuscript. Financial support was provided by grants from the National Institutes of Health and the American Heart Association National Center. LAB is an Established Investigator of the American Heart Association.

[17] Volume Measurements in Confocal Microscopy

By GARY C. SIECK, CARLOS B. MANTILLA, and Y. S. PRAKASH

Introduction

The confocal microscope is a convenient tool for obtaining spatially registered two-dimensional (2D) optical sections of three-dimensional (3D) objects (for reviews on confocal microscopy, see Matsumoto,[1] Pawley,[2] Schild,[3] Stevens *et al.*,[4] and White *et al.*[5]). The ability of the confocal microscope to eliminate out-of-focus information is dependent on the presence of an aperture limit before the detector, which determines the thickness of the optical plane imaged for a given objective lens. By restricting the aperture of the detector, out-of-focus light is unable to pass the restricted aperture and thus does not reach the detector, limiting the signal to the focal plane.[6–10] Image processing software can then be used to reconstruct a series of optical sections in 3D, thereby making direct volume and surface area measurements of an object of interest.[4,11,12] When using the confocal technique, a match between optical section thickness and the spacing between optical sections establishes optimal sampling, i.e., the least amount of over- or undersampling. However, a perfect match between optical section thickness and stepper motor control of the Z axis is rarely achieved. Moreover, the small optical section thickness obtained when using objective

[1] B. Matsumoto, ed., "Cell Biological Applications of Confocal Microscopy." Academic Press, San Diego, 1993.

[2] J. B. Pawley, ed., "Handbook of Biological Confocal Microscopy." Plenum Press, New York, 1995.

[3] D. Schild, *Cell Calcium.* **19,** 281 (1996).

[4] J. K. Stevens, L. R. Mills, and J. E. Trogadis, eds., "Three-Dimensional Confocal Microscopy: Volume Investigation of Biological Specimens." Academic Press, San Diego, 1994.

[5] J. G. White, W. B. Amos, and M. Fordham, *J. Cell Biol.* **105,** 41 (1987).

[6] G. J. Brakenhoff, H. T. van der Voort, E. A. van Spronsen, and N. Nanninga, *J. Microsc.* **153**(Pt 2), 151 (1989).

[7] S. Inoue, *in* "Handbook of Biological Confocal Microscopy" (J. B. Pawley, ed.), p. 1. Plenum Press, New York, 1995.

[8] T. Wilson and C. J. R. Sheppard, "Theory and Practice of Scanning Optical Microscopy." Academic Press, London, 1984.

[9] T. Wilson, *J. Microsc.* **153**(Pt 2), 161 (1989).

[10] T. Wilson, *in* "Handbook of Biological Confocal Microscopy" (J. B. Pawley, ed.), p. 113. Plenum Press, New York, 1995.

[11] Y. S. Prakash, K. G. Smithson, and G. C. Sieck, *NeuroImage* **1,** 95 (1993).

[12] J. N. Turner, J. W. Swann, D. H. Szarowski, K. L. Smith, W. Shain, D. O. Carpenter, and M. Fejtl, *Int. Rev. Exp. Pathol.* **36,** 53 (1996).

lenses with high numerical aperture (NA) and magnification necessitates collecting a large number of optical sections. This large data set makes subsequent 3D reconstruction both computer and labor intensive. Accordingly, stereological methods may be applied to reconstruct 3D images from a set of serial optical sections, e.g., the Cavalieri principle.[13–15.] These stereological methods, combined with confocal microscopy, lessen the analytical burden of large image sets greatly and offer an unbiased approach to 3D rendering without any necessary assumptions about object shape, size, or orientation.

Advances in the speed and sensitivity of image acquisition of confocal microscopy have made it an ideal tool for the qualitative and quantitative assessment of cellular dynamics in 3D. Until recently, limitations in detector sensitivity and the need for frame averaging, as well as computer hardware and software limitations, had allowed enhanced 3D acquisition only at the expense of considerable reductions in image acquisition rates, especially when obtaining repetitive 3D images of cellular events over time (4D). Accordingly, realistic and reliable 4D confocal microscopy had been possible only under conditions where cellular dynamics were slow enough to be captured repetitively through the volume of a cell. However, confocal systems can now achieve high-sensitivity, faster-than-video rates of acquisition in one optical plane and, by combining this high-speed 2D imaging with rapid 3D optical sectioning, are capable of relatively rapid 4D acquisitions. Certainly, new investigations on a variety of cellular dynamics across a wide range of spatial and temporal response patterns can be expected in the near future.

The goal of this article is to focus on the techniques and limitations of using confocal microscopy for volume measurements. Practical solutions are offered for calibration and estimation of errors, particularly in the Z axis. Finally, specific examples are provided for the use of confocal microscopy in volume measurements.

Selection of Optics in Confocal Microscopy

The selection of the objective lens is particularly important when subsequent analysis of the optical sections includes 3D reconstruction of an object. It is beyond the scope of this article to provide a complete description of objective lenses (for a review, see Keller.[16]) However, several issues

[13] Y. Geinisman, H. J. Gundersen, E. van der Zee, and M. J. West, *J. Neurocytol.* **25,** 805 (1997).
[14] H. J. Gundersen and E. B. Jensen, *J. Microsc.* **147,** 229 (1987).
[15] Y. S. Prakash, K. G. Smithson, and G. C. Sieck, *NeuroImage* **1,** 325 (1994).
[16] H. E. Keller, *in* "Handbook of Biological Confocal Microscopy" (J. B. Pawley, ed.), p. 77. Plenum Press, New York, 1995.

regarding the selection of the objective lens have real significance, especially when imaging cellular structures requires a minimal loss of transmitted signal (e.g., fluorescence), minimal distortion, and maximal signal-to-noise ratio throughout the image field. A basic understanding of the microscope optics design is needed to optimally match the properties of the selected optical components.

When using confocal microscopy for volume measurements, NA is an important determinant in the selection of the objective lens. Generally, the higher the NA, the greater the XY spatial resolution and the thinner the optical section (see later). Another important consideration is the working distance of the objective lens, defined as the distance from the outermost part of the lens to the deepest focal plane in the specimen that can be achieved without compression of the specimen. In general, the working distance is shorter with high NA lenses, although it varies significantly among lenses and manufacturers. For example, an Olympus UV 40×/0.85 NA has a working distance of 0.25 mm, whereas an Olympus UV 40×/ 1.00 NA has only 0.16 mm. When focusing deep into a thick specimen, even lenses with excellent correction properties lose their optimal performance and images of lesser intensity and greater blur are obtained.[7] In many cases, specimen thickness can be optimized to the working distance of the objective lens.

Currently, there is no *one* fully standard, independent test to evaluate the performance of an objective lens. However, manufacturer-provided test procedures, such as the use of fluorescently labeled microspheres and microscopic grids, allow empirical assessment of lens performance under various experimental conditions. Most currently available lenses are near diffraction limited (i.e., the corrective properties of the objective lens have been adjusted to match the theoretical limits imposed by the medium refractive index and wavelength of light in use),[16] at least in the center of the field. However, the long-term use of laser illumination may deteriorate lens performance. In addition, lens aberrations can introduce deviations from the "theoretical" diffraction-limited image. Therefore, regular empirical assessment of lens performance is highly recommended.

Lens aberrations can be grouped as either monochromatic (wavelength independent) or chromatic. Monochromatic aberrations include spherical aberrations and usually can be avoided easily by the use of lenses with appropriate corrective optics (see manufacturer brochures for details). In particular, dry or water-immersion lenses of high NA become very sensitive to imaging conditions, such as the coverslip thickness and unmatched refractive indices.[17] In addition, when focusing deep into a specimen, oil-

---

[17] S. W. Hell and E. H. K. Stelzer, *in* "Handbook of Biological Confocal Microscopy" (J. B. Pawley, ed.), p. 347. Plenum Press, New York, 1995.

immersion lenses will no longer have an adequate spherical correction, with consequent loss of signal and resolution. In general, factors that increase spherical aberration also worsen other monochromatic aberrations, particularly when imaging off axis. Curvature of field is especially important for the spatial reconstruction of specimens imaged from thick materials because of the narrow depth of field of the confocal image and the dish-shaped distortion introduced by noncorrected objectives (nonplan).[18] Reconstruction of the 3D stack of optical sections could include correction for the curvature of the field off axis, but would make calculations more complex.

Chromatic aberrations arise from the fact that the refractive index (η) of every optical medium is dependent on the wavelength of incident light (dispersion), thus affecting the lens focal length for each wavelength (λ). If the objective lens is not properly corrected for the specific light wavelengths in use, the image along the Z axis at one wavelength will be different from that at another. Obviously, this is particularly important for double- or triple-fluorescence labeling of 3D objects, where registration along the Z axis is important, e.g., for colocalization of subcellular structures. Achromat objective lenses include optical materials of linear dispersion and are usually corrected for only two wavelengths. Apochromat or Semiapochromat lenses have excellent correction for chromatic aberrations at three or more wavelengths, but require glass of nonlinear dispersion, which unfortunately may be unsuitable for UV imaging (because of more rapid changes in the η at this end of the spectra).[16] Overall, the best results, in terms of fluorescence transmission and image resolution, are obtained when the emission and excitation wavelengths have similar corrections.[18]

In conclusion, a basic understanding of the major components of the optical pathway of the objective lens in confocal systems can help improve on image acquisition and resolution. Most importantly, the adequacy of individual components should be fully evaluated before using any confocal system for specific imaging protocols.

Optical Sectioning Using Confocal Microscopy

By virtue of its restricted aperture, the confocal microscope offers a narrow depth of field, reducing problems associated with light scattering. The significant reduction in out-of-focus information and the inherent spatial registration of sequential optical sections make the confocal microscope an excellent tool for direct analysis of 3D structures.[4,6–9] Thin optical sections of physically thick specimens can thus be obtained with complete spatial registration. Furthermore, attenuation of out-of-focus information

[18] D. R. Sandison, R. M. Williams, K. S. Wells, J. Strickler, and W. W. Webb, in "Handbook of Biological Confocal Microscopy." (J. B. Pawley, ed.), p. 39. Plenum Press, New York, 1995.

reduces the need for various image processing and enhancement techniques commonly used with standard light microscopy to improve the quality of images by mathematically removing out-of-focus information *post hoc.*[19-23]

The strength of confocal imaging thus lies in optimal sampling with optical sections and in the faithful reproduction of shape and size along the Z axis. It is, therefore, important to validate the confocal system in terms of the fidelity of optical sectioning and with regard to the distortions introduced, particularly along the Z axis.

In addition, an important consideration is the potential for photobleaching during optical sectioning, especially when image intensity is also a measured parameter and is critical in determining which parts of the image belong to an object and which do not. In particular, investigators commonly rely on threshold definitions in the calculation of object volume, and bleaching will render different estimations of the same object (or others along the illuminated column). Because an entire column of the specimen (along the Z axis) is illuminated by the confocal laser when acquiring images of a single plane, bleaching will occur outside of the focal plane.[24] Fortunately, the rate at which bleaching occurs is inversely proportional to the square of the distance from the focal plane. Nonetheless, as with all fluorescent imaging, limiting the exposure of the specimen to incident light only to that necessary for image acquisition will allow maximal preservation of the sample.

The selection of the most appropriate confocal imaging system for a particular application involves making several choices which may impact spatial and temporal resolution (i.e., the ability to distinguish two points of an object or two temporally separate events). It is well beyond the scope of this article to describe the theoretical aspects of confocal imaging and the advantages and limitations of currently available systems (for reviews, see Prakash *et al.,*[25] Wilson,[10] and Koester[26]). Nonetheless, a general under-

[19] D. A. Agard, Y. Hiraoka, P. Shaw, and J. W. Sedat, *Methods Cell Biol.* **30,** 353 (1989).

[20] D. A. Agard, *Annu. Rev. Biophys. Bioeng.* **13,** 191 (1984).

[21] H. Chen, J. R. Swedlow, M. Grote, J. W. Sedat, and D. A. Agard, *in* "Handbook of Biological Confocal Microscopy" (J. B. Pawley, ed.), p. 197. Plenum Press, New York, 1995.

[22] P. J. Shaw, *in* "Handbook of Biological Confocal Microscopy" (J. B. Pawley, ed.), p. 373. Plenum Press, New York, 1995.

[23] P. Shaw, *Histochem. J.* **26,** 687 (1994).

[24] J. B. Pawley, *in* "Handbook of Biological Confocal Microscopy" (J. B. Pawley, ed.), p. 19. Plenum Press, New York, 1995.

[25] Y. S. Prakash, M. S. Kannan, and G. C. Sieck, *in* "Fluorescent and Luminescent Probes" (W. T. Mason, ed.). Academic Press, London, 1998.

[26] C. J. Koester, *in* "Handbook of Biological Confocal Microscopy" (J. B. Pawley, ed.), p. 525. Plenum Press, New York, 1995.

standing of the basic differences among the various available confocal systems is required for selection of the optimal system for each particular study. Therefore, a brief description of the general classes of confocal systems is included with emphasis on their suitability for volume measurements.

The confocal principle is based on the introduction of an aperture limit in the light illumination pathway and a corresponding one in the transmitted light path such that only information from a single optical plane is obtained (by convention, Z-axis resolution).[3,7,27] The use of laser illumination permits coherent illumination of high intensity, allowing for better correction of possible chromatic aberration (see Selection of Optics). The illuminated specimen reflects light or emits fluorescence, which is registered by a sensitive detector, usually a photomultiplier tube or a cooled charge-coupled device. The signal over time is then digitized and stored in a computer.[21]

Scanning microscopes successively illuminate finite sections of the object, therefore increasing the lateral resolution (XY resolution) by reducing the light scattered by neighboring parts of the specimen.[24] Scanning introduces a temporal delay in acquisition of an image, especially when using a *pinhole* (circular) aperture. Different scanning methods are available, including stage and beam scanning systems, albeit with limited scan rates.[3,7,27,28] In order to improve temporal acquisition, the tandem scanning microscope[29] and acousto-optical modulator systems have been developed. A different solution is the use of a rectangular *slit* aperture to produce a confocal image.[30] The slit design is ideally suited for rapid acquisition rates with a reduced signal-to-noise ratio,[31,32] although it suffers from reduced axial (Z axis) resolution.[26] Therefore, depending on the fluorescence and intensity distribution in the specimen, the required XY and Z axis resolutions, and even perhaps on the required temporal resolution, a judicious choice must be made to allow for 3D measurements.

[27] S. J. Wright, V. E. Centonze, S. A. Stricker, P. J. DeVries, S. W. Paddock, and G. Schatten, *in* "Cell Biological Applications of Confocal Microscopy" (B. Matsumoto, ed.), p. 1. Academic Press, San Diego, 1993.

[28] J. J. Art and M. B. Goodman, *in* "Cell Biological Applications of Confocal Microscopy" (B. Matsumoto, ed.), p. 47. Academic Press, San Diego, 1993.

[29] G. S. Kino, *in* "Handbook of Biological Confocal Microscopy" (J. B. Pawley, ed.), p. 155. Plenum Press, New York, 1995.

[30] T. Wilson and S. J. Hewlett, *J. Microsc.* **160**(Pt 2), 115 (1990).

[31] W. B. Amos and J. G. White, *in* "Handbook of Biological Confocal Microscopy" (J. B. Pawley, ed.), p. 403. Plenum Press, New York, 1995.

[32] C. J. R. Sheppard, X. Gan, M. Gu, and M. Roy, *in* "Handbook of Biological Confocal Microscopy" (J. B. Pawley, ed.). Plenum Press, New York, 1995.

XY Axis Resolution

In light microscopy, the NA of the objective lens and the wavelength of the incident light (λ) determine the XY (horizontal plane, parallel to the stage by convention) spatial resolution as given by Eq. (1)[7]:

$$XY \text{ spatial resolution} = 0.61 \frac{\lambda}{\text{NA}} \qquad (1)$$

Currently available confocal systems and objective lenses approximate the theoretical diffraction limit (0.2 μm; described by Abbe[33]) using visible light sources. However, aberrations in the optical path usually limit actual XY spatial resolution to ~0.3 μm.

Available confocal imaging systems digitize the optical image within an optical section into picture elements (or pixels).[34] The pixel dimension along any axis is given by the ratio of the total length scanned to the total number of pixels along that axis. In accordance with the practical limit of objective lenses, the finest useful pixel dimension is obviously ~0.3 μm (pixel resolution). However, it must always be kept in mind that pixel dimension need not necessarily match optical resolution (see Webb and Dorey[34] for a detailed discussion). For example, when using low magnification lenses such as a 10 or 20×, the scanned areas are large, but given the fixed total number of pixels (typically 256 to 1024), the pixel resolution is likely to be worse than the optical resolution, and there is considerable lost data. With lenses of increased magnification, pixel resolution may approximate the optical resolution, optimizing the digitization process. Magnifications beyond this point result in empty magnification (oversampling), where pixels of decreased dimensions do not provide additional morphological information. The actual minimal resolved area is not changed, but is now represented by a greater number of pixels. However, under certain conditions, this oversampling may actually be used advantageously by limiting the scanned area but maintaining the total number of pixels (hardware zoom), particularly when imaging objects containing periodic data, e.g., Z lines in skeletal or cardiac muscle sarcomeres.

The optimal spatial resolution for acquisition is dictated by the Nyquist theorem, such that sampling should be at least twice the spatial frequency of the feature being imaged within an object.[24,34] In practice, this means obtaining images where the pixel resolution is adjusted to be less than half

[33] E. Abbe, *J. R. Microsc. Soc.* **4,** 348 (1884).
[34] R. H. Webb and C. K. Dorey, *in* "Handbook of Biological Confocal Microscopy" (J. B. Pawley, ed.), p. 55. Plenum Press, New York, 1995.

the dimension of any periodicity within the specimen.[34,35] For example, Z lines are typically separated by 2.5 μm in skeletal muscle fibers. Thus, the optimal spatial resolution of acquisition needs to be at least 1.25 μm. These spatial acquisition limits do not apply only to periodic features within objects, but dictate the ability to resolve differences between features (two-point discrimination). Thus, in the example provided earlier, a 10× objective lens with a 0.5 NA is barely sufficient to resolve Z line periodicity using visible light (XY resolution of ~0.62 μm at 515 nm). However, the Nyquist theorem only establishes the minimal requirements for spatial resolution. Oversampling offers distinct advantages, particularly in resolving finer features. In the example given earlier, using the 10× objective lens permits a hardware zoom of approximately two times with no loss of spatial resolution. In contrast, neuronal dendrites are frequently 1.0 μm in diameter and bundled with an intervening space of ~1.0 μm. Thus, in order to discriminate between two adjacent dendrites, the spatial resolution has to be at least 0.5 μm. Using a 20× objective lens with a 0.7 NA the XY spatial resolution is ~0.44 μm, which is barely sufficient, but would not allow for hardware zoom. In contrast, a 60× objective lens with a 1.3 NA provides an XY spatial resolution ~0.24 μm, approximating the theoretical limit of light microscopy. The use of the 60× objective lens in this specific example also offers the advantage of applying a two times hardware zoom, which will resolve finer details in dendritic structure.

Z-Axis Resolution

The major source of error in volume measurements using confocal microscopy is related to a mismatch between optical section thickness (Z-axis resolution) and the sampling interval along the Z axis (step size). Optical section thickness depends on the NA of the objective lens, the wavelength of incident light, the size of the confocal aperture, and the refractive index (η) of the surrounding medium and tissue, as given by Eq. (2)[3,7]:

$$Z\text{-axis resolution} = \frac{0.45\lambda}{\eta(1 - \cos[\sin^{-1}(NA)/\eta])} \qquad (2)$$

Thus, optical section thickness is related inversely to the NA of the objective lens and related directly to the wavelength of incident light. Decreasing aperture size increases Z-axis resolution, but significantly re-

[35] V. Centonze and J. Pawley, *in* "Handbook of Biological Confocal Microscopy" (J. B. Pawley, ed.), p. 549. Plenum Press, New York, 1995.

duces the amount of light reaching the detector with consequent loss of information and worsening of the signal-to-noise ratio.[10,24] Therefore, in selecting the aperture setting, a compromise must be reached between optical section thickness and signal-to-noise ratio. In general, the better the match among the refractive indices of the objective lens, coverslip glass (if any), and the tissue mounting medium, the greater the Z-axis resolution.[17] The theoretical Z-axis resolution can be calculated using Eq. (2) e.g., with a $60\times/1.3$ NA oil-immersion objective lens at 515-nm emission wavelength, the theoretical Z-axis resolution is $\sim$0.3 μm. However, several factors, such as lens properties (chromatic aberration and flatness of field) and mismatches in refractive indices, reduce the actual Z-axis resolution. Therefore, in several practical applications involving volume measurements, the major goal cannot be to achieve the best (theoretical) Z-axis resolution, but to minimize or account for distortions along the Z axis.

Based on its theoretical determinants,[7,24] the Z-axis resolution is at best twice the XY resolution, and given the number of factors that can affect practical Z-axis resolution, it is difficult to calculate exactly. However, the overall Z-axis distortion can be determined empirically for any particular confocal system.[11] For example, fluorescently labeled latex microspheres of varying diameters embedded in tissue mounting medium can be imaged under different conditions. Microsphere diameters can be measured in both the XY and the XZ planes and compared to manufacturer-specified values. The ratio of specified diameter to measured XZ diameter serves as an estimate of overall Z-axis distortion under the specific conditions of tissue preparation, optics, and Z-axis step size. Although not the focus of the present discussion, this approach can be utilized to estimate the relative contribution of different factors to the overall distortion. For example, the distortion introduced solely by the mismatch between the refractive indices of embedding medium and the objective lens immersion oil can be estimated by mounting the fluorescent microspheres in the immersion oil rather than tissue mounting medium.[11]

Under a given experimental condition, the actual Z-axis resolution can also be determined from the point-spread function (PSF) of the objective lens (for a detailed review, see Hell and Stelzer,[17] Agard et al.,[19] and Hiraoka et al.[36]). Intensity profiles along the Z axis are obtained when imaging objects of finite size (e.g., 0.2-μm fluorescent beads), which can be averaged to estimate a mean PSF. The full-width half-maximum (FWHM) of the mean profile is then taken to represent the optical section thickness (Fig. 1).

[36] Y. Hiraoka, J. W. Sedat, and D. A. Agard, *Biophys. J.* **57,** 325 (1990).

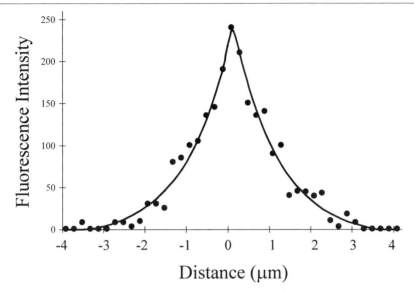

FIG. 1. Empirically determined mean point-spread function (PSF) for a 40×/1.3 oil-immersion objective lens (Nikon Instruments). Fluorescently labeled microspheres (0.2 μm in diameter) were sectioned optically at 0.18-μm step size. The average PSF of the individual microspheres was obtained after alignment along the maximal axis (individual data points) and was then fitted in a modified four-parameter Gaussian regression curve (line). The full-width half-maximum was estimated to be 0.8 μm.

Matching Optical Section Thickness to Step Size

The match between the optical section thickness and the increment between optical sections, as determined by the stepper motor (step size), determines the optimal sampling conditions in the Z axis and, ultimately, the reliability of the volume measurements.[10] A number of commercially available confocal systems are equipped with stepper motors that control the fine focus knob of the microscope. Most stepper motors have a step size resolution of 0.05 to 0.20 μm, which is more than sufficient when compared to the Z-axis resolution of light microscope objective lenses. Accordingly, the use of a step size smaller than the optical section thickness results in repeated sampling of more or less the same optical plane. In contrast, use of a step size larger than the optical section thickness results in an undersampling of the specimen with a consequent loss of information. An example of the impact of over- and undersampling in the Z axis is provided in Fig. 2. Using a Bio-Rad (Richard, CA) MRC 600 confocal imaging system and a 40×/1.3 NA Fluor objective lens (Nikon Instruments), confocal images were obtained of a 10.3-μm fluorescent (488/515 nm) latex

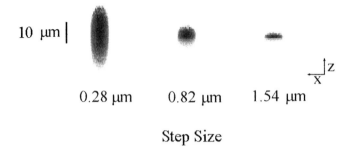

10 µm |

0.28 µm   0.82 µm   1.54 µm

Step Size

FIG. 2. Two-dimensional images of fluorescently labeled latex microspheres (FluoSpheres, Molecular Probes, OR) are shown. Measurement of the microsphere diameter on the XY plane correlates closely with the manufacturer's specification (10.3 µm). Measured diameters on the XZ plane, however, show the effect of varying the sampling interval (step size).

microsphere suspended in immersion oil (to minimize distortions due to refractive index mismatches). In the XY plane the measured diameter of the microsphere matched the specifications of the manufacturer closely. At the smallest pinhole aperture size, the step size was adjusted in 0.18-µm increments (stepper motor resolution). As can be seen, the measured diameter of the microsphere in the XZ plane depended on step size. At 0.28-µm step size, microsphere volume was overestimated by ~70% (if no interpolation is performed). In contrast, at 1.54-µm step size, microsphere volume was underestimated by ~45% (without interpolation).

A common mistake is to assume that step size is per manufacturer specification. Accuracy of the stepper motor can be affected severely by backlash and hysteresis. The direction of the focus motor movement determines the size of the step (hysteresis). In addition, changes in direction of the focus motor step can introduce movement of the stage or objective turret (backlash). Hysteresis and backlash in the stepper motor can be estimated by moving the motor from a known initial position to a final position and back, measuring the error in placement. Clearly, the accuracy of optical sectioning is critically dependent on the performance of the Z-axis stepper motor. Errors in the step size introduced by the stepper motor would not directly affect each optical section, but would generate discrepancies between the optical section thickness and the actual step. Furthermore, the specimen thickness may also be inferred incorrectly. As described earlier, this could lead to under- or oversampling and incorrect 3D rendering. Performance of the stepper motor can be determined empirically with the use of large (>10 µm diameter) fluorescent microspheres immersed in oil. The number of steps needed to optically section through the microsphere provides an estimate of actual step size.[11]

Estimation of Errors in Confocal Volume Measurements

Errors Introduced by Tissue Compression

Tissue preparation and mounting are important considerations in volume measurements using confocal microscopy.[37] Most importantly, the possibility of tissue compression during optical sectioning should be evaluated, especially with high NA lenses with short working distances. One simple approach is to determine the number of optical sections required to scan through a sample of known thickness, e.g., a 50-μm-thick section of densely stained tissue. In cryosections, sample thickness can be calibrated independently using optical densitometry.[38]

Errors Introduced by Z-Axis Distortion

The greatest source of error in confocal volume measurements is introduced by Z-axis distortion due to a number of factors (see earlier). Although the contribution of each of these factors could be evaluated independently, the most practical approach is to determine empirically the overall distortion in the Z axis by assessing confocal images of fluorescent microspheres of known diameter.[11] Even under optimized conditions matching optical section thickness and step size, some Z-axis distortion will be present. For example, in the illustration shown in Fig. 2, the optical section thickness was 0.84 μm as calculated by the point-spread function of the 40×/1.3 NA oil-immersion objective lens. At step sizes of 0.82 and 1.00 μm, the XZ diameter of the 10.3-μm fluorescent (488/515 nm) latex microsphere suspended in immersion oil was 10.6 and 9.5 μm, respectively. In contrast, in the XY plane the measured diameter of the microsphere matched the specifications of the manufacturer closely. The differences in measured diameter of the microsphere in the XZ plane reflected Z-axis distortion. These types of measurements can be repeated for microspheres located at different depths within the working distance of the objective lens. Generally, Z-axis distortion worsens at the distal part of the working distance of the objective lens, i.e., deeper in the specimen. For example, if fluorescent microspheres were imbedded in mounting medium, rather than immersion oil, and imaged, the Z-axis distortion is more pronounced and approximates tissue conditions more closely. In this case, we have estimated an ~16% overestimation of microsphere volume.[11] Such volume measurement errors

[37] R. Bacallao, K. Kiai, and L. Jesaitis, *in* "Handbook of Biological Confocal Microscopy" (J. B. Pawley, ed.), p. 311. Plenum Press, New York, 1995.
[38] C. E. Blanco, M. Fournier, and G. C. Sieck, *Histochem. J.* **23**, 366 (1991).

488/515   568/590

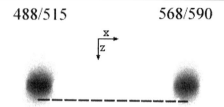

Fig. 3. Double-labeled fluorescent microspheres (4 μm in diameter; MultiSpeck, Molecular Probes Inc., OR) were imaged using a two-channel Bio-Rad MRC 600 confocal system with an Olympus 40×/1.3 NA objective lens. Correct alignment of the images at the two different wavelengths (emission: 488 and 568 nm, excitation: 515 and 590 nm, respectively) provides confirmation of the adequacy of instrument selection and calibration.

attributed to Z-axis distortion are fairly consistent and therefore can be appropriately corrected.

Error Introduced by Multicolor Imaging

An additional error can be introduced in volume measurements of multicolor images because different light wavelengths are focused by the objective lens at different planes (dispersion). Therefore, image registration becomes a problem when using different fluorescent indicators, e.g., the use of double or triple labeling to colocalize subcellular structures. To avoid this problem, the effects of dispersion introduced with each objective lens should be established. This can be accomplished by obtaining XZ sections of double- or triple-labeled fluorescent microspheres (e.g., MultiSpeck microspheres with excitation/emission wavelengths of 505/515 and 560/580 nm; Molecular Probes Inc.). In the example shown in Fig. 3, a two-channel Bio-Rad MRC 600 confocal system was used to image 4-μm microspheres (MultiSpeck) at two different wavelengths (excitation light: 488 and 568 nm, emission light: 515 and 590 nm, respectively). Registration error can be estimated by the XZ depth of microsphere at each wavelength. In addition, the difference between XY and XZ diameters can be used as an index of the overall distortion along the Z axis for each individual wavelength.[39]

Practical Limitations of Using Confocal Microscopy for
  Volume Measurements

Size of Data Set

As with other imaging technologies, improvements in confocal imaging have been accompanied by increased data content. Developments in com-

---

[39] Y. S. Prakash, K. G. Smithson, and G. C. Sieck, *J. Neurocytol.* **24,** 225 (1995).

puter technology have considerably expanded the capabilities of data acquisition hardware and software at much reduced cost. However, investigators are still faced with the immense task of probing through large numbers of optical sections. Furthermore, given the 2D nature of computer displays, manipulation and presentation of a set of optical sections can be a daunting task.[21,40,41]

Several unbiased, stereological techniques have been used to estimate 3D variables (such as volume) from a limited set of 2D images. These techniques make no assumptions of object shape or orientation; therefore, they are applicable to any situation where random samples are selected.[42-45] The Cavalieri principle is an example of such an unbiased stereological technique that has been verified extensively.[13,14,46] The tissue sectioning protocol of the Cavalieri principle systematically obtains parallel sections of 3D objects. Starting with a random section at one end of the object, uniformly spaced sections are sampled for measurement. Areas from regions of interest are measured in each section and are multiplied by the section spacing to obtain an estimate of volume for these areas. The uniformly spaced sections of fixed thickness are then used to estimate object volume by linearly interpolating between sampled sections.[44] The method of optical sectioning in confocal microscopy also obtains a series of parallel sections of a 3D object; however, unlike the Cavalieri principle, the spacing between sections can be zero with complete confocal optical sectioning, i.e., no interpolation is performed. The optical sectioning technique of confocal microscopy can thus be considered a limiting case of the Cavalieri principle, and therefore the Cavalieri protocol can be applied to digitized images from a confocal system.[11] In addition, in the Cavalieri principle, there is no need to match section thickness to the spacing between sections. Therefore, by interpolating between adjacent sections, the Cavalieri principle offers the advantage of reliable estimation of volume from a reduced number of confocal sections. Application of such an interpolation technique to a large set of confocal optical sections reduces data size and makes computations and measurements easier.

In the example shown in Fig. 4, a combination of confocal microscopy and the Cavalieri principle was applied to measurements of phrenic motoneuron somal volumes. In rats, the phrenic motoneuron pool was labeled

[40] L. Lucas, N. Gilbert, D. Ploton, and N. Bonnet, *J. Microsc.* **181**(Pt 3), 238 (1996).
[41] S. Sabri, F. Richelme, A. Pierres, A. Benoliel, and P. Bongrand, *J. Immunol. Methods* **208,** 1 (1997).
[42] L. M. Cruz-Orive and E. R. Weibel, *Am. J. Physiol.* **258,** L148 (1990).
[43] T. M. Mayhew, *Exp. Physiol.* **76,** 639 (1991).
[44] T. M. Mayhew, *J. Neurocytol.* **21,** 313 (1992).
[45] J.-P. Royet, *Prog. Neurobiol.* **37,** 433 (1991).
[46] R. P. Michel and O. L. Cruz, *J. Microsc.* **150,** 117 (1988).

A   B

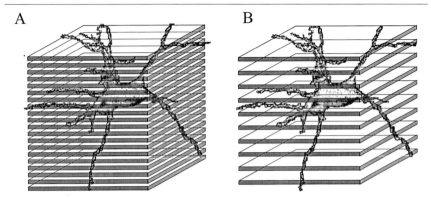

FIG. 4. Application of the Cavalieri principle to estimate object volumes from confocal image sets. Starting with a randomly determined section within the first n sections, every nth successive optical section is selected. A phrenic motoneuron was optically sliced, and the complete stack of optical sections is seen (A). The Cavalieri principle allows manipulation of a much reduced set of optical sections (B), with minimal loss of information.

retrogradely by injection of cholera toxin B fragment into the diaphragm muscle 2 days prior to tissue collection.[15] The spinal cord was removed, fixed, and immunostained with Cy5-conjugated antibody to cholera toxin B fragment. Optical sections of the rat cervical spinal cord at 0.6 μm were obtained using a Bio-Rad MRC 600 laser confocal system mounted on an Olympus (BH2) upright microscope (with an Olympus ApoUV 40×/NA 1.3 oil-immersion lens). The long axis of phrenic motoneurons ranges from 30 to 50 μm. Step size was set at different sampling intervals, ranging from 0.6 to 3.0 μm. Thus, the number of sampled optical sections containing segments of a phrenic motoneuron varied from 10 to 80, depending on the sampling interval. Previous studies have suggested that reliable volume estimates can be obtained from approximately 10 sections using the Cavalieri protocol.[45] There was no difference in the mean values of phrenic motoneuron somal volumes across the different step sizes (Fig. 5). However, it should be stressed that the size of the set of optical sections using a 3.0-μm step size was only 20% of that using a 0.6-μm step size.

Time Constraints in Data Acquisition

Spatial resolution is dependent only on the inherent properties of the optics involved. However, acquisition of the image results in an interdependency between spatial and temporal resolution. Depending on the confocal system, there are temporal limitations in image acquisition in both XY and XZ planes. The introduction of acousto-optical control has markedly enhanced temporal resolution in the XY plane. Confocal systems are cur-

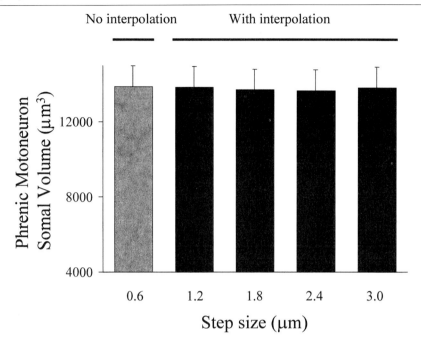

FIG. 5. Bar graph of motoneuron somal volume estimates using different step size intervals with the Cavalieri principle. Noninterpolated confocal measurements correspond to a step size of 0.6 μm. The bar represents mean somal volumes and the error bar the standard error. There is no difference in the mean somal volumes or variance when imaging 20% of the image set (step size 3.0 μm) using the Cavalieri principle.

rently available and have full-frame (640 × 480 pixels) acquisition at video rates (30 frames/sec) and limited frame (640 × 16 pixels) at up to 480 frames/sec. Of course, with the faster acquisition rates, pixel dwell time is reduced and either laser intensity must be increased or frame averaging must be applied to provide an adequate signal-to-noise ratio.

Even with the improvements in frame acquisition rate, the main limitation for repeated volume measurements in time (4D analysis) is control of the Z axis. With control of the movement of the microscope stage, the mass of the stage introduces significant inertia, causing marked delays for optical sectioning. For example, under unloaded conditions (with motor not attached to the focusing knob of the microscope or attached to the microscope but with no lenses or specimen stage), the focus motor of a Noran Odyssey XL confocal imaging system can achieve a single step in ~20 msec. However, as with most other focus motors, the Odyssey system also has extremely fine step resolutions (50 nm). Because optical section thickness is optimal at ~0.6–1.0 μm, several single motor steps are required. Therefore, ~400 msec is required to achieve a 0.8-μm step size. In addition, delays in software control and communication with the focus motor may prolong this time even further. Although it is difficult to directly determine the temporal resolution for an imaging system during 3D volume measurements, an average time per image can be obtained by measuring the time taken for collecting a set of optical sections. For example, collecting a set of 0.8-μm optical sections through a 10-μm distance requires ~6 sec using the Odyssey system, representing a ~500-msec delay for each individual step. The time for image acquisition also contributes to the overall delay in 4D imaging, but compared to the delay introduced by Z-axis control, this delay is relatively insignificant. For example, at video rate (30 frames/sec), frame acquisition requires ~33 msec. Increasing the acquisition rate to 480 frames/sec saves only ~30 msec.

The temporal resolution of 4D volume measurements may be improved by using a faster stepper motor. However, faster motors are also likely to suffer from oscillation artifacts due to the rapid starts and stops ("ringing"), particularly if massive stages are controlled. Therefore, considerable delay may be introduced waiting for the oscillations to dampen. An alternative approach is use of piezoelectric control of the microscope objective lens. For example, when using piezoelectric control of the objective lens on a Noran Odyssey confocal system at 480 frames/sec and 1-μm step size, 10 optical slices can be obtained in ~67 msec. The major cause of this delay results from the fact that image acquisition is noncontinuous. An optical section is obtained at a set depth, then after the next Z-axis position is reached, another optical section is obtained. This incremental adjustment of the Z-axis position is time-consuming because of the mechanical linkages involved. Improvements in 4D acquisition rate could be achieved if Z-axis control was continuous and image acquisition was synchronized to the beginning and end of objective lens movement. The advantage of this approach is that the interaction between image acquisition and Z-axis control is minimal. However, the image set would contain a motion-blurred representation of the object, but it would be possible to deblur the images using mathematical transforms that are currently being used to correct motion artifacts in MRI and CT images.[47]

Biological Applications of 4D Analysis

Obviously many intracellular events occur in space, e.g., within localized compartments of the cell. These events can be sampled repeatedly during

---

[47] J. C. Russ, "The Image Processing Handbook." CRC Press, Boca Raton, FL, 1995.

the collection of a set of 3D images. As mentioned earlier, the main limitation currently is control of the Z axis. In the collection of sequential sets of 3D optical sections, if there is no time delay between sequential 3D sets, the time elapsed before a given optical plane is sampled again is simply the total time taken to collect one full stack of 3D images. Under these conditions, the 4D resolution obviously depends on the number of 3D sections. When the different factors contributing to time delays in sequential 3D image acquisition are considered (see earlier), it may be quite difficult to attain a 3D temporal resolution better than ~600 msec. For example, we repeatedly imaged a 3D set of optical sections of an isolated porcine airway smooth muscle cell. With the Noran Odyssey XL confocal system, 10 optical sections with a 1-μm step size can be obtained at 15 frames/sec, i.e., every 667 msec. The Nyquist theorem also applies to 4D analysis of biological intracellular events; therefore, only cellular events occurring with a frequency of every 1334 msec or more can be detected reliably. From previous studies,[25] the temporal profile of Ca^{2+} sparks in a porcine airway smooth muscle showed a ~200-msec peak within a very localized area of the cell. Therefore, these imaging conditions can fortuitously detect the occurrence of sparks in airway smooth muscle cells and would not allow the characterization of the spark (Fig. 6).

In addition to the study of localized cellular events, 4D confocal imaging holds promise for the study of events where the primary result is the actual change in morphology (i.e., shape and/or size).[41,48,49] In particular, studies relating to cell division can benefit from relatively rapid 3D image acquisition rates, whereas studies of cell differentiation may require more prolonged intervals between sampling. For example, in an article by Errington et al.,[50] the authors studied volume regulation of living chondrocytes in response to osmotic stress. This and other applications are of great interest in the biological sciences.

Limitations in Image Processing and Display

Innovative solutions for the limitations imposed by the rate of data storage include resizing of the image to increase the rate of data acquisition from a smaller frame. However, this can result in important loss of spatial information. Obviously, more research has to focus on methods to improve data collection.[21] Fortunately, the impressive improvements in video hardware and software and the ever-faster computer display terminals are likely

[48] S. W. Paddock, *Proc. Soc. Exp. Biol. Med.* **213,** 24 (1996).
[49] W. M. Petroll, J. V. Jester, and H. D. Cavanagh, *Int. Rev. Exp. Pathol.* **36,** 93 (1996).
[50] R. J. Errington, M. D. Fricker, J. L. Wood, A. C. Hall, and N. S. White, *Am. J. Physiol.* **272,** C1040 (1997).

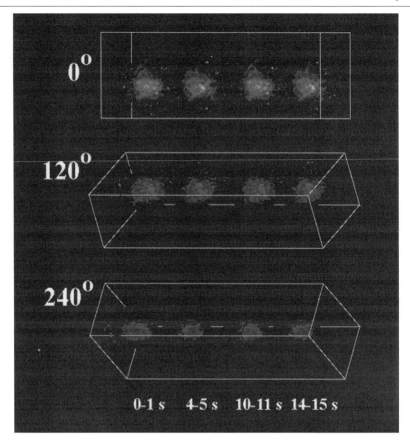

Fig. 6. Rotational views of temporal changes in the 3D volume distribution of Ca^{2+} within a spark in a TSM cell. Such rotational views of 3D reconstructions allow representation of 4D data sets such that temporal changes in all three spatial dimensions can be evaluated. In this example, the volume distribution of the Ca^{2+} spark does not appear to change considerably over time.

to reduce problems relating to rapid display greatly. Furthermore, access to extremely rapid, inexpensive data storage devices will also alleviate some of the data storage problems. For example, with new video display technology and computer hardware, single images of reasonable size (e.g., 1024×1024 pixels) can be displayed easily and saved within 30–50 msec. However, because of the huge amounts of image data that can be stored with 4D confocal imaging, it is essential that the rate of image acquisition be chosen based on the process of interest, e.g., 4D registration of intracellular dynamics or changes in cellular morphology.

After the image set has been collected, the display of the optical section information should allow easy, accessible interpretation of the data content.[40,51] For 3D objects, this usually means a 2D representation, using either a computer display or paper. Most image processing software packages with capabilities for 3D reconstruction employ special algorithms such as voxel gradient shading, depth gradient shading, surface shading, and maximum intensity projections (see Russ[47] for a review). Using these algorithms, the 3D reconstructions can be viewed at any desired angle and resectioned if necessary to view the interior of the specimen. The processing of these large amounts of information, imported into image processing packages such as ANALYZE, VoxelView, and NIH Image, still represents a very time and effort consuming task. While the visualization environment for 3D data sets appears to be well under control, the adequate representation of 4D data is still an unresolved issue, although some impressive options exist.[25] One option is to display a 2D view of the 3D reconstruction at one defined angle. The time sequence of 3D reconstructions at a certain angle would appear as a movie (when the successive images are played back as video frames). However, when displaying only the view from only one angle, the representation has eliminated the 4D character of the data and made it only 3D (2D spatial over time). However, this can be obviated by the simultaneous display of multiple angles of view, which would prove very valuable, particularly for the spatial analysis of intracellular dynamics over time. In addition, rotational sets can be constructed for successive 3D data sets, where the entire 4D data set rotates as the time sequence is displayed. Obviously, the choice of an appropriate visualization procedure will depend on the desired point to emphasize. For example, in order to show the temporal changes in the volume occupied by an object (or even compartments within the object), temporal sequences of 3D reconstructions at one or multiple angles would yield the most information in a user-friendly manner. The impressive improvements that are being made constantly in computer technology will surely help in this regard, and evidently, constant attention to optimal data acquisition could limit the amount of data collected to that necessary to address the question(s) of interest.

Acknowledgments

This work is supported by Grants HL34817, HL37680, GM57816-01, and HL57498 from the National Institutes of Health and the Mayo Foundation.

[51] N. S. White, *in* "Handbook of Biological Confocal Microscopy" (J. B. Pawley, ed.), p. 211. Plenum Press, New York, 1995.

Section IV

Imaging of Processes

[18] Quantitation of Phagocytosis by Confocal Microscopy

By GEORGE F. BABCOCK

Quantification of Phagocytosis by Confocal Microscopy

Phagocytosis has traditionally been defined as being mediated by leukocytes and functioning as the first line of defense against infection. These phagocytic leukocytes, which include neutrophils, monocytes, and tissue macrophages, are considered professional phagocytic cells.[1] It is now clear that many other cell types are also capable of phagocytosing various types of particles.

The phagocytic process is the engulfment of particles, which involves the expansion of the plasma membrane to engulf the particle. When performing assays, it is important to separate the events by which the plasma membrane is extended (phagocytosis) from the events that involve the inversion of the plasma membrane (endocytosis). These two processes are often difficult to separate experimentally. Because the endocytic process is only capable of capturing smaller particles, this difference in particle size can be used to differentiate the two processes.

Phagocytic cells are capable of engulfing particles of less than 1 μm in size to large particles up to many micrometers in size. In fact, there have been reports of cells phagocytosing other cell types, which are larger actual phagocytosing cells. Phagocytosis may be divided conceptually into a two-step process.[2,3] The first step is the attachment of the particle to receptors on the plasma membrane of the phagocytotic cell. The second step involves internalization of the bound particle. Problems associated with the quantification of phagocytosis are related to the nature of this two-step process. The number of particles or relative amount of material phagocytosed must be quantified. If the cell type being examined is a professional phagocyte, during and following phagocytosis, the cells produce large amounts of reactive oxygen intermediates.[4] In conjunction with digestive enzymes discharged into the phagosomes, these compounds may result in the quenching

[1] R. Bjerknes, *in* "Phagocyte Function" (J. P. Robinson and G. F. Babcock, eds.), p. 187. Wiley Liss, New York, 1998.

[2] S. C. Silverstein, R. M. Steinman, and Z. A. Cohn, *Annu. Rev. Biochem.* **46,** 669 (1977).

[3] S. J. Klebanoff and R. A. Clark, "The Neutrophil: Function and Clinical Disorders." North-Holland, Amsterdam, 1978.

[4] B. M. Babior, *Blood* **64,** 959 (1984).

TABLE I
OPSONIN BINDING RECEPTORS ON PHAGOCYTIC CELLS

Receptor	Cluster designation	Ligand	Phagocytic cell expression
FcγI	CD64	IgG_1, IgG_3	Monocytes, neutrophils
FcγII	CD32	IgG_1, IgG_2, IgG_3, IgG_4	Monocytes, neutrophils
FcγIII	CD16	IgG_1, IgG_3	Monocytes, neutrophils, natural killer cells
CR1	CD35	C3	Monocytes, neutrophils
CR3	CD11b/CD18	IC3b	Monocytes, neutrophils, natural killer cells
Various lectin receptors	None	Mannose among other carbohydrates	Most cell types

of some fluorochromes and/or the destruction of the phagocytosed particle itself. This is further complicated by the fact that the phagocytosed particles, which must be quantified, are intracellular, so in addition to quenching or destruction of the particles, intracellular organelles and/or granules may obscure the particle. An additional complication relates to the fact that particles bind to the outside of cells before being internalized. This external binding, which may be the stronger of the two signals, must be separated from internalization. Confocal microscopy allows the user to focus on various planes within the cell, which permits the investigator to separate truly internalized particles from externally bound particles.

Opsonins and Receptors

Some particles in their native form may be ingested by cells without further treatment. However, the phagocytosis of most microorganisms and many other types of particles requires or is enhanced greatly by serum factors termed opsonins.[2,3] Opsonins form a bridge between the particle and the phagocytic cell.[5] The major opsonins include immunoglobulins of the IgG class and complements components of the C3 class. These opsonins bind to several important receptors on phagocytic cells and initiate adherence and uptake of the opsonin-coated particle (Table I). Even in model systems where opsonins are not required, the assay benefits from their inclusion.

[5] E. J. Brown, *BioEssays* **17,** 109 (1995).

Methods Used to Measure Phagocytosis

Many procedures and variations of these procedures have been used to monitor phagocytosis. Some of the common methods are quantitative whereas others are qualitative. Almost any particle of the proper size can be used to measure phagocytosis. However, most of the published phagocytosis procedures have utilized bacteria, yeast, red blood cells, or beads (polystyrene). Particles used for measuring phagocytosis are usually labeled with some type of reagent, which allows for easier detection and more sensitivity. Particles have been labeled with opaque dyes, fluorescent dyes, and radionuclides such as ^{3}H or ^{57}Cr among others.[6-8] In some cases, the quantification of phagocytosis has been accomplished using viable organisms and determining the number of internalized colony-forming units (cfu).[9]

Confocal microscopy and flow cytometry have become the methods of choice for analyzing phagocytosis, as these methods can provide quantifiable data in a relative short time compared with most other methods. Although confocal microscopy can be utilized for phagocytosis with a particle, which can be visualized by ordinary light microscopy, cells or particles labeled with fluorescent dyes are the most popular. By exciting the fluorescent particle with the appropriate wavelength of light, the particles become much easier to detect and quantify.

Choosing Proper Particle for Phagocytosis

Each investigator must decide which type of particle is optimal for their particular model. Some of the properties, as well as the pros and cons, for each of several types of particles are discussed in Table II. It should be noted that Table II is not all inclusive and represents only a partial list.

Fluorochrome Labeling of Bacteria

Latex particles can be purchased with fluorescent labels incorporated either into the particle or on the surface of the bead. However, the investigator must label the bacteria or yeast. Many dyes can be used to label bacteria; however, fluorescein is the most commonly used. Although it is not neces-

[6] R. C. Bjerknes, C. F. Bassoe, H. Sjursen, O. D. Laerum, and C. O. Solberg, *Rev. Infect. Dis.* **11,** 16 (1989).
[7] M. E. Scheetz, L. J. Thomas, D. K. Allemenos, and M. R. Schinitsky, *Immunol. Comm.* **5,** 189 (1976).
[8] E. M. Goodell, S. Bilgin, and R. A. Carchman, *Exp. Cell Res.* **114,** 57 (1978).
[9] A. A. Amoscato, P. J. Davies, and G. F. Babcock, *Ann. N.Y. Acad. Sci.* **419,** 114 (1983).

TABLE II
PARTICLES USED FOR PHAGOCYTIC ASSAYS

Particle	Detection	Advantages	Disadvantages
Polystyrene beads	Can be observed without further labeling; can be purchased with broad-spectrum fluorochrome incorporated into bead; specific fluorochromes can be bound covalently to bead surface	Can be obtained in many sizes; incorporated fluorochromes are not quenched easily; resistant to destruction with most cellular enzymes; specific ligands can be attached to surface of beads to examine receptor binding	Nonviable, nonbiological particle; multiple receptor–ligand interactions difficult to examine; may bind inherently to lectin receptors
Bacteria	Can be labeled with variety of fluorochromes	Can be used to measure microbiocidal activity and phagocytosis	Difficult to see without labeling due to small size; fluorochrome labeling may change adherence and phagocytosis; can be destroyed by professional phagocytes and fluorescence quenched
Yeast	Can be labeled with variety of fluorochromes; can be observed without labeling (externally)	Same as for bacteria	Same as for bacteria; difficult to observe internally with labeling
Red blood cells	Can be observed without labeling; autofluorescent at certain excitation wavelengths	Difficult to quantify by confocal microscopy other than by manual counting	Can be destroyed by cellular enzymes, making it difficult to detect; rather large particle

sarily the best reagent to use for this purpose, it is reasonably priced and easy to use. Two methods are commonly employed to label bacteria or yeast with fluorescein. The simplest is to add fluorescein isothiocyanate (FITC) directly to the culture medium.[10] To 100 ml of the appropriate culture medium (trypticase soy broth, brain heart infusion, etc.), 25 μg of FITC is added and stirred gently. All remaining procedures should be performed in reduced lighting, especially fluorescent lighting. The medium is then inoculated with the organism (bacteria or yeast) and incubated with shaking for 18 hr at 37°. Bacteria are then collected by centrifugation at 2000g for 20 min at 4°. The organisms should be washed three times in a large volume of phosphate-buffered saline (PBS, 250 ml), recollected, and resuspended in Hanks' balanced salt solution (HBSS) to the appropriate

[10] C. White-Owen, J. W. Alexander, R. M. Sramkoski, and G. F. Babcock, **30,** 2071 (1992).

concentration. Organisms prepared in this manner appear to retain their normal growth and attachment properties. Organisms can be frozen without further processing and stored in the dark at $-70°$ for at least 6 months. Labeling by this method produces bright and generally stable fluorescence. However, some release of fluorescein over time has been noted. If nonviable organisms are desired, they can be inactivated by heating at $60°$ for 30 to 60 min following the labeling and washing procedure. Chemical treatment such as formaldehyde has also been used. It should be noted that inactivating bacteria, especially by chemical means, may change the binding properties of the bacteria.

A second labeling method couples the ε-amino group of lysine on the fluorochrome to the bacterial proteins through a thiourea bond.[11] Bacteria are washed twice and resuspended in 2 ml of carbonate/bicarbonate buffer (pH 9.5, 8.6 g Na_2CO_3 and 17.2 g $NaHCO_3$ in 1 liter of H_2O). The FITC is dissolved in dimethyl sulfoxide at 10 mg/ml. Two hundred microliters of this solution is added to the bacterial suspension slowly under constant stirring. The mixture should be rotated end over end for 90 min at room temperature. Bacteria should then be washed three times in 100 ml of PBS, aliquoted, and frozen as described earlier. Fluorescein bound by this procedure is very stable if used or stored in the dark. However, the properties of some strains of bacteria and yeast are altered.

Equipment

The minimum equipment required to quantify phagocytosis is a confocal microscope. However, quantification using only confocal microscopy would be tedious at best. An epifluorescence attachment allows for the use of fluorochrome-labeled particles. A confocal microscope with an epifluorescence attachment should be considered the practical minimum. Excitation of fluorochomes with laser light and collection of light emissions using photomultiplier tubes make quantification rather simple.[12] Other desirable features include laser scanning to automate the process, a temperature-controlled stage for "real time" phagocytosis measurements, and a color charge-coupled device video camera. In addition, an integrated computer with software to analyze the fluorescent signals or the color intensity is also desirable. If adherent cells are used, an inverted confocal microscope is usually necessary.

[11] J. W. Goding, "Monoclonal Antibodies: Principles and Practice." Academic Press, London, 1986.
[12] D. A. Rodeberg, R. E. Morris, and G. F. Babcock, *Infect. Immun.* **65,** 4747 (1997).

Separation of External from Internal Particles

The use of confocal microscopy by itself allows the investigator to separate surface-adherent particles from phagocytosed particles. Although this procedure can be performed with unlabeled particles, fluorescence labeling improves the ability to detect the phagocytosed particle. For quantification, the individual fluorescent particles detected by fluorescence microscopy must be observed in several planes below the cell surface and then must be counted manually. Particles in a minimum of 100 cells (preferably 200) should be counted. A more efficient method of separating internal particles from external particles is to mark the external particles in a manner such that they can be separated from the internal ones. The methods used most commonly can be divided into two groups. One group of methods involves quenching or resonance energy transfer of the fluorochrome on the external particle to a different emission wavelength, which can be separated from the emission wavelength of the internal particle. Dyes such as ethidium bromide or trypan blue have commonly been used for this purpose.[13,14] These dyes are added immediately prior to reading the assay at final concentrations of 3 mg/ml for trypan blue and 50 μg/ml for ethidium bromide. Another quenching method is the use of Immuno-Lyse (Coulter Corp., Hialeah, FL), which can be used when phagocytic assays are performed on whole blood.[10] This reagent quenches the fluorescence of the external particles and lyses red blood cells.

The second set of methods is to use an antibody, which reacts with the particles being used in the assay.[15] Particles, which are phagocytosed, are isolated and do not react with the antibody, whereas external particles bind the antibody. The antibody can be fluorochrome labeled directly or a fluorochrome-labeled antiimmuonglobulin can be used to detect the primary antibody. It is important to choose an antibody that reacts only with the external particle and that is labeled with a fluorochrome that produces an emission wavelength that can be separated from the fluorochrome associated with the particle. The choice of the fluorochrome for the antibody depends on not only the fluorochrome chosen for the particle, but also the wavelengths available for excitation and the filters available for emission. A list of commonly used fluorochromes appears in Table III.

[13] A. R. Fattorossi, R. Nisini, J. G. Pizzolo, and R. D'Amelio, *Cytometry* **10,** 320 (1989).
[14] C. F. Bassoe, *in* "Handbook of Flow Cytometry" (J. P. Robinson, ed.), p. 177. Wiley Liss, New York, 1993.
[15] K. L. Hess, G. F. Babcock, D. S. Askew, and J. M. Cook-Mills, *Cytometry* **27,** 145 (1997).

TABLE III
COMMON FLUOROCHROMES USEFUL IN QUANTIFYING PHAGOCYTOSIS

Fluorochrome	Excitation[a] (common wavelengths used)	Emission[a] (common wavelengths collected)
Fluorescein	450–490[b], 488[c]	520–530
Alexa488	450–490[b], 488[c]	520–530
Phycoerythrin	450–490[b], 488[c]	560–590
Rhodamine	510–560[b], 514[c]	610–620
Cy3	510–560[b], 514[c]	610–620
Texas Red	540–580[b], 568[d]	600–660
Cy5	633[e]	675
Allophycocyanin	633[e]	675
Cy7	752[d]	800

[a] These represent the excitation and emission wavelengths used most commonly by typical laser-driven or epifluorescence microscopes. The optimal wavelengths are often different.
[b] Excitation with a mercury arc lamp.
[c] Excitation with an argon ion laser.
[d] Excitation with a krypton ion laser.
[e] Excitation with a helium–neon laser.

Quantification of Phagocytosis by Confocal Microscopy

Many cell types besides the "professional phagocytes" can phagocytose particles. However, the same procedures can be used to quantify phagocytosis for all cell types. Adjustments must be made to the general procedure depending on the cell type being used. Adherent cells in particular require slight modifications in basic procedures. These modifications are mentioned at the appropriate points in the protocol. The incubation period required for maximal phagocytosis varies greatly, depending on the type of cells being analyzed. This difference can be from minutes to many hours, especially for "nonprofessional" phagocytes such as endothelial cells, which may require especially long incubation periods for maximal phagocytosis to occur. Each investigator must perform a time course experiment to determine when optimal phagocytosis occurs in their particular system. The following procedure is written for neutrophils, which are nonadherent cells that phagocytose particles rapidly. Differences in the technical aspects of the procedures, which must be adjusted when using adherent cells, are stated. The particles used as the phagocytic reagent in this protocol are fluorescein-labeled *Staphylococcus aureus*.

Neutrophils are obtained from whole blood. Phagocytosis can be performed directly on whole blood or on isolated neutrophils. If whole blood is used the neutrophils must be separated from other leukocytes either visually or by the use of a specific marker. In the following protocol, isolated neutrophils are used.

Sufficient numbers of cells can be obtained from 5 ml of blood using ethylenediaminetetraacetic acid (EDTA) or heparin as the anticoagulant. The blood is separated into neutrophils on mononuclear leukocytes by centrifugation on neutrophil isolation media.[16] Five milliliters of blood is layered over 5.0 ml of modified Ficoll–Hypaque solution in 15-ml conical centrifuge tubes. The sample is centrifuged at 550g for 30 min at 25°. The neutrophil band should be collected. This band is the second band going from top to bottom.

The neutrophils are washed twice by centrifugation and resuspended in 5-ml volumes of HBSS. The washed cells are then resuspended to 5 × 10^6 cells/ml in HBSS or cell culture medium. Cell culture medium is preferred and RPMI 1640 works well. One hundred microliters of this cell suspension is added to 12 × 75-mm round-bottom polystyrene tubes. Ten microliters of serum as a source of opsonin is added to each tube. Autologous serum should be used when possible, but the sera of certain species are low in opsonic ability and should probably be avoided. If adherent cells are used, they are plated in microwell tissue culture plates (12-well) or Lab-Tek chamber slides (4-well) and grown to confluency. Before use, 100 μl of serum as a source of oponsin is added. *Note:* If fetal bovine serum (FBS) is used, the medium must still be supplemented with opsonins as FBS is usually deficient in opsonic activity. The assay should be performed in triplicate with three tubes or wells being utilized for *each* time point.

The labeled particles, in this case fluorescein-labeled *S. aureus,* are added to each well. The *S. aureus* should be resuspended in culture medium or HBSS to a total volume of 10 μl (100 μl for the adherent cell cultures). Several ratios of particles/phagocyte should be used to determine the optimal ratio. Generally the optimal concentration ranges from a ratio 2–3 : 1 particles/phagocytic cell for endothelial cells to 25 : 1 or even more for neutrophils. In this particular procedure the 10 : 1 ratio has been determined to be optimal. The bacteria should be sonicated for 5 sec in a bath-type sonicator to break up clumps immediately before using.

Because neutrophils phagocytose bacteria rapidly, time points (incubation times at 37°) of 0, 2, 5, 10, 15, 20, and 30 min. are used. Following the incubation period, 10 μl of cytochalasin D (5 μM final concentration) is added to the sample, which is chilled immediately to 4°. For the time-zero point, 10 μl of cytochalasin D is added to the sample before the addition

---

[16] G. F. Babcock, *in* "Handbook of Flow Cytometry" (J. P. Robinson, ed.), p. 22. Wiley Liss, New York, 1993.

of *S. aureus* to prevent any phagocytosis. This time point allows for an examination of binding prior to phagocytosis. Following the proper incubation period and the treatments listed previously, all the samples are washed twice in cold HBSS and resuspended in 200 μl of HBSS. In adherent cultures, enough HBSS is added to just cover the monolayers.

A monoclonal antistaphylococcal antibody or a quenching agent (ethidium bromide or trypan blue) is then added. This step is important to separate internal from external particles. Following a 30-min incubation at 4°, the cells are washed once in HBSS and a fluorochrome-labeled goat antimouse antibody is added. Following an additional 30-min incubation at 4°, the cells are washed twice and fixed with 1% (w/v) paraformaldehyde.

Controls should include duplicate cultures for each time point, which are incubated at 4° instead of 37°. Additional controls to monitor nonspecific antibody binding should include tubes containing immunoglobulins, which do not react with either the cells or the bacteria (particles) and are of the same isotype and labeled with the same fluorochrome as the specific antibody. The cells are then held at 4° in the dark until analyzed. They can be stored for several days. If a quenching reagent is used, then a slightly different procedure is followed. After the addition of the quenching agent, the cells should be analyzed within a short period of time. These samples should be held at 4° in the dark until analysis by confocal microscopy.

Data Analysis and Interpretation

Phagocytosis can be quantified in a number of different ways, depending on the type of data collected. The simplest method is to determine the percentage of cells positive for phagocytosis. Any cells containing internal particles are considered phagocytosed. In the example given earlier, green particles would indicate positive phagocytosis whereas green/orange particles are external and not indicative of phagocytosis. Determining the percentage of cells positive for phagocytosis is usually performed even if other quantification methods are used.

A second method involves determining the number of bacteria phagocytosed per cell. Collecting these data is very labor-intensive as each phagocytosed particle must be counted individually. To minimize variations, it is recommended that the particles be counted in at least 200 cells. Data are then expressed as the percentage of cells phagocytosing a mean number of particles.

Most investigators find it easier to determine the relative phagocytosis per cell rather than the actual number of particles per cell. This method requires the computerization of data for analysis. Fluorescence can be collected either by the use of photomultiplier tubes or by a video camera.

Data from photomultiplier tubes can be channelized according to intensity
and a mean and/or median fluorescence can be obtained. Data captured
by the video camera can be contoured according to pixel numbers. The
intensity of fluorescence within the pixel contours can then be calculated.
Simply stated, the greater the fluorescence intensity per cell, the more
phagocytosis per cell. In the protocol given earlier, only the green fluorescence, which represents phagocytosis, is analyzed. Dual-labeled cells, green/
orange in the protocol described earlier, are not analyzed as this represents
binding. The green fluorescence of cells incubated consecutively with bacteria at 4° and then stained with the appropriate antibody (same as for the
experiment) represents the background or nonspecific fluorescence control,
as the cells should not phagocytose at 4°. Only green fluorescence signals
exceeding this value should be included or alternatively subtracted from
the experimental values. The isotypic antibody control is used to ensure
that the dual fluorescence is indeed specific for organisms (particles) on
the outside of cells and is not caused by the nonspecific binding of immunoglobulin molecules.

Summary

Confocal microscopy is an excellent tool to quantify phagocytosis. Depending on the particle used, phagocytosis can be determined by the simple
manual counting of internalized particles. If a fluorescence probe is utilized,
an analysis of fluorescence intensity can be used for quantification. The
basic procedure can be altered in a number of areas to conform with the
scientific needs of the investigator. This includes the use of different particles, cell types, fluorescence dyes, and even the degree of sophistication of
the instrumentation. The major pitfall encountered when trying to quantify
phagocytosis is the inability to separate external from internal (phagocytosed) particles. If not determined properly, data will be erroneous, usually
indicating a much higher degree of phagocytosis than actually occurred.

[19] Measurement of Secretion in Confocal Microscopy

By Akihisa Segawa

Introduction

Secretion is a dynamic biological activity, seen widely in glandular and
nonglandular tissues, accomplished by the spatially and temporally organized movement of molecules and organelles within and between the cells.

Secretion of cellular products (hormones, enzymes, and neurotransmitters, etc.) relies on the cascade of the biosynthetic pathway with the final release step exocytosis,[1] whereas secretion of body fluids (sweat, tears, digestive juice, etc.) comprises the transepithelial passage of serum components across epithelial cells (transcellular pathway) or the intercellular junction (paracellular pathway)[2] (Fig. 1). Studies on secretion have been done successfully using biochemical and electrophysiological approaches, and several crucial events involved in secretion have been unraveled. Of exocytosis, these studies resolved the molecular and kinetic events at the level of single secretory granules, e.g., granule docking, priming, triggering, fusion/release, and removal,[3] each of which constitutes the rate-limiting process of secretion.[4] For epithelial transport, they also demonstrated the existence of a "leaky" tight junction, not a "tight" seal as recognized previously, that alters its permeability in response to physiological stimuli to allow passage of water, ions, nonelectrolytes,[5–7] and even macromolecular substances[8] through the paracellular pathway. Morphologically, light and electron microscopy on fixed cells have elucidated such processes in detail,[9–12] but only in static images. An important challenge is to clarify their dynamic aspects and to integrate morphology directly with the physiological events that occur during secretion. Confocal microscopy of living cells is expected, and indeed has offered opportunities, to answer such demand.[13–22] This article

[1] G. Palade, *Science* **189,** 347 (1975).
[2] J. A. Young, D. I. Cook, E. W. van Lennep, and M. Roberts, *in* "Physiology of the Gastrointestinal Tract," 2nd ed., p. 773. Raven Press, New York, 1987.
[3] T. F. J. Martin, *Trends Cell Biol.* **7,** 271 (1997).
[4] Y. Ninomiya, T. Kishimoto, T. Yamazawa, H. Ikeda, Y. Miyashita, and H. Kasai, *EMBO J.* **16,** 929 (1997).
[5] E. Frömter and J. Diamond, *Nature New Biol.* **235,** 9 (1972).
[6] J. L. Madara, *Cell* **53,** 497 (1988).
[7] S. Citi, *J. Cell Biol.* **121,** 485 (1993).
[8] J. R. Garrett, *in* "Glandular Mechanisms of Salivary Secretion," Vol. 10 of Frontiers of Oral Biology, p. 153. Karger, Basel, 1998.
[9] W. W. Douglas, *Br. J. Pharmacol.* **34,** 453 (1968).
[10] A. Amsterdam, I. Ohad, and M. Schramm, *J. Cell Biol.* **41,** 753 (1969).
[11] N. Takai, Y. Yoshida, and Y. Kakudo, *J. Dent. Res.* **62,** 1022 (1983).
[12] M. R. Mazariegos and A. R. Hand, *J. Dent. Res.* **63,** 1102 (1984).
[13] A. Segawa, S. Terakawa, S. Yamashina, and C. R. Hopkins, *Eur. J. Cell Biol.* **54,** 322 (1991).
[14] Y. Kawasaki, T. Saitoh, T. Okabe, K. Kumakura, and M. Ohara-Imaizumi, *Biochim. Biophys. Acta* **1067,** 71 (1991).
[15] A. Nakamura, T. Nakahari, T. Senda, and Y. Imai, *Jpn. J. Physiol.* **43,** 833 (1993).
[16] A. Segawa, *J. Electr. Microsc.* **43,** 290 (1994).
[17] M. Terasaki, *J. Cell Sci.* **108,** 2293 (1995).
[18] T. Whalley, M. Terasaki, M.- S. Cho, and S. Vogel, *J. Cell Biol.* **131,** 1183 (1995).
[19] A. Segawa and A. Riva, *Eur. J. Morph.* **34,** 215 (1996).

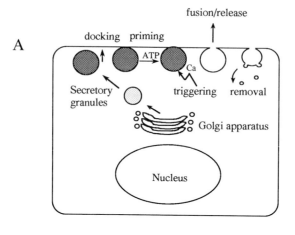

Exocytosis

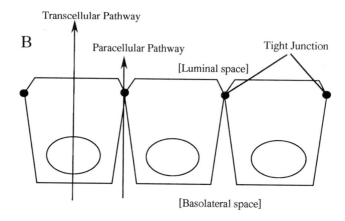

Epithelial Transport

FIG. 1. Two major events in secretion: Exocytosis (A) and epithelial transport (B).

describes the measurement of exocytosis and the paracellular pathway in rat salivary glands, using confocal microscopy combined with the fluorescent tracer technique. Salivary acini, the secretory end piece of this gland, are composed of highly polarized acinar cells capable of secreting enzymes and fluid; the secretion is regulated distinctively under different autonomic receptor control.[2,12,16,19,22,23] Methods of cell dissociation, selection of fluorescent tracers, and analytical procedures are described. A general method of confocal microscopy of living cells has been described previously by other authors,[24] and is not described here in detail.

Preparation of Cells

Probably the ideal way to visualize glandular secretion is to observe the gland *in situ*. However, the technique of confocal microscopy for this purpose has not been established and, more importantly, it is difficult for such a preparation to obtain a transmitted light or Nomarsky image, which is necessary to correlate with the confocal fluorescence image to locate the sites of secretory events exactly in the cells and tissues. Use of tissue slices is thus a better choice, but this requires skilled techniques to obtain clear images. Slices must be cut intactly as thin as possible, preferably less than 150 μm. Practically, cell dissociation is the commendable way to image cells with high resolution.

Medium

Eagle's minimal essential medium (MEM) is supplemented with 25 mM HEPES (N-2-hydroxyethylpiperazine-N'-2-ethanesulfonic acid), pH 7.3. Bubble the medium with O_2 gas at least 10 min before use to avoid anoxia of the cell. In rat salivary acinar cells, anoxia induces a "vacuole" formation in the cytoplasm.[16,25]

Cell Dissociation

Dissect salivary glands after sacrifice of the rats, and place the excised glands on a plastic dish. Remove capsules, connective tissues, and lymph

[20] C. B. Smith and W. J. Betz, *Nature* **380**, 531 (1996).

[21] A. Segawa, *Bioimages* **5**, 153 (1977).

[22] A. Segawa and S. Yamashina, *in* "Glandular Mechanisms of Salivary Secretion," Vol. 10 of Frontiers of Oral Biology, p. 89. Karger, Basel, 1998.

[23] B. J. Baum, *Ann. N.Y. Acad. Sci.* **694**, 17 (1993).

[24] M. Terasaki and M. E. Dailey, *in* "Handbook of Biological Confocal Microscopy," 2nd ed., p. 327. Plenum Press, New York, 1995.

[25] R. L. Tapp and O. A. Trowell, *J. Physiol* **188**, 191 (1967).

nodes with fine forceps and a pair of scissors under the binocular mac-
roscope. Cut tissues into small pieces (approximately 1–2 mm^2) with scal-
pels, put them into a tissue culture flask (bottle size 25 ml with double seal
cap, Iwaki Glass, Japan), and incubate with 10 ml medium containing
20–30 mg of collagenase (Wako Pure Chemical Industries, Osaka, Japan).
Hyaluronidase and bovine serum albumin (BSA) may be added to the
medium to increase the cell viability. After capping tightly with the screw
cap, incubate the culture flask in a water bath kept at 37° for 45 min
with constant shaking at 130 oscillations per minute. Every 15 min during
digestion, pipette tissues gently with a Pasteur pipette to facilitate mechanically the cell dissociation. Following digestion, centrifuge the cell suspension
at 1000 rpm for 1 min, discard medium, and resuspend the cells in MEM.
Wash twice. This procedure yields cell aggregates having a well-preserved
acinar structure. They exhibit a high secretory response as intact tissue
slices.[26] Keep the cell suspension on ice until use. Prepared cells should be
used within 1 hr. Later, secretory response decreases and many "vacuoles" appear.

Fluorescent Tracers

For the measurement of paracellular pathway (tight junctional permeability), fluid-phase fluorescent tracers of varying molecular weights are
used.[16] These include Lucifer yellow (molecular weight 457), dextrans labeled with FITC, RITC, or Texas Red (molecular weight 3, 10, 40, 70 and
500 K are available from Molecular Probes Inc., Eugene, OR). They are
dissolved in MEM at concentrations of 0.5–2 mg/ml. When perfused, these
tracers flood the extracellular space of the tissue and, under laser excitation,
reveal bright fluorescence against the nonfluorescent acinar cell cytoplasm.
As will be described, fluorescence is detectable in the basolateral (interstitial) space, but not the luminal space, of salivary acini unless a tight junction
permeates the tracer into the lumen.

The measurement of exocytosis includes four different categories of
fluorescent approaches. One is the use of fluid-phase tracers to stain extracellular space as described earlier.[13–15,17–19,21,22] If exocytosis occurs, the
exocytosed granule space is flooded with the tracer and reveals bright
fluorescence. The second approach is to stain all the cytoplasm. On exocytosis, granules lose their fluorescence. BCECF-AM [2′,7′-bis(2-carboxy-
ethyl)-5,6-carboxyfluorescein, Dojin, Japan] has been used at 3 μM for this
purpose.[15] The third approach is to stain the contents of secretory granules.
As in the second approach, exocytosed granules lose fluorescence. Use of

[26] A. Segawa, N. Sahara, K. Suzuki, and S. Yamashina, *J. Cell Sci.* **78,** 67 (1985).

acridine orange (1 μM) has been reported.[14] The fourth approach is to stain plasma membrane. Following exocytosis, the secretory granule membrane becomes continuous with the plasma membrane, thus revealing ring-shaped fluorescence. FM1-43 (2 μM)[17,18,20] and TMA-DPH (1 μM)[14] have been used. For the second and third approaches, preincubation of the cells with fluorochromes is necessary to load the dye into the cell. The loaded cells are washed, placed in the chamber, and perfused with MEM. The first and fourth approaches do not require preincubation. Instead, cells are placed in the chamber and perfused with MEM containing the fluorescent dyes. The combined use of different fluorescent tracers is possible when the emission wavelengths of tracers are discriminated distinctively by the detector.

Perfusion

Coat the cover slides (24 × 60 mm, for the observation with inverted microscope) or the slide glass (for the upright microscope) with CELL TAK (Collaborative Biomedical Products, Two Oak Park, Bedford). Put the cell suspension on the slide and allow it to settle for several seconds for cells to adhere onto the glass surface (Fig. 2). Wipe the medium with filter paper so that nonadherent cells are removed. If cells do not adhere sufficiently, centrifuge specimens with Cytospin (Shandon, Cheshire, UK) at 200 rpm for a few minutes. For observation with the inverted microscope, specimens can be seen without a cover slide. Simply drop the perfusion medium on the specimen. Wipe it using filter paper. This application method is especially recommended when the response occurs quite rapidly, in the order of milliseconds to seconds, after the addition of secretagogues.[27] For the longer observation, apply medium at least every 2–3 min to avoid anoxia of the cells. Observation with the upright microscope and observation of tissue slices need a cover slide. To make the perfusion space, strips of vinyl tape are attached on lateral sides of the glass. Place the specimen in the center, cover with cover slides (24 × 24 mm), and seal with petrolium jelly along the vinyl tape. Pour medium from one side into the chamber and wipe it from the other side with the filter paper. Adjust the thickness of the chamber so as to allow passage of perfusion medium while retaining the specimens (tissue slices) in a fixed position.

[27] M. Yamamoto-Hino, A. Miyawaki, A. Segawa, E. Adachi, S. Yamashina, T. Fujimoto, T. Sugiyama, T. Furuichi, M. Hasegawa, and K. Mikoshiba, *J. Cell Biol.* **141,** 135 (1998).

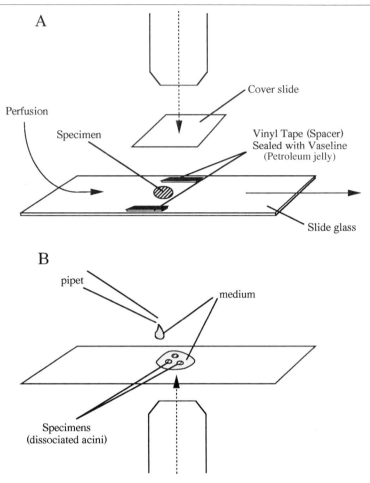

Fig. 2. Methods of perfusion for an upright microscope (A) and an inverted microscope (B).

Confocal Microscopy

Place the sample on the microscope stage warmed to 37°. Using the ordinary microscope mode, confirm the location, shape, and visibility of cells. Choose the best cells. Change to the confocal microscope mode. Adjust brightness and the plane of focus (height of optical sectioning) by the simultaneous fluorescence and transmitted light or Nomarsky imaging. Select the box (image) size. A large box allows imaging with high resolution but requires longer scanning time. A small box enables rapid acquisition of images, up to three to four frames per second (128 × 128 pixel) for Bio-

Rad MRC 1024. For more rapid observation, a real confocal microscope (Nikon RCM 8000, Noran Oz and Yokogawa CSU10) obtains images faster than the video rate (33 msec/frame). To increase the signal-to-noise ratio, average several images. Averaging three improves the image quality greatly. Before starting the image aquisition, decide the time interval. The rapid acquisition of images consumes a lot of computer memory, which limits the observation period to a short time. The acquisition of large images also consumes memory and reduces the temporal resolution. Thus for ordinary observation, we collect images every 1.5 to 5 sec with a moderate image size (768 × 512 pixel).

Secretory Stimulation

Rat parotid acinar cells possess two distinct secretion mechanisms regulated by different receptor-signaling systems: (1) enzyme release by exocytosis through the activation of β-adrenergic receptor generating cyclic AMP and (2) fluid secretion activated by muscarinic, α-adrenergic, and substance P receptors via an increase in cytosolic calcium.[2,23] For stimulation of exocytosis, DL-isoproterenol (1–20 μM) is dissolved in the perfusion medium to activate β receptor. Fluid secretion is stimulated by carbachol (10 μM), which activates the muscarinic receptor, a major regulator of fluid secretion.

Measurement of Tight Junctional Permeability:
  Paracellular Pathway

The basic structure of salivary acini is described first to evaluate properly the results of experiments. As mentioned previously, a tight junction adjoins neighboring cells to separate the lumen from the interstitial (basolateral) space (Fig. 1). In parotid acini, the configuration of lumen is not simple but exhibits complex narrow (approximately 1 μm in diameter) canalicular extensions called the intercellular canaliculi (Fig. 3). In the sectioned image, they reveal a very small ring or tubular appearance within the acini. Intercellular canaliculi also provide the exclusive site of exocytosis in salivary acini. Therefore, it is very important to define the location of intercellular canaliculi in the confocal images (Fig. 4).

Apply medium containing the fluid-phase fluorescent tracers. In untreated cells, bright fluorescence can be detected in the basolateral space but not in the luminal space (Fig. 4). This indicates that the tight junction of parotid acini does not allow permeation of fluorescent tracers into the lumen. Stimulate cells by adding secretagogues in the perfusion medium and observe if the fluorescence appears in the lumen. If it appears, then

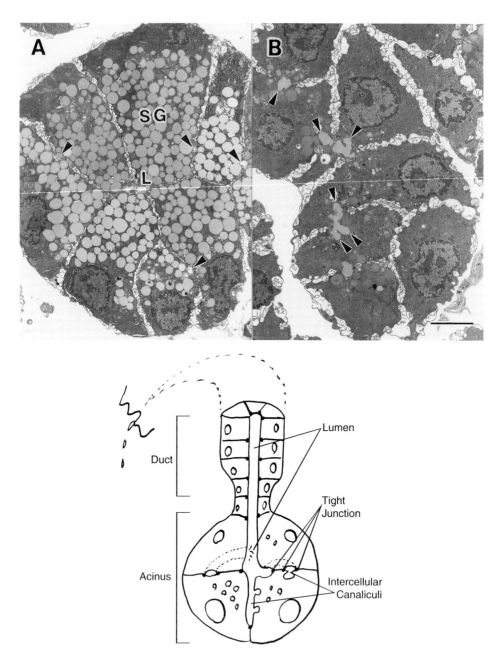

FIG. 3. Parotid acini shown by electron micrograph [modified from A. Segawa, *J. Electr. Microsc.* **43**, 290 (1994)]. (A) Unstimulated. (B) Stimulated with isoproterenol *in vitro* for 30 min. Arrowheads denote the intercellular canaliculi. Secretory granules (SG) are numerous in A but depleted in B, where the exocytotic profiles are seen at the intercellular canaliculi. L, lumen. Bar: 5 μm. (C) Schematic representation.

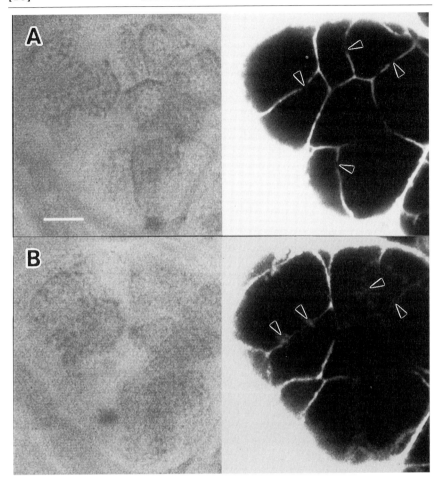

FIG. 4. Dynamic changes in tight junctional permeability shown by confocal microscopy. Reproduced from A. Segawa, *J. Electr. Microsc.* **43,** 290 (1994). Tissue slices of rat parotid glands were perfused with Lucifer yellow and stimulated with isoproterenol for 1 (A) and 5 (B) min. Simultaneous transmitted light (left) and confocal fluorescence (right) imaging. Arrowheads indicate the intercellular canaliculi. The intensity of fluorescence in the intercellular canaliculi is weak in A but high in B. Bar: 10 μm.

examine the size-selectivity characteristics. Apply medium without fluorescent tracers to wash out the luminal fluorescence and reintroduce medium containing tracers having larger molecular sizes. In rat parotid acini, a tight junction exhibits different size-selectivity characteristics in response to different secretory stimuli: We detected up to molecular weight 40 K for isoproterenol and 10 K for carbachol when fluorescent dextrans were used as a monitoring tracer.[16] This observation coincides with the results of previous experiments carried out by biochemical and morphological approaches.[8,12,28] The existence of size-selectivity characteristics also rules out the possibility that the luminal fluorescence originates by the retrograde diffusion of tracers from the duct.

Advantages of this method include (1) applicability to the glandular epithelia whose lumen is too narrow to allow for the insertion of electrode, which is necessary for the electrophysiological approach; (2) distinction of para- from transcellular pathways involved in transepithelial transport, which is difficult in the biochemical approach; and (3) consecutive observation of morphological changes occurring in the same cell, which is impossible for electron microscopy.

Note: Enzymatic digestion sometimes damages salivary tight junction; luminal fluorescence often appears in the dissociated acini, even though they do not receive secretory stimulation.[16] Thus the use of tissue slices is preferable to the study of tight junction dynamics. Anoxia also seems to increase the permeability of tight junction as anoxic "vacuoles" were found to exhibit fluorescence.[16] To avoid this, carefully keep perfusing tissues with a well-oxygenated medium.

Measurement of Exocytosis

Secretion by exocytosis involves the cycles of fusion and the removal of secretory granule membranes at the cell surface (exocytosis–removal cycle). This section describes the visualization of the exocytosis–removal cycle of single secretory granules using fluid-phase fluorescent tracers.[21]

Place dissociated acini on a slide glass. Perfuse cells with MEM containing fluorescent tracers without secretagogues. Using simultaneous fluorescence and transmitted light or Nomarsky imaging, observe for a few minutes to confirm that cells exhibit little changes. Start acquisition of images (Fig. 5). Apply fluorescent medium containing isoproterenol. Ten to 15 sec later, omega-shaped fluorescent spots appear abruptly along the intercellular canaliculi. This represents the exocytotic response. It is difficult to predict the sites exhibiting exocytosis until the cells receive secretory stimuli. Therefore, it is recommended to image a wide area with a relatively large box size (768 × 512 pixels) to capture as many morphological events as possible. Continue image collection for at least 2 min to follow the fate of fluorescent spots. Granule movement and the dynamics of membrane reorganization during exocytotic secretion are analyzed later.

Image analyses are performed by merging the confocal fluorescence

---

[28] L. C. U. Junqueira, A. M. S. Toledo, and R. G. Ferri, *Arch. Oral Biol.* **10,** 863 (1965).

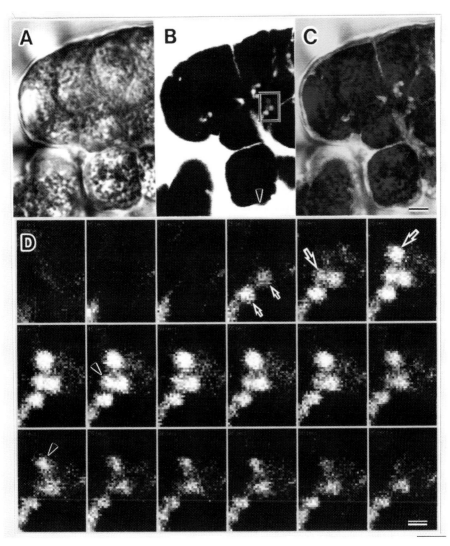

FIG. 6. Analytical confocal images showing the exocytosis–removal cycle of single secretory granules, reproduced from A. Segawa, *Bioimages* **5,** 153 (1997). (A and B) A merged time series image. (B) Fluorescent spots are demonstrated in various colors: white (unchanged spots during observation), yellow (red plus green, appeared at 25 sec), and red (appeared at 35 sec). (C) A merged fluorescence (red) and transmitted light (green) image taken at 35 sec. Bar: 3 μm. (D) Enlarged images, indicated by rectangle in B, displayed sequentially from the top left to the right bottom. Appearing fluorescent spots are indicated by small and large arrows, whereas disappearing spots are shown by arrowheads. Bar: 1 μm.

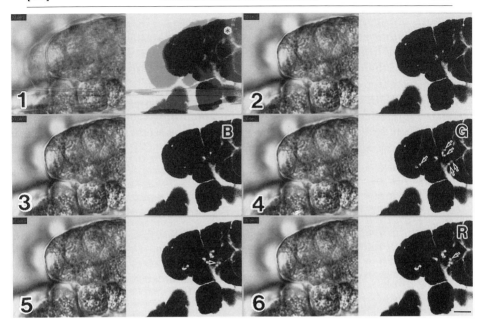

FIG. 5. Exocytosis in living cells shown by confocal microscopy. Dissociated rat parotid acini were perfused with FITC–dextran (molecular weight 3000). Simultaneous fluorescence and transmitted light images were taken every 5 sec and displayed consecutively. Isoproterenol was added at image 1, so that acini moved (asterisk). Arrows indicate the appearance of fluorescent spots. Some images are pseudocolored in blue (B), green (G), and red (R) for image analyses (see Figs. 6A and 6B). Bar: 10 μm.

image with transmitted light or Nomarsky image (Fig. 6C, see color insert). Confirm that the size and shape of the fluorescent spots are comparable to those of single secretory granules. Analyze granule movement and membrane dynamics by displaying images in a time sequence (Fig. 5). Find critical images showing dynamic changes. Merge them to make the time-resolved changes clearer. MRC 1024 makes it possible to merge images obtained at three different time points. Pesudocolor each image in blue, green, and red and merge into one (Figs. 6A and 6B, see color insert). The unchanged area summates all three colors and is displayed in gray scale. In Fig. 6B, only the granule-shaped fluorescent spots are displayed in the color image. Other areas are displayed mostly in gray scale. This is important as it indicates that the observed changes in fluorescent spots are not caused by the movement of the specimen as a whole. The transmitted light image also shows that the granules are immobile before exhibiting the exocytotic response. This indicates the presence of docked granules in salivary exocytosis, as in neurons and neuroendocrine cells.[3,22] Select an area where the exocytotic response is recognized (Fig. 6B). Enlarge and display the area sequentially (Fig. 6D). This demonstrates the appearing and disappearing changes of fluorescent spots, representing the exocytosis–removal cycle of a single secretory granule. Fluorescent spots appear 10–15 sec after the addition of isoproterenol. This latency period is likely to represent the time required for priming and triggering. Following their appearance, the fluorescent spots maintain their round shape for several seconds to tens of seconds, diminish gradually, and finally disappear, with a total detectable period of 40–70 sec in many instances. Animated demonstration of these changes helps recognize the dynamic image.[29] There are many softwares, such as "Confocal assistant," which present a motion picture on the personal computer and also enables the animated demonstration of merged transmitted and fluorescence image.

[29] A. Segawa and M. Ono, in "Image Analysis & 3D Reconstruction CD-ROM Series, Vol. 1," Purdue University Cytometry Laboratories, West Lafayette, USA (1998).

[20] Receptor–Ligand Internalization

By GUIDO ORLANDINI, NICOLETTA RONDA, RITA GATTI, GIAN CARLO GAZZOLA, and ALBERICO BORGHETTI

Introduction

The interaction between biologically active compounds and target cells has been studied extensively by various quantitative and qualitative approaches as it is the crucial first step in the chain of events leading to the final effect. Additionally it is a suitable process to be studied for pharmacological purposes.[1] The visualization of receptor–ligand binding and internalization has been studied by immunocytochemistry techniques for both light and electron microscopy on fixed, permeabilized tissues or cells. The information achievable by these methods is limited and static, as the localization observed in fixed samples might not correspond to the actual binding site

[1] H. Lodish, D. Baltimore, A. Berk, S. L. Zipursky, P. Matsudaira, and J. Darnell, in "Molecular Cell Biology" (J. Darnell, ed.), 3rd ed. Scientific American Books, New York, 1995.

in living cells and the time course of the interaction cannot be evaluated with satisfactory precision.[2,3]

A technique is now available that allows the visualization of the interaction between directly fluoresceinated ligands and living cells and enables one to follow their possible internalization. In living adherent cells, such a methodological approach was hindered previously by the relatively poor resolution of the conventional fluorescence microscope and the unfavorable signal-to-noise ratio. As a consequence, the visual information has not been exhaustive and subcellular localization has been limited to main cell compartments.[4]

More recently, confocal imaging has provided new insights in the observation of fluorescent specimens. The virtual absence of out-of-focus blurring allows a much better definition of probe localization at the subcellular level together with the possibility of exploiting the three-dimensional reconstruction capability of most confocal systems.

We have coupled a self-constructed flow chamber to an inverted confocal scanning laser microscope that allows long-term observation of adherent cells under controlled microenvironmental conditions.[5] This method not only provides images of intact, nonfixed cells, but also allows one to change culture conditions and to observe living cell responses directly or to perform two-step staining to identify subcellular structures involved in the observed processes.

The reliability of this procedure has been verified on a well-known model of receptor–ligand internalization, that of insulin and insulin receptor. The same technique has also been applied to studying the interactions between natural human antibodies, circulating in healthy subjects, and living human endothelial cells, fibroblasts, and proximal tubular epithelial cells.[6]

[2] P. Jackson and D. Blythe, in "Immunocytochemistry" (J. E. Beesley, ed.), p. 22. Oxford Univ. Press, Oxford, 1993.

[3] P. Monaghan, D. Robertson, and E. J. Beesley, in "Immunocytochemistry" (J. E. Beesley, ed.), p. 47. Oxford Univ. Press, Oxford, 1993.

[4] H. Lodish, D. Baltimore, A. Berk, Z. S. Lawrence, P. Matsudaira, and J. Darnell, in "Molecular Cell Biology" (J. Darnell, ed.), 3rd ed. Scientific American Books, New York, 1995.

[5] V. Dall'Asta, R. Gatti, G. Orlandini, P. A. Rossi, B. M. Rotoli, R. Sala, O. Bussolati, and G. C. Gazzola. *Exp. Cell Res.* **231,** 260 (1997).

[6] N. Ronda, R. Gatti, G. Orlandini, and A. Borghetti, *Clin. Exp. Immunol,* **109**(1), 211 (1997).

Materials

Flow Chamber

Despite the ever-increasing number of available fluorescent probes for living cells acting as vital, almost real-time indicators for a series of functional parameters, several technical problems arise in trying to exploit fully confocal laser scanning microscopy of viable cell monolayers. This kind of system is extremely sensitive to changes of the focal plane, the distance between lens and specimen must be very short, and high numerical aperture lenses are mandatory for optimal resolution. Moreover, three major general requisites must be satisfied in order to achieve improvement over the techniques previously available: (1) perturbation of culture conditions must be kept at a minimum; (2) the system should allow one to extend observation as long as desired; and (3) the relevant stimuli, used in the experimental protocol, should be applied easily and then washed out easily.

The device described in this article is simple and inexpensive but fulfills all of the just described criteria. It was developed to fit a Multiprobe 2001-Molecular Dynamics computer scanning laser microscope whose optical "conventional" side is based on a Nikon Diaphot inverted microscope (Sunnyvale, CA).

The upper portion of the flow chamber is a transparent polyacetate block (Figs. 1, 2). According to the type of experiment to be carried out a hollow, whose depth can vary from 0.1 to 0.5 mm, has been milled on the bottom. When the cover slide (5 × 2.5 cm) bearing the cell culture is applied to the bottom of the block, the chamber is completed, with the

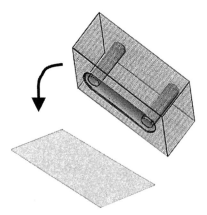

FIG. 1. Schematic representation of the flow chamber. The bottom face of the polyacetate block has to be sealed to the cover slide bearing the cell monolayer.

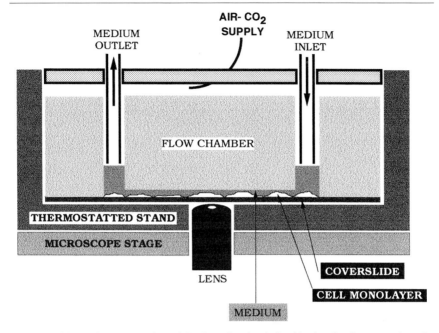

FIG. 2. Schematic cutaway view of the flow chamber lodged in the aluminum stand on the microscope stage.

cover slide and the block serving as the floor and the vault, respectively. According to the depth of the hollow the volume of the flow chamber ranges between 60 and 300 μl. Two vertical tunnels (diameter 4 mm) opening on the upper face of the block allow the introduction of small sylastic catheters (crossing the lid) for inflow and outflow of medium. The inlet and outlet tubes are usually connected to a micropipette and a vacuum apparatus, respectively, but a peristaltic pump for continuous replacement can also be used.

Various sealers, provided they are nontoxic to cells, can be adopted in order to secure the cover slide to the polyacetate block and to prevent medium leakage. In our experience the best results can be achieved with silicon vacuum grease or, better, with a special silicon-based liquid gasket for a high-performance engine (Motorsil D, Arexons, Cernusco, Milano, Italy). In any case, a thin, narrow layer of the sealer must be applied around all the block edges in order to avoid untoward spreading into the chamber once the block itself is pushed on the cover slide.

The flow chamber is lodged in a thermostatted stand (Fig. 3), a cylindrical aluminum block (diameter 14 × 2 cm) whose base has been opportunely mill finished so as to adhere tightly to the round opening in the microscope

FIG. 3. View of the stand in place. Note the small tubes for medium substitution and air–CO₂ supply. The cable from the aluminum stand is connected to the temperature control unit. The clear Perspex lid allows field lighting from the above condenser. Microscope stage translation (for field selection) carries about the whole system.

stage. A parallelepipedal niche (5 × 2.5 × 1.4 cm) is carved within the stand in order to contain precisely the flow chamber. The bottom of the niche shows an ellipsoidal slit (3 × 1 cm) devoted to objective apposition, wide enough to allow the observation of a large area of the culture by stage translation. The flow cell niche is closed tightly by a Perspex lid with three openings for a medium inlet and outlet and for atmosphere conditioning, respectively. The pressure of the lid also secures the two parts of the system fixed in place. Temperature in the flow chamber is controlled by a resistance coil embedded in the stand. A thermorelay probe is located at the inner surface of the slot; this configuration grants the widest surface for contact between the radiating element and the flow chamber.

Confocal Microscope

The confocal scanning laser microscope employed is a Molecular Dynamics Multiprobe 2001 (inverted) equipped with an argon ion laser. Samples are observed through either a 60 or a 100× oil-immersion objective (Nikon PlanApo, NA 1.4), allowing vertical resolution around 1 μm, and the step size is set accordingly. The confocal aperture (pinhole) is set at 50 and 100 μm for 60 and 100× lenses, respectively. For practical purposes, the first setting is chosen whenever possible, as it provides the best vertical resolution (0.8 μm) but, due to the lower brightness, it requires optimal signal intensity.

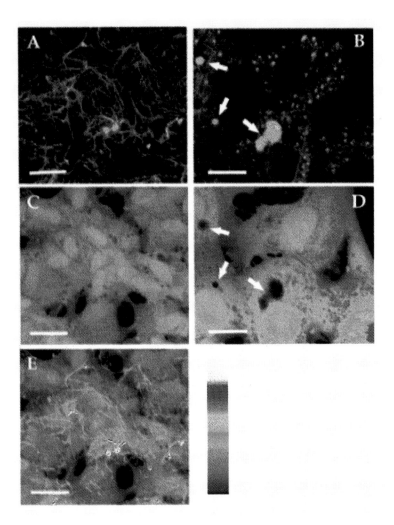

Fig. 5. Insulin–FITC internalization by living endothelial cells. (A) After 2 min of incubation, the insulin–FITC signal appears as filaments. (C) The same field as in (A) after counterstaining with calcein. Cytoplasm and nuclei are now evident. Acquisition parameters fit for calcein, with a higher signal than that of insulin–FITC, do not allow the visualization of intracellular insulin. (E) Overlapping of (A) and (C) shows the actual intracellular localization of insulin. The image shown in (A) was added to that in (C), the intensity of which had been reduced by 30%. (B) In a distinct experiment, the image was acquired after 10 min of incubation with insulin–FITC; the distribution pattern is now mostly granular. Insulin–FITC also accumulates in large bodies, like the one in this field (arrow). (D) Counterstaining demonstrates that granules are cytoplasmic and correspond to structures inaccessible to calcein. Note in particular the large body in this field (arrow); the insulin–FITC it contains is not detectable at the low sensitivity of this acquisition setting (see text). Bar (A, C, and E): 20 μm; Bar (B and D): 10 μm. For gray-scale palette, see Fig. 4.

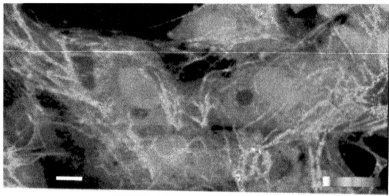

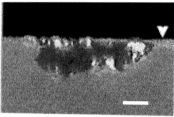

FIG. 6. IgG–FITC internalization by living endothelial cells. (Top) After 15 min of incubation, the IgG–FITC signal was distributed with a cytoplasmic fibrillar pattern. Following calcein addition (which shows cell bodies and demonstrates cell membrane integrity), there was no need to change the acquisition setting or to perform image overlapping because of the high intensity of the IgG–FITC signal. (Bottom) Vertical section of a cytoplasmic protrusion showing intracellular IgG–FITC in a cell not counterstained with calcein. Remnants of extracellular IgG–FITC in culture medium provide contrast for the cell. Arrowhead: cover slide. Bars: 5 μm. Reproduced with permission from N. Ronda, R. Gatti, G. Orlandini, and A. Borghetti, *Clin. Exp. Immunol.* **109**(1), 211 (1997).

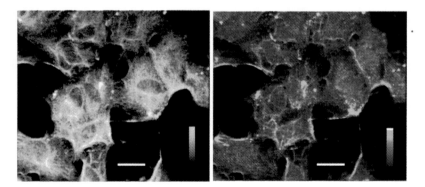

FIG. 7. Simultaneous visualization of endothelial cell microtubules and internalized IgG–FITC. Following internalization of IgG–FITC, endothelial cells were fixed and microtubules were stained by indirect immunofluorescence using a rhodamine TRITC-conjugated secondary antibody. Images were then acquired simultaneously with a double-channel system (see text) to show the localization of microtubules (left) and IgG–FITC (right). All IgG–FITC signals correspond to some of the microtubular network. Bars: 10 μm.

Because of its high quantum yield, fluorescein can be detected with a very low excitation power, which is crucial when imaging living cells. Settings exceeding 1 mW produce phototoxic events in the culture (such as apoptosis or cell detachment) that are probably related to local energy delivery and consequent heating. Another advantage of using low excitation power is that photobleaching and background noise are negligible.

In postfixation experiments with two tracers (see later), a secondary beam splitter is placed after the pinhole aperture and before the barrier filters. Signals from fluorescein and the second chromophore, usually tetramethylrhodamine isothiocyanate (TRITC) or Texas Red, are then acquired concurrently by two different photomultipliers that can be set independently in order to correct the unevenness in the quantum yield of the fluorophores.

Once the apparatus is ready for image acquisition, it is important that the field of observation is chosen in bright-field microscopy. This allows one to take time in selecting the best field according to the experimental requirements without delivering useless or even noxious high power.

Fresh medium or the relevant study solutions can be replaced without shifting along the x, y, or z axis. At each experimental step, a section series is acquired along the whole thickness of the cell (for endothelial cells, 5–6 sections). Complete scanning yields information about the whole cell, allows one to choose the most representative section, and provides the material for accurate three-dimensional reconstruction.

Whenever it is important to give a global representation of an internalization process, the image series are smoothed with a Gaussian $3 \times 3 \times 3$ kernel filter and three-dimensional reconstruction is performed according to a maximum intensity algorithm. In other words, for all pixels of a given x–y coordinate in the series, the one with the highest intensity is chosen for final image rendering. Image processing is performed on a Silicon Graphics Personal Iris workstation (Image Space Software, Molecular Dynamics).

Preparation of Ligand–Fluorescein Conjugates

Human insulin (Sigma, St. Louis, MO), purified normal human IgG (pooled normal IgG, Sandoglobulin, Sandoz, Basel, Switzerland), and IgG purified from single healthy donors[6] are coupled to fluorescein isothiocyanate (FITC),[7] purified by chromatography on a Sephadex G-25 column, and dialyzed extensively against phosphate-buffered saline (PBS) using a 3500-kDa cutoff membrane (Spectra/Por, Spectrum Medical Industries, Inc., Los Angeles, CA) to allow total elimination of free FITC, as detected

---

[7] A. Johnstone and R. Thorpe, "Immunochemistry in Practice," p. 258. Blackwell, Oxford, 1982.

by spectrophotometric analysis of the dialysis medium at 495 nm. Fluoresceinated ligands are finally dialyzed in Dulbecco's modified Eagle's medium (DMEM) containing L-glutamine, 50 U/ml penicillin, and 50 mg/ml streptomycin, and filtered with 0.22-μm filters into sterile tubes. The FITC to protein ratio is 6.3 for IgG and 5.5 for insulin, and final protein concentrations are 8 and 6 mg/ml, respectively. The ligand concentrations used, as indicated in the description of each experiment, are obtained by diluting ligand solutions with DMEM with the addition of L-glutamine, penicillin–streptomycin, and fetal calf serum at a final concentration of 5% (v/v). The presence of FCS does not modify ligand–cell interactions, as demonstrated by comparison with serum-free experiments, but is associated with better preserved cell morphology and greater adhesion to the cover slide, especially in the case of endothelial cells. Aliquots of the last dialysis medium of ligand–FITC are filtered and saved for incubating with cells at least 2 hr before the beginning of the experiments to further exclude contamination of the ligand–FITC solutions by free FITC (see later). When indicated, possible effects due to lipopolysaccharide (LPS) contamination of the ligand solutions are excluded by repeating the experiments with the addition of 5 μg/ml polymyxin B to all incubation media. Fluoresceinated ligands are kept sterile at 4° without preservatives for up to 2 months.

Cell Culture

Human umbilical cord vein endothelial cells (EC),[8] human fibroblasts,[9] and human proximal tubular epithelial cells (PTEC)[10] at the first passage are grown to subconfluence without attachment factors (to reduce background signal) on a glass cover slide fitting the flow chamber. In order to obtain good cell density within 24 hr, endothelial cells (which proliferate slowly on glass and in the absence of attachment and growth factors) are seeded carefully, placing 0.5 ml of cell suspension (10^5 cells/ml) on the cover slide, letting the cells adhere for 4 hr in the incubator, and finally adding the proper amount of culture medium for incubation. All the experiments are performed 24 hr after seeding the cells.

Membrane Binding of Ligands

The system described allows one to observe membrane binding of a ligand to living cells. In fact, we have been able to show that purified normal IgG binds to the cell membrane of cultured living fibroblasts and is not

[8] S. Oravec, N. Ronda, A. Carayon, J. Milliez, M. D. Kazatchkine, and A. Hornych, *Nephrol. Dialys. Transplant.* **10,** 796 (1995).
[9] G. Gazzola, V. Dall'Asta, and G. Guidotti, *J. Biol. Chem.* **255,** 929 (1980).
[10] C. J. Detrisac, M. A. Sens, A. J. Garvin, S. S. Spicer, and D. A. Sens, *Kidney Int.* **25,** 383 (1984).

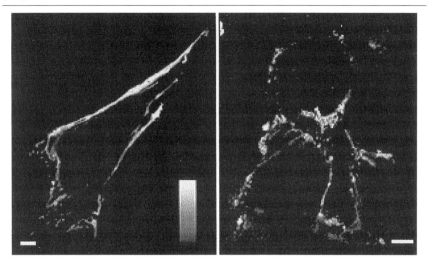

Fig. 4. IgG–FITC membrane binding to a living fibroblast (left-hand side) and to endothelial cells (right-hand side). Bars: 10 μm (left) and 5 μm (right). Reproduced with permission from N. Ronda, R. Gatti, G. Orlandini, and A. Borghetti, *Clin. Exp. Immunol.* **109**(1), 211 (1997).

internalized (Fig. 4). Before starting a specific experiment, it is advisable to obtain a basal image of the cells that had been incubated for 2 hr with the dialysis medium saved after the final dialysis of IgG–FITC in DMEM. The complete absence of a fluorescent signal excludes free FITC contamination of the fluoresceinated ligand. Such a preliminary test should be performed before every experiment described in this article. We then incubate fibroblasts in the confocal flow chamber for 30 min with IgG–FITC at 2 mg/ ml in standard culture conditions, wash them with culture medium, and observe the cells at 10- to 15-min intervals for 2 hr. A fluorescent signal is detectable with IgG diluted out to 0.5 mg/ml.

We did not observe IgG–FITC binding to PTEC under the same conditions, even using an IgG concentration of 8 mg/ml. In contrast, normal IgG entered living EC within minutes (see later). In order to show membrane binding of IgG to endothelial cells, we have inhibited cell energy-dependent processes by setting the flow chamber temperature to 27° (the lowest temperature tolerated by EC without cell damage in our system) and observed cells after 5 min of incubation with 2 mg/ml of IgG–FITC (Fig. 4).

Receptor–Ligand Internalization

Whenever a signal from the ligand under investigation is detected, it is necessary to check its localization within the cell and it is advisable to identify the nature of the structure or compartment involved. Moreover, it is important to monitor the morphological evolution of the process. In order to achieve these goals, it is useful to counterstain the cells. This is possible, with no need to change the field of observation, by flushing 1 μM calcein AM through the flow chamber medium. This neutral dye is converted by intracellular esterases into a fluorescent anionic compound, optimally excited at 488 nm with a peak emission around 520 nm, which stains the nucleus and the cytoplasm (nucleus/cytoplasm signal intensity ratio, 2:1), except for cationic compartments. Calcein allows one to check cell vitality as the fluorescent signal immediately disappears in the presence of membrane damage, despite residual cell esterase activity.[11] Because the calcein signal often overwhelms that emitted by the internalized ligand (see later), once counterstaining is carried out, possible subsequent changes in the distribution pattern of the internalized ligand cannot be visualized. Therefore, it is possible to program the counterstaining at various times of incubation in different samples or to track the relevant changes in a single microscopic field and to delay counterstaining until appropriate. In order to verify the actual intracellular nature of the signal, when the raw and the counterstained images of the same field are overlapped digitally, usually the intensity of the latter must be reduced evenly. However, comparison of the two separate images also can provide additional useful information, as will be illustrated later.

As an example of the information achievable related to morphology, timing, and specificity (i.e., receptor involvement) of the internalization of a ligand, we first describe the visualization of receptor–ligand internalization in a well-known model (insulin and endothelial cells) and then in the case of a previously unknown ligand–cell interaction (IgG in endothelial cells).

Confocal observation of living endothelial cells incubated with insulin–FITC allows the direct visualization of insulin internalization, the morphology and timing of which are consistent with previous knowledge of the process. Insulin binding and internalization are almost immediate, with a fluorescent cytoplasmic fibrillar signal evident after only 2 min of incubation with 1 mg/ml insulin–FITC followed by washing with culture medium (Fig. 5A, see color insert).

The initial fibrillar aspect evolves rapidly, and after 10 min the fluorescence is distributed almost entirely in cytoplasmic granules, some of which are larger bodies of 2–8 μm in diameter (Fig. 5B, see color insert).

The signal from intracellular insulin–FITC is relatively weak as compared to that of calcein and is no longer appreciable after counterstaining

[11] P. Moore, I. MacCoubrey, and R. Haughland, *J. Cell Biol.* **111,** 58 (1990).

(Figs. 5C and 5D, see color insert). For overlapping images, it is necessary to attenuate the calcein signal evenly by 25–35% to show insulin localization together with cytoplasmic staining (Fig. 5E, see color insert). This is particularly evident from the comparison between Fig. 5B and Fig. 5D. For the acquisition of the latter, the sensitivity of the photomultiplier, which was appropriate for calcein, was too weak to detect insulin–FITC. Therefore the large granules containing insulin and excluding calcein appear as negative bodies. The preservation from calcein loading indicates that these are acidic compartments, likely to correspond to late endosomes involved in insulin catabolism[12] (Fig. 5D).

As stated earlier, the incubation of living endothelial cells with IgG–FITC at 2 mg/ml is followed by the internalization of IgG, detectable after 10–15 min of incubation and most evident after 20–30 min. For experiments excluding the presence of contaminants other than IgG that could be responsible for the intracellular fluorescence, see Ronda *et al.* [6] The intracellular localization of fluorescence can be demonstrated first through a vertical section of a protrusion of a cell (Fig. 6, see color insert) and by calcein loading, as shown earlier, but without a need to reduce counterstain intensity, as the IgG–FITC signal is higher. The fluorescence pattern is that of a fibrillar network, particularly abundant in peripheral areas of the cytoplasm and in protrusions that apparently connect adjacent cells (Fig. 6, top). After 1 hr, most of the fluorescence is localized in cytoplasmic granules and it is reduced greatly after 2 hr. The time course, morphology, and inhibition by the low temperature of the IgG internalization process in endothelial cells are consistent with a receptor-mediated mechanism rather than with pinocytosis, which is a slow process, with poor quantitative efficiency, leading to a nonspecific uptake of extracellular medium. The demonstration that pinocytosis was not responsible for our observation came from the lack of internalization of fluoresceinated IgG fragments by endothelial cells under the same conditions. Proximal tubular epithelial cells, whose nonspecific reabsorption of proteins from preurine is well known, show cytoplasmic granules of IgG after only 48 hr of incubation with IgG–FITC or fluoresceinated IgG fragments.

As compared to insulin–FITC, the process of internalization appears morphologically similar, but insulin internalization is faster and, as noted earlier, requires a higher power of excitation and magnification of the signal. Such a difference in fluorescence intensity is likely to be due mainly to the smaller number of FITC molecules per molecule of the ligand in the case of insulin (5600 molecular weight) as compared to IgG (150,000), rather than determined by differences in cell receptor number or affinity.

[12] J. Carpentier, *Diabetologia* **37**, S117 (1994).

Two-Step Staining for Identification of Subcellular Structures

The fibrillar pattern of cytoplasmic fluorescence shortly following internalization of IgG–FITC and insulin–FITC, together with the well-known function of microtubules in molecular/vesicular intracellular trafficking, suggests the possibility that microtubules are involved in the receptor–ligand internalization systems. We have thus designed an experimental procedure to demonstrate the possible overlapping of signals from internalized fluoresceinated ligands and microtubules stained with a different probe. As a preliminary test we ensured that the cytoplasmic IgG–FITC signal remains unmodified after cell fixation with methanol. We then induced internalization of IgG–FITC in endothelial cells in the confocal flow chamber as described, turned off the thermostat, fixed the cells by flushing 100% methanol at 4° through the chamber for 2 min, and washed the cells with PBS at room temperature. We then performed indirect immunofluorescence at room temperature using a mouse monoclonal anti-α-tubulin antibody (Sigma) and an antimouse IgG TRITC (λ_{ex} = 552 nm; λ_{em} = 570 nm)-conjugated antibody (Sigma). Images were then acquired using a double-channel system placing a secondary beam splitter (565 nm) after the pinhole and 535-nm (± 15) bandpass and 570-nm long-pass barrier filters before two separate photomultipliers. After acquisition, barrier filters were inverted to check for contamination of fluorescein image by TRITC; indeed the absolute negativity of the field excluded such a possibility. The actual purity of each signal was also enhanced using the "separation enhancement" routine of the software, which subtracts a chosen percentage of the signal of each channel from the other one.

The images obtained show a perfect correspondence between internalized ligand–FITC localization and some of the microtubular filaments (Fig. 7, see color insert). It is known that microtubules, intermediate filaments, and parts of the endoplasmic reticulum often colocalize,[13] but the actual involvement of microtubules in the internalization of IgG has been demonstrated by the total inhibition of the process obtained by pretreating endothelial cells for 20 min with 100 μg/ml colchicine before incubation with IgG–FITC.[6]

Acknowledgments

This work was funded by the Department of Clinical Medicine, Nephrology and Health Sciences and partly by CNR target project "Biotechnology." The confocal apparatus is a facility of the Centro Interfacoltà Misure of the University of Parma.

[13] H. Lodish, D. Baltimore, A. Berk, S. L. Zipursky, P. Matsudaira, and J. Darnell, in "Molecular Cell Biology" (J. Darnell, ed.). Scientific American Books, New York, 1995.

[21] Quantitative Imaging of Metabolism by Two-Photon Excitation Microscopy

By David W. Piston and Susan M. Knobel

Introduction

Many of the important biological discoveries made with fluorescence microscopy have resulted from experiments on fixed samples. Unlike other experimental approaches, such as electron microscopy, fluorescence microscopy offers the possibility of working with living specimen. Since the early 1980s, significant developments have allowed fluorescence microscopy assays of processes in living cells (e.g., Ca^{2+}, membrane potential, vesicular transport). Most of these techniques depend on the addition of an extrinsic fluorescence reporter, which can introduce difficulties in the interpretation of results. Recently introduced fluorescent reporters based on the green fluorescent protein (GFP) have the potential to alleviate some of these problems, but considerable work remains before these will be of general use.[1] Still, any external indicator dye may alter the process under observation. Instead of using extrinsic probes, we have used intrinsic cellular fluorescence, which offers several advantages for the investigation of cellular metabolism. Because these autofluorescent compounds are natural constituents of every cell, there is no problem of uniform loading of the dye. In addition, these probes can be active participants in cellular processes. However, intrinsic fluorophores are typically not as bright or photostable as artificial probes, so they are more difficult to measure in the microscope. This article describes the use of the naturally occurring reduced nicotinamide adenine dinucleotide(phosphate) [NAD(P)H] as a monitor of cellular metabolism. To image these UV-absorbing fluorophores in living cells, we have utilized two-photon excitation microscopy, which minimizes the photodamage associated with NAD(P)H imaging.

This article describes the use of NAD(P)H as a metabolic indicator and discusses the two-photon excitation microscopy methods that we use to image its activity. It also details the instrument that we use for these experiments, with emphasis on the important design criteria for this demanding application. Finally, this article presents the application of two-photon exci-

---

[1] A. Miyawaki, J. Llopis, R. Heim, J. M. McCaffery, J. A. Adams, M. Ikura, and R. Y. Tsien, *Nature* **388,** 882 (1997).

tation imaging of NAD(P)H to assay glucose-stimulated metabolism in both pancreatic and muscle cells.

Use of NAD(P)H Autofluorescence for Metabolic Imaging

Fluorescence from naturally occurring NAD(P)H can be used as an indicator of cellular respiration and therefore as an intrinsic probe to study cellular metabolism.[2] Under normal conditions, roughly 25% of the reduced pyridine nucleotides are phosphorylated. Because this amount varies among different cells and because the spectra of NADH and NADPH are indistinguishable in our imaging experiments, we simply refer to both as NAD(P)H. NAD(P)H fluorescence is normally excited with light of ~360 nm and emits in the 400- to 500-nm region. Because the fluorescence yield of reduced forms [NAD(P)H] is significantly greater than for oxidized forms [NAD$^+$(P)], fluorescence intensity can be used to monitor the cellular redox state. Measurement of the NAD(P)H/NAD$^+$(P) ratio has been developed into a noninvasive optical method to monitor cellular respiration, called redox fluorometry,[3] which has been widely used.[4,5] Biochemical experiments have shown good qualitative agreement between the rise in NAD(P)H levels and the changes in fluorescence intensity measured on the addition of cyanide.[6]

Because NAD(P)H has a small absorption cross section and a low quantum yield, it is difficult to measure and has the potential to cause considerable photodamage. Furthermore, it absorbs in the UV, which is also more biologically damaging than visible or infrared light. Thus, a researcher interested in imaging metabolic dynamics would never choose NAD(P)H as the fluorophore, except that it is an active participant in cellular metabolic events. This means that it can be used without perturbation of the events under study. Despite the difficulties of imaging this dim, UV-absorbing fluorophore, two-dimensional images of the fluorescence intensity from NAD(P)H have been obtained with a conventional fluorescence microscope from several cellular systems.[7–10] These images have also

[2] B. Chance and B. Thorell, *J. Biol. Chem.* **234**, 3044 (1959).
[3] B. Chance and M. Lieberman, *Exp. Eye Res.* **26**, 111 (1978).
[4] C. Ince, J. M. Coremans, and H.A. Bruining, *Adv. Exp. Med. Biol.* **317**, 277 (1992).
[5] S. A. French, P. R. Territo, and R. S. Balaban, *Am. J. Physiol.* **275**, C900 (1998).
[6] B. R. Masters, A. K. Ghosh, J. Wilson, and F. M. Matschinsky, *Invest. Ophthalmol. Vis. Sci.* **30**, 861 (1989).
[7] J. Eng, R. M. Lynch, and R. S. Balaban, *Biophys. J.* **55**, 621 (1989).
[8] W.-F. Pralong, C. Bartley, and C. B. Wollheim, *EMBO J.* **9**, 53 (1990).
[9] W. Halangk and W. S. Kunz, *Biochim. Biophys. Acta* **1056**, 273 (1991).
[10] J. M. Coremans, C. Ince, H. A. Bruining, and G. J. Puppels, *Biophys. J.* **72**, 1849 (1997).

been used to assess metabolic dynamics as a function of pharmacological or electrical stimulation of the cells, but the experiments are limited to whole cell measurements (e.g., of isolated cardiac myocytes) or surface imaging (e.g., of an intact perfused heart).

To measure events with subcellular resolution or within a single cell in an intact tissue accurately, it is necessary to use an optical sectioning microscope. While use of the confocal microscope for optical sectioning is well established, confocal imaging of autofluorescence from living cells is problematic. Unfortunately, UV confocal microscopy is degraded by optical system problems, especially chromatic aberration between UV excitation and visible fluorescence. The introduction of commercial UV confocal microscopes has permitted confocal imaging of cellular autofluorescence dynamics.[11,12] However, these UV confocal observations were still limited severely by photobleaching of the autofluorescence and could not be performed in thick tissues. Many of the limitations of confocal microscopy for the imaging of NAD(P)H in living tissue can be overcome using two-photon excitation microscopy.[13–15]

Two-Photon Excitation Microscopy

Background and Concepts

The effective sensitivity of fluorescence microscopy measurements is often limited by out-of-focus flare. This limitation is reduced greatly in a confocal microscope, where the out-of-focus background is rejected by a confocal pinhole to produce thin (<1 μm), unblurred optical sections from within thick samples. A new alternative to confocal microscopy is two-photon excitation microscopy, which excels at imaging of living cells.[16] Two-photon excitation arises from the simultaneous absorption of two photons in a single quantized event, which is dependent on the square of the excitation intensity. Because the energy of a photon is inversely proportional to its wavelength, the two photons should be about twice the wavelength required for single-photon excitation. For example, NAD(P)H that normally absorbs

[11] B. R. Masters, A. Kriete and J. Kukulies, *Appl. Opt.* **32**, 592 (1993).
[12] A. L. Nieminen, A. M. Byrne, B. Herman, and J. J. Lemasters, *Am. J. Physiol.* **272**, C1286 (1997).
[13] D. W. Piston, B. R. Masters, and W. W. Webb, *J. Microsc* **178**, 20 (1995).
[14] B. D. Bennett, T. L. Jetton, G. Ying, M. A. Magnuson, and D. W. Piston, *J. Biol Chem.* **271**, 3647 (1996).
[15] D. W. Piston, S. M. Knobel, C. Postic, K. D. Shelton, and M. A. Magnuson, *J. Biol Chem.* **274**, 1000 (1999).
[16] W. Denk, J. H. Strickler, and W. W. Webb, *Science* **248**, 73 (1990).

ultraviolet light ($\sim$350 nm) can also be excited by two red photons ($\sim$700 nm). In the case of fluorescence, the emission after two-photon excitation is the same as would be generated in a typical biological fluorescence experiment.[17] To obtain sufficient two-photon absorption events for an imaging application, very high laser powers are required. These powers are achieved practically using mode-locked (pulsed) lasers, where the power during the peak of the pulse is high enough to generate significant two-photon excitation, but the average laser power is fairly low ($<$10 mW, just slightly greater than what is used in confocal microscopy).

The application of two-photon excitation to laser scanning microscopy is very powerful. In the microscope, two-photon excitation microscopy is made possible not only by concentrating the photons in time (by using the pulses from a mode-locked laser), but also by crowding the photons spatially (by focusing in the microscope). As a laser beam is focused in the microscope, the only place where the photons are crowded enough to generate an appreciable amount of two-photon excitation is at the focal point. The localization of excitation yields many advantageous effects.[17] Most importantly, the use of two-photon excitation minimizes photobleaching and photodamage—the ultimate limiting factors in fluorescence microscopy of living cells and tissues. In addition, it is not necessary to use a pinhole to obtain optical sectioning, so flexible detection geometries can be used. For instance, it is now possible to develop a high efficiency direct detection scheme (where the fluorescence does not pass back through the scanning system as it must in a confocal microscope) as described later. Further details about two-photon excitation microscopy are presented elsewhere (for an introductory review, see Piston[18]; for a more advanced and comprehensive description, see Denk et al.[17])

Two-photon excitation microscopy is the only currently available method capable of yielding high-resolution images of NADH autofluorescence throughout an extended sample such as the pancreatic islet.[14,15] The two-photon technique also allows for increased signal detection, and that translates into reduced photobleaching, which in turn leads to a better effective subcellular resolution. Most importantly, though, is the reduction in photodamage, which preserves sample viability and permits extended time-course measurements from living cells. Because the time scale of metabolic events in mammalian tissue may be up to several hours, increased

[17] W. Denk, D. W. Piston, and W. W. Webb, in "The Handbook of Biological Confocal Microscopy" (J. Pawley, ed.), 2nd ed., p. 445. Plenum, New York, 1995.
[18] D. W. Piston, Trends Cell Biol. 9, 66 (1999).

sample viability is paramount. Even for less demanding studies, however, minimization of cellular photodamage is always a worthwhile goal.

Two-Photon Excitation Microscope Optimized for Quantitative NAD(P)H Imaging

Because NAD(P)H has a small absorption cross section and a low quantum yield, it is very important to minimize photodamage and to maximize the fluorescence collection. The use of two-photon excitation minimizes photodamage because of its inherent three-dimensional localization. Although there is still a chance of photodamage associated with NAD(P)H excitation in the focal plane, this type of photodamage does not occur out of the focus. To optimize the fluorescence collection system for NAD(P)H, modifications must be made to the optical system. Because normal fluorescence is red shifted and the wavelengths used in confocal microscopy are in the visible, most commercial laser scanning microscopes are designed to increase fluorescence collection in the red wavelengths. This usually means that they are not optically efficient in the deep blue/near-UV range, and use of one of these systems for two-photon excitation requires the addition of an external nondescanned detector. Furthermore, even UV confocal systems [which are more efficient optically in the NAD(P)H fluorescence wavelengths] usually have red-enhanced optics and photomultiplier tubes (PMTs) as well. For two-photon excitation imaging, it is necessary not only to optimize collection, but also to reject the red excitation light, which may be up to 10,000-fold greater than the excitation in a normal confocal system. Thus, to obtain sufficient excitation light rejection, a barrier filter used with two-photon excitation must have four orders of magnitude better rejection in the excitation band than a barrier filter used with confocal microscopy. Part of this rejection can be achieved by replacing the red-enhanced PMTs with red-blind PMTs. Inexpensive, bialkali photocathode PMTs are near ideal for this purpose because they offer high quantum efficiencies in the blue, but are insensitive to wavelengths above 650 nm. The system described in this article uses a nondescanned detection pathway with minimal optics and the correct PMTs to maximize the collection of NAD(P)H autofluorescence. Even though this system is entirely custom-built, this type of external detection pathway can be added easily to any confocal microscope system.

A schematic diagram of the two-photon excitation laser scanning microscope is shown in Fig. 1. To produce illumination for two-photon excitation, an all-lines argon ion laser (Coherent Innova 310, Santa Clara, CA) is used to pump a Coherent Mira mode-locked Ti:sapphire femtosecond laser using the X wave mirror set [the mirror set allows tuning from 690 to 960

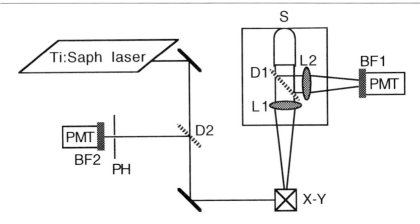

FIG. 1. A schematic diagram of the two-photon laser scanning microscope with optimized nondescanned detection, plus confocal detection. The incoming laser light is raster scanned (X–Y scan mirrors) and is focused onto the sample (S) through the tube lens (L1) and the objective lens. Fluorescence returns down the same path until it is reflected to the detection system by D1 (550 DCLP). The emitted signal is then refocused by a transfer lens (L2) so that the back aperture of the objective is conjugate to the front face of the PMT. A custom 700- to 720-nm blocked low-pass filter (BF1, Chroma Technology, Brattleboro, VT) is used to collect the NAD(P)H fluorescence. For confocal detection, D1 is removed and D2 is used. The scanning is done from below the microscope (bottom port of the Zeiss Axiovert 135TV), which allows for access to the stage from both sides for perfusion, temperature control, or microinjection experiments.

nm, but for NAD(P)H we use ~705 nm]. The output of the Mira laser at this wavelength is a pulse train of ~150-fsec pulses, running at 78 MHz, with an average output power of ~300 mW. For our NAD(P)H imaging experiments, the laser power must be attenuated considerably because we only require ~3 mW at the focal plane. This attenuation can be done with a variable neutral density filter (such as the Newport Corp. 925B, Irvine, CA) or with a low dispersion Pockel's cell modulator designed specifically for ultrafast laser applications (FastPulse Technology 5026, Saddlebrook, NJ). The pulses can be chirped negatively by extracavity prisms to maintain transform limited pulses at the sample, which will give the highest two-photon excitation. However, we find dispersion compensation to be a highly unnecessary procedure with our optical system (obviously for fiber-coupled systems such compensation is a must). The pulse train is directed onto two orthogonal galvanometer scanners (Cambridge Technology 6350, Cambridge, MA), and the resulting scanned beam is focused onto the sample through the basement port of a Zeiss Axiovert microscope. For the objective lens, we use either a 40× 1.3 NA F-Fluar (to obtain the highest possible signal collection) or a 40× 1.3 NA plan-Neofluar (when a larger flat field is needed). More highly corrected lenses, such as plan-Apochromats, should not be used for NAD(P)H imaging because they have lower transmission efficiency and the chromatic corrections are not really needed with two-photon excitation of a single fluorophore.

Because the confocal spatial filter is not needed to obtain confocal properties with two-photon excitation, it is best to collect and measure the fluorescence as near to the sample as possible (i.e., before it reaches the scanning mirrors). This eliminates losses at optical surfaces that are not necessary for fluorescence filtering. The entire generated fluorescence signal is collected because no spatial filter is used, nor is it needed because two-photon excitation is confined inherently to the focal volume. Therefore, nondescanned detection allows increased detection efficiency without any loss of three-dimensional discrimination. The fluroescence is split from the reflected signal by a dichroic mirror (with high reflectivity from 380 to 550 nm) to the detection unit that contains a single barrier filter (with high transmittance from 380 to 550 nm, and 10 OD blocking at 700 to 720 nm) for NAD(P)H autofluorescence. Both of these components are custom made (Chroma Technologies, Brattleboro, VT) for optimal NAD(P)H signal collection. A single transfer lens (antireflection coated from 380 to 550 nm, Melles Griot, Carlsbad, CA) is used to optically map the back aperture of the objective lens to the detection surface of the PMT (Hamamatsu R268, Bridgewater, NJ). Because the PMT is optically conjugate to the back aperture, which is a stationary pivot point of the scanning beam, there should be no effects of any spatial heterogeneities of the photocathode in the resultant image. The use of a bialkali photocathode PMT is best suited for this application. This nondescanned pathway contains only four optical elements (objective lens, dichroic mirror, transfer lens, and barrier filter), and each of these elements is optimized for transmission (or in the case of the dichroic mirror, reflection) of NAD(P)H fluorescence. To obtain the highest gain in the PMT, the supply voltage is set near its maximum (1150 V) throughout the experiments. In our system, the PMT signals are amplified (Hamamatsu C1053), integrated for the pixel duration, and then directed to a frame store card (Data Translation 3852, Marlboro, MA). It should be emphasized that such an external nondescanned detection system can be added easily to any commercial confocal microscope.

Applications to Quantitative Metabolic Imaging

General Quantitative Imaging Considerations

The instrument just described allows for extended dynamic studies of many cells simultaneously, thereby permitting observation of the temporal and spatial organization of metabolic activity within intact tissues. To perform quantitative experiments there are some important calibration experiments that must be performed first. These include viability controls, determination of the linearity of detection, and photobleaching controls.

When imaging living cells, sample viability is perhaps the major issue to address. This becomes even more important for NAD(P)H because it is not only a poor fluorophore, but it is also a participant in the metabolic events under investigation. For instance, it may be possible to photoinactivate NAD(P)H by excessive fluorescence excitation, which could alter the cellular redox state. To assay cellular viability in the experiments described here, we check the cellular response to glucose before and after a given laser imaging exposure. Laser irradiation of ~3 mW (average power at the sample) generated signals sufficient for imaging without evidence of cellular damage (i.e., the autofluorescence response to glucose was the same after irradiation as it was before irradiation) even after 1 hr of continuous laser scanning irradiation. However, laser irradiation of 5 mW resulted in an immediate slow rise in autofluorescence and led to a significant reduction in the glucose-induced autofluorescence response.[14] We further determined that this laser-induced photodamage was due to two-photon excitation, not just the incident red light. To show this, islets were exposed to intense laser illumination that focused into the coverslip (not focused in the islet, so there was no two-photon excited fluorescence generated in the sample). In this case, even 15 mW of unfocused red light did not affect the subsequent glucose-induced NAD(P)H response.

Because laser scanning microscopy uses PMT detectors, which offer a high dynamic range, low noise, and excellent linearity, quantitation of image data is straightforward. However, to make sure that data will be acquired in a linear fashion that will allow quantitation, there are two considerations. First, the analog-to-digital converter that is used to get the PMT signal into the computer must not be saturated (i.e., for a typical 8-bit system, the PMT gain must be set so that all pixel values are <255). Once the correct gain is set, it should not be changed; in our system we always use 1150 V (near the maximum allowable voltage) on the PMT. As described earlier, we are limited to 3 mW of illumination by viability considerations, and with this excitation level we never observe saturated pixels in our NAD(P)H images. Second, the amplifier offset must be set so that the zero pixel value corresponds to the zero fluorescence signal. This can be done using a standard fluorescent sample (we use pure NADH for this) as follows. Image four concentrations (in a ratio of $1:2:3:4$) of NADH that each give a good signal but for which the maximum pixel value is <255 at the gain used for cellular imaging. Determine the mean pixel value of each image using the histogram command and plot the mean pixel value versus the concentration.

The resulting plot should be a straight line, and the Y intercept will go through zero when the black level is correctly set. If the Y intercept is positive, the offset should be reduced; if the Y intercept is negative, the offset should be increased. Although it may require several trials to achieve the optimum gain/offset combination for a given sample, this calibration only needs to be done once (but it is a good idea to check it occasionally to assure accurate quantitation).

To determine the photobleaching correction, the laser is scanned continuously, and images are generally collected every 10th scan. The whole images are then analyzed as described at the end of this article and the results are plotted out. This plot usually shows a decay of fluorescence, which can be fit to an exponential curve. This curve can then be used to correct the results from a time-course experiment. Surprisingly, the photobleaching controls in our two-photon excited NAD(P)H experiments have shown that no bleaching correction is needed. At the excitation levels determined by the viability controls above, there was no measurable photobleaching, even after 1 hr of continuous laser scanning of a single optical section. While we cannot be sure why there is no measurable photobleaching, it is likely due to a combination of the low laser intensities used and the excitation localization of the two-photon technique. The small amount of photobleached NAD(P)H may be replaced by natural cellular mechanisms, and as long as the bleaching rate is slow enough, the cells can keep the level of NAD(P)H + NAD$^+$(P) constant during an experiment.

Image Analysis and Quantitation

For the quantitation of NAD(P)H autofluorescence data, we perform most of our digital image analysis on Macintosh Power PC computers running NIH Image 1.61 (Bethesda, MD). This free program is good and easy to use and it is available from the NIH via the World Wide Web at http://rsb.info.nih.gov/nih-image/Default.html. For single-cell analysis, we usually use a 25 pixel circular region of interest (ROI) that does not include the cell nucleus. It is also important to use the same ROI for all measurements on that cell (in images acquired at different times or with different glucose concentrations). We also collect data on several 25 pixel ROI that are not within any cells to use as a background standard. These values are averaged and subtracted from the autofluorescence values. Because we have set our acquisition system to have zero offset, this residual background represents the small portion of the excitation light that is detected. The background value in our images has pixel values of 2 or 3, compared with peak autofluorescence pixel values ~100.

Quantitative Metabolic Imaging of β Cells in Intact Pancreatic Islets

Insulin secretion from pancreatic β cells is tightly coupled to glucose metabolism. At physiological glucose levels, glucose is phosphorylated by the high K_m hexokinase, glucokinase (GK).[19] The kinetic features of GK are quite distinct from the other three mammalian hexokinases: K_m for glucose of ~8 mM, sigmoidal kinetics, and the lack of significant inhibition by glucose 6-phosphate. As glucose signaling proceeds in the β cell, intermediate metabolism results in an increase in the ATP/ADP ratio, which is reflected in a concomitant change in the reduced-to-oxidized NAD(P)H/NAD(P)$^+$ ratio. Thus, the measurement of NAD(P)H autofluorescence is a powerful tool for investigating the glucose response of β cells.

β cells normally exist within the islet of Langerhans, a quasi-spherical microorgan in the pancreas consisting of ~1000 cells. When β cells are isolated from their natural environment, marked variability in their metabolic responses to glucose has been observed using NAD(P)H autofluorescence as an index of the cellular redox state. These studies have been performed using flow cytometry[20] and fluorescence microscopy.[8,14] In order to extend these measurements to the intact islet, we used two-photon excitation microscopy to image glucose-induced NAD(P)H autofluorescence. Using the following protocol for two-photon excitation microscopy to image NAD(P)H levels quantitatively, we have shown that β cells within the islet form a much more uniform population than isolated β cells.[14]

1. Islets are isolated from mice by distention of the splenic portion of the pancreas followed by collagenase digestion.[21] The splenic portion of the pancreas is preferable because it yields islets that are enriched in β cells,[22] which are the focus of our current work.

2. Isolated islets are maintained in petri dishes (Falcon) in culture media consisting of RPMI 1640 (Life Technologies, Inc., Grand Island, NY) with 5 mM glucose (basal), supplemented with 10% fetal bovine serum (FBS, Life Technologies, Inc.) containing 100 units/ml penicillin and 100 μg/ml streptomycin (Life Technologies, Inc.) at 37° in an atmosphere of 5% (v/v) CO_2. Generally, islets can be maintained in culture dishes for up to 1 week. For longer culture periods, islet viability can be increased by culturing them on extracellular matrix (ECM).[23] However, islets cultured this way should be on ECM-

[19] F. M. Matschinsky, B. Glaser, and M. A. Magnuson, *Diabetes* **47**, 307 (1998).

[20] D. Pipeleers, R. Kiekens, Z. Ling, A. Wilikens, and F. Schuit, *Diabetologia* **37**, S57 (1994).

[21] D. W. Sharp, C. B. Kemp, M. J. Knight, W. F. Ballinger, and P. F. Lacy, *Transplantation* **16**, 686 (1973).

[22] Y. Stefan, P. Meda, M. Neufeld, and L. Orci, *J. Clin. Invest.* **80**, 175 (1987).

[23] G. M. Beattie, D. A. Lappi, A. Baird, and A. Hayek, *J. Clin. Endocrinol. Metab.* **73**, 93 (1991).

coated coverslips because they cannot be removed from the ECM without damage.

3. Prior to imaging or fixing, islets are attached to the bottom of a MatTek dish (MatTek Corp.) using Cell-Tak (Collaborative Biomedical Products, Bedford, MA). A 0.5-μl drop of Cell-Tak is placed in the center of a MatTek dish and dried for 30 sec at 42°; the dish is rinsed with Hanks' balanced salt solution (HBSS, Life Technologies, Inc.) and then 100 μl HBSS is added so that it covers the dried drop of Cell-Tak. We find the MatTek dishes very easy to use for most experiments and they fit into our microscope stage incubator. For experiments that require perifusion of solutions, we can either use these dishes with a input and aspirated outflow or use one of many home-built chambers, which hold a 1-inch diameter coverslip. In the case of chambers, we dry the Cell-Tak onto the coverslip, place the coverslip in the chamber, and then add the islet as described next.

4. Using a dissecting microscope, islets are picked up by a pipette, washed quickly in HBSS, and then placed directly onto the circle of dried Cell-Tak by pipetting through the HBSS. Islets must be washed in serum-free buffer because serum inhibits the attachment to Cell-Tak. We often place two or more islets contiguously in the dish. For many experiments (e.g., comparing islets from normal and transgenic mice), it is extremely useful to perform the imaging of different islets simultaneously. This provides an internal control for each experimental observation.

5. The islet should immediately attach firmly to the Cell-Tak, and then 2 ml of BMHH buffer (125 mM NaCl, 5.7 mM KCl, 2.5 mM CaCl$_2$, 1.2 mM MgCl$_2$, 10 mM HEPES, and 0.1% bovine serum albumin, pH 7.4) with 1 mM glucose is added to the dish. It is important to add the buffer gently to avoid dislodging the islets.

6. The dish containing the islet(s) is then transferred to the microscope stage and allowed to equilibrate for 15 min at 1 mM (basal) glucose. During the imaging experiments, the islets are held at 37° using a commercial microincubator (TLC-MI, Adams & List Associates, Westbury, NY). An air stream incubator (Nicholson Precision Instruments, Gaithersburg, MD) is also used to heat the objective to eliminate heat transfer through the glass–oil–objective interface.

7. NAD(P)H autofluorescence measurements can now be made in response to changes in glucose or other treatments. Each image is formed using a 9-sec scan. Because we use the external nondescanned detector, all measurements must be made in the dark. NAD(P)H–glucose dose–response images are generally acquired after a 5-min eqilibration at each glucose concentration. For extended time course

measurements, we usually take one scan every 10 sec; several hundred images can be taken with no noticeable degradation in islet viability.

Figure 2 shows a typical image of β cell NAD(P)H autofluorescence within an intact islet, where signals from both the cytoplasm and mitochondria can be observed. The outlines of single cells are visible as are the nuclei, both of which appear dark. These NAD(P)H imaging experiments simply cannot be performed by confocal microscopy due to the limitations imposed by photobleaching and UV-induced photodamage. In addition, these data open up the possibility of subcellular resolution to investigate differences between cytoplasmic and mitochondrial metabolism. Figure 3 shows a NAD(P)H–glucose dose–response curve averaged from the responses of 80 cells within four different islets. The apparent K_m of this dose response is ~8 mM glucose, indicative of the central role of GK in β-cell glucose-stimulated metabolism.

We have also developed protocols for the indirect immunostaining of glucokinase, insulin, glucagon, somatostatin, and pancreatic polypeptide in intact islets. These immunofluorescence methods allow us to correlate the results of dynamic metabolic imaging with enzyme and hormone levels in

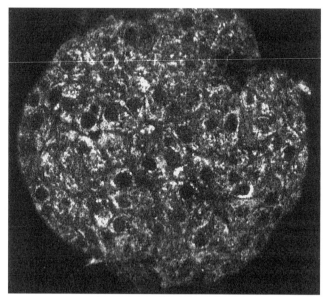

FIG. 2. Optical section of NAD(P)H autofluorescence from an intact islet. The NAD(P)H signal arises from both the cytoplasm and mitochondria, the latter of which sometimes can be seen as punctate bright spots. Cell outlines and nuclei [where there is little or no NAD(P)H] appear dark.

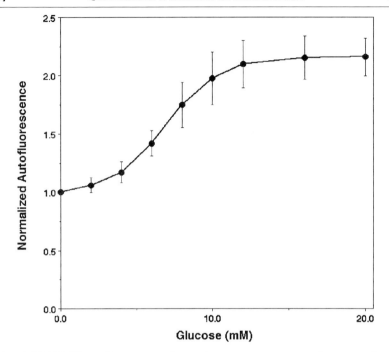

Fig. 3. NAD(P)H autofluorescence glucose dose response, average of responses from 80 β cells in intact islets. Each NAD(P)H level was measured after a 5-min equilibration with glucose solutions. The inflection point of this curve is consistent with the kinetics of glucose binding to glucokinase.

the cells. Because the islet is a thick multicellular sample, it is necessary to use low antibody concentrations and long incubation times. Any attempt to rush these steps will lead to brighter immunostaining in cells on the periphery and poor staining of cells in the middle of the islet.

1. Islets used for immunofluorescence staining are fixed in 2 ml of cold (4°) 4.0% (v/v) paraformaldehyde (Electron Microscopy Sciences, Ft. Washington, PA) in 10 mM phosphate-buffered saline (PBS), and then incubated at 4° for 45 min. This should be added to the dish gently, taking care not to dislodge the islets.
2. The dish is then washed gently three times with cold (4°) 10 mM PBS. At this point the sample can be stored at 4° in 10 mM PBS for several days before staining.
3. The islets are permeabilized by first pipetting off the PBS and then adding 10 mM PBS with 0.2% Triton X-100; incubate this for 2.5 hr at room temperature. All of the staining solutions must be added

to and removed carefully from the glass coverslip in the center of the dish with 100 μl volumes for each incubation.

4. Replace the permeabilization solution with the blocking solution (10 mM PBS + 0.2% Triton X-100 with 5% normal donkey serum) and incubate for 2.5 hr at room temperature.

5. Replace the blocking solution with fresh antibody dilution buffer [10 mM PBS + 0.2% Triton X-100 + 1.0% bovine serum albumin (Sigma, St. Louis, MO)] and incubate for 20 min at room temperature.

6. This is then replaced with the primary antibody diluted in antibody dilution buffer at a final immunoglobulin G (IgG) concentration of about 10 μg/ml. The dish is then incubated undisturbed overnight at room temperature. This should be placed in a humidified chamber made from a P-100 size petri plate with a circle of damp paper towel placed inside the bottom of the dish. Place the MatTek dish on top of this paper towel carefully.

7. The following day, remove the primary antibody and wash the dish four times carefully with 10 mM PBS + 0.2% Triton X-100 for 20 min each time.

8. Replace the final wash with antibody dilution buffer and incubate for 1 hr at room temperature.

9. Replace this with the secondary antibody in antibody dilution buffer, return the dish to the humidified chamber, and incubate overnight undisturbed in the dark at room temperature. Because the secondary antibody is light sensitive, it is helpful to perform these washes with minimal light in the room.

10. The following day, remove the secondary antibody and wash four times, 20 min/wash, with 10 mM PBS + 0.2% Triton X-100.

11. Prior to mounting the sample, rinse it once in 10 mM PBS without Triton X-100. For most sample mounting, we use a few drops of Aqua-Polymount (Polysciences, Inc.). This dries into a solid sample that will last for several months (at least). The Aqua-Polymount must fill the cut-out section in the center of the MatTek disk to the rim, as it will shrink a little as it hardens. For the most demanding high-resolution imaging, we dehydrate the samples and mount them in methyl salicylate, which has the same index of refraction as objective lens immersion oil.[24]

12. This dish is now ready for analysis by quantitative confocal microscopy and comparison with the two-photon excitation metabolic im-

[24] R. G. Summers, S. A. Stricker, and R. A. Cameron, in "Methods in Cell Biology" (B. Matsumoto, ed.), Vol. 38, p. 265. Academic Press, San Diego, 1993.

aging. It can be stored at room temperature in the dark until imaging is complete.

Quantitative Imaging of Muscle Metabolism

Glucokinase is not expressed in peripheral tissue, such as muscle, so in these cells glucose metabolism is mediated by other hexokinases. Because the K_m for glucose of the other hexokinases is much lower than that of GK, this difference can be seen readily in the NAD(P)H responses to glucose. Of the several hexokinase isoforms, hexokinase II (HKII) is thought to be most important in muscle. HKII has been shown by biochemical methods to bind mitochondria,[25] where it has better access to the ATP needed to phosphorylate glucose and where it is less sensitive to inhibition by glucose 6-phosphate, the reaction product. Further, this mitochondrial binding of HKII is thought to be regulated by insulin. We therefore would like to determine the relationship among HKII distribution, insulin action, and glucose-stimulated metabolism. Redox fluorometry based on NAD(P)H autofluorescence has been used extensively to image metabolism in cardiac muscle[4,7] and should be quite useful for other muscle types as well. We are using two-photon excitation microscopy to image cultured L6 muscle cells in order to assess the glucose-stimulated changes in NAD(P)H autofluorescence in the entire cell and its mitochondria. For this application, two-photon excitation microscopy offers the unique ability to follow these changes in many individual cells in real time. Similar to the experiments with pancreatic islets, the cells could be fixed and stained immediately after the metabolic experiment, and the precise distribution of HKII can be determined by immunofluorescence. Currently, we are developing methods to image the HKII distribution in living cells using a fusion protein of HKII and the green fluorescent protein.

Through the two-photon excitation imaging procedure described here, we can derive the kinetics of glucose-dependent changes in mitochondrial and cellular NAD(P)H autofluorescence and attempt to correlate these results with HKII location in the cells. We can also determine whether insulin treatment (either through long-term culture or rapid application) contributes to enhanced cellular glucose metabolism. Once the metabolic effects are well characterized, they can be correlated with the binding of HKII to the mitochondria.

The procedures used are similar to those for islets, but because the cultured cells are not as thick, we can use less input power. We have found that only 2 mW at the sample gives a good signal with no obvious cellular

[25] D. X. Sui and J. E. Wilson, *Arch. Biochem. Biophys.* **345**, 111 (1997).

Fig. 4. Optical section of NAD(P)H autofluorescence from a field of L6 myotubes. In this case the NAD(P)H signal arises mostly from the mitochondria, which can be seen as brighter punctate objects. In the more highly differentiated cells, fluorescence is evident in mitochondria collimated between muscle fiber striations. As in Fig. 2, cell nuclei appear dark.

photodamage; even after several minutes of continuous imaging, the morphology of the cell and its NAD(P)H response to glucose are unchanged. The gain and offset of the microscope are the same as used for the islet imaging, and again there was no measurable photobleaching during control experiments.

1. Partially differentiated L6 myotubes[26] are plated onto coverslip-bottom dishes (MatTek) that have been coated with mouse collagen, type IV (Life Technologies, Grand Island, NY), and cultured overnight in RPMI 1640 medium supplemented with 100 nM insulin and 4.5 mM D-glucose. For many cell cultures, it is important to grow them on some kind of matrix to which they will stick. We have used Cell-Tak or collagen in many experiments, but other matrices (e.g., polylysine) may work better for certain cell types.

2. Cells are washed and incubated in insulin- and glucose-free medium for 4 hr at 37°. It is important to begin with a baseline (in this case zero glucose) that will allow you to ratio the observed changes in autofluorescence.

3. At the start of the experiment, a plate of cells is positioned on the microscope stage of the two-photon instrument, where the cells are

[26] H. M. Blau and C. Webster, *Proc. Natl. Acad. Sci. U.S.A.* **78,** 5623 (1981).

maintained at 37°. The cells must be focused by eye and then be left to equilibrate to the temperature on the stage for at least 10 min. The temperature-controlled stage fits both MatTek dishes and several custom-built chambers that use 1-inch round coverslips. Failure to equilibrate the sample to the stage will result in motions during the experiments.

4. After the cells have equilibrated on the stage, a single image is taken at zero glucose concentration. As discussed earlier, a reliable baseline is very helpful in data interpretation. Because some cells may have more mitochondria in the optical section than others, the absolute value of the autofluorescence intensity is not of much value.

5. Images are taken 5 min after varying the medium glucose to each concentration. The glucose concentration can be changed by perfusing media with the desired concentration or by adding a small amount of 1 M glucose to the media in order to reach the desired concentration. In either case, we have confirmed that for L6 cells the fluorescence changes are complete within 2–3 min of changing the glucose concentration.

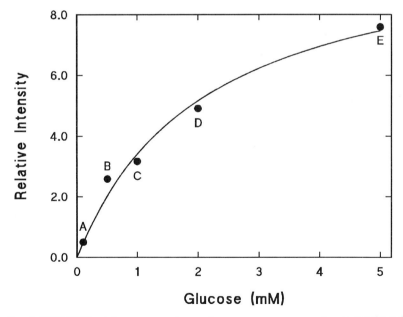

FIG. 5. NAD(P)H autofluorescence glucose dose response, average of responses from five L6 cells from Fig. 4. Each NAD(P)H level was measured after a 5-min equilibration with glucose solutions. Unlike the dose–response curves from β cells shown in Fig. 3, the inflection point of this curve is consistent with the kinetics of hexokinases other than glucokinase.

Figure 4 shows the NAD(P)H autofluorescence pattern in several cells at 0.2 mM glucose. The pattern reflects primarily NAD(P)H in mitochondria, which is punctate, but there is also some cytoplasmic distribution of fluorescence. In differentiated cells, the bright mitochondrial fluorescence is evident as columns of mitochondria between muscle fiber striations. With increasing glucose concentration in the medium, an increased fluorescence was apparent, which is shown in Fig. 5 averaged over the same cells at each glucose concentration. A saturable increase in averaged fluorescence is apparent, with an apparent K_m that is in between that of the transporter (6 mM) and the endogenous hexokinase (0.1 mM). The response of various cells was quite homogeneous (i.e., all cells show a similar change). The addition of insulin to the overnight culture medium in step 1 increased the NAD(P)H autofluorescence signals significantly at all glucose concentrations (including zero glucose). This suggests persistent effects of the hormone on the cellular redox state. Overall, these results show that we can define the kinetics of glucose utilization in real time from single cells using two-photon excitation microscopy.

Acknowledgments

The authors thank Drs. James May, Dick Whitesell, and Daryl Granner for help with the L6 cell experiments and Mr. Guangtao Ying for help in constructing the two-photon excitation microscope. This work has been supported by an NIH grant to DWP (DK53434) and the Vanderbilt Cell Imaging Resource (supported by CA68485 and DK20593).

[22] Localization of Proteases and Peptide Receptors by Confocal Microscopy

By CELIA J. SNYMAN, DESHANDRA M. RAIDOO, and KANTI D. BHOOLA

Introduction

Historical Overview

In the 1970s, a new generation of confocal microscopes was introduced to the scientific world simultaneously in Oxford and Amsterdam. These technical innovations were combined in 1977 and comprised the theoretical background for the geometry of confocal imaging by Wilson and Sheppard,[1]

---

[1] T. Wilson and C. J. R. Sheppard, "Theory and Practice of Scanning Optical Microscopy." Academic Press, London, 1994.

both members of the Oxford group. These authors also pioneered the description of the nonlinear relationship between light and atoms of an illuminated object and the theory of Raman spectroscopy using laser scanners.

The first convincing practical demonstration of a confocal microscope with improved resolution was performed by Brakenhoff and colleagues[2] at the University of Amsterdam using a microscope equipped with high numerical aperture lenses designed for imaging in the transmission mode. Soon after that, the new technique was ready for practical application in various fields of medicine and biology. Clinical examination techniques in ophthalmology immediately profited from the new technique. With confocal microscopy it became possible to routinely inspect the endothelial cells at the inner corneal side of a patient's eye. For this particular application, Koester designed a confocal slit scanner for which he obtained a U.S. patent in 1979 (USP 4,170,398). Another application was added later, which allowed the three-dimensional (3D) imaging of the retina.

Carl Zeiss developed the first prototype of a laser scanning microscope in 1982. However, Wijnaendts van Resandt et al.[3] presented the first successful demonstration of optical sectioning of fluorophore-labeled biological material in 1985, who were also the first to perform a section along the Z axis. This marked the time when all the key technologies necessary for the production and digital recording of three-dimensional data sets became available. Among the prerequisites were a set of perfectly designed optical lenses, high sensitivity detectors, versatile fluorophore markers, and last but not least, powerful lasers capable of producing coherent light of the correct wavelength required for the desired excitation of a given fluorophore. Furthermore, integrated computer systems were developed for controlling the scanning mechanics and for storing and managing the vast amount of data required for reconstructing three-dimensional images. Van der Voort et al.[4] reported the first description of such a completely integrated version of a confocal microscope. These instruments, now referred to as the first generation and equipped with a laser as the light source, had been designed as object scanners, meaning that the object stage was moved, whereas the beam was fixed in place.

The alternative version with a beam scanner was more suitable for the study of living cells, as it uses rapidly scanning mirrors to scan a laser beam

[2] G. J. Brakenhoff, P. Blom, and C. Bakker, *Proc ICO* **11,** 215 (1978).

[3] R. W. Wijnaendts van Resandt, H. J. B. Marsman, R. Kaplan, J. Davoust, E. H. K. Stelzer, and R. Stocker, *J. Microsc.* **138,** 29 (1985).

[4] H. T. M. Van der Voort, G. J. Brakenhof, J. A. C. Valkenburg, and N. Nanninga, *Scanning* **7,** 66 (1985).

across the specimen. This construct only became available in 1985, but it is now a standard feature of the confocal microscope. Currently, there are several different versions of confocal microscopes on the market, one of which is illustrated in Fig. 1 (color insert).

The trend toward linking even more powerful computer workstations with confocal microscopes is already evident. With increased computational power it is now possible to view an object from all possible spatial orientations and combine the morphological information with quantitative physiological data such that a full representation of the structures and dynamic changes of a cell during physiological events can be obtained.[4a]

Nature of Confocal Imaging

For more than 200 years, light microscopy has provided biologists with a powerful system to unravel the structure of microorganisms, to analyze cell and tissue structure, and to study the dynamics of cell function. Further, with conventional light microscopes, researchers have always been limited by the preparation procedures necessary to produce suitable contrast for visualizing structures. With the advent of the electron microscope came the ability to decode structures of biological objects much smaller than those observed by light microscopy.

Although light microscopy has played a crucial role in characterizing tissue structure and electron microscopy in providing powerful visuals of cell ultrastructure, the images are static. Research into the functions of nonfixed living cells requires an experimental approach that permits observation of live cells. Living specimens offer unique imaging problems due to their opacity, motion, and photoabsorption. Using noninvasive techniques, advancements in fluorescence microscopy have permitted real time analysis of free ions, second messengers, retrieval of peptide receptors, and many other molecular events in living cells and molecules on the surface of nonfixed cells or membranes with considerable precision using novel and specific probes and ligands.

The confocal microscope provides optimal operating conditions for the study of living tissue. What distinguishes confocal microscopy from conventional light microscopy is its ability to optically slice tissue sections or cells. In the confocal microscope, all structures being out of focus are suppressed at image formation, providing images free of out-of-focus information. This is achieved because the object is not illuminated and imaged as a whole at the same time, but in sequence at one focal point after the other, with the

---

[4a] J. Engelhard and W. Knebel, "Confocal Spectrum." Leica Lasertechnik GmbH, Im Neurnheimer Feld 518, Heidelberg, Germany.

optical scanning of each plane. An additional advantage is the ability to create, with increased image resolution, three-dimensional constructs.

Principal Elements

The following components are typically found in a confocal microscope (Fig. 2, color insert). 1. The laser light source has the versatility that allows suitable wavelengths to be selected either individually or as a combination. Currently, a number of lasers are available: HeNe 543 nm, HeNe 633 nm, and Ar 458, 488, 514, and 566 nm for the visible light range and the KrAr 351, 364 nm for the ultraviolet range. 2. It should have a scan unit for moving the illuminating beam across the object. 3. Detectors to record the amount of photons coming from the object are usually designed for detecting fluorescence and reflected light. Optionally an additional detector can be installed that would allow viewing of the object in a nonconfocal transmission mode. 4. A central processing unit to control the hardware and data storage unit to store and manipulate image data. 5. A scanning system attached to a research microscope.

With these components, a multifunctional imaging system is assembled capable of performing all major techniques of microscopical analysis, e.g., confocal multiple fluorescence detection and reflection, as well as basic nonconfocal transmission techniques such as absorption, phase, and differential interference contrast. The combination of these imaging techniques and image processing allow the manipulation of all aspects of an image, including the rotation of the object in space. Therefore, a much more detailed view of the specimen is possible. This explains why confocal microscopes have become extremely important in medicine and biological science.

Basic Steps

1. Illumination of a spot by confocal microscopes which achieve high resolution of a selected plane in a specimen in three basic steps. The light is focused by an objective lens into an hourglass-shaped beam so that the bright "waist" of the beam strikes one spot at some chosen depth in a specimen.
2. Once focused, the laser light passes through a pinhole to the detecting device.
3. The light emitted from that spot is split into its spectral components by a prism, focused to a point, and allowed to pass in its entirety through a slit formed by two mirrors, a pinhole aperture.
4. Light passing through the slit is detected by a photomultiplier. The light reflected by the mirrors is directed to subsequent pairs of mir-

rors, forming slits for the second and third detector as well. Up to four detectors can be supplied with custom spectral bands simultaneously.

5. The opaque regions around the pinhole black out most of the rays that would tend to obscure the resulting image, namely those rays reflected by illuminated parts of the specimen lying above and below the plane of interest.

6. The confocal uses a variable size detection pinhole for two primary reasons: (a) If the fluorescence is very weak the number of available photons emitted from the specimen is very small (1/10,000 of excitation intensity). To increase the number of photons available for detection, one can open the detection pinhole. However, opening the pinhole increases the signal-to-noise ratio of the image at the expense of optical sectioning performance. (b) In order to achieve optimal optical section thickness, one needs to match the pinhole diameter to the diffraction pattern in the intermediate plane.

7. Scan the entire plane. Finally, in order to produce a two-dimensional image, it is necessary to move the illuminating light spot rapidly from point to point and in consecutive lines along the object until the entire plane has been scanned. The optical path depicted in Fig. 3 describes the geometry underlying the formation of just a single image spot. Alternatively, in slit scanners, a scan line can be produced at once in a single process and consecutive lines can be added to it until the image is completed.

8. Another important feature is that serial optical sections may be obtained by moving the focal plane progressively through the specimen, as the objective lens is positioned closer to or further from the sample. To obtain a three-dimensional image of an object, successive planes in a specimen are scanned. It produces a stack of images, each of which is an optical selection; such selections are analogous to images of fine slices cut physically from a specimen.

Merits of the Confocal Microscope

Advantages of the confocal microscope over a conventional microscope and the major improvements offered by the confocal microscope on performance may be summarized as follows: 1. Light rays from outside the focal plane will not be recorded. 2. Defocusing does not create blurring, but gradually cuts out parts of the object as they move away from the focal plane. The practical consequence is that these parts become darker and eventually disappear. The feature is called optical sectioning. 3. True, three-dimensional data sets can be recorded. 4. Scanning the object in the xy direction as well as in the z direction (along the optical axis) allows one

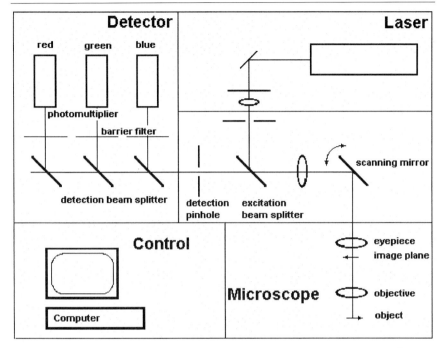

FIG. 3. Detailed organogram of the laser beam light path that complements the diagrammatic representation in Fig. 2.

to view the object from all sides. 5. Due to the small dimension of the illuminating light spot in the focal plane, stray light is minimized. 6. By image processing many slices can be superimposed, giving an extended focus image that can only be achieved in conventional microscopy by reduction of the aperture and thus sacrificing resolution. 7. It is therefore clear that quantitative measurements are operator dependent because until all intensities above and below the specimen are physically gathered, analysis of the images cannot be performed.

Optical Sectioning

Similar to the optics in a conventional microscope, the depth of the focal plane in the confocal microscope is determined in particular by the numerical aperture of the objective used and the diameter of the diaphragm. What is the thickness of an optical section for a given pinhole size? This question is not simple to answer. Optical sections do not have well-defined edges like a section taken from a microtome. In fact, the z-intensity profile is shaped like a bell. The thickness of an optical section is defined as the full-width half-maximum (FWHM) of this bell-shaped profile. The exact profile depends on many factors, such as the correction of the optical components, lens type, refractive index, and matching of indices. However, the most important factor is the specimen itself. Taking all these factors into consideration, the empirical section thickness can deviate from theoretical section thickness by one-third or more. At a wider detection pinhole the confocal effect can be reduced. This means that by widening the pinhole to a maximum width, an image can be achieved that is similar to that observable in a conventional microscope, where blurred parts surrounding a particular object or structure, arising from the out-of-focal plane, contribute to the image.

Technical Aspects: Size of Detector Pinhole

Due to diffraction, a point of light in the focal plane is imaged as a bright disk surrounded by increasingly dimmer rings. The bright center is called the Airy disk (G. B. Airy, 1801–1892, British astronomer). The diameter of the Airy disk depends on several optical parameters, such as wavelength, numerical aperture, objective magnification, and additional magnification factors of the system. Setting the size of the detector pinhole to the size of the Airy disk achieves optimal performance. Due to diffraction, setting the detection pinhole diameter less than the diameter of the Airy disk does not result in a thinner optical section.

The performance in optical resolution of a microscope is expressed as FWHM, which is two times the distance from a point-like object at which its luminosity drops by half the maximum light intensity measured in the center of the object. Two point-like objects separated by a distance of one FWHM from each other are just recognizable as two disparate structures, i.e., they can be resolved optically. The FWHM is defined by the wavelength of light used for illuminating the object and by the aperture of the objective lens. Although this is true for conventional as well as confocal microscopes, the actual resolution is still better by a factor of 1.4 in the confocal setup. This is described by the formula:

$$FWHM_{lat} = 0.4\lambda/NA$$

The wavelength is designated λ and the numerical aperture (NA) is $n \sin \alpha$, where n is the refractive index of the immersion medium and α is the aperture angle of the entrance lens of the objective. The immersion medium is placed between the objective lens and the object.

It is noteworthy that the axial resolution is much more dependent on the numerical aperture than the lateral one; also, objectives that allow air as the immersion medium have a better axial resolution than those that require liquid media at the small numerical aperture.

Axial resolution can be deduced from the following formula:

$$\text{FWHM}_{ax} = 0.45/[n(1 - \cos \alpha)]$$

The actual values for axial and lateral resolution measured in commercial high-performance lenses come close to the theoretical values within the range of a few percentage points.

Optimal Image Quality

As in other microscopes, the useful wavelength range of light is limited by the design of the objective lens. A very careful correction of chromatic lens aberrations is necessary for high-quality fluorescence images. Demands on the quality of chromatic correction become even more pressing as modern applications of multiple color fluorescence need even wider ranges of chromatic correction in order to increase the number of fluorophores when applied simultaneously in a single experiment. This in turn is only possible if lasers can be provided that generate the suitable spectrum of excitation wavelengths. To this end, modern mix gas lasers have been introduced that can be tuned to generate lines over the entire spectrum of visible light. However, the intensity for excitation illumination in a microscope should not exceed certain levels because of light saturation phenomena within the object, which can have adverse effects such as a reduction of optical resolution and inaccurate readings in quantitative measurements. Another important parameter in obtaining optimal images of quality is the signal-to-noise ratio, which depends on the type of detector and the design of the scanner. Optical discrimination between image points within and outside the focus is realized best in single beam scanners equipped with small confocal pinholes. Slit scanners and multiple spot scanners produced by a Nipkov-type disk suffer from imperfect optical discrimination and hence reduced contrast. For the detection of fluorescence signals in single beam scanners the signal-to-noise ratio is in practice determined only by the number of photons. This is why confocal imaging systems must be designed for optimal light transmission and maximum yield of photons at the detector, both of which depend heavily on the quality of lenses and the width of the apertures in the light path. Consequently, the signal-to-noise ratio is also the major limiting parameter in determining the scan rate, i.e., the time necessary to build a single image.

Engineering Aspects of Scanners

The following types of scanners may be distinguished.

1. The speed with which an object can be moved relative to the beam in object scanners is limited by mass inertia of the moving parts and by viscosity and surface interactions of the immersion media with lens and object. The advantage, however, is that because of the fixed position of the beam, object scanners require less sophisticated designs for the optical path. Mechanical work in object scanners is performed by galvanometers, piezo crystals, and electromotors fitted with suitable gears. Objective scanners have an instrument design similar to that of object scanners except that the object itself is not moved. However, the obvious restrictions imposed on the capability to perform high frequency movements has prevented a widespread application of both types of scanners. Object movement is still used but only for the slower movements along the z axis.
2. Beam scanners, however, are sufficiently fast to satisfy the requirements of a confocal system. They operate as single or multiple beams or slit scanners. Between these, the single beam scanners appear to bring together the best combination of features. Although slit scanners have the fastest rates of image buildup, the signal-to-noise ratio is inferior compared to single point scanners because stray light from out-of-focus areas is blocked off less effectively. In addition, slit scanners have a built-in asymmetry in their resolving power. Disk scanners also have disadvantages in that they have a lower fluorescence yield compared to single point scanners because in the latter, pinhole position and geometry can be adapted more easily to the experimental needs. In slit scanners, movements of the beam are realized by mounting polygonal mirrors on a motor axis or by galvanometers. Nipkov disks are rotated by electromotors. A very fast deflection of mirrors is achieved in acousto-optical devices, which, however, are curbed in their performance by the fact that the crystals used in these devices reflect light differently depending on the wavelength. Their use is therefore limited to semiconfocal slit scanners in fluorescence applications.

Generally the scan rate must be set as a compromise between speed and image quality and this in turn is dictated by the signal-to-noise ratio defined by the statistical profile of the photons collected. The latter is a result of the level of fluorescence output and conditions set by other experimental prerequisites. In principle, a higher fluorescence output can be produced using more powerful lasers; however, the intensity that can

FIG. 1. Confocal microscope assembly. The image visualization and capturing system of the confocal microscope comprises a fluorescent microscope and a controlling computer workstation linked to a 3D image analyzer.

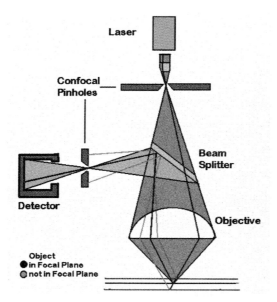

FIG. 2. Confocal optics. After it passes through a pinhole, the laser light beam widens and becomes concentrated by the objective that focuses the beam to a discrete point in a specified focal plane. Photons emitted from the excited fluorophore are directed through a second pinhole for capture by the detector system. Observe that the brown and red lines showing light paths that produce the blurred image in a conventional microscope are not passed through the second pinhole.

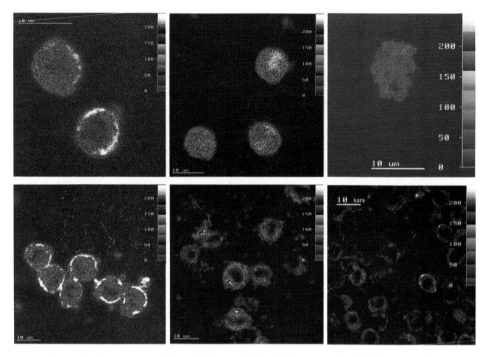

FIG. 4. Confocal images of tissue kallikrein (TK) and kinin moiety in the kininogen molecule on the external surface of human neutrophils. A1, TK; A2, hKLK1 (TK) mRNA antisense; and A3, hKLK1 (TK) mRNA sense. B1, Kinin moiety in the kininogen molecule on the external surface of the neutrophil membrane; B2, loss of kinin moiety from the neutrophil membrane when incubated with bacterial enzymes for 15 min; and B3, for 30 min.

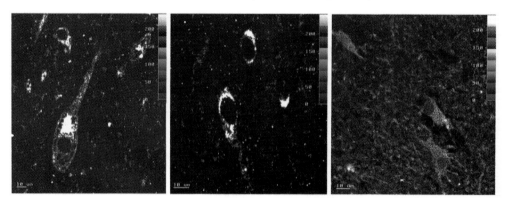

FIG. 5. Confocal images of tissue kallikrein (TK) and kinin receptors in the human brain. A1, TK in the cell body and axons of the neurons; A2, kinin B2 receptors on neurons; and A3, kinin B1 receptors on astrocytic tumor cells.

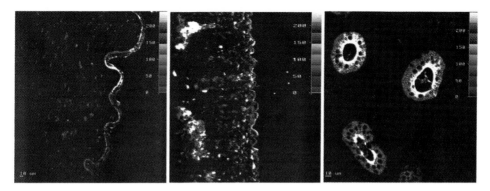

FIG. 6. Confocal images of tissue kallikrein (TK) in human blood vessels and salivary gland. A1, TK in normal artery; A2, TK in atheromatous plaque; and A3, TK in ductal cells of the salivary gland (positive control).

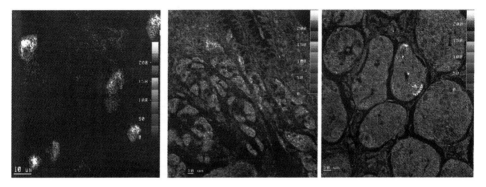

FIG. 7. Confocal images of tissue kallikrein (TK) and kinin B2 receptors in human gastric glands. A1, TK in parietal cells of the glands; A2, TK in parietal cells and deep glands in gastritis; and A3, kinin B2 receptors on parietal cells and cells of the deep gastric glands.

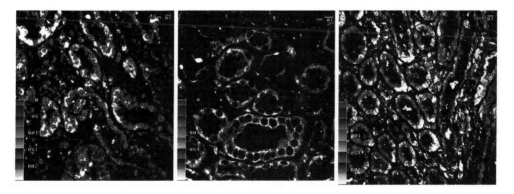

FIG. 8. Confocal images of tissue kallikrein (TK) and kinin B2 receptors in the kidney. A1, TK in distal connecting tubule cells and kinin B2 receptors in the normal kidney; A2, mild loss of TK in acute transplant renal rejection; and A3, kinin B2 receptors in the convoluted portion of distal and collecting ducts of normal kidney.

be applied maximally is limited by a saturation effect occurring in the process of fluorophore excitation. Scan rates in fluorescence applications, therefore, do not usually exceed one image per second. If image quality is not a limiting factor, images may be captured at the much faster rate of three to five images per second.

Confocal Image

Confocal images can typically be recorded at variable pixel densities with the highest resolution at a setting of 1024 × 1024 pixels. Depending on the size of the object there may be from 16 to several hundred optical sections compiled in a single 3D data set. In addition to the digitized information on the 3D position of an image pixel, the data set contains intensity values for each pixel recorded by 8-bit leading of 256 gray values. Such data sets, therefore, encompass several megabytes of information. This amount of data requires a fairly powerful computer system equipped with large memory storage media. Hence, it is obvious why technical realization of confocal microscopy was historically closely related to the developments in the computer field.

Three-Dimensional Constructs of Images

One of the most pressing problems in medical research is cancer. When cancer develops within a patient, cells of the body suffer subtle internal deviations from the normal developmental pattern, leading to more pronounced changes in cell shape and organization of tissues. Monitoring these changes as early as possible and describing the unfolding features of the many types of cancer may be achieved by preparing three-dimensional constructs of confocal images of the morphology of cancer cells.

Looking deeper inside the cell, it is becoming more important to study the microarchitecture of the cell nucleus with the aim of understanding the processes underlying cell growth and differentiation. Modern techniques of cell and molecular biology, such as *in situ* hybridization and immunofluorescence, provide the means for the study of spatial order in the nucleus, particularly by visualizing chromosomal domains at interface or in dividing nuclei, when the genome is split and partitioned to the progeny cells. Kinetochores are important structures in the cell nucleus, which are thought to be of instrumental importance during mitosis. In order to quantify structural changes that occur in the course of these processes, complex three-dimensional segmentation must be applied, which eventually leads to the exact determination of kinetochore positions.

As far as actual changes in the tissue go, there is great interest in applying confocal microscopy to monitoring the progression or remission of bone tumors. Particularly important are lesions in the bone caused by the activity of special cells called osteoclasts. These cells are also active during normal bone development, and the disintegrating bone structures caused by them have long been known from electron microscope studies. However, it has been difficult to determine the degree of damage inflicted on the bone. Using the confocal microscope, it is possible to determine the size of the lesion in a bone quantitatively and thus come to a realistic estimation of the scale of damage inflicted on the bone structure.

Simultaneous Detection of Separate Fluorophores

A great number of suitable fluorophores are known to differ in their emission and excitation spectra (Table I). These differences can be used to simultaneously visualize two or more cell structures labeled with specific fluorophore-conjugated markers. Such an application is beginning to gain importance in the localization of several genes in the nuclear genome as well as in studies on the spatial arrangement of the cytoskeleton within the microarchitecture of cells.

Confocal Microscopes

The essentials of confocal microscopy, namely optical sectioning, multiple lasers in the visible light and ultraviolet ranges, multidimensional imaging, and improved contrast and resolution of images, are provided by all of the current models on the market (Zeiss LSM 510, Leica TCS SP, Olympus FLUOVIEW, Noran OZ CLSM).

Value of Confocal Microscopy in Cell Research

Confocal microscopy is a new imaging technique allowing new insight into the morphology and dynamics of cells and tissues. In contrast to the performance of conventional microscopy, larger tissues or densely packed structures are no longer obstacles in arriving at high-quality images. The capability to produce consecutive optical sections with a thickness of less than 0.5 μm enables direct recording of three-dimensional data sets. Suitable programs to determine surface features, volume, near-neighbor analysis of given structures, and views of the total structure from any angle in space can process these data.

Studying dynamic aspects of cell activities is facilitated greatly by confocal microscopy in conjunction with suitable fluorophores. New applications that are becoming possible are the quantitative estimation of local ion concentrations and membrane potentials in cells and tissues. Similar experiments using conventional microscopy techniques have produced ambiguous

TABLE I

FLUOROPHORES USED COMMONLY IN BIOMEDICAL SCIENCE[a]

Fluorophore	Absorbance (nm)	Emission		Use/function
		nm	Color	
7-Aminoactinomycin D	546	647	Yellow	Labels nucleic acids in intact or permeabilized cells
Cy3	550	570	Red	Detection of antigen–antibody complexes in immunohisto-chemistry
DAPI	358	461	Blue	Labeling and counterstaining of chromosomes
Ethidium bromide	518	605	Brown/red	Detects breaks in nucleic acids
FITC	494	518	Green	Detection of antigen–antibody complexes in immunohisto-chemistry and FISH
Fluo-3	506	526	Green	Fluorescent indicator for cal-cium during signal transduc-tion in excitable cells
Mito Tracker Red	578	599	Red	Labels active mitochondria selec-tively
Texas Red	595	615	Red	Flow cytometry, detection of an-tigen–antibody complexes in immunohistochemistry, and la-beling unpolymerized G-actin
SNARF-1	575	635	Red/orange	Label to study cell morphology, adhesion, communication, and migration
SYTO 17	621	634	Red	Fluorescent labeling of DNA and RNA in live cells

[a] The range of fluorophores with their absorption and emission wavelengths used in confocal research are listed, together with their use in identifying cellular structures, dynamics of cell function, and intracellular transport of ions.

results in the past. The fact that the geometrically defined focal spot can be positioned precisely anywhere in the specimen provides ideal precondi-tions for a number of additional interesting experiments with the micro-scope.

Confocal microscopy involves the collection and computation of numer-ous data related to structure and physiological activity of the cell or tissue under study. Most data, such as multiple fluorophores and transmission signals, fluorescence recovery course, and color and Raman spectroscopy, can be recorded simultaneously. The integration of such information pro-vides an insight that is much closer to reality than has been ever possible to obtain before confocal microscopes became available to researchers in cell and tissue medical research and in material science.

Medical and Biological Applications of Confocal Microscopy

Due to the strength, versatility, and little limitation of the instrument in the choice of objects or preparation techniques, the applications for confocal microscopy are numerous. Material science was one of the first areas that took advantage of the three-dimensional imaging capability in the reflection mode. Because it is not necessary to place the object into a vacuum chamber, such as that used previously for scanning electron microscopy, handling and processing of samples are much less time-consuming. Because of this and the fact that imaging in a confocal microscope is essentially noninvasive and nondestructive, medical and biological research has begun to rely heavily on this new type of microscopical instrumentation. In particular, fluorescence techniques in the modern version of highly specific multifluorochrome labeling are experiencing a renaissance in current cell structure research.

Visualization of Images of Living, Nonfixed Cells

Apart from the immunofluorescence techniques already mentioned, cell structure research depends on the availability of fluorophores that can be used to stain specific cell components in the living cell. The lipophilic dicarbocyanine dye ($DiOC_6$) (Molecular Probes, Inc.) is one that accumulates in the endoplasmic reticulum (ER) and, because it can be monitored over time, provides information on the structural dynamics and functional importance of ER. However, it was also found that immobile, fixed sites exist that determine the location and overall shape of the ER over time. The same dye also stains the mitochondria of cells.

Spontaneous calcium release (SCR) from the sarcoplasmic reticulum (SR) initiates contractile waves in both uni- and multicellular cardiac preparations. The dynamics of contractile waves, originating from sites in the SR and visualized as propagating bands of shortened sacromeres, have been studied extensively. When loaded with the esterified derivative fluor-3-acetoxymethyl ester (Fluo-3-AM, Sigma Chemicals, St. Louis, MO), cell suspensions of isolated ventricular myocytes have been used for intracellular release studies of calcium. With the development of Ca^{2+}-sensitive fluorescent probes it has been possible to confirm that local changes in intracellular Ca^{2+} levels, described as "Ca^{2+} waves," are associated with the spontaneous contractile activity of myocytes.

Methods and Procedures

Ethics

Ethical permission for all of these studies has been obtained from the ethics committee of the Faculty of Medicine, University of Natal.

Collection and Processing of Neutrophils

Whole blood is obtained by venipuncture through a precision glide 21-gauge Vacutainer needle (Becton-Dickinson Vacutainer Systems, UK) from healthy volunteers and patients presenting with infection. Neutrophils isolated by careful layering of blood on 3.5 ml Histopaque 1119/1077 (Sigma, UK) on centrifugation at 1500*g* for 30 min at 20° form a band. Neutrophils from this layer are harvested with a Pasteur pipette, mixed with an equal volume of phosphate-buffered saline (PBS), centrifuged (500*g*), and the process repeated twice.

Tissue Collection

Brain, blood vessels, and control kidney tissue are obtained at autopsy: Individuals declared dead on arrival as a result of trauma not involving the head or sudden unexplained death are immediately refrigerated and maintained at 4° in the hospital morgue. The age, sex, cause, and time of injury and death are noted. The age ranges from 24 to 40 years. These patients sustained gunshot wounds to the chest or abdomen, acute myocardial infarction, or carbon monoxide poisoning. At postmortem, performed within 24 hr of death, the jugular veins are opened first to allow free drainage of blood from the brain. Immediately following its removal, the brain is divided sagittally along the midline. One-half is suspended in 5% formal saline [41% formaldehyde in 0.9% NaCl, 1:8 (v/v)] and is dissected after 48 hr. Topographical distribution of the various brain areas is established from three neuroanatomical texts,[5–7] as well as maintaining orientation by marking appropriate areas with India ink (Reeves, London) and 1% Alcian blue (pH 2.5, Sigma, St. Louis, MO). Blood vessel and control kidney samples are fixed similarly in 5% formal saline for 24 hr.

[5] J. D. Fix, *in* "Atlas of the Human Brain and Spinal Cord." Waverly Press, 1974.
[6] M. Roberts, J. Hanaway, and D. K. Morest, "Atlas of the Human Brain in Section." Lea and Febiger, Philadelphia, 1987.
[7] M. L. Barr and J. A. Kiernan, *in* "The Human Nervous System: An Anatomical Viewpoint." Lippincott, Philadelphia, 1993.

Brain Tumor Tissue. Samples of human brain tumor are obtained at surgery.

Gastric Tissue. Ten patients with dyspepsia are endoscoped, and two biopsy specimens each are taken from the fundus and antrum. Control tissue is obtained surgically by antrectomy from the antrum of a single patient. Patient selection is restricted to males and nonpregnant females between 18 and 75 years old undergoing routine evaluation for dyspepsia. Patients exhibiting any severe abnormality of the renal, respiratory, or cardiovascular system are excluded. Patients who had undergone recent esophageal or stomach surgery for an ulcer and those who presented with pyloric stenosis, perforation, or arterial bleeding are also omitted.

Renal Tissue. Renal tissue samples from 18 patients with deteriorating renal function were obtained by ultrasound-directed needle biopsy.

Tissue Processing

All of the tissue sections are fixed routinely in 5% (v/v) buffered formaldehyde–saline for 24 hr at room temperature. After fixation the tissue samples are dehydrated in a graded series of ethanol and embedded in paraffin wax.

Histology

Sections (4 μm thick) are cut on a rotary microtome (Leica, Germany) and adhered onto glass slides. These are stained with hematoxylin and eosin for histology and grading of disease; Giemsa, to determine bacterial presence; as well as periodic acid–Schiff, Luxol fast blue, methamine silver, Masson's trichrome, and elastic von Gieson. The control tissue shows "normal" histology.

Antibodies

Recombinant TK is obtained from Dr. M. Kemme (Institute for Biochemistry, Technical University of Darmstadt, Darmstadt, Germany). The antibody is produced in a goat host using 100 μg of recombinant tissue kallikrein conjugated to 125 μl TiterMax adjuvant injected intramuscularly at 4-week intervals over a 4-month period. Isolation and characterization of IgG are performed according to the method described by Johnston and Thorpe.[8] Aliquots (30 μl) are stored at $-20°$ and reconstituted when required.

---

[8] A. Johnston and R. Thorpe, *in* "Immunocytochemistry in Practice," p. 41. Blackwell Scientific, Oxford, 1982.

Eight polyclonal antibodies, fully characterized for specificity, have been raised to synthetic peptides of the amino-terminal and loop regions encoded by the rat B2 receptor cDNA and based on the homology between these regions with the human receptor. However, only four are shown to react strongly with human epithelial cells and neutrophils.[9] Matches between the peptide sequences of intracellular domains one (ID1, LHK) and two (ID2, DRY) and the fourth extracellular domain (ED4, DTL) in the rat and human receptors are 80, 100, and 75%, respectively. It was therefore decided to perform experiments with a combination of antibodies directed to these regions. The antibodies were kindly donated by Werner Muller-Esterl (Department of Pathobiochemistry, Johannes Gutenberg-University of Mainz, Duesbergweg 6, D-6500 Mainz, Germany). Fred Hess (Merck Research Laboratories, R80M-213, Rahway, NJ) kindly supplied an antibody directed at the C terminus of the peptide of the kinin B1 receptor (I-S-S-S-H-R-K-E-I-[F-Q-L]-F-W-R-N), and characterized for specificity.

Immunocytochemistry

Nonfixed Cells. The pellet of neutrophils is resuspended in oxygenated (95% O_2 and 5% CO_2, v/v) PBS cotaining 5 mM glucose and is then dispensed into microfuge tubes for incubation with kinin-releasing enzymes. Control neutrophils are incubated without the enzyme. Following an incubation period of 20 min, the cells are pelleted (1000g for 10 min at room temperature) and spotted onto glass slides. Cells are washed twice with PBS containing 1% human IgG, 1% bovine serum albumin, and 0.2% (w/v) sodium azide. Cells are treated with this buffer in order to block nonspecific binding of the antisera to Fc receptors and to prevent internalization of the receptors. Cells are first immunolabeled with the appropriate primary antibody (antibody against kinins, tissue kallikrein, and kininogen fragments or kinin receptors) under humid conditions and then labeled with fluorescein isothiocyanate (FITC)-conjugated specific anti-species secondary antibody for 30 min. For method controls, the primary antibody is replaced by PBS buffer. The slides are mounted with 10% PBS/90% (v/v) glycerol and viewed on the confocal microscope.

Fixed Cells. Isolated neutrophils are floated onto glass slides and fixed with 4% (w/v) paraformaldehyde in PBS for 2 min. The labeling procedure is similar to that used for nonfixed cells prior to viewing on the confocal microscope.

[9] M. Haaseman, C. D. Figueroa, L. Henderson, S. Grigoriev, S. Abd Alla, C. B. Gonzales, I. Dunia, E. L. Benedetti, J. Hoebeke, K. Jarnagin, J. Cartaud, K. D. Bhoola, and W. Muller-Esterl, *Braz. J. Med. Bio. Res.* **27**, 1739 (1994).

Fixed Tissue. Sections (3–4 μm thick) are cut from wax-embedded tissue on a rotary microtome (Leica, Germany), adhered onto poly(L-lysine)-coated glass slides, and probed for the presence of TK, B1, and B2 receptors using standard immunolabeling techniques detected with FITC viewed by confocal microscopy. Sections are dewaxed using xylene, rehydrated with increasingly dilute ethanol solutions (100, 90, 70, and 50%) and distilled H_2O as the final rehydrant, and trypsinized with 0.01% trypsin (Sigma) for 3 min. The tissue is then boiled at 80° in 0.1 M sodium citrate buffer for 10 min in a microwave oven (25% power, Sharp R-4A52) for antigen retrieval and allowed to cool at room temperature for 20 min.

Prior to incubation with the primary antibodies the sections are incubated with 1% (w/v) human immunoglobulin G (IgG) (Sigma) for 20 min to block nonspecific sites. Incubation of the tissue sections with specific, primary antibodies (100 μl), diluted in 1 mM phosphate buffer, pH 7.2, is performed at 4° for 18 hr. These antibodies comprise (1) goat anti-human recombinant TK IgG diluted 1:1000, (2) rabbit anti-B2 receptor peptide IgGs (a combination of the following three antibodies: ED4 DTL, 283; ID1 LKH, 280; and ID2 DRY, 277) diluted 1:200, and (3) rabbit anti-B1 receptor peptide IgG diluted 1:100 (v/v).

All incubations are carried out in a humidified chamber maintained at 21°. Between incubations the sections are washed by submerging in 0.01 M PBS, pH 7.4, for 10 min. Binding of the primary antibody is detected with a sheep antigoat or sheep antirabbit IgG coupled to a fluorescein probe, either fluorescein isothiocyanate (525 nm) or Cy3 (fluorophore 552 nm), at room temperature for 30 min.

Replacement of the primary antibody with nonimmune IgG serves as a method control. The loss of immunolabeling following preabsorption of the specific primary antibody with its respective antigen, namely (1) recombinant TK, (2) the kinin B2 receptor peptides (D-T-L-L-R-L-G-V-L-S-G-C for ED4, L-H-K-T-N-C-T-V-A-E for ID1, and D-R-Y-L-A-L-V-K-T-M-S-M-G-R-M for ID2), and (3) the kinin B1 receptor peptide (I-S-S-S-H-R-K-E-I-[F-Q-L]-F-W-R-N), demonstrates absolute specificity.

The labeled slides are finally mounted with 90% (v/v) glycerol–phosphate. Immunofluorescence is observed with a Leitz DM IRB confocal microscope (Leica, Germany), and fluorescent emission is analyzed using the Leica TD4 confocal microscopy system (Leica, Germany).

Image Analysis

Confocal images are typically recorded at a pixel density of 225 × 225 pixels. The gray scale ranges from 0 to 256 and is divided into eight equal phases (POLI Look-Up-Table), with each phase having a lower and upper gray-scale value threshold. With image analysis using the Analysis 2.1 Pro system (Soft-Imaging Software GmbH, 1996, Germany), the regions of interest in each image are encircled. This information is used to calculate the number of pixels falling within each phase, as well as the area circled for analysis. Therefore, the mean high intensity of immunolabeling in (n) number of cells is quantified in pixel/μm^2.

Images of Tissue Kallikrein (Protease) and Kinin Receptors (Peptide Receptors) Captured by Confocal Microscopy

Literature Review

There has been a paucity of research on the imaging of receptor populations by confocal microscopy. So far only a few citations have been captured on literature survey. Studies using the confocal microscope have been performed mainly on neutrophils or on cultured cells, but also on some fixed tissue.

Bacterial N-formyl peptides chemotactically attract neutrophils to sites of infection. The binding of N-formyl peptide to its G-protein-coupled receptor on the neutrophil membrane triggers signal transduction accompanied by the release of second messengers. Using a fluorescent probe, the distribution and possible internalization of the receptor were examined by confocal microscopy. Activated N-formyl peptides receptors seem to aggregate and bind to the membrane prior to internalization by a process sensitive to cytochalasin.[10]

The neutrophil CD14 receptor for lipopolysaccharide (LPS) regulates the chemotactic response of these cells to gram-negative bacteria. In cells so attracted the surface-expressed CD14 receptor becomes internalized, and this process can be visualized by confocal microscopy.[11]

Receptors to chemokine proteins are members of the seven transmembrane G-protein-coupled family and play a role in inflammation. The receptor-mediated endocytosis of chemokines was investigated by Solari *et al.*[12] using both fluorescently coupled chemokine agonists and antagonists. The distribution of the fluorescent label was followed confocally, and analysis of the results indicated that internalization had been triggered by agonists and not by antagonists. The question of the internalization of high-affinity

[10] B. Johansson, M. P. Wymann, K. Holmgren-Peterson, and K. E. Magnusson, *J. Cell Biol.* **121**(6), 1281 (1993).

[11] D. A. Rodeberg, R. E. Morris, and G. F. Babcock, *Infect. Immun.* **65**(11), 4747 (1997).

[12] R. I. Solari, R. E. Offord, S. Remy, J. P. Aubry, T. N. Wells, E. Whitehorn, T. Oung, and A. E. Proudfoot, *J. Biol Chem.* **272**(15), 9617 (1997).

neurotransmitters has been addressed by Grady *et al.*[13] Receptors for substance P and neurokinin 1 show internalization and recycling of the receptors. This is clearly a phenomenon common to all peptide receptors once activation of the cell has occurred and signal transduction is initiated.

Additionally, laser scanning confocal microscopy has been used in developmental biology. It has been particularly useful for the visualization of cellular actin together with chromatin in human arrested preimplantation embryos and in unfertilized oocytes obtained by *in vitro* fertilization.[14]

Visualization of Protease and Receptor Images on Nonfixed Cells by Confocal Microscopy

Images of Contact-Phase Factors on the Human Neutrophil Membrane

The contact-phase assembly comprises the serine proteases, plasma prekallikrein (PpreK) and tissue prokallikrein (TproK), the endogenous substrates for these enzymes, H- and L-kininogens, and the clotting factors, XI and XII.[15] In humans the two kallikreins are different proteins with the common capacity to generate the vasoactive peptides known as kinins. Tissue kallikrein, found in the epithelial cells of many organs, was discovered a decade ago in granules of the neutrophil.[16] In contrast, plasma kallikrein is synthesized as a precursor enzyme in hepatocytes together with H-kininogen. Both proteins are secreted into the circulation where they form a complex. The proenzyme (PpreK) docks on domain 6 of H-kininogen, in equilibrium with factor XI. On discovery of tissue kallikrein in human neutrophils by Figueroa and Bhoola,[17] these authors developed a concept that components of the kallikrein–kinin may be linked or spatially orientated on the neutrophil membrane. The unique optical feature of the confocal laser scanning microscope enabled the very first mapping of the essential proteins that comprise the contact-phase assembly on the external surface of the neutrophil cell membrane.[18] Additionally, L-kininogen[18] and

---

[13] F. Grady, P. D. Gamp, E. Jones, P. Baluk, D. M. McDonald, D. G. Payan, and N. W. Bunnett, *Neuroscience* **75**(4), 1239 (1996).

[14] R. Levy, M. Benchaib, H. Cordonier, C. Souchier, J. F. Guerin, and J. C. Czyba, *Ital. J. Anat. Embryol.* **102**(3), 141 (1997).

[15] K. D. Bhoola, C. D. Figueroa, and K. Worthy, *Pharmacol. Rev.* **44**(1) (1992).

[16] C. D. Figueroa, A. G. MacIver, P. Dieppe, J. C. Mackenzie, and K. D. Bhoola, *Adv. Exp. Med. Biol. Kinins V* **247B**, 207 (1989).

[17] C. D. Figueroa and K. D. Bhoola, *in* "The Kallikrein-Kinin System in Health and Disease— Leucocyte Tissue Kallikrein: An Acute Phase Signal for Inflammation" (H. Fritz, J. Schmidt, and G. Dietz, eds.), p. 311. Limbach-Verlag, Braunschweig, 1989.

[18] L. M. Henderson, C. D. Figueroa, W. Muller-Esterl, and K. D. Bhoola, *Blood* **84,** 474 (1994).

TproK[19] also appear to have binding sites on the membrane. Specific antibodies to each of these proteins, linked on incubation with FITC-labeled F(ab)$_2$ secondary antibodies, provided a powerful new imaging technique that established that these proteins were bound to the external surface of the neutrophil membrane (Fig. 4,A1, color insert). Images of tissue kallikrein mRNA are illustrated in Fig. 4,A2-3, color insert. The precise molecular sequences involved in the binding process have not as yet been fully elucidated for some of these proteins.

Images of Kinin B2 Receptor on External Surface of the Neutrophil Membrane

The cloning of kinin B2 and B1 receptors has shown them to be members of the superfamily of G-protein-coupled receptors linked to the G$_q$ subtype of the family. The receptor comprises seven α helices that span the membrane. Both receptors, through the G-protein link, activate phospholipase C, which results in the hydrolysis of phosphotidylinositol, the release of second messengers, and a rise in calcium transients within the cell that initiates cellular function.

The kinin B2 receptors are constitutive whereas the cell surface expression of the B1 receptors is mainly induced. The amino acid sequence of the membrane core and the transmembrane loops of the two receptors were determined from their cDNA profile. The subsequent raising of antibodies to the external and internal loop sequences of the receptor enabled the visualization of the receptors on the cell membrane of epithelial cells. Having established the presence of kinin-releasing enzymes and their substrates on the neutrophil membrane, insight clearly suggested the possible occurrence of autoreceptors to kinins also on the external surface. Experiments were designed therefore to examine this possibility. With antibodies raised to its external loops, the kinin B2 receptor was immunolocalized by confocal microscopy on the external surface of nonfixed human neutrophils.[9]

Images of Kinin Moiety Translocated from Kininogen Molecule on the Surface of the Neutrophil Membrane

The kinin nona- and decapeptides are primary mediators of the cellular response to tissue injury. For this reason they are important therapeutic targets. The kinin moiety resides in domain 4 of the kininogen molecule. Once released from kininogen by the enzymatic action of the kallikreins,

---

[19] Y. Naidoo, C. Snyman, D. M. Raidoo, K. D. Bhoola, M. Kemme, and W. Muller-Esterl, *Brit. J. Haematol.*, in press.

kinins cause vascular dilatation and, by increasing capillary permeability, promote the traffic of neutrophils into sites of tissue injury and infection. We have therefore examined the question of whether kinins are released from the kininogen molecule on the external surface circulating neutrophils when incubated with bacterial enzymes.

Using confocal microscopy linked to an image processing program, we captured the fluorescence emitted from an FITC-labeled F(ab)$_2$ secondary antibody complexed to a monoclonal primary antibody to bradykinin (Fig 4,B1, color insert). The images provided a qualitative estimate of the amount of antigen visualized on each optical plane. There was an almost complete loss of the kinin moiety in a dose-dependent manner from the surface membrane of the neutrophils incubated with either nagarse or serratiopeptidase (Fig 4,B2-3, color insert) and the kallikreins. Similarly, neutrophils harvested from patients with infections also showed a loss of the peptide, with the segments on either side of the kinin molecule, namely domains 1-3 and 5-6, still attached to the neutrophil membrane. In confirmation electron microscopy studies, immunoreactive kinin, seen as clusters of gold particles on the surface of the membrane, was absent from the circulating neutrophils of patients with systemic infection.[20]

Visualization of Protease and Receptor Images on Fixed Cells by Confocal Microscopy

Images of Tissue Kallikrein in Human Brain

Tissue kallikrein is a serine protease that has the capacity to generate kinins that are vasoactive peptides with multiple functions. Except for the localization of tissue kallikrein in human prolactin-secreting adenomas, its apparent deficiency in Alzheimer's disease, and its presence in human cerebrospinal fluid (CSF), the regional distribution of TK in human brain has not been defined. The study by Raidoo et al.[21] was the first attempt at mapping the regional distribution of tissue kallikrein in the human brain. The presence of this enzyme in several areas of the human brain provided a basis for investigating the localization by confocal microscopy and the expression of kinin receptors in the human brain. Further, such a study may aid in the understanding of the pathophysiological role of these receptors and their cell surface orientation to other peptide receptor systems in the human brain.

[20] Y. Naidoo, S. Naidoo, R. Nadar, and K. D. Bhoola, *Immunopharmacology* **33**, 387 (1996).
[21] D. M. Raidoo, R. Ramsaroop, S. Naidoo, and K. D. Bhoola, *Immunopharmacology* **32**, 39 (1996).

The regional distribution and the cellular localization of TK in human brains, collected within 24 hr of death, were determined by labeling with specific antibody to recombinant TK. Thus far this enzyme has been visualized in neurons of the hypothalamus, thalamus, cerebral gray matter, and reticular areas of the brain stem, as well as in cells of the anterior pituitary and choroid plexus by light and confocal microscopy (Fig 5,A1, color insert). In the central nervous system, a greater amount of TK exists in the proenzyme form and is located preferentially in the neuronal cell bodies and their processes in both the cerebral cortex and the brain stem. The cellular distribution of TK in specific areas suggests a role for TK in the neurons and epithelial cells of the brain. The question whether the functional importance of TK may relate to a particular cell type remains to be elucidated.

Images of Kinin Receptors on Human Neurons

Knowledge of the distribution of kinin receptors in the human brain should enhance our understanding of the neurophysiological role of kinins. Furthermore, induction of the kinin B1 receptor may be important in the pathogenesis of neural diseases. Using polyclonal antibodies directed to specific regions of the B1 and B2 kinin receptors and standard immunolabelling techniques, we have determined the localization of these receptors on neurons of specific areas of the human brain by confocal microscopy.[22] Kinin B2 receptors were identified in neurons of the brain stem, basal nuclei, cerebral cortex, thalamus, and hypothalamus (Fig 5,A2, color insert). Kinin B2 receptor immunolabeling was also observed in the endothelial lining of the superior sagittal dural sinus and ependyma of the lateral and third ventricles, whereas kinin B1 receptors were localized on neurons of the thalamus, spinal cord, and hypothalums. Although the binding of radiolabeled bradykinin to neuronal membranes has been demonstrated, this study provided the first conclusive evidence for the existence of immunoreactive B1 and the further confirmation of B2 receptors on human neurons.

The presence of immunoreactive kinin B2 receptors in neurons of the human brain implies that the receptor-mediated interaction of kinins may occur in the human nervous system and further supports a neuronal or humoral role for kinins. The localization of tissue kallikrein in choroidal epithelial cells and the occurrence of immunoreactive B2 receptors in the ependyma of the lateral and fourth ventricles, as well as in the endothelial lining of the superior sagittal dural sinus, suggest that kinins may regulate the homeostatic balance between the cerebral vasculature and brain parenchyma.

[22] D. M. Raidoo and K. D. Bhoola, *J. Neuroimmunol.* **77,** 39 (1997).

Kinin B1 receptors in the hypothalamus and thalamus may also modulate the cellular actions of kinins in neural tissue. The presence of B1 receptors on neurons in the substantia gelatinosa and some interneurons of the spinal cord, as well as specific thalamic nuclei in all of the brains studied, suggests a role for kinins in nociception.

Our knowledge of the localization of kinin receptors will be valuable, especially since a new generation of potent antagonists could be targeted to specific cellular functions.

Identification of Tissue Kallikrein and Kinin Receptors in Astrocytomas of Human Brain

Some of the major diagnostic and prognostic proteinases associated with cancer include the cathepsins (D, B, and L), collagenase, and urokinase-type plasminogen activator. The serine protease tissue kallikrein is present in tumors of the breast (ductal breast cancer cells), lung (Lewis lung tumor cells), stomach (gastric carcinoma cells), and pituitary (pituitary prolactin-secreting adenomas). By degrading components of the extracellular matrix, these enzymes facilitate tumor proliferation and invasion. In addition, the vasodilator effect of the bioactive peptides, bradykinin and kallidin, generated by tissue kallikrein would increase vascular permeability, thereby enhancing metastasis as well as providing additional nutrients important for tumor growth.

In order to determine whether tissue kallikrein has a role in the pathogenesis of tumors, its localization was carried out by confocal and electron microscopy and confirmed by *in situ* hybridization. Because kinins act by signal transduction mechanisms linked to B1 and B2 kinin receptors, it was necessary to elucidate the occurrence of B1 and B2 receptors in normal human cerebrum and astrocytic tumors.

Our studies have established that tissue kallikrein and kinin B1 receptors are abundant in proliferating tissue, whether of inflammatory or cancer origin. This observation is supported by the demonstration of tissue kallikrein and kinin B1 receptors in the astrocytes showing malignant transformation,[23] as illustrated in Fig. 5,A3 (color insert). These findings also suggest that, like other proteases, components of the kallikrein–kinin system may have diagnostic and prognostic relevance and that specific inhibitors of tissue kallikrein, kinin B1, and B2 receptors or their gene expression may be of therapeutic value in tumors of the central nervous system.

---

[23] D. M. Raidoo, R. Ramsaroop, S. Naidoo, and K. D. Bhoola, *Microsc. Soc. South. Afr. Proc.* **27,** 100 (1997).

Images of Tissue Kallikrein and Kinin Receptors in Human Blood Vessel and Atheromatous Plaques

Kinins are very potent dilators of arterioles, constrict veins, and contract endothelial cells. The specificity of pharmacological responses to kinin agonists is determined by the kinin receptor. Whereas kinin B2 receptors are stimulated by bradykinin and lysyl-bradykinin (kallidin), kinin B1 receptors are activated by the desArg analogs of these two peptides.

The first step was to visualize tissue kallikrein as the preferred kinin-releasing enzyme in blood vessels by confocal microscopy. Immunoreactive tissue kallikrein was localized in the cytoplasm of endothelial cells and, with lesser intensity, in the smooth muscle cells of muscular arteries and arterioles (Fig. 6,A1, color insert). In atheromatous plaques (Fig. 6,A2, color insert) of larger blood vessels the enzyme was present in endothelial cells and foamy macrophages.[24]

Immunoreactive B1 and B2 kinin receptors were also observed on endothelial cells and smooth muscle cells of the tunica media of regional blood vessels. In contrast, only B2 kinin receptors were located on the endothelium of the renal vein. The higher density conical images of these receptors on blood vessels suggest that they may be uniquely placed to homeostatically regulate blood pressure in humans.

Images of Tissue Kallikrein in Helicobacter pylori-Associated Gastric Ulcer Disease

The association of *H. pylori* (*Hp*) with ulcer disease is a common form of gastric disorder involving mucosal damage and invasion of the mucosa by polymorphic inflammatory cells with concomitant changes in the epithelial cell structure. Bacteria are thought to adhere by specific junction zones to the epithelial cell surface, resulting in the degeneration of the mucosal layer. The relative status of tissue kallikrein in antral and fundic biopsies, obtained endoscopically from 10 patients suspected of having gastric disorders, was examined by Naidoo *et al.*[25] For cellular evidence of inflammation, the tissue was stained with hematoxylin and eosin and classified as mild, active, chronic, and chronic active gastritis. The presence of the infective agent *Hp* was determined by Giemsa staining. For the localization of tissue kallikrein, slide-mounted tissue sections were subjected to peroxidase–antiperoxidase (PAP) and immunofluorescent staining using a goat IgG

[24] D. M. Raidoo, R. Ramsaroop, S. Naidoo, W. Muller-Esterl, and K. D. Bhoola, *Immunopharmacology* **36,** 153 (1997).

[25] S. Naidoo, R. Ramsaroop, R. Bhoola, and K. D. Bhoola, *Immunopharmacology* **36,** 263 (1997).

antibody to recombinant human tissue kallikrein. Confocal microscopy results revealed that tissue kallikrein in antral control tissue removed during partial resection of the stomach was immunovisualized along the luminal border of the deep pyloric glands (Fig. 7,A1, color insert). The surface epithelia and superficial glands showed no labeling. The fundic control tissue showed an absence of TK in the superficial and surface epithelial glands, but was positive in the parietal cells. Fundic biopsy specimens showed similar immunoreactivity in these areas. In contrast, in the inflamed pyloric mucosa, there was a shift of TK localization to the basal part of the glandular cells and there was also expression of TK in the superficial glands that showed cellular evidence of regeneration (Fig. 7,A2, color insert). In fundic biopsies, there was no change observed in the sites of TK localization (similar to control tissue). It was noted that even though infection by *Hp* could be demonstrated in 8 of the 10 subjects, the inflamed mucosa showed no discernible difference in staining patterns between infected and noninfected tissue sections.

Images of Kinin Receptor Status in Normal and Inflamed Gastric Mucosa

No documented studies have been reported on the occurrence of B1 and B2 kinin receptors in the mammalian gastric mucosa. This first study attempted to immunolocalize by confocal microscopy sites of B1 and B2 kinin receptors in the human pyloric gastric mucosa and to evaluate its role in gastritis.[26] Biopsies were obtained from patients with dyspepsia during endoscopic examination of the patient. Diagnosis and grading of the gastritis were performed following histological examination. In gastritis there is destruction of the normal mucosal glandular architecture with subsequent regeneration of the epithelial cells.

Kinin B2 receptors are known to mediate the physiological action of kinins, especially with regard to mucous production. Schachter *et al.*[27] postulated that tissue kallikrein may play a role in goblet cells by processing mucoproteins. During inflammation there is destruction of mucous-secreting cells, which would explain the loss of the B2 receptors.

Because the antrum is affected more by inflammation, the kinin receptor loss is more readily discernible in this region. Inflammation involves the migration of neutrophils, macrophages, and lymphocytes to the antrum. There is transmigration of the neutrophils through the basement membrane

---

[26] R. Bhoola, R. Ramsaroop, S. Naidoo, W. Muller-Esterl, and K. D. Bhoola, *Immunopharmacology* **36,** 161 (1997).

[27] M. Schachter, D. J. Longridge, G. D. Wheeler, J. G. Metha, and Y. Uchida, *J. Histochem. Cytochem.* **34,** 927 (1986).

of the deep glands, resulting in cryptitis. Additionally, the acute inflammatory cells infiltrate the pyloric glands, resulting in the formation of abscesses in the crypts, with the concomitant loss of epithelial cell membranes. This finding may explain the reduction in the confocal localization of the B2 kinin receptors (Fig. 7,A3, color insert). Later there is a loss of the glands themselves and replacement by fibrous tissue.

The inflammatory process is also characterized by the release of cytokine that causes the induction of the kinin B1 receptors in active gastritis. Whether the stem cells normally carry the gene for these receptors or whether they are synthesized *de novo* is open to further study by molecular probes. This first study on the identification of kinin receptors on the gastric mucosal cells indicates a role for kinin B1 receptors in gastritis and may provide a new pathway for treatment of gastritis with kinin receptor antagonists. Follow-up studies after treatment of the inflammation with a combination of B1 and B2 receptor antagonists are therefore indicated. Further, the occurrence of B2 receptors on acid-producing (parietal) cells points to a novel role for kinin B2 receptor antagonists in the management of patients with similar potency against both kinin receptors.[26]

Images of Tissue Kallikrein in Transplant Kidney

Literature survey has established a decrease in the excretion of urinary tissue kallikrein (TK) in transplant patients with a further reduction of the enzyme during episodes of acute rejection. The localization of tissue kallikrein in biopsies of the transplant kidney was compared at cellular and subcellular levels to autopsy-derived normal renal tissue.[28] Renal biopsies from 18 transplant patients with deteriorating renal function were obtained. Immonolabeling for tissue kallikrein, using a polyclonal goat antibody raised against recombinant human tissue kallikrein, was performed following routine enzymatic, confocal immunofluorescence, and electron microscopic techniques. In normal kidney tissue, tissue kallikrein was immunolocalized in the distal connecting tubules and collecting ducts (Fig. 8,A1, color insert). By comparison, the renal transplant tissue showed a reduction in the intensity of label, but maintained the sites of localization (Fig. 8,A2, color insert). Although tissue kallikrein was confined mainly at the luminal side of the cell on electron micrographs, some label was noted along the basolateral membranes. In the transplant kidneys, there was a reduction in the overall number of gold particles counted, which correlated with the decreased intensity observed on confocal immunocytochemistry. In addition, there was a shift to a basolateral orientation of the immunolabel.

[28] R. Ramsaroop, S. Naicker, T. Naicker, S. Naidoo, and K. D. Bhoola, *Immunopharmacology* **36,** 255 (1997).

Edema, tubulitis, and vasculitis characterize acute rejection. Destruction of the tubule cells and leakage of tissue kallikrein into the interstitial tissue space and the resultant effect of the formed kinins on renal capillary vasculature could explain the observed renal parenchymal edema and transplant rejection.

Images of Kinin B2 Receptors in Acute Renal Transplant Rejection

The mechanisms of renal rejection are complex and involve cell-mediated immunological reactions. The question of whether the immediate acute rejection process is initiated by inflammatory molecules such as kinins prompted us to examine the status of kinin receptors in patients undergoing acute transplant rejection.[29] In the normal kidney the B2 kinin receptors observed by confocal microscopy are confined mainly to the convoluted portion of the distal tubules and the collecting duct (Fig. 8,A3, color insert). These images were lost in the transplant kidney, indicating considerable downregulation of the receptors.

Acknowledgments

We thank the Foundation for Research Development (FRD), Medical Research Council of South Africa (MRC, SA), Cancer Research Association of South Africa (CANSA), and the University of Natal Research Fund (NURF) for generous financial support.

[29] S. Naidoo, R. Ramsaroop, Y. Naidoo, and K. D. Bhoola, *Immunopharmacology* **33,** 157 (1996).

[23] Observation of Microcirculatory Kinetics by Real-Time Confocal Laser Scanning Microscopy

By KAZUHIRO YAMAGUCHI

Importance of Real-Time Confocal Scanning Microscopy in Studying Cell Kinetics in Microcirculation

Studies of microcirculation have made rapid progress due to the development of various methods allowing precise estimation of hemodynamics and blood cell kinetics in microvessels of various organs and tissues. The relationship between blood flow and metabolism, measured with a variety of physiological tools, has been traditionally utilized for elucidating blood flow distribution in connection with oxidative metabolism in whole-organ preparation or in the intact organism. Magnetic resonance spectroscopy has come into greater use for specifying regional metabolic states in organs. Furthermore, the introduction of electron microscopic techniques has brought great improvements in clarifying the fine structure, enzyme distribution, and localization of substances related to metabolic activity within the cells constituting microvascular walls in fixed tissue sections. The available methods, however, provide limited information on the details of dynamically changing microcirculatory behavior under a wide variety of diseased conditions. One of the most important points in the qualitative and quantitative assessment of events occurring in the microcirculation of a given organ or tissue is to establish a methodology that allows the direct estimation of blood cell kinetics in the microvasculature under physiologically maintained flow conditions and also detects rapidly changing concentrations of key substances that regulate subcellular functions, such as intracellular calcium ion (Ca^{2+}), hydrogen ion (H^+), and membrane potential in the living cell. To meet the former requirement with certainty, conventional intravital microscopy was introduced in the 1960s in order to estimate the blood cell kinetics at the microcirculatory level under conditions equivalent to those in living organs. Subsequently, a number of photographing optical instruments have been developed, including the silicon-intensified target camera (SIT), the charged-coupled device camera (CCD), the CCD camera with image intensifier (I.I. camera), the photon counting tube, and the photomultiplier. These have distinctly accelerated, in association with the utilization of various fluorescent dyes, the study of microcirculatory blood cell kinetics by means of conventional intravital microscopy. Although intravital microscopy coupled with fluorescent dyes (i.e., epiluminescence microscopy) has been widely used for analyzing blood cell kinetics in the microcirculation of a given organ, its application is largely confined to that located in a single plane, such as mesenteric, pial, or mucosal microcirculation.[1-4] Epiluminescence microscopy is not always suitable for examining cell behavior in a microcirculatory network in which microvessels are positioned in different planes and superimposed on each other. This is especially true in pulmonary microcirculation in which the capillary network conforming to the shape of alveoli is densely intertwined and piled up on precapillary

[1] K. Ley and P. Gaehtgens, *Circ. Res.* **69,** 1034 (1991).
[2] K. Ley, and G. Linnemann, M. Meinen, L. M. Stoolman, and P. Gaehtgens, *Blood* **81,** 177 (1993).
[3] M. A. Perry and D. N. Granger, *J. Clin. Invest.* **87,** 1798 (1991).
[4] M. Suematsu, H. Suzuki, T. Tamatani, Y. Iigou, F. A. DeLano, M. Miyasaka, M. J. Forrest, R. Kannagi, B. W. Zweifach, Y. Ishimura, and G. W. Schmid-Schoenbein, *J. Clin. Invest.* **96,** 2009 (1995).

arterioles and postcapillary venules.[5] Therefore, images of capillaries are frequently superimposed on those of arterioles and venules. This creates the possibility that cell behavior in capillaries is frequently confused with that in adjacent arterioles or venules, when observation is performed by applying a conventional epifluorescent system. In addition, with an epifluorescent microscope, illumination of the entire field of view excites fluorescent emissions in the full depth of the specimen, rather than in the focal plane only. Much of the emitted light coming from regions above and below the focal plane is collected by the objective lens, thereby producing an out-of-focus blur in the final image of the specimen and markedly diminishing the contrast and sharpness of the image.[6] When the specimen is sufficiently thin and flat, as with the mesentery, the drawback mentioned previously presumably has little impact on the measurement of blood cell kinetics in small vessels.[5]

To conquer the obstacles presented by classical intravital microscopy, confocal scanning optical microscopic techniques (CSOM) were introduced in the late 1980s in the microcirculation field. These newly elaborated confocal systems allow both precise discrimination of individual microvessels from neighboring vessels in the $X-Y$ plane and direct measurement of axial velocities of various blood cells, including leukocytes, erythrocytes, platelets, and artificial small particles, in complicated microcirculation where microvascular architecture is closely interwoven. While CSOM in combination with laser-activated fluorescent probes emitting a variety of wavelengths [i.e., confocal laser scanning luminescence microscopy (CLSM)] is presently the most important optical scanning microscope imaging mode for biological specimens, it may also be used in other optical modes, such as confocal reflection contrast and transmission imaging. The former is particularly useful in detecting the light reflected by colloidal gold immunoconjugate labels. In application of CLSM to the investigation of dynamic process occurring in living cells or in microcirculation or to analysis of large areas of cytological or hematological specimens, the issue warranting careful consideration is the scan speed over the two-dimensional X (horizontal axis)–Y (vertical axis) plane, i.e., temporal resolution in the $X-Y$ plane. Although CSOM instruments (including CLSM) provide imaging data with high spatial resolution, most of them generally require nearly 1 sec or more to collect a high-quality image. The relatively slow scan rate means that the results of refocusing the image are not seen instantaneously. For instance, fast scanning at a rate of more than 30 frames/

[5] K. Yamaguchi, K. Nishio, N. Sato, H. Tsumura, A. Ichihara, H. Kudo, T. Aoki, K. Naoki, K. Suzuki, A. Miyata, Y. Suzuki, and S. Morooka, *Lab. Invest.* **76,** 809 (1997).
[6] D. M. Shotton, *J. Cell Sci.* **94,** 175 (1989).

sec is indispensable for precise discrimination between leukocytes moving slowly along the microvascular wall due to interaction with the endothelium (rolling leukocytes) and those flowing at the center line without any interruption because the center line linear velocity of leukocytes in microvessels, exclusive of the capillaries, is over 2–3 mm/sec.[3,5] This limitation is removed when using a CLSM system with a higher scanning frame rate. Instruments allowing the objective to be scanned at standard video rates (25 or 30 frames/sec) may be tentatively defined as fast CLSM, whereas those with scan speeds exceeding video rates may be called ultrafast CLSM. Both may be joined together and defined as real-time CLSM. Although CSOM, including CLSM, may possibly construct a three-dimensional image by storing all the optical tomographic information accumulated from a series of focal planes along the Z axis, such a procedure cannot be completed in a very short span, implying that real-time CLSM accelerates the scan speed in the two-dimensional X–Y plane at the expense of taking stereoscopic portraits. Subsequent discussion in this article will focus on the details of real-time CLSM as it is applicable to the study of events in the X–Y plane occurring in the microcirculation of various organs and tissues.

Real-Time Confocal Laser Scanning Microscopes:
  Principle and Characteristics

   In classical CLSM, scanning of a unitary beam emitted from a laser source is conveniently brought about either by lateral movement of the specimen in the focal plane relative to a stationary optical path (stage scan) or by the angular movement of the illuminating beam causing the focused light beam to move laterally in the focal plane relative to a motionless specimen (beam scan). Classical CLSM has the important advantage of constant axial illumination with little optical aberration, a feature that is desirable in constructing high-quality images and subsequent image processing. Because of physical limitations, however, these CLSM instruments generally have a relatively slow scan rate, which precludes their application in the real-time measurement of dynamic processes in the microcirculation or in living cells, although the time required for scanning may be reduced proportionally by decreasing the scan field. Such instruments physically resemble scanning electron microscopes, lacking the alternative capability for direct full-field viewing of the specimen through an eyepiece. However, the ingenious modification of classical types of CLSM since the early 1980s has resulted in the development of confocal scanning microscopes usable in real-time measurement.[6]
   Real-time CLSM is divided into four fundamentally distinct types, de-

TABLE I

REAL-TIME CONFOCAL SCANNING OPTICAL MICROSCOPES CURRENTLY AVAILABLE[a]

Type of CSOM	Laser beam	Aperture form	Scan system	Imaging rates[b] (frames/sec)	Image formation	Model (manufacturer)
Resonance type	Unitary	Variable pinhole	Resonance vibrator	30	Photomultiplier	RCM8000 (Nikon)
Slit type	Unitary	Variable slit	Galvanometer scanner	120	Eyepiece, CCD camera	Insight plus-1Q (Meridan) DVC-250 (Bio-Rad)
AOD type	Unitary	CCD line sensor equivalent to slit[c]	AOD scanner	30	CCD line sensor	2LM31 (Lasertec)
TSRLM type	Multiple	Pinholes	Nipkow disk	360–1000	Eyepiece, CCD camera, I.I. camera	CSU10 (Yokogawa) K2-Bio (Tec. Inst.) VX100 (Newport)

[a] AOD, acousto-optic deflector; TSRLM, tandem scanning reflected light microscope.

[b] Image formation rates in the two-dimensional X–Y plane.

[c] This is not an actual slit, but shows characteristics equivalent to those of a slit. See text for further details.

pending on the method of image formation (Table I). Except for the tandem scanning reflected light microscope (TSRLM) in which the specimen is scanned simultaneously by numerous focused light beams (multiple beam CLSM), other forms of instruments are basically categorized as unitary beam CLSM as scanning is achieved by a single light beam in these instruments. Depending on the aperture used to focus an illuminating laser beam on a single limited point or an in-focus volume element (voxel) within a three-dimensional specimen, CLSM with a unitary beam is further classified into two different types: pinhole and slit.

The pinhole type requires two-dimensional (horizontal and vertical directions) rapid scanning of an expanded light beam filling the pinhole of the back focal plane of the objective in order to obtain real-time confocal images. Scanning frequency along the horizontal axis is the most important factor in actually determining the image construction rate. A scanning frequency of more than 16 kHz in the horizontal direction realizes real-time confocal image formation of over 30 frames/sec. Although the beam scan is generally achieved by rotating or vibrating mirrors with the aid of a galvanometer, causing angular scanning of the laser beam along the horizontal axis, a high-frequency beam scan of over 16 kHz cannot be produced with the customary galvanometer, by which scanning frame rates are limited to those of less than 1 frame/sec (e.g., MRC-1024, Bio-Rad Microscience, Richmond, CA; LSM410, Carl Zeiss Co., Thornwood, NY). High-frequency scanning in the horizontal direction is accomplished successfully using a special vibrator such as a resonance galvanometer (e.g.,

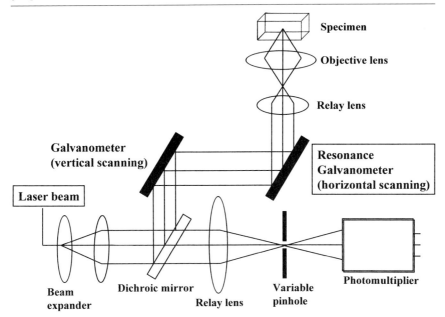

FIG. 1. Diagram of resonance-type CLSM. High-frequency beam scanning in the horizontal direction is achieved by a resonant galvanometer, whereas relatively slow scanning in the vertical direction is made by a classical galvanometer. The pinhole is used for focusing the illuminating laser beam.

RCM8000, Nikon Co., Tokyo, Japan) (Fig. 1). In CLSM with a resonant mirror scanner, the confocal image may be distorted significantly as the scanning frequency increases. This restraint is eliminated considerably by decreasing the resonant mirror size, although it then incidentally suffers from the disadvantage of a reduced optical field of view. By way of suggestion, the RCM8000 (Nikon Co.) scans a light beam at a rate of 8 kHz to and fro in the horizontal direction (but at only 60 Hz along the vertical axis), thus achieving a scan frequency equivalent to 16 kHz, which is required for taking real-time confocal images at a video rate.

However, slit-type CLSM, scanning laser beam and descanning fluorescence emission through a slit rather than the circular aperture of a pinhole, can realize real-time confocal observation without rapid scanning in the horizontal direction (e.g., Insight plus-1Q, Meridian Instruments; DVC-250, Bio-Rad Microscience) (Fig. 2). Instead of performing rapid scanning along the horizontal X axis, the laser beam in slit-type CLSM is split by a cylindrical lens into a slender slit-shaped light, functioning comparably to a resonant scanner with high frequency. The slit light is in turn scanned at a low frequency, with the mirror driven by the galvanometer in the vertical

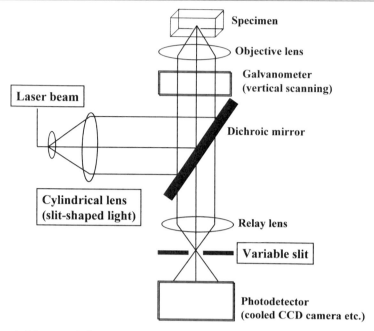

FIG. 2. Diagram of slit-type CLSM. This type of CLSM does not require rapid scanning along the horizontal axis but does slow scanning along the vertical axis with a classical galvanometer. Laser beam with a slit shape produced by a cylindrical lens functions equivalently to rapid scanning along the horizontal axis in other types of real-time CLSM. In-focus emission signals are descanned through a slit-shape aperture.

Y direction.[7] Only the emission signals from the in-focus voxel pass through the imaging aperture (which has a slit shape arranged in front of a photodetector), thus forming optical sectioning images. However, in slit-type CLSM, confocality and out-of-focus blur rejection of the image are expected to be low in comparison with pinhole-type CLSM because slit-type CLSM does not scan the specimen by the diffraction-limited spot of a laser beam, but rather scans it only in a one-dimensional, vertical direction, thus deviating from the most important principle of confocal sectioning.[8] Despite these limitations, slit-type CLSM instruments permit practically instantaneous confocal images at a rate of 120 frames/sec.

Another alternative method for achieving rapid beam scanning along the horizontal axis is the use of acousto-optic deflectors (AOD).[9] AOD

[7] C. J. Koester, *Appl. Opt.* **19,** 1749 (1980).
[8] T. Wilson, *J. Microsc.* **154,** 143 (1989).
[9] A. Draaijer and P. M. Houpt, *Scanning* **10,** 139 (1988).

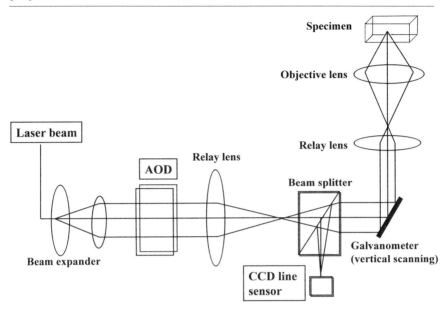

FIG. 3. Diagram of AOD-type CLSM. Rapid beam scanning along the horizontal axis is accomplished by adding AOD-generated ultrasonic waves to the laser beam perpendicularly. Generally, pinhole and slit are not used in this type of CLSM. Instead, descanned emission signals are projected directly onto an array of CCD line sensors.

generates ultrasonic waves that are then added perpendicularly to the light source. The result is that the laser beam is diffracted at a high frequency, thus allowing it to be scanned rapidly (e.g., 2LM31, Lasertec Co.). The scanning speed of AOD-associated CLSM reaches a video rate (Fig. 3). However, CLSM with AOD has a few problems, such as geometric distortions and decreased intensity of emitted light.[10] The former is introduced by the distortion of the laser beam passing through the AOD, causing the limited spot of the laser light to be significantly distorted on or in the specimen. At present, this is corrected for only partially. The latter, i.e., decreased intensity of emitted light, poses a serious problem because it explicitly indicates that the intensity of the reflected and/or transmitted light is reduced significantly when passing through the same route as the illuminating beam. This leads to trouble, especially with luminescent indicators, because fluorescent signals with longer wavelengths generally have very low intensity, indicating that they cannot be descanned back through the wavelength-specific acousto-optic modulator.[10] These drawbacks may

[10] Y. Horikawa, M. Yamamoto, and S. Dosaka, *J. Microsc.* **148**, 1 (1987).

be lessened considerably by scanning and descanning along the vertical axis with mirrors driven by a conventional galvanometer, allowing the partially descanned fluorescence emission signal to be projected onto a linear (one-dimensional) array of high-sensitivity photodetectors, such as a CCD line sensor that exhibits characteristics functionally similar to those of the slit. The descanned signal is then read out synchronously with the information of the horizontal scan. Another way to avoid the decrease in emission intensity is to image the descanned fluorescent signal through a slit, rather than a pinhole, that is arranged in parallel with the scan direction of the AOD-generated illuminating beam.

The final possibility for constructing real-time confocal images is that of scanning a numerous array of illuminating apertures relative to a stationary optical beam, a stationary objective, and a stationary specimen. This type of imaging system has been known as the tandem scanning reflected light microscope (TSRLM).[11,12] With this instrument, scanning is achieved by high-speed rotation of two Nipkow disks containing numerous pairs of diametrically opposed pinhole apertures in the primary image plane of the objective (e.g., CSU10, Yokogawa Electric Co.; K2-Bio, Technical Instruments; VX100, Newport Co.). Thus, when rotated together, these disks cover the entire field of view of the objective and largely eliminate any image distortion (Fig. 4). The arrangement of pinholes is extremely important in order to gain a uniform image. Although a fixed-angle helical pattern was traditionally used due to its simplicity, it had the disadvantage of uneven illumination because light intensity at the outside of the disk is significantly lower than at the inside. With a tetragonal pinhole pattern, there is little imbalance in light intensity between the two ends of the disk, but the scanning pitch is not kept constant during disk rotation. Meanwhile, an arrangement of pinholes in a constant-pitch helical pattern is designed to yield uniform light illumination and an invariable scanning pitch independent of the disk radius (CSU10, Yokogawa Electric Co.[13]).

In the original design of TSRLM, the illuminating light passes down to the specimen through one set of apertures, whereas the reflected or emitted fluorescent light, separated from the illuminating light by a beam splitter, returns along a different route through a diametrically opposed set of apertures on the opposite side of the Nipkow disk.[14] This caused significant image distortions. However, Xiao et al.[15] have developed a TSRLM instru-

[11] M. D. Egger and M. Petran, Science 157, 305 (1967).

[12] A. Boyde, Science 230, 1270 (1985).

[13] A. Ichihara, T. Tanaami, K. Isozaki, Y. Sugiyama, Y. Kosugi, K. Mikuriya, M. Abe, and I. Uemura, Bioimages 4, 57 (1996).

[14] M. Petran, M. Hadravsky, M. Egger, and R. Galambos, J. Opt. Soc. Am. 58, 661 (1968).

[15] G. Q. Xiao, T. R. Corle, and G. S. King, Appl. Phys. Lett. 53, 716 (1988).

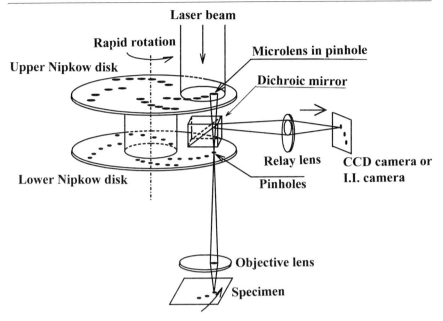

Fig. 4. Diagram of modified TSRLM used in our laboratory. The essentially important point in this type of CLSM is that rapid scanning along the horizontal axis is achieved by numerous focused light beams passing through rapidly rotating two Nipkow disks having numerous pairs of diametrically opposed apertures. The optical efficiency is increased by installing microlenses in each of the apertures located in the upper Nipkow disk. See text for further details.

ment in which both illuminating and reflected light beams pass through the same set of apertures, thus increasing the alignment between the two. A TSRLM with a Nipkow-disk scanner differs fundamentally from unitary beam CSOM in a variety of ways.[6] Because beam scanning is performed many times a second, the TSRLM produces a stable, continuous image that can be viewed via an eyepiece in real time as with any conventional light microscope. The scanning rate of the TSRLM can be approximately 1000 times as fast as that of a conventional confocal scanning microscope with a unitary beam, although this depends on the rotation speed of the Nipkow disk.[5,13] Because noncoherent white light may be used as the illuminating beam source instead of a monochromatic laser, the TSRLM may permit construction of images with natural color. As with a conventional optical microscope, the visual image is not immediately available as an electronic signal, although this is accomplished readily by installing a high-sensitivity camera. The major drawback of classical TSRLM is that the Nipkow disk situated in the optical path typically transmits only 1% of the incident light.[16] Thus, when used in combination with fluorescent indicators, the TSRLM requires that image intensifiers should be attached to achieve a satisfactory signal-to-noise ratio. This drawback to TSRLM has been partially overcome by installing microlenses in each of the apertures located in the upper Nipkow disk, greatly augmenting the efficiency of collecting the light incident on the upper disk.[5,13]

Real-Time Confocal Laser Scanning Luminescence
  Microscope Applied to Observation of Dynamic Events
  in Thick Specimens

In view of the distinctive features of the several types of CLSM mentioned earlier that allow real-time measurement, we have applied a modified type of TSRLM (CSU10, Yokogawa Electric Co.) in investigating a wide variety of microcirculatory events[5,17–21] (Fig. 5). In our confocal system, each Nipkow disk has 20,000 pinhole apertures 50 μm in diameter placed 250 μm apart, allowing 1000 beams of light to be transmitted onto an 8 × 8-mm visual field at any one time. Thus, our confocal unit yields a resolution velocity approximately 1000 times greater than a conventional confocal scanning optical microscope with a unitary beam. Because the disk is designed to rotate at 5000 rpm and to scan 12 frames in one rotation, our system actually generates a continuous image at a rate of 1000 frames/sec. In order to enhance the efficiency of light transmission through the Nipkow disk, 20,000 microlenses are installed in each of the apertures located on the upper disk, thereby enhancing optical efficiency up to 40%. The collimated beam from the light source impinging on the upper disk is focused by microlenses on the lower disk, which has an array of apertures corresponding exactly to that of the upper disk. Subsequently, the light beam passing through the disk apertures is focused by an objective lens onto a

[16] M. Petran, M. Hadravsky, J. Benes, R. Kucera, and A. Boyde, *Proc. R. Microsc. Soc.* **20,** 125 (1985).
[17] K. Yamaguchi, K. Suzuki, K. Naoki, K. Nishio, N. Sato, K. Takeshita, H. Kudo, T. Aoki, Y. Suzuki, A. Miyata, and H. Tsumura, *Circ. Res.* **82,** 722 (1998).
[18] K. Yamaguchi, K. Nishio, T. Aoki, Y. Suzuki, N. Sato, K. Naoki, K. Takeshita, and H. Kudo, *Histol. Histopathol.* **13,** 1089 (1998).
[19] T. Aoki, Y. Suzuki, K. Nishio, K. Suzuki, A. Miayata, Y. Iigou, H. Serizawa, H. Tsumura, Y. Ishimura, M. Suematsu, and K. Yamaguchi, *Am. J. Physiol.* **273,** H2361 (1997).
[20] K. Nishio, Y. Suzuki, T. Aoki, K. Suzuki, A. Miyata, N. Sato, K. Naoki, H. Kudo, H. Tsumura, H. Serizawa, S. Morooka, Y. Ishimura, M. Suematsu, and K. Yamaguchi, *Am. J. Respir. Crit. Care Med.* **157,** 599 (1998).
[21] K. Suzuki, K. Naoki, H. Kudo, K. Nishio, N. Sato, T. Aoki, Y. Suzuki, K. Takeshita, A. Miyata, H. Tsumura, Y. Yamakawa, and K. Yamaguchi, *Am. J. Respir. Crit. Care Med.* **158,** 602 (1998).

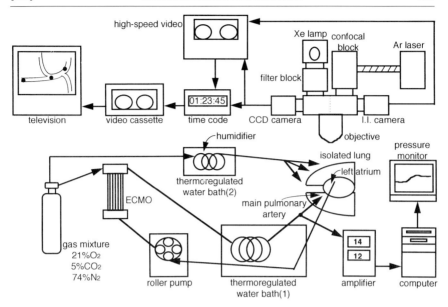

FIG. 5. Confocal and nonconfocal luminescence microscopes equipped with a high-speed video analysis system currently used in our laboratory. The isolated lung is fixed on a microscopic stage in the supine position and perfused at a constant flow rate (10–20 ml/min) in a recirculating manner with a roller pump. Krebs–Henseleit solution with 3% bovine serum albumin and a small quantity of whole blood is used as the perfusate. Gas exchange is maintained with an extracorporeal membrane oxygenator (ECMO). A warmed and humidified gas mixture containing the same composition of gases as those used for the ECMO is supplied continuously to the lung surface in order to maintain a temperature of 37° and to avoid the desiccation of the lung surface. Pulmonary artery pressures are monitored by force displacement of the pressure transducer. See text for further details concerning the practical way of obtaining confocal and nonconfocal images of pulmonary microvessels. From K. Yamaguchi, K. Nishio, N. Sato, H. Tsumura, A. Ichihara, H. Kudo, T. Aoki, K. Naoki, K. Suzuki, A. Miyata, Y. Suzuki, and S. Morooka, *Lab. Invest.* **76**, 809 (1997).

single diffraction-limited point or an in-focus volume element within a three-dimensional specimen. The same objective is then used, in conjunction with an appropriate dichroic beam splitter (cutoff wavelength: 508 nm) and a relay lens (01LAO-028, Melles Griot, CA), to image the reflected or emitted fluorescent light from the specimen onto an image intensifier with a high-sensitivity CCD camera (EktaPro Intensified Imager VSG, Kodak, CA) that can detect even very low fluorescence signals, indicating that our TRSLM allows both illuminating beam and reflected or emitted fluorescent light to go through the same path. Significant elimination of out-of-focus information, flare, and scattered light was achieved by impinging the excitation beam through the lateral side of the confocal unit with a selective barrier filter (Asahi Spectra Co., Tokyo) such that the light passing through the dichroic mirror is not scattered within the unit. In addition, all glass parts, the array of apertures with microlenses, and the dichroic mirrors are made of synthetic quartz, a material free of autofluorescence, thus improving the signal-to-noise ratio by minimizing background flare. By incorporating an excitation wavelength of 488 nm emitted from a low-power air-cooled argon (Ar) ion laser (532-BSA04, output power: 10 mW/V, Omnichrome, CA) with appropriate fluoresceins, the present system allows us to obtain apparently instantaneous fluorescent images at 1000 frames/sec, with an optical sectioning depth of 0.75 μm (using an objective with a magnification of 100$\times$ and 0.9 NA) and to achieve a two-point spatial resolution of 0.2 μm. The optical sectioning depth, i.e., depth of focus, was defined as the distance in the Z direction between the focal planes on either side of the focus at which the intensity of the image falls to half-maximum. In our series of experiments, we generally used a long-distance objective at magnification of either 20$\times$ (M Plan ELWD 20$\times$/0.4 NA, Nikon, Tokyo) or 40$\times$ (M Plan ELWD 40$\times$/0.5 NA, Nikon, Tokyo) such that the depth of focus was 3.5 or 1.8 μm, respectively, while observing blood cell kinetics in the microcirculation of living organs. Employment of high-magnification, high-numerical aperture objectives is important when attempting confocal observation, but because these have a limited working distance, they are not practically suitable for looking deep into thick specimens. Because the objectives used in our study have a 10-mm working distance, they are a good compromise for measuring events occurring in the microcirculation of living organs or in tissue slices, both of which are very thick, by means of a confocal laser scanning luminescence microscope. By combining the magnification achieved by the objective lens (20 or 40$\times$) and the confocal unit (2.5$\times$) with electric magnification, the final magnifying power of our system reached either 484$\times$ (with the 20$\times$ objective) or 968$\times$ (with the 40$\times$ objective) on the video screen. If necessary, magnification could be enhanced easily by changing the objective. The size of the resulting field of view was 210 $\times$ 210 μm with the 20$\times$ objective and 105 $\times$ 105 μm with the 40$\times$ objective. Our confocal system yielded stable, actually continuous images in the X–Y plane in real time. However, the optical design does not permit simultaneous scanning in the X and Z directions, necessitating the independent construction of three-dimensional images. If continuous portraits in the X–Y plane are not required, however, such images are obtained readily from a series of focal planes along the Z axis.

Although our real-time TSRLM was found to be useful in studying the dynamic processes in living cells and analyzing the rapid movement of cytological or hematological specimens in a given organ, the issue of how to record real-time images obtained from TSRLM still needs to be resolved. To overcome this difficulty, we attempted to record all TSRLM information by means of a high-speed video analysis system (EktaPro 1000 Processor, Kodak, CA) connected to a image-intensified CCD camera. We registered confocal images at a rate of either 250 or 500 frames/sec. All obtained images were monitored on color video television (PVM-1444Q, Sony, Tokyo) and stored in a video cassette recorder (SVQ-260, Sony, Tokyo). To facilitate evaluation of the dynamic video images, a time code generator (VTG-55B, FOR. A, Tokyo) was superimposed onto the video screen.

To compare confocal scanning microscopy data with data obtained from nonconfocal observation, we also attached classical oblique illumination and conventional epifluorescence units (i.e., nonconfocal units) to our confocal microscopic system. Oblique illumination with a 150-W halogen lamp [Techno Light KTS-150 (15V), Kenko, Tokyo] was applied simply to examine a bright field of view. For conventional fluorescence microscopy, epiillumination was achieved with a 100-W xenon (Xe) lamp (XPS-100, Nikon, Tokyo) through a filter cassette consisting of an excitation filter (420–490 nm), a dichroic mirror (DM-510, Nikon, Tokyo), and a barrier filter (cutoff wavelength: 520 nm). An additional neutral-density (ND) filter was inserted between the filter cassette and the objective, which resulted in reducing the intensity of the Xe light by 94% (1/16 ND filter, Nikon, Tokyo). Images obtained by the nonconfocal microscopic units were recorded using a cooled CCD camera with high sensitivity (TEC-470, Optronics, CA; shutter speed: 1/60) and transferred to the video analysis system. The same objective lens as used in TSRLM yielded comparable magnification in nonconfocal microscopic observation: magnifications of 375 and 750× were attained with the 20 and 40× objective lenses, respectively.

Practical Application of Real-Time Confocal Laser
  Scanning Luminescence Microscope to Observation of
  Microcirculatory Kinetics in Various Organs and Tissues

As mentioned previously, the measurement of blood cell kinetics in the microcirculation of sufficiently thin, flat tissues (or organs) may be performed successfully, with marginal error, using conventional epiluminescence, nonconfocal microscopes. However, most organs do not meet these requirements, as they generally have complicated shapes with various thicknesses. For instance, microcirculatory architecture in the pulmonary acinus is fundamentally intertwined such that the alveolar capillaries are piled densely onto the precapillary arterioles and postcapillary venules, suggesting that the nonconfocal observation of pulmonary microvasculature with conventional epiluminescence microscopes would greatly diminish position-dependent contrast and sharpness so that events occurring in arterioles or venules could possibly be mistaken for those in capillaries, and vice versa.[5,6] In conventional nonconfocal microscopy, the integrated intensity of the light emitted from the various points in the specimen is not altered when changing the focus, indicating that the light is merely redistributed in the final image, eventually contributing to a uniform background light intensity and thus reducing specimen contrast.

In order to precisely assess blood cell kinetics in the acinar microcirculation of the lung under a wide range of experimental conditions, the behavior of fluorescence-labeled leukocytes and erythrocytes was investigated using the modified TSRLM presented earlier in an isolated rat lung preparation perfused with Krebs–Henseleit solution containing 3% bovine serum albumin as well as a small quantity of whole blood, thus adjusting the perfusate hematocrit (Ht) to about 5%[5,17–21] (Fig. 5). Leukocytes were stained with carboxyfluorescein diacetate succinimidyl ester (CFDASE, Molecular Probes, Eugene, OR). CFDASE solution diluted with 1.5 ml of normal saline was injected into the femoral veins of donor rats [*in vivo* concentration: 1 μg/g (body weight)]. CFDASE is a nonfluorescent precursor that diffuses into cells and forms a stable fluorochrome called carboxyfluorescein succinimidyl ester (CFSE), which is present predominantly in leukocytes and platelets but is not found in erythrocytes.[22] After 30 min of *in vivo* incubation, blood was aspirated gently from the left ventricle. A blood sample containing stable CFSE-stained leukocytes was administered into the reservoir of the perfused circuit connected to the isolated lung so that the final perfusate concentration of leukocytes with and without CFSE labeling reached about 500 cells/μl. Leukocytes and platelets stained with CFSE were readily distinguishable during luminescence microscopic measurement because of their distinctive size difference. Erythrocytes were stained with fluorescein isothiocyanate (FITC, Sigma, St. Louis, MO). Fresh rat blood was centrifuged at 1000 rpm for 5 min and the buffy coat was discarded. The same procedure was repeated three times in the presence of phosphate-buffered saline (PBS), reducing leukocyte contamination to 0.1% or less. The packed erythrocyte solution thus prepared was diluted with PBS to adjust the Ht to 10%, and FITC was added to give a final concentration of 0.1 mg/ml. This solution was then incubated at 37° for 30 min. The FITC-labeled erythrocyte samples were washed three times in PBS and 1 ml was added to the reservoir after the measurement of leukocyte kinetics had been completed. In replaying the video tapes photographed by the high-speed video analysis system at the normal video rate, we estimated the FITC–erythrocyte velocity by measuring the distance traveled

[22] M. Bronner-Fraser, *J. Cell. Biol.* **101,** 610 (1985).

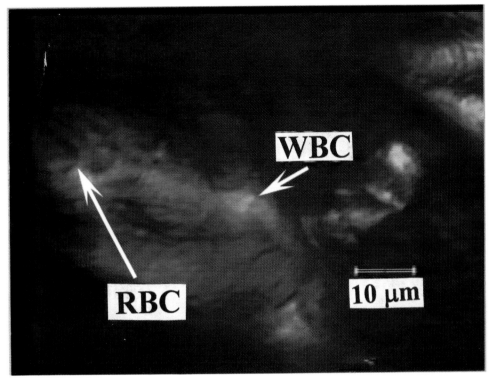

Fig. 15. Confocal image of blood cells in rat alveolar capillaries (×40 objective). Photograph is taken by means of an AOD-type CLSM with no fluorescent dyes. In this experiment, three laser beams are used simultaneously as light sources, allowing real-time confocal "color" images to be obtained. From T. Nakamoto, personal communication (1998).

between two or more successive video frames. In reply, one frame indicates events during a 2- or 4-msec interval. The mean flow velocity and wall shear rate in each microvessel were calculated from the velocity of the FITC–erythrocytes flowing at the center line portion without any interruption. In addition, the measurement of erythrocyte flow direction permitted reliable discrimination between precapillary arterioles and postcapillary venules. The microvessel through which the FITC–erythrocytes entered the capillary networks may be defined as the arteriole, whereas the microvessel into which FITC–erythrocytes flowed from the capillary networks is taken as the venule. To precisely determine the diameters and architecture of microvessels in which leukocyte and erythrocyte behavior was examined, we added 200 μl of 5% FITC-dextran with a molecular weight of 145,000 (Sigma) to the reservoir. Vessel diameters were estimated precisely by processing a confocal video image by a computer-assisted digital image analyzing system (NIH image software operated on Quadra 840AV/Image 1.58, Apple, CA), composing an image of 552 × 436 pixels. Each pixel had a gray-scale value ranging from zero (white) to 255 (black), which was proportional to the fluorescence intensity. The microvascular edge was defined as the portion having an abruptly changing gray-scale value (Fig. 6).

In images obtained from the same region by changing the focal depth using TSRLM, precapillary arterioles and postcapillary venules in the pulmonary acinus were clearly distinguished from capillary images. However,

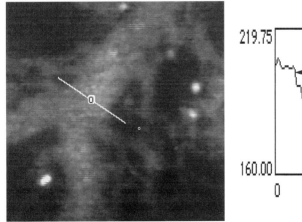
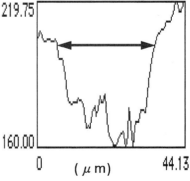

FIG. 6. Precise estimation of microvascular diameter from confocal images (×20 objective). (Left) Confocal view of a microvessel. (Right) Changes in gray-scale values over the line connecting two points located outside the microvascular edges. The microvascular edge can be defined as the portion at which the gray scale changes steeply.

with the nonconfocal epiluminescence microscope, they were largely super-imposed on each other, and the edges of individual microvessels were extremely difficult to judge in images taken from exactly the same region. This indicates the absolute necessity of confocal measurement when attempting to estimate events in a microcirculatory network with a complicated nature, such as in pulmonary microcirculation (Fig. 7). Briefly, wall shear rates in pulmonary arterioles were twice as low as those in venules

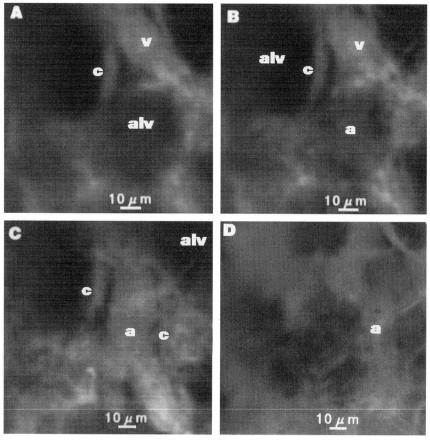

FIG. 7. Confocal and nonconfocal images of rat pulmonary microcirculation (×20 objective). Microvessels are stained with FITC. (A) Image of in-focus plane at a 5-μm depth from the lung surface. (B) Image of in-focus plane at a 15-μm depth of the surface. (C) Image of in-focus plane at a 25-μm depth from the surface. (D) Nonconfocal image obtained from exactly the same region as that shown in (A) to (C). alv, alveolus; a, arteriole; v, venule; c, capillary. From K. Yamaguchi, K. Nishio, N. Sato, H. Tsumura, A. Ichihara, H. Kudo, T. Aoki, K. Naoki, K. Suzuki, Y. Suzuki, and S. Morooka, *Lab. Invest.* **76,** 809 (1997).

with a comparable size.[5,20] This result is highly inconsistent with findings in systemic microcirculation in which shear force applied to arterioles is approximately three times that in venules with nearly the same size.[3] Unstimulated leukocyte kinetics in intact acinar microvessels examined by TSRLM exhibited the following tendencies[5,18] (Fig. 8): (1) leukocytes do not adhere firmly to any of the acinar microvessels, including arterioles and venules; (2) leukocyte rolling is found to occur preferentially in arterioles, mediated in part by an intercellular adhesion molecule-1 (ICAM-1)- and P-selectin-independent by L-selectin-dependent mechanism; (3) spontaneous rolling in pulmonary venules is not related to adhesion molecules; (4) pulmonary leukocyte sequestration occurs mainly in alveolar capillaries rather than in postcapillary venules; and (5) adhesion molecules are not involved in leukocyte tethering to the intact capillary endothelium. These findings differ qualitatively from those for systemic microcirculation in several respects: (1) in systemic microvessels, rolling cells show movement

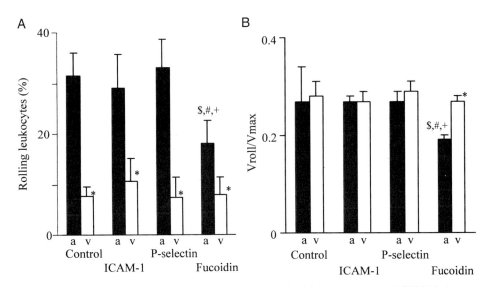

FIG. 8. Leukocyte rolling in arterioles (a) and venules (v) in intact rat acini. (A) Relative frequency of rolling leukocytes. (B) Rolling velocity of leukocytes (V_{roll}) relative to center line velocity of erythrocytes (V_{max}). Control: group with no medications. ICAM-1: group in which ICAM-1 is inhibited by 1A29. P-selectin: group in which P-selectin is inhibited by ARP2-4. Fucoidin: group in which P- and L-selectins are inhibited by fucoidin. *, significantly different from the values obtained for arterioles; \$, smaller than the values obtained for the control group; #, smaller than the values observed for the ICAM-1 group; +, smaller than the values obtained for the P-selectin group. From K. Yamaguchi, K. Nishio, N. Sato, H. Tsumura, A. Ichihara, H. Kudo, T. Aoki, K. Naoki, K. Suzuki, A. Miyata, Y. Suzuki, and S. Morooka, *Lab. Invest.* **76**, 809 (1997).

that can be described as "clumsy," but such caterpillar-like movement was not observed in pulmonary microvessels[19]; (2) although leukocyte–endothelium interactions leading to rolling are confined mainly to the venular segment in systemic microcirculation,[1,2,4,23] they were predominant in arterioles in pulmonary microcirculation; (3) L-selectin has been demonstrated convincingly to be the decisive factor in causing spontaneous leukocyte rolling in systemic venules,[2,24] whereas it played no role in pulmonary acinar venules; and (4) in capillaries of systemic organs, leukocyte entrapment occurs when microcirculation is exposed to morbidly low shear conditions,[25] whereas pulmonary capillaries could entrap leukocytes at physiological shear rates. Leukocyte behavior unique to the pulmonary microvasculature was exaggerated in lungs with endothelial injury due to hyperoxia exposure. Exposure to a hyperoxic environment caused upregulation of ICAM-1 in both venules and capillaries, but with scattering augmentation of P-selectin expression in arterioles.[20] Hyperoxia exposure significantly increased the relative frequency of rolling leukocytes in pulmonary venules, which largely exceeded that in arterioles (Fig. 9). Increased numbers of rolling venular leukocytes can be certainly explained based on the pathway related to endothelial ICAM-1 and leukocyte L-selectin. Leukocyte entrapment in capillaries was additionally augmented in lungs activated by hyperoxia exposure.[20] In opposition to the finding for intact lungs, however, this enhancement was almost fully diminished by inhibiting ICAM-1. Absorbing results were obtained in experiments in which leukocytes were activated with rat cytokine-induced neutrophil chemoattractant (CINC/gro), a peptide possessing biological activities analogous to those of human interleukin-8 (IL-8).[26] FACScan flow cytometry revealed that L-selectin was markedly downregulated but CD18 was inversely upregulated on the leukocyte surface in response to CINC/gro stimulation.[19] The behavior of CINC/gro-activated leukocytes was investigated in the pulmonary microcirculation as well as in mesenteric venules,[19] with the result that CINC/gro-stimulation distinctly diminished rolling leukocyte flux along the mesenteric venular wall (Fig. 10), while it contradictorily augmented entrapped leukocytes in pulmonary capillaries (Fig. 11). These findings strongly suggest that L-selectin is fundamentally important in initiating leukocyte rolling in systemic venules, whereas ICAM-1 exerts

[23] U. H. von Andrian, J. D. Chambers, L. M. McEvoy, R. F. Bargatze, K. E. Arfors, and E. C. Butcher, *Proc. Natl. Acad. Sci. U.S.A.* **88,** 7538 (1991).

[24] E. Tuomanen, *J. Clin. Invest.* **93,** 917 (1994).

[25] E. Mori, G. J. del Zoppo, J. D. Chambers, B. R. Copeland, and E. Arfors, *Stroke* **23,** 712 (1992).

[26] K. Watanabe, M. Suematsu, M. Iida, K. Takaishi, Y. Iizuka, H. Suzuki, M. Suzuki, M. Tsuchiya, and S. Tsurufuji, *Exp. Mol. Pathol.* **56,** 60 (1992).

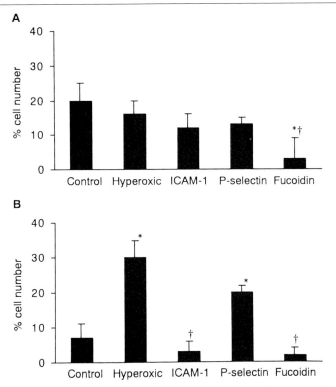

FIG. 9. Relative frequency of leukocytes rolling along arteriolar (A) or venular (B) walls in the rat pulmonary microcirculation exposed to normoxic or hyperoxic conditions. Control: normoxia-exposed lungs. Hyperoxic: hyperoxia-exposed lungs with no medications. ICAM-1: hyperoxia-exposed lungs treated with monoclonal antibody to ICAM-1 (1A29). P-selectin: hyperoxia-exposed lungs treated with monoclonal antibody to P-selectin (ARP2-4). Fucoidin: administration of fucoidin-treated leukocytes into hyperoxia-exposed lungs. *, differing from the control group; †, smaller than the value obtained for the hyperoxic group. From K. Nishio, Y. Suzuki, T. Aoki, K. Suzuki, A. Miyata, N. Sato, K. Naoki, H. Kudo, H. Tsumura, H. Serizawa, S. Morooka, Y. Ishimura, M. Suematsu, and K. Yamaguchi, *Am. J. Respir. Crit. Care Med.* **157**, 599 (1998).

a significant impact on the enhanced margination of activated leukocytes in pulmonary capillaries.

To confirm the importance of various adhesion molecules in abnormal leukocyte behavior in the pulmonary microvasculature, constitutively expressed ICAM-1 and P-selectin distribution along pulmonary microvessel walls was examined by applying TSRLM. The monoclonal antibody against ICAM-1 [1A29: 4 μg/g (body weight)] was injected into an anesthetized ventilated rat through the femoral vein.[5,18,19] Five minutes later, an isolated

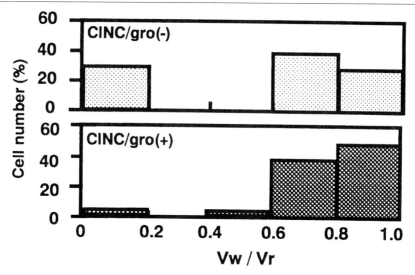

FIG. 10. Linear velocity of leukocytes relative to that of erythrocyte (V_w/V_r) in rat mesenteric venules. CINC/gro ($-$) and CINC/gro ($+$): data collected from animals injected with CINC/gro-unstimulated and -stimulated cells, respectively. On the basis of Wilcoxon's signed ranking test, there is a significant difference in mean values between the two groups. From T. Aoki, Y. Suzuki, K. Nishio, K. Suzuki, A. Miyata, Y. Iigou, H. Serizawa, H. Tsumura, Y. Ishimura, M. Suematsu, and K. Yamaguchi, *Am. J. Physiol.* **273**, H2361 (1997).

perfused lung was prepared from the animal and was injected with 0.5 ml of FITC-labeled antimouse IgG antibody (Sigma), the secondary antibody to 1A29, over 45 sec. Subsequently, flow was halted for 15 min so as to allow conjugations among endothelial ICAM-1, 1A29, and FITC–IgG antibody. Thereafter, perfusion was resumed for another 15 min. After compounds of 1A29 and FITC–IgG antibody flowing freely in pulmonary circulation had been thoroughly washed out, endothelial ICAM-1 bound to 1A29 and FITC–IgG antibody was determined under confocal scanning luminescence microscopy and displayed by a high-sensitivity SIT camera (C2400-08, Hamamatsu Photonics, Tokyo). Because 1A29 consists of mouse IgG, mouse IgG [4 μg/g (body weight) (Sigma)] and FITC-labeled anti-mouse IgG antibody were used as controls. Arterioles, venules, and capillaries were distinguished by administering both FITC-dextran and FITC-labeled erythrocytes at the end of each experiment. Procedures qualitatively the same as those used for endothelial ICAM-1 detection were applied to examine endothelial P-selectin expression in pulmonary microvessels. A monoclonal antibody against P-selectin [ARP2-4: 4 μg/g (body weight)] and a FITC-labeled antimouse IgG antibody were used. Because ARP2-4 is composed of mouse IgG as well, mouse IgG [4 μg/g (body weight)] and

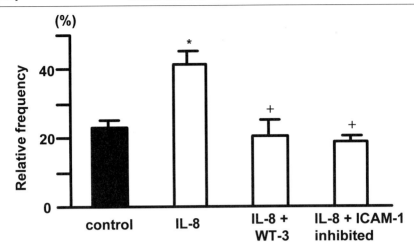

(%)

FIG. 11. Relative frequency of CINC/gro-stimulated leukocytes entrapped in rat alveolar capillaries. Control, unstimulated leukocytes; IL-8, leukocytes stimulated with CINC/gro; IL-8 + WT-3, CINC/gro-stimulated leukocytes treated with monoclonal antibody to CD18 on leukocyte surface (WT-3); IL-8 + ICAM-1 inhibited, kinetics of CINC/gro-stimulated leukocytes is observed under conditions in which pulmonary endothelial ICAM-1 is inhibited by 1A29. *, Higher than the controls group; +, lower than the IL-8 group but not differing from the control group. From T. Aoki, Y. Suzuki, K. Nishio, K. Suzuki, A. Miyata, Y. Iigou, H. Serizawa, H. Tsumura, Y. Ishimura, M. Suematsu, and K. Yamaguchi, *Am. J. Physiol.* **273,** H2361 (1997).

its FITC-labeled conjugate served as controls. Although P-selectin was not detected in any of the examined microvessels, ICAM-1 was demonstrated clearly along postcapillary venules and capillary networks but not along precapillary arteriolar walls (Fig. 12). By applying qualitatively the same method as ours, Iigo et al.[27] showed that postcapillary venules were the major portion expressing ICAM-1 constitutively in the mesenteric microcirculation, whereas the liver expressed ICAM-1 abundantly in sinusoids to an extent similar to that in central venules. These authors further examined heterogeneity in ICAM-1 distribution among various mesenteric venules and ascertained that postcapillary venules with a 25-μm diameter exhibited the greatest density of ICAM-1 on their endothelial surfaces, suggesting that venular segments of a particular size preferentially express ICAM-1 molecules.

Pulmonary acinar microvessel responses to various physiological stimuli, including hypoxia, hypercapnic acidosis, and isocapnic acidosis, are essen-

[27] Y. Iigo, M. Suematsu, T. Higashida, J. Oheda, K. Matsumoto, Y. Wakabayashi, Y. Ishimura, M. Miyasaka, and T. Takashi, *Am. J. Physiol.* **273,** H138 (1997).

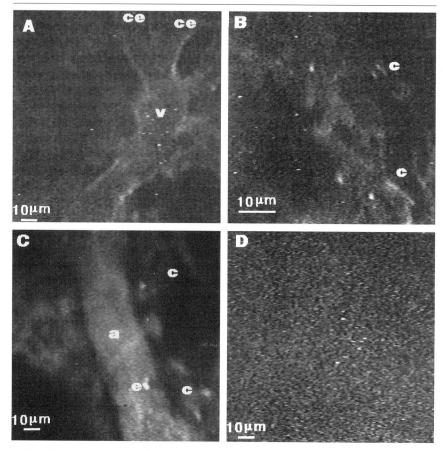

FIG. 12. Confocal views of ICAM-1 distribution along rat pulmonary microvessel walls (×20 objective). (A) ICAM-1 expression along venular walls and entrances of capillaries connected to the venule. (B) ICAM-1 expression in capillaries. (C) Microvascular architecture of the region photographed in (B). The vascular architecture is delineated by administering FITC-dextran and FITC-labeled erythrocytes. An obliquely running arteriole, not visible in (B), is observed. (D) Negative control (mouse IgG and FITC-labeled antimouse IgG antibody) showing no fluorescence emission. a, arteriole; v, venule; c, capillary; ce, capillary entrance connected to a venule; e, erythrocyte. From K. Yamaguchi, K. Nishio, N. Sato, H. Tsumura, A. Ichihara, H. Kudo, T. Aoki, K. Naoki, K. Suzuki, A. Miyata, Y. Suzuki, and S. Morooka, *Lab. Invest.* **76,** 809 (1997).

tially important in maintaining gas exchange efficiency in the lung. However, due to technical difficulty, they have not been evaluated critically thus far.[17,21] Applying TSRLM to isolated rat lungs, we estimated acinar microvessel responsiveness to O_2, CO_2, and H^+ in connection with enzymes producing vasoactive mediators, including nitric oxide (NO) and prostaglandins (PG) in lungs, especially those injured by hyperoxia exposure.[17,21] Our experimental results revealed that alveolar hypoxia constricted precapillary arterioles (Fig. 13), whereas alveolar hypercapnia evoked postcapillary venular dilatation in intact acini under medication-free conditions (Fig. 14). These phenomena, however, disappeared in hyperoxia-injured acini, strongly suggesting that microvascular response to physiological stimuli is

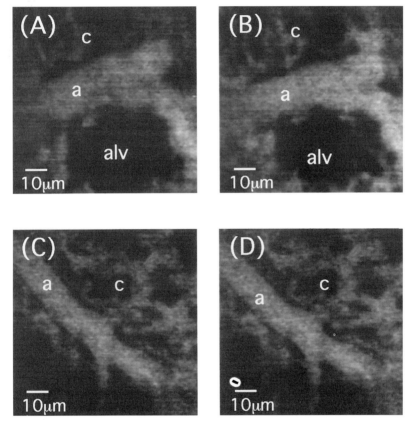

FIG. 13. Confocal views of rat intraacinar arterioles before and after hypoxic stimulation (×20 objective). (A) and (B) Arteriole before and after hypoxic stimulation in a normoxia-exposed lung, respectively. (C) and (D) Arteriole before and after hypoxic stimulation in the hyperoxia-exposed lung. Arteriole in the normoxia-exposed lung is constricted, whereas that in the hyperoxia-exposed lung is not on hypoxic stimulation. alv, alveolus; a, arteriole; c, capillary. From K. Suzuki, K. Naoki, H. Kudo, K. Nishio, N. Sato, T. Aoki, Y. Suzuki, K. Takeshita, A. Miyata, H. Tsumura, Y. Yamakawa, and K. Yamaguchi, *Am. J. Respir. Crit. Care Med.* **158,** 602 (1998).

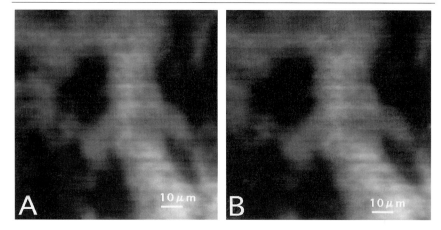

FIG. 14. Confocal images of the intraacinar venule before (A) and after (B) hypercapnic gas breathing in the intact rat lung (×20 objective). Intact venule is dilated at hypercapnic conditions. However, the venule injured by hyperoxia exposure fails to be dilated on hypercapnic stimulation.

significantly distorted in injured lungs. Inhibition of either NO synthase (NOS) or cyclooxygenase (COX) allowed arterioles to regain contractility during hypoxic stimulation in hyperoxia-exposed lungs. Blunted hypercapnia-associated venular dilation in hyperoxia-injured lungs was restored unexpectedly by NOS inhibition to the level of intact venules, but this peculiar phenomenon disappeared when NOS and COX were inhibited concurrently, with the result that the NO-related pathway negatively modulates the COX-dependent pathway under hypercapnic conditions, i.e., there is cross-talk between cGMP and cAMP pathways.

Although several groups of investigators have attempted to measure microcirculatory behavior in various organs and tissues by applying confocal laser scanning microscope with high spatial resolution, most of their experiments have not been aimed at real-time observation but instead drew conclusions from stationary confocal images, including three-dimensional portraits.[28–33] Of course, stereoscopic images with high spatial resolution

[28] U. Dirnagl, A. Villringer, R. Gebhardt, R. L. Haberl, P. Schmiedek, and K. M. Einhaupl, J. Cereb. Blood Flow Metab. 11, 353 (1991).
[29] F. A. Merchant, S. J. Aggarwal, K. R. Diller, and A. C. Bovik, J. Microsc. 176, 262 (1994).
[30] V. Rummelt, L. M. G. Gardner, R. Folberg, S. Beck, B. Knosp, T. O. Moninger, and K. C. Moore, J. Histochem. Cytochem. 42, 681 (1994).
[31] P. He and R. H. Adamson, Microcirc. 2, 267 (1995).
[32] M. Nakano, Y. Nakajima, Y. Tsuchiya, S. Kudo, H. Nakamura, and O. Fukuoka, Int. J. Microcirc. 17, 159 (1997).
[33] M. Jirkovska, L. Kubinova, and I. Krekule, Anat. Embryol. 197, 263 (1998).

are extremely valuable in analyzing in detail the stationary structures constituting a given cell. However, this is a digression from the main subject of
this article, which is obtaining two-dimensional images in real time. Villinger
et al.[34] and Barfod et al.[35] estimated blood cell velocity in rat brain cortical
capillaries by applying a single-line imaging approach, allowing them to
take one-dimensional images at an acquisition time of 2 msec. In their
experiments, sodium fluorescein injected intravenously served as the vascular marker because it was present only in the plasma. Because erythrocytes
were not stained with fluorescent dyes, they were visualized in negative
contrast (gray) between labeled plasma gaps (white). However, the serious
drawback in their method is that the single-line imaging approach permits
construction of one-dimensional images only during a very short period of
time so that it sacrifices the advantage of confocal scanning microscopy,
i.e., temporal resolution is gained at the expense of spatial resolution.
Chauhan and Smith[36] analyzed the validity of confocal scanning laser Doppler flowmetry (SLDF), a new noninvasive technique for measuring retinal
and optic nerve head hemodynamics originally developed by Webb et al.[37]
SLDF contains 780-μm diode laser confocal optics and a photodiode to
measure the intensity of backscattered light from the funds. Lights scattered
by stationary structures and by moving blood cells interfere with each
other, leading to beating at a specific frequency, which is detected by
the photodiode as intensity variations in the backscattered light. In this
apparatus, each of 64 horizontal lines is scanned 128 times with a line
repetition rate of 4 kHz, requiring about 2 sec as total image acquisition
time. Although SLDF does not allow real-time measurement of microcirculatory hemodynamics, it detects flow velocities of less than 1 mm/sec within
a marginal error.[36] Applying AOD-associated CLSM (2LM31, Lasertec,
Tokyo), Kato and colleagues[38] have succeeded in investigating the dynamic
movement of blood cells, including erythrocytes, leukocytes, and platelets,
in rat alveolar septal microvasculature in real time. In their experiments,
the blood cells were not stained with fluorescent probes. Instead, reflected
light from the blood cells was detected through an AOD-type CLSM, in
which three lasers (red: He/Ne 633 nm, green: Ar 515 nm, and blue: Ar 488
nm) were used concurrently as the illuminating light source, thus permitting
"pseudo-color image" formation in real time (Fig. 15, color insert).

[34] A. Villringer, A. Them, U. Lindauer, K. Einhaupl, and U. Drinagl, Circ. Res. 75, 55 (1994).
[35] C. Barfod, N. Akgoren, M. Fabricius, U. Dirnagl, and M. Lauritzen, Acta Physiol. Scand. 160, 123 (1997).
[36] B. C. Chauhan and F. M. Smith, J. Glaucoma 6, 237 (1997).
[37] R. H. Webb, G. W. Hughes, and F. C. Delori, Appl. Opt. 26, 1492 (1987).
[38] S. Kato, N. Ohnuma, K. Ohno, K. Takasaki, S. Okamoto, T. Asai, M. Okada, T. Nakamoto, and M. Iizuka, Int. J. Microcirc. 17, 290 (1997).

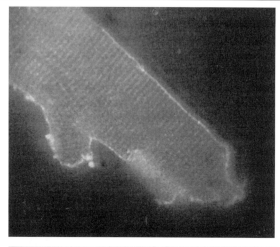

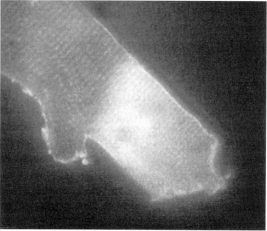

FIG. 16. Confocal images of Ca^{2+} wave in the cardiac myocyte harvested from rat ventricles in terms of modified TSRLM (x100 objective). The intracellular Ca^{2+} wave is detected using fluo-3 AM (10 μM) loaded into the cell. (Top) Myocyte before exposure to fluo-3, in which only t-tubules stained with di-8-ANEPP (10 μM) are observed. (Bottom) The same myocyte 30 min after exposure to fluo-3 at 37° in which t-tubules and Ca^{2+} propagation within the cell are evidently observed. From H. Ishida, personal communication (1998).

In addition to measurements of various cell kinetics in microcirculation, it is also important to investigate subcellular functions in cells constituting the microvascular wall. Incorporating a confocal imaging system with calcium indicator dyes of fluo-3 and Fura Red, Yip and Marsh[39] succeeded

[39] K. Yip and D. J. Marsh, *Am. J. Physiol.* **271**, F1004 (1996).

TABLE II
LASER LINEUP DEPENDING ON POWER SUPPLY

Type of laser	Monochromatic beam emitted (nm)
Ar (25 mW)	488 (blue), 514 (green)
Ar (100 mW)	457 (blue), 488, 514
Ar (250 mW)	351 (violet), 363 (violet), 488, 514
Ar/Kr (15–60 mW)	488, 568 (yellow), 647 (red)
He/Ne (1 mW)	543 (green)
He/Ne (10 mW)	633 (red)
He/Cd (10 mW)	442 (blue)

TABLE III
ARGON LASER (488 nm)-ACTIVATED FLUORESCENCE PROBES AVAILABLE FOR REAL-TIME
OBSERVATION OF MICROCIRCULATORY KINETICS

Fluorescence probe	Excitation wavelength (nm)	Fluorescence wavelength (nm)	Remarks[a]
FITC	495	520	Cell frame (L, E, P)
CFSE	490	520	Cell frame (L, P)
Bodipy-FL	503	512	Cell frame
Rhodamine 123	505	534	Organelles, MC
Acridine orange	502 (DNA), 460 (RNA)	526 (DNA), 650 (RNA)	DNA, RNA
Ethidium bromide	482	616	DNA, RNA
Propidium iodide	536	617	Dead cell (DNA, RNA)
Di-4-ANEPPS	496	705	MP
Di-8-ANEPPS	498	680	MP
Fluorescein-PE	497	521	Cell surface pH
SNAFL	506 (acid), 538 (base)	535 (acid), 620 (base)	Intracellular pH (ER: 540/630, acid sensitive)
SNARF	518 (acid), 579 (base)	587 (acid), 635 (base)	Intracellular pH (ER: 580/630, base sensitive)
BCECF	457 (acid), 508 (base)	520 (acid), 531 (base)	Intracellular pH (ER: 450/500 or 490/540)
Calcium green	506	534	Intracellular Ca^{2+}
Fluo-3	506	526	Intracellular Ca^{2+}
Fura Red	472 (low), 436 (high)	645 (low), 640 (high)	Intracellular Ca^{2+}
Fura Red/fluo-3	503	526, 640–645	Intracellular Ca^{2+} (ER: 526/640)

[a] L, leukocyte; E, erythrocyte; P, platelet; MC, mitochondria; MP, membrane potential (voltage-dependent); ER, emission ratio.

in detecting concentration changes in Ca^{2+} in vascular smooth muscle and endothelial cells of microperfused afferent arterioles isolated from rat juxtamedullary nephrons. When excited at 488 nm, fluo-3 exhibits an increase in green fluorescence (525 nm) in Ca^{2+} binding, whereas Fura Red shows a decrease in red fluorescence (640 nm). The result is that the emission ratio of these two fluorescent signals precisely indicates intracellular Ca^{2+} changes, independent of the loaded dye concentrations. However, because green and red fluorescence images were acquired on two separate photomultipliers at 1 Hz, their method required about 0.4 sec to obtain each image, indicating that the calcium images taken in their study did not reflect real-time Ca^{2+} kinetics within the cell. An attempt has been made to acquire real-time calcium images in a cardiac myocyte isolated from the rat ventricle.[40] Ishida[40] has succeeded in capturing Ca^{2+} waves within the cardiac myocyte at a video rate by means of a modified type of TSRLM (CSU10, Yokogawa Electric Co.). In his experiment, a myocyte was loaded simultaneously with fluo-3 and di-8-ANEPPS (voltage-sensitive fluorescent indicator for t-tubules). Spontaneously changing local Ca^{2+} wave propagation was clearly demonstrated against a systemic array of t-tubules as a landmark (Fig. 16).

Although FITC, rhodamine, or acridine orange has been generally used as the fluorescent dye for investigating microcirculatory kinetics, numerous luminescence probes may be used, in combination with appropriate monochromatic laser beams, for detecting events occurring in microcirculation. Information on the kinds of laser beams usable under various experimental conditions and on the excitation wavelengths (by laser beam) and fluorescent wavelengths (emitted from each probe) of representative luminescence dyes is summarized in Tables II and III. In Table III, fluorescent dyes activated by a 488-nm laser beam were included only for simplicity. These probes can be utilized in accordance with different experimental purposes.

Acknowledgments

The author is grateful to Dr. T. Nakamoto, First Department of Internal Medicine, School of Medicine, Dokkyo University, Tochigi, Japan, and Dr. H. Ishida, Department of Physiology, School of Medicine, Tokai University, Kanagawa, Japan, for generously providing the excellent photographs taken via a real-time CLSM.

[40] H. Ishida, personal communication (1998).

Section V

Imaging of Ions

[24] Confocal Imaging Analysis of Intracellular Ions in Mixed Cellular Systems or *in Situ* Using Two Types of Confocal Microscopic Systems

By Hisayuki Ohata, Masayuki Yamamoto, Yosuke Ujike, Gousei Rie, and Kazutaka Momose

Introduction

Intracellular Ca^{2+} is involved in the regulation of various cellular functions as an intracellular messenger system. Since the development of digital video imaging of intracellular free Ca^{2+} concentrations ($[Ca^{2+}]_i$) using Ca^{2+}-sensitive fluorophores, new findings such as Ca^{2+} oscillations[1,2] and Ca^{2+} waves[3] have been reported in many different cell types. However, most of these results have been obtained using monolayers of cultured cells or freshly isolated cells because of restrictions in the optical axis (Z axis) resolution in conventional fluorescence microscopy. To determine changes in $[Ca^{2+}]_i$ in mixed cellular systems such as *in situ* and exact spatial heterogeneity of the Ca^{2+} response within single cells, microscopic techniques with a high Z-axis resolution are needed. In this connection, the thin optical sectioning capability of laser scanning confocal microscopy, which rejects light from out-of-focus planes, permits imaging of $[Ca^{2+}]_i$ in individual cells *in situ* in optical sections about 1 μm thick.[4]

Ca^{2+} Imaging Using Confocal Microscopy

Figure 1 shows fundamental principles of confocal microscopy that can be applied to Ca^{2+} imaging using Ca^{2+}-sensitive fluorophores such as fluo-3. Excitation laser light is reflected by the dichroic mirror and is focused by the objective lens to a diffraction limited spot at the focal plane within the thick specimen loaded with Ca^{2+}-sensitive fluorophores. Fluorescence emissions originating from the focal plane are passed through a small pinhole to a photomultiplier detector. However, fluorescence emissions from above and below the focal plane are not focused on the pinhole and

[1] M. J. Berridge and A. Galione, *FASEB J.* **2,** 3074 (1988).
[2] M. J. Berridge, *J. Biol. Chem.* **265,** 9583 (1990).
[3] M. J. Berridge, *Nature* **361,** 315 (1993).
[4] J. J. Lemasters, E. Chacon, H. Ohata, I. S. Harper, A.-L. Nieminen, S. A. Tesfai, and B. Herman, *Methods Enzymol.* **260,** 428 (1995).

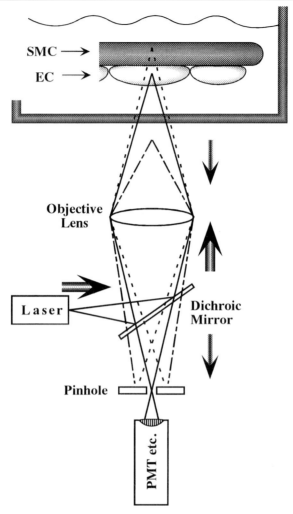

FIG. 1. Fundamental principles of confocal microscopy. Excitation laser light is reflected by the dichroic mirror and is focused by the objective lens to a diffraction-limited spot at the focal plane (endothelial cells, EC) within the thick specimen. Fluorescence emission originating from the focal plane (EC) is passed through a small pinhole to a photomultiplier tube (PMT) detector. However, fluorescence emission from above (smooth muscle cells, SMC) and below the focal plane is not focused on the pinhole and is not transmitted to the detector.

are not transmitted to the detector. By this simple optical principle, a confocal microscopy permits Ca^{2+} imaging of individual cells in optical sections about 1 μm thick, although the beam must scan the specimen using vibrating mirrors to get two-dimensional images.[5] This principle can also be applied to Ca^{2+} imaging in individual cells in tissue whose depth is less than about 30 μm.

Practical Considerations for Confocal Ca^{2+} Imaging in Situ

Photodamage. Because confocal microscopy creates images from a thin optical section using the more complicated optics of scan generator, confocal microscopy requires brighter excitation light than conventional wide field fluorescence microscopy. Brighter excitation light also improves signal-to-noise ratios. However, living cells are damaged by high excitation light, especially in tissue preparations *in situ*. Because the susceptibility of cells to the excitation light may be dependent on the kind of cells, laser power needs to be adjusted for each cell type and objective lens used. Further, photodamage increases with total time to acquire images, especially during continuous imaging.

Lens. Because Z-axis resolution and brightness of confocal imaging are dependent on the numerical aperture (NA) of the objective lens, a high NA objective lens should also be used in *in situ* experiments. However, oil-immersion lens should not be used when a bath solution exists between the specimen and the objective lens because differences in the refractive index between the immersion oil and the object medium decrease image formation capability. Water-immersion lens with high NA should be used to get confocal imaging *in situ*. However, this requires adjustment with a correction collar to spherically correct for thickness of coverslip for fine focusing.

Pinhole. Most commercial point-scanning confocal systems have adjustable pinhole apertures. Decreasing the diameter of the pinhole to the size of the diffraction-limited Airy disk projected on the pinhole improves Z-axis resolution, but further decreases in the diameter decrease image brightness without improvement of Z-axis resolution. When examining dim specimens, increasing pinhole diameter is effective for improving the signal-to-noise ratio, although Z-axis resolution decreases slightly. Full-width at half-maximum as an index of Z-axis resolution should be determined for each objective lens used at the suitable pinhole size.

[5] J. G. White, W. B. Amos, and M. Fordham, *J. Cell Biol.* **105,** 41 (1987).

Confocal Imaging Analysis of Ca^{2+} Response in Individual Endothelial Cells of Artery *in Situ*

It is important to characterize the $[Ca^{2+}]_i$ response of intact endothelial cells to various stimuli *in situ* because Ca^{2+} mobilization in endothelial cells may alter pathological states such as hypertension and diabetes as well as by isolation with enzymatic treatments or culture procedures.[6] However, little is known about the $[Ca^{2+}]_i$ in individual endothelial cells because of the difficulty in separating fluorescence from endothelial cells and smooth muscle cells *in situ* using a digital video-imaging system and conventional microscopy. We have established an imaging analysis of $[Ca^{2+}]_i$ in individual endothelial cells of guinea pig artery loaded with fluo-3/AM *in situ* using laser scanning confocal microscopy.[7] The method permitted us to distinguish signals from individual endothelial and smooth muscle cells. In the present study, to characterize properties of agonist-induced changes in endothelial $[Ca^{2+}]_i$ *in situ*, we have examined the effect of ACh, a physiological major vasorelaxant, on $[Ca^{2+}]_i$ in individual endothelial cells of rat aortic strips, a blood vessel used widely for the study of vascular mechanical function, using the confocal imaging analysis.

Tissue Preparation and Loading of Fluorophores. Thoracic aorta is excised carefully from male Wistar rats weighing 200–300 g in physiological saline solution (PSS) and carefully cleaned of the surrounding connective tissue. The aorta is cut open carefully into 2×5-mm (width $\times$ length) strips. Aortic strips are incubated with 5 μM calcein/AM, an ion-insensitive fluorophore, as a marker of intracellular space,[8] or 5 μM fluo-3/AM, Ca^{2+}-sensitive fluorophore,[9] with 0.03% cremophor EL in PSS for 30 min at room temperature. The aortic strips are then rinsed several times with PSS. Because loading of excess fluorophores to cells may cause compartmentalization of fluorophores[10–12] or may modify physiological changes in $[Ca^{2+}]_i$, loading of excess fluorophores is avoided. Loading time and concentration of fluorophores are largely dependent on cell and fluorophore type.

[6] G. De Nucci, R. J. Gryglewski, T. D. Warner, and J. R. Vane, *Proc. Natl. Acad. Sci. U.S.A.* **85**, 2334 (1988).

[7] H. Ohata, Y. Ujike, and K. Momose, *Am. J. Physiol.* **272**, C1980 (1997).

[8] E. Chacon, J. M. Bond, J. M. Reece, G. Zahrebelski, A.-L. Nieminen, B. Herman, and J. J. Lemasters, *Toxicologist* **12**, 371 (1992).

[9] A. Minta, J. P. Y. Kao, and R. Y. Tsien, *J. Biol. Chem.* **264**, 8171 (1989).

[10] M. W. Roe, J. J. Lemasters, and B. Herman, *Cell Calcium* **11**, 63 (1990).

[11] A.-L. Nieminen, A. K. Saylor, S. A. Tesfai, B. Herman, and J. J. Lemasters, *Biochem. J.* **307**, 99 (1995).

[12] H. Ohata, E. Chacon, S. A. Tesfai, I. S. Harper, B. Herman, and J. J. Lemasters, *J. Bioenerg. Biomembr.* **30**, 207 (1998).

Collection of Fluorescence Images in Situ. Endothelial $[Ca^{2+}]_i$ *in situ* is measured as described previously.[7] The adventitia side of the aortic strip loaded with fluorophores is attached to a plastic plate with double-sided sticky tape. The plastic plate is mounted on a glass coverslip-bottomed experimental chamber (volume 3 ml) with the endothelium surface facing the objective as shown in Fig. 1. The distance between the endothelial surface and the coverslip mounted at the bottom of the chamber is about 1 mm. The bath solution passes freely between them. Fluorescence images are collected using a Bio-Rad (Richmond, CA) MRC-500 confocal laser scanning attachment mounted on a Nikon Diaphot inverted microscope with a water-immersion objective (40× W, NA 0.75, Zeiss). An excitation wavelength of 488 nm is provided by an argon–krypton laser and is attenuated with a 3% neutral density filter to minimize photobleaching and photodamage. The green fluorescence of fluorophore is collected using a 510-nm long-pass dichroic reflector and a 515-nm long-pass emission filter. *XY*-scanning images composed of 384×256 pixels (319×212 μm) are acquired every 0.90 sec. In this setting, because the distance between an objective lens and the specimen is more than 1 mm, an objective lens with a long working distance and medium NA is used. If the distance is less than 0.25 mm, a water-immersion objective lens with an NA of more than 1.1 should be used.

Confocal Imaging of Endothelial Cells and Smooth Muscle Cells in Aortic Strip. To confirm whether fluorescence signals in endothelial cells can be distinguished from those in smooth muscle cells *in situ,* aortic strips are loaded with calcein/AM and fluorescence images are collected. Figures 2A and 2B are calcein fluorescence images taken at different focal planes, the endothelial and smooth muscle layer, respectively, in the same $X–Y$ region of an aortic strip. Oval-shaped endothelial cells disperse in the bloodstream at the surface of the aortic strip, whereas spindle-shaped smooth muscle cells disperse at right angles to the bloodstream directly under the endothelial cells. These clear confocal images confirm the optical sectioning capability of confocal microscopy, permitting imaging of fluorescent probes in individual cells within the endothelial or smooth muscle layer.

Using the same confocal imaging settings, changes in $[Ca^{2+}]_i$ in individual endothelial cells loaded with fluo-3 to ACh, a physiologically important stimulator of endothelial cells, are examined *in situ.* We do not convert the intensity change of fluo-3 fluorescence to $[Ca^{2+}]_i$ because it is difficult to truly determine using nonratiometric dyes such as fluo-3. The mean intensity of fluo-3 fluorescence in the selected area of the image after stimulation is divided by resting fluorescence intensity, the mean intensity in the same area three frames before stimulation, after subtraction of background fluo-

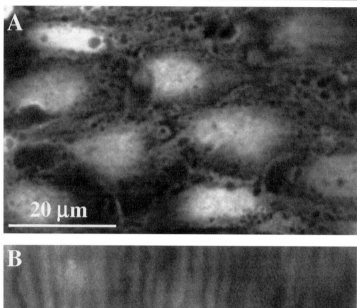

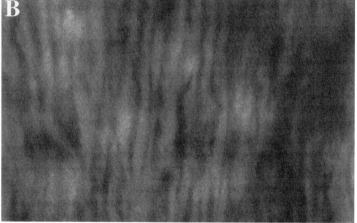

Fig. 2. Confocal imaging of endothelial cells and smooth muscle cells in an aortic strip loaded with calcein/AM. A calcein fluorescence image was collected from the endothelial layer (A) and smooth muscle layer of rat aortic strips (B) in the same X–Y region *in situ.*

rescence. The relative fluorescence intensity is used as an indicator for $[Ca^{2+}]_i$. An approximate value of resting $[Ca^{2+}]_i$ can be estimated by *in situ* calibration based on the following equation:

$$[Ca^{2+}]_i = K_d[(F - F_{min})/(F_{max} - F)]$$

where K_d is the dissociation constant for Ca^{2+} (316 nM) of fluo-3, and F_{max} and F_{min} are fluorescence intensities obtained by the addition of 10 μM ionomycin and by subsequent addition of 10 mM EGTA, respectively.[9]

Figure 3 shows changes in fluo-3 confocal images (Fig. 3A) *in situ,* the typical patterns of changes in $[Ca^{2+}]_i$ in individual endothelial cells (open and closed circles in Fig. 3B), and the averaged change in $[Ca^{2+}]_i$ of 84 cells during the cumulative addition of acetylcholine (ACh, 0.1, 1.0, and 10 μM). Fluo-3 confocal images show that endothelial cells of oval shape disperse in the bloodstream at the surface of the aortic strip, and the fluorescence signals could be distinguished from the signals derived from smooth muscle cells (Fig. 3A). At resting state, endothelial $[Ca^{2+}]_i$ is stable at low level without any oscillatory change in $[Ca^{2+}]_i$. The cumulative addition of ACh causes a concentration-dependent increase in endothelial $[Ca^{2+}]_i$. Cells that respond to 10 μM ACh are 89% of the cells that respond to 10 μM ionomycin. The oscillatory increase in $[Ca^{2+}]_i$, as shown by the open circle in Fig. 3B, is observed in 86% (72 cells out of 84 cells) of cells that respond to ACh. Although a step-like increase in $[Ca^{2+}]_i$, as shown by the closed circle in Fig. 3B, is observed in only 10% (8 of 84 cells) of cells, the average change in $[Ca^{2+}]_i$ of all cells that respond to ACh increases in a step-like manner (dotted line of Fig. 3B), suggesting that the oscillation of each endothelial cell does not synchronize in this microscopic area. As shown by the open circle in Fig. 3B, the frequency of $[Ca^{2+}]_i$ spikes is very high in most oscillating cells and is in the range of 10 to 20 spikes/min. It is very important that the phenomenon is demonstrated *in situ,* suggesting the physiological significance of $[Ca^{2+}]_i$ oscillations in endothelial cell functions. However, our results are different from those in experiments using rat aorta *in situ* by Carter and Ogden.[13] They also observed that ACh induces large $[Ca^{2+}]_i$ oscillations, but the frequency was less than 2 spikes/min. This discrepancy may be due to the fact that they measured $[Ca^{2+}]_i$ by image analysis with a conventional microscope using low-affinity Ca^{2+} indicator furaptra. Therefore, it is possible that they did not detect small changes. Additionally, their simultaneous measurement of membrane potential using the whole-cell patch-clamp configuration might affect changes in $[Ca^{2+}]_i$. In this respect, it is clear that our results are reliable in regard to physiological response, although careful setting of the experimental conditions is required to detect this phenomenon *in situ.*

High-Speed Imaging Using Multipinhole Confocal Scanner

Most point-scanning confocal systems require 0.1–1 sec to acquire a two-dimensional image. This temporal resolution is not sufficient to detect precise changes in $[Ca^{2+}]_i$, even in one focal plane. A temporal resolution greater than the video rate (33 frames/sec) is required to measure changes

[13] T. D. Carter and D. Ogden, *Pflueg. Arch.* **428,** 476 (1994).

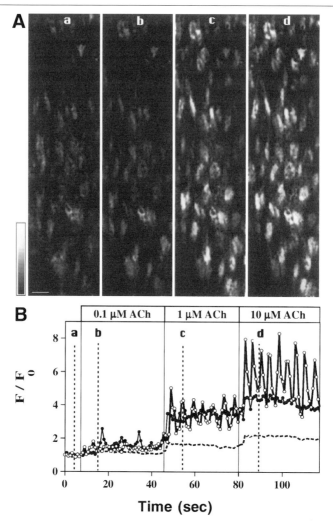

FIG. 3. Typical fluo-3 confocal images and time course of ACh-induced $[Ca^{2+}]_i$ oscillations in individual endothelial cells in an aortic strip. Confocal images (A, a–d) in endothelial cells loaded with fluo-3/AM were collected every 0.9 sec during the cumulative addition of ACh (0.1, 1, and 10 μM). The images are pseudocolored using a color table of NIH Image. Horizontal bar represents 20 μm. (B) Time course of changes in fluo-3 fluorescence in the two different cells (open and closed circles) and averaged value (dotted line) of 84 cells in image (A). The lowercase letters (a–d) in (B) indicate the time at which the frames in image (A) were taken.

in $[Ca^{2+}]_i$ in different focal planes sequentially. An add-on type scanner consisting of an innovative Nipkow disk with microlens array developed as a multipinhole confocal scanner (CSU-10) by Yokogawa Electric Co. (Tokyo, Japan) enabled us to acquire a two-dimensional image at a few milliseconds per frame without any decrease in spatial resolution, although a high-speed and high-sensitive camera system was required to detect and to record the images.

Principles of Multipinhole Confocal Scanner. Figure 4 shows the principles of the scanner consisting of a Nipkow disk with microlens array (CSU-10Z). Excitation laser light is focused by microlens on the upper microlens disk to the focal point on the lower pinhole disk, where 20,000 pinholes of 50 μm diameter are formed with the same pattern as that of the microlenses. This microlens array provides one of the brightest advances in confocal technology currently available. The laser light passing through the pinhole is focused by the objective lens to a diffraction-limited spot at the focal plane within a thick specimen loaded with appropriate fluorophores. Fluorescence emissions originating from the focal plane are passed through the objective lens and the same pinhole, and the light reflected by the dichroic mirror focuses on a camera detector through relay lens. Unlike point-scanning confocal systems, a large number of light spots illuminate the specimen simultaneously and are rotated at 1000–3600 rpm. The multipinhole confocal disk scanner can form a two-dimensional image effectively within a few milliseconds. Therefore, although the pinhole size is not adjustable, the confocal image can be observed through the eyepiece and a cooled charged-coupled device (CCD) can be used as detector, which has quantum efficiency for photon detection higher than that of the photomultipliers used for a point-scanning confocal systems.[14]

System Configuration Using Multipinhole Confocal Disk Scanner for High-Speed Three-Dimensional Confocal Imaging. Figure 5 is a diagram of system configuration for the high-speed three-dimensional confocal measurement of $[Ca^{2+}]_i$. The rotating speed of the confocal scanner consisting of a Nipkow disk with a microlens array (CSU-10Z) is controlled by a synthesized function generator (FG120, Yokogawa Electric Co.) to synchronize with the exposure time of the high-speed cooled CCD camera controlled by an ARGUS HiSCA system (Hamamatsu Photonics K.K., Hamamatsu, Japan). Z-axis scanning is performed with a Microscope Objective NanoPositioner (P-721.10, PI Polytec K.K., Tokyo, Japan) mounted between the objective lens and the objective lens holder of an inverted microscope (Axiovert 100, Carl Zeiss Co., Tokyo, Japan) controlled by a LVPZT Position Serve Controller (E-662, PI Polytec K.K.)

[14] K.-H. Marien and E. Pitz, *Opt. Engin.* **26,** 742 (1987).

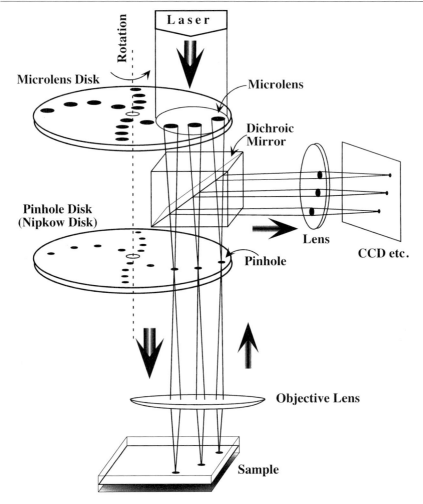

FIG. 4. Principles of multipinhole confocal disk scanner. Important optical and mechanical parts for the multipinhole confocal disk scanner are labeled and described in the text.

and is regulated by a synthesized function generator triggered by electrical signals from the ARGUS HiSCA system at the start of the exposure to synchronize it with the exposure interval of the CCD camera.

CCD Camera. Although CSU-10Z can form 700 full-frame images per second, practical time resolution and *X–Y* resolution of the imaging system are dependent on the performance of the camera and recording systems

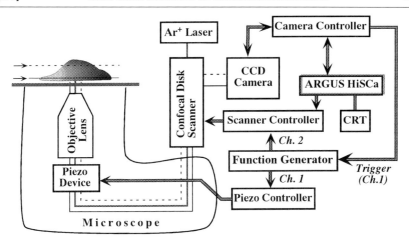

Fig. 5. Diagram of system configuration for a high-speed three-dimensional confocal imaging system. The rotating speed of a confocal disk scanner is controlled by a synthesized function generator to synchronize with exposure time of a high-speed cooled CCD camera controlled by the ARGUS HiSCA system. Z-axis scanning is performed by the Microscope Objective NanoPositioner (piezo device) mounted between an objective lens and an objective lens holder of an inverted microscope controlled by a LVPZT Position Serve Controller and is regulated by a synthesized function generator that is triggered by electrical signals originating from the ARGUS HiSCA system to synchronize it with the exposure interval of the CCD camera.

used. The choice of camera is also important for determining characteristics of the confocal imaging system. A high-speed cooled CCD camera controlled by ARGUS HiSCA is used as an interline transfer CCD using on-chip lens technology to improve the open area ratio and has high flexibility. In this CCD, the full-well capacity is 30,000–50,000 electrons/pixel, the maximum quantum efficiency is 42% at 520 nm, and the dynamic range is 12 bits at a readout rate of 300 Kpixels/sec with a readout noise of 5–7 electrons/pixel or 10 bits at a readout rate of 10 Mpixels/sec with a readout noise of 20–30 electrons/pixel. The full frame area image of the CCD with 659 × 494 pixels can be taken at 1 frame/sec with a 12-bit dynamic range or 27.5 frames/sec with a 10-bit dynamic range. Furthermore, using the binning technique, images with 20 × 15 pixels with 32 × 32 binning can be taken at 500 frames/sec with a 10-bit dynamic range, although X–Y resolution decreases. Thus, the CCD camera system used here can fully utilize the full potential of the CSU-10Z, especially for high-speed confocal imaging.

Z-axis Scanning. The Microscope Objective NanoPositioner consists of piezoelectric translator ceramics that directly convert electrical energy into motion. The closed loop versions we used allowed for absolute position control, high linearity, and repeatability based on an integrated position feedback sensor and provided a positioning and scanning range of up to 100 μm with nanometer resolution. Z-axis resolution of the objective lens was tested by Z-axis scanning of 0.5-μm fluorescent microspheres (Molecular Probes, Inc., Eugene, OR) using the Microscope Objective NanoPositioner at a 488-nm excitation wavelength with an argon ion laser. As shown in Fig. 6, Z-axis resolution shown as full-width at half-maximum (FWHM) was dependent on the NA of the objective lens used. Maximum Z-axis resolution was obtained using UPlanApo 60× W (NA 1.2, Olympus, Tokyo,

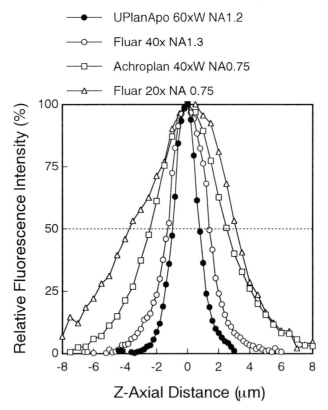

FIG. 6. Z-axis resolution of the high-speed three-dimensional confocal imaging system with each different objective lens. Z-axis resolution using each objective lens, including oil-immersion objective lens and water-immersion objective lens, was tested by Z-axis scanning of 0.5-μm fluorescent microspheres using the Microscope Objective NanoPositioner at a 488-nm excitation wavelength with an argon ion laser.

Japan), a water-immersion objective lens, among some object lenses used, and the FWHM was estimated as 1.3 μm. This value was comparable to that obtained using Bio-Rad MRC-500 confocal microscopy with the same kind of objective lens on a small pinhole setting.[4]

To obtain three-dimensional images of living cells, high-speed Z-axis scanning synchronized with image acquisition is required. The closed loop versions of the Microscope Objective NanoPositioner used here provide a Z-axis positioning and a scanning range of up to 100 μm with a setting time of about 8 msec as shown in Fig. 7A. Therefore, sequential confocal images of two different focal planes can be acquired with the video rate of 30 frames/sec when the exposure time of each focal plane was 8 msec, although an oil-immersion objective lens could not be used in this condition. When an oil-immersion lens was used, the setting time of the Z-axis positioning required more than 20 msec because of the high viscosity of the immersion oil.

Application of High-Speed Three-Dimensional Confocal Imaging to Measurement of [Ca²⁺]ᵢ in Mixed Cellular Systems

Using the high-speed three-dimensional confocal imaging system described previously, changes in $[Ca^{2+}]_i$ in each cell in a mixed culture of hippocampal neuronal cells and astroglial cells loaded with 5 μM fluo-4/ AM, a new Ca^{2+}-sensitive fluorophore improved by Molecular Probes, Inc. (Eugene, OR), were measured by the experimental protocol shown in Fig. 7A. In this mixed cell culture, neuronal cells were grown on astroglial cells[15] and the cells could be discriminated morphologically by transparent infrared light images.[16] Images of 60 × 60 pixels with 4 × 4 binning were acquired with a 17.3-msec exposure at a 12-bit dynamic range using UPlanApo 60× W. Z-axis scanning with a closed loop version of the Microscope Objective NanoPositioner was controlled by an electrical square wave with a 69.2-msec/cycle and a 0.4-V amplitude with 0.2 V offset, which corresponds with 4 μm Z-axis scanning. Scanning was synchronized with exposure of the camera by the TTL level signal output from the ARGUS HiSCA system at the start of each exposure. Therefore, in this experimental setting, each fluo-4 fluorescence image of neuronal cells and astroglial cells could be acquired reciprocally at a 69.2-msec interval with an $X-Y$ resolution of 0.67 μm/pixel and a Z-axis resolution of 1.3 μm.

[15] K. Inoue, K. Nakazawa, K. Fujimori, T. Watano, and A. Takanaka, *Neurosci. Lett.* **134,** 215 (1992).
[16] M. Yamamoto, T. Kawanishi, T. Kiuchi, M. Ohta, I. Yokota, H. Ohata, and K. Momose, *Life Sci.* **63,** 55 (1998).

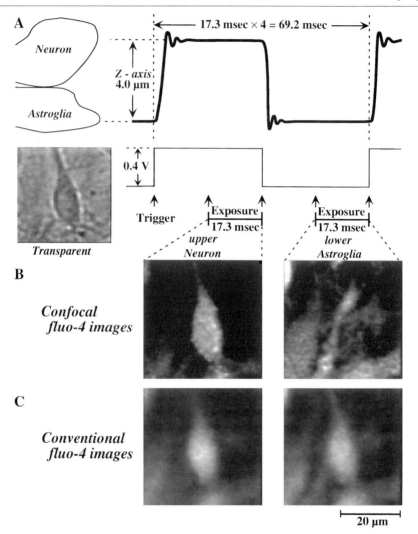

FIG. 7. Simultaneous confocal imaging of a mixed cellular system of neuronal cells and astroglial cells using the high-speed three-dimensional confocal imaging system. (A) Relationship between changes in Z-axis position with the Microscope Objective NanoPositioner controlled by electrical square waves and camera exposure. The detailed protocol is described in the text. (B) Confocal fluorescence images of fluo-4-loaded neuronal cell (upper) and astroglial cells (lower) obtained by this confocal system. (C) Conventional fluorescence images of the same cells as those in (B) by the system without the confocal disk scanner (CSU-10Z).

Confocal fluo-4 fluorescence images of neuronal and astroglial cells obtained are shown in Fig. 7B. The images indicate clearly that this method permits discriminative imaging of neuronal and underlying astroglial cells. However, without CSU-10Z, this system could not discriminate between neuronal and astroglial cells, as shown in Fig. 7C, although the other experimental setting and the cells observed were the same as that in the confocal imaging using CSU-10Z. These results indicate clearly the usefulness of the confocal system for the Ca^{2+} imaging of mixed cellular systems such as neuronal cells and astroglial cells.

It is well known that the presynaptic glutamate receptor in neuronal cells plays an important physiological role in central nervous systems.[17] In addition, hypoxic–ischemic cell injury has been linked to excessive activation of the postsynaptic glutamate receptor, followed by an influx of extracellular Ca^{2+}.[18] However, the functional relationship between neuronal cells and astroglial cells is unclear. The effect of glutamate on the changes in $[Ca^{2+}]_i$ in the neuronal and astroglial cells was then examined simultaneously by high-speed three-dimensional confocal imaging. The application of 30 μM glutamate increased $[Ca^{2+}]_i$ in both neuronal and astroglial cells, but the time course of the changes was quite different, as shown in Fig. 8. In the neuronal cell, $[Ca^{2+}]_i$ increased markedly right after the addition of glutamate in all three regions measured. The rate of increase in $[Ca^{2+}]_i$ was highest in the beginning and then decreased gradually. These results show clearly that the increases in $[Ca^{2+}]_i$ in neuronal cells induced by glutamate are due to Ca^{2+} influx, probably through activation of the NMDA receptor.[19] However, in astroglial cells, $[Ca^{2+}]_i$ increased from region (a) shown in Fig. 8 after a lag time of about 1 sec. $[Ca^{2+}]_i$ increases spread as a wave from region (a), near the one process, to the other regions along the long axis of the astroglial cell. The propagation rate of the wave of the $[Ca^{2+}]_i$ increase induced by glutamate in astroglial cells was about 50 $\mu m/sec$. These results show that the increase in $[Ca^{2+}]_i$ in astroglial cells induced by glutamate is due to Ca^{2+} release from intracellular stores through the activation of metabotropic receptors for glutamate.[20] In addition, the rate of increase in $[Ca^{2+}]_i$ was slow in the beginning and increased with increases in $[Ca^{2+}]_i$ up to the submaximal level of $[Ca^{2+}]_i$. This pattern of Ca^{2+} increase can be explained by a Ca^{2+}-dependent positive feedback control of inositol 1,4,5-trisphosphate-induced Ca^{2+} release.[21] Thus, the high-speed three-di-

[17] M. L. Mayer and G. L. Westbrook, *Prog. Neurobiol.* **28,** 197 (1987).

[18] D. W. Choi, *Trends Neurosci.* **11,** 465 (1988).

[19] A. B. MacDermott, M. L. Mayer, G. L. Westbrook, S. J. Smith, and J. L. Barker, *Nature* **321,** 519 (1986).

[20] S. Nakanishi, *Science* **258,** 597 (1992).

[21] M. Iino and M. Endo, *Nature* **360,** 76 (1992).

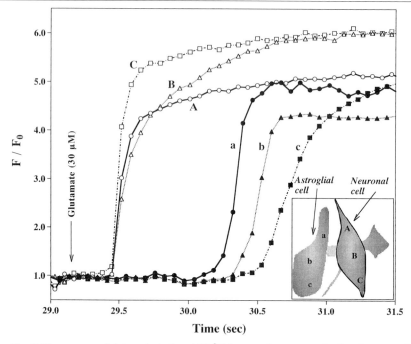

FIG. 8. Time course of glutamate-induced $[Ca^{2+}]_i$ increase in a neuronal cell and an astroglial cell. Time courses of changes in fluo-4 fluorescence induced by 30 μM glutamate for three different regions (3.3 × 4.0 μm–4.0 × 6.7 μm) in neuronal and astroglial cells shown in the inset were obtained by analyzing the confocal images shown in Fig. 7B.

mensional confocal Ca^{2+} imaging system permits simultaneous imaging of cells in different focal planes with high spatiotemporal resolution and should be useful for the study of interactions between these cells.

Conclusion

Confocal imaging analysis permits measurement of $[Ca^{2+}]_i$ in individual cells in tissue preparations, such as blood vessels, *in situ*. Accordingly, it should be useful in the characterization of Ca^{2+} mobilization in vascular endothelial cells in pathological states such as hypertension and diabetes, which cannot be examined using cultured cells with conventional fluorescence imaging systems. In addition, a high-speed three-dimensional confocal imaging system using a multipinhole confocal disk scanner and a high-speed cooled CCD camera permits simultaneous imaging of cells in different focal planes at video rates and should be useful for the study of interactions between these cells. Also, confocal imaging technology can apply analysis of other intracellular ions such as Mg^{2+}, Na$^+$, K$^+$, H$^+$, and Cl$^-$ as well as Ca^{2+} using each appropriate ion-sensitive fluorophore. In the near future, improvement of confocal imaging technology with further development of these ion-sensitive fluorophores may permit visualization of activation of an ion channel with high spatiotemporal resolution comparable to those of electrophysiological recording.

Acknowledgments

We are grateful to Dr. T. Kawanishi (National Institute of Health Sciences) for insightful comments. This work was supported in part by a grant-in-aid for Drug Innovation Science Project (to T. Kawanishi and K. Momose) from the Japan Health Science Foundation.

[25] Confocal Ca^{2+} Imaging of Organelles, Cells, Tissues, and Organs

By DAVID A. WILLIAMS, DAVID N. BOWSER, and STEVEN PETROU

Introduction

Laser scanning confocal microscopy (LSCM) has much to offer in the study of cellular physiology, anatomy, and pathology because of its ability to remove most of the out-of-focus information from two-dimensional images. Light emanating from locations in a biological sample adjacent to the point of focus or that are above or below the plane in which the point lies is excluded from detection by pinhole apertures placed in the emission light path (see Fig. 1). This feature has allowed for improvements in spatial resolution (z depth and x–y) and, in essence, has enabled the production of optical sections of living cells in isolation or within tissues. In particular, in motile cell systems, such as isolated muscle cells, the confounding effects of cell contraction (or movement), and out-of-focus information, on measured fluorescence levels can be minimized through the ability to confine data acquisition to restricted and well-defined volumes within the cell. These volumes may be represented by complete two-dimensional slices of the cell or by specific regions of interest within the cell, both of which will be featured in some of the methods and studies described in this article. As the combination of confocal microscopy and fluorescent calcium sensors is no longer new, details of the potential advantages and disadvantages of

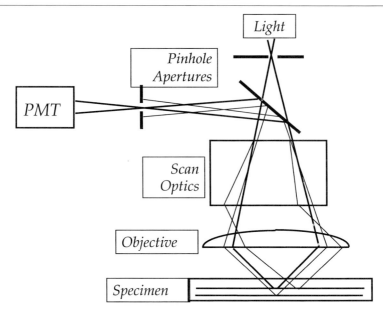

FIG. 1. Schematic representation of the light path of a generic confocal microscope. Only light that emanates from a point of focus within a desired plane of the biological specimen (bold lines) is brought to a focus at an adjustable pinhole aperture placed in front of the detector (photomultiplier tube, PMT). Light that is out of focus or adjacent to the focal point (thin lines) has alternative focal points and contributes little to light that is detected by the PMT.

various combinations of fluorescent and confocal imaging paradigms are largely known and have been reviewed extensively.[1]

Ca^{2+} Sensors for Confocal Microscopy

Chemically Synthesized Sensors

Fluorescent sensors that show a spectral response on binding Ca^{2+} have enabled researchers to investigate changes in intracellular-free Ca^{2+} concentrations using fluorescence and confocal microscopy, flow cytometry, and fluorescence spectroscopy. These fluorescent sensors, most of which are variations of the nonfluorescent Ca^{2+} chelators EGTA and BAPTA, have evolved largely through the efforts of Tsien and colleagues[2] and, more

---

[1] P. A. Dilberto, X. F. Wang, and B. Herman, in "Methods in Cell Biology" (R. Nuccitelli, ed.), Vol. 40, p. 243. Academic Press, San Diego, 1994.
[2] R. Y. Tsien and A. Waggoner, in "Handbook of Biological Confocal Microscopy" (J. B. Pawley, ed.), p. 267. Plenum Press, New York, 1995.

recently, through those of scientists at private chemical synthesis companies such as Molecular Probes.[3] These Ca^{2+} sensors are essentially fluorescent Ca^{2+} buffers and hence the chemical constraints governing their characteristics and limitations are intuitively clear. A 1:1 stoichiometry of sensor Ca^{2+}-binding (buffering) dictates a sigmoidal relationship between the proportion of sensor bound to Ca^{2+} (and hence the fluorescence produced) and the local (solution) $[Ca^{2+}]$. This relationship also dictates the effective buffering (and Ca^{2+}-reporting) range of each sensor. The "working" estimate of this range of $[Ca^{2+}]$ encompasses one order of magnitude above and below the association (K_a) or dissociation (K_d) constant, representing the range from 9 to 91% saturation of sensor with Ca^{2+}.

Almost exclusively, this working range and the spectral characteristics of the fluorophore govern the experimental choice of sensor. The expected range of experimentally measured Ca^{2+} levels dictates the working range of the desirable sensor, whereas the hardware characteristics of the recording system, in concert with the time course of Ca^{2+} measurements that need to be made, determine the appropriate spectral characteristics of the sensor. Although compromises have to be made between these features on occasion, the range of useful sensors available is growing rapidly, leading to closer approximations to ideal combinations. Estimates of $[Ca^{2+}]$ working ranges, using representative values of K_d largely determined in this laboratory, are listed in Table I for some of the Ca^{2+} sensors used most commonly with confocal microscopy and for some sensors that have potential for use in this area. An additional readily accessible source for technical and application information can be found in Haugland.[3]

In particular the Ca^{2+} sensors fluo-3, Calcium Green, and rhod-2 (and a structurally dissimilar probe Fura Red) (Molecular Probes) have been widely employed over the last decade for quantitative measurements of calcium concentrations with confocal microscopy. While UV-light excitable sensors such as Indo-1 have found some use in confocal Ca^{2+} studies, visible light-excitable Ca^{2+} sensors offer several advantages and are used more commonly. The utilization of argon ion or krypton–argon lasers in most commercially available confocal systems limits excitation wavelengths to the major bands emitted by such lasers, generally 488, 514, and 568 nm. As such the main interest has been in the use of BAPTA-based Ca^{2+} sensors that are related structurally to fluorescein or rhodamine and can therefore be excited in the visible wavelength spectrum. These sensors offer reduced interference from the cell or tissue autofluorescence that is induced by UV-light illumination. In addition, less cellular photodamage is evident and the

---

[3] R. P. Haugland, "Handbook of Fluorescent Probes and Research Chemicals" (M. T. Z. Spence, ed.), 6th ed. Molecular Probes, Eugene, OR, 1996.

TABLE I
COMMONLY USED AND NEW ALTERNATIVE Ca^{2+} SENSORS FOR CONFOCAL MICROSCOPY[a]

Ca^{2+} sensor	Confocal mode (ex–em)	Dissociation constant $(K_d)^b$ (nM)	Effective working range (μM)
Fluo-3	Single–single	430	0.04–4.3
Calcium Green-1	Single–single	200	0.02–2.0
Calcium Green-2	Single–single	515	0.05–5.1
Calcium Orange	Single–single	200	0.02–2.0
Calcium Green/ Texas Red	Single–dual	370	0.04–3.7
Rhod-2	Single–single	600	0.06–6.0
Oregon Green 488 BAPTA-1	Single–single	170	0.02–2.0
Oregon Green 488 BAPTA-2	Single–single	580	0.06–6.0
Oregon Green 488 BAPTA-5	Single–single	20,000	2–200
Indo-1	Single–dual	230	0.02–2.0
Fura Red	Dual–single	140	0.01–1.4
Fura Red/fluo-3	Single–dual^c	c	c
BTC	Dual–single	7000	0.7–70
Fluo-4	Single–single	345	0.35–3.5
X-rhod-1	Single–single	700	0.07–7.0

[a] "Confocal mode" represents whether the sensor is used in a single excitation–single emission wavelength (single–single) mode, single excitation–dual emission wavelength (single–dual) mode, or dual excitation–single emission wavelength (dual–single) mode. Temperature is 20–22°

[b] The K_d for BAPTA-based Ca^{2+} sensors is very sensitive to a number of extracellular and environmental factors, including temperature, pH, ionic strength, and potential interactions of the sensor with proteins. The values of K_d determined within cells can be severalfold different (generally higher) than values determined in vitro.

[c] Denotes mode when used in dual-loaded combination with fluo-3.[31] Estimation of K_d and working range, which represent the weighted average of the K_d for each sensor, require knowledge of the proportion of each sensor in the calibrated compartment.

higher absorbance of these molecules permits the use of lower sensor concentrations with a resultant lower phototoxicity for live cells. Finally, these sensors allow compatibility with UV-light photoactivatable ("caged") compounds, increasing options for multiparameter measurements in experiments.

Genetically Encoded Ca^{2+} Sensors

A development in molecular biological synthesis now provides sensors that offer a viable alternative to chemically synthesized Ca^{2+} sensors. Protein-based sensors of Ca^{2+} concentration are now finding application in confocal microscopy. There are two types of genetically encoded Ca^{2+} sensors, both of which use photoproteins from the jellyfish *Aequorea victoria*. While sensors based on the luminescent and Ca^{2+}-sensitive protein aequorin have been in use for many years,[4] those incorporating green fluorescent protein (GFP) and calmodulin, or "cameleons," have been only recently developed and used.[5] We will primarily describe the use and manipulation of cameleons for the confocal measurement of cell Ca^{2+} levels.

Cameleons are genetically engineered sensors of free Ca^{2+} composed of tandem fusions of blue or cyan mutants of fluorescent protein (GFP), calmodulin, calmodulin-binding protein M13, and an enhanced green or yellow emitting GFP.[5] By monitoring the resonance energy transfer (RET) of the two fused GFPs, the Ca^{2+}-dependent conformational change of calmodulin as it binds Ca^{2+} can be determined. The degree to which RET occurs then represents a direct measure of free Ca^{2+} concentration. RET is the nonradiative and nondestructive transfer of the energy of an absorbed photon by one fluorophore to a nearby fluorophore and is a common spectroscopic method used to monitor the proximity and molecular orientation at high spatiotemporal resolution.[6] The advantage of this method over Ca^{2+}-evoked aequorin chemiluminescence is that the calmodulin-binding reaction is reversible, and apart from the inevitable photobleaching inherent in any fluorescent sensor, the concentration in a cell can be maintained for the duration of the experiment. When cameleons were first introduced they were available in several forms, with different RET partners, varying Ca^{2+} affinities, and targeting to cytoplasm, endoplasmic reticulum, and the nucleus. These cameleons appear better suited than targeted recombinant aequorin, which requires coenzymes, gives limited signals at the single cell level, and is exhausted during long-term measurements.

The ability to "tune" the affinity of cameleons for Ca^{2+} according to the targeted location is vital for the successful application of these sensors. A cameleon that possesses high affinity for Ca^{2+} would make a poor choice for monitoring endoplasmic reticulum (ER) Ca^{2+} levels. Mutant forms of the calmodulin domain can be created with altered affinities for Ca^{2+}. The potential for fine-tuning of Ca^{2+} affinities for a particular application is great and is amenable to standard molecular genetic techniques. We have had excellent results performing site-directed mutagenesis using Stratagene's (LaJolla, CA) QuickChange mutagenesis protocol. This is a non-

[4] T. T. DeFeo and K. G. Morgan, *J. Physiol.* **369,** 269 (1985).
[5] A. Miyawaki, J. Llola, R. Hein, J. M. McCaffery, J. A. Adams, M. Ikura, and R. Y. Tsien, *Nature* **388,** 882 (1997).
[6] L. Stryer, *Annu. Rev. Biochem.* **47,** 819 (1978).

PCR-based protocol and relies on the ability of the restriction enzyme *Dpn*I to only restrict methylated DNA, thus newly synthesized nonmethylated DNA, created using mutagenic primers and DNA polymerase, will not be digested while the methylated parental DNA is digested. This allows for simple selection of mutated plasmids.

The advantages of molecular genetic and RET approaches over chemical synthetic approaches are numerous. As sensor creation relies on genetic techniques, genetically encoded sensors are less expensive to create than chemically synthesized sensors. The sensor can be produced *in situ* by the gene transfer of plasmid DNA into a number of cell types using a variety of well-established transfection techniques. It is a simple matter to substitute improved forms of GFP into existing sensors as new spectral variants are developed. Using known targeting sequences, probes can be targeted to subcellular locations—endoplasmic/sarcoplasmic reticulum, mitochondria, nucleus, plasma membrane, or cytosol—and targeted by fusing to localized host proteins. This article describes methods to create "custom" cameleons with any combination of RET partners, affinities, and desired targeting by using readily available molecular genetic techniques.

Introduction of Ca^{2+} Sensors into Biological Samples

The ultimate aim in the use of a Ca^{2+} sensor is to obtain a measurement of the average concentration or intracellular distribution of Ca^{2+} under conditions where the result itself is not influenced by the imaging methodology or through the presence of the sensor. Several general problems are encountered in trying to minimize the error in such measurements, including (i) balancing the need to have detectable levels of the sensor while minimizing cell damage, by-product liberation, and buffering of Ca^{2+}; (ii) limiting the distribution of the sensor to the intracellular sites of interest (e.g., cytosol or individual organelles); and (iii) determining the effects of the intracellular environment (e.g., viscosity, binding) on the behavior (and calibration constants) of the sensor.

Loading Protocols for Sensors

A number of well-established and fully evaluated techniques exist for introducing sensors into organelles, cells, and tissues. These include acetoxymethyl (AM) ester loading, microinjection, iontophoresis, and patch pipette perfusion and techniques that transiently increase the permeability of membranes for fluorophore molecules. This last category includes techniques such as ATP-induced permeabilization, electroporation, hypo-osmotic shock, and a new pinocytotic cell-loading procedure.[7] Reviews of some of these techniques have been published, and a comparative summary may be found in Haugland.[3] We have had continued success with AM ester loading of sensors for most cell or tissue types and will describe typical protocols for these processes. In addition, molecular biological techniques have been employed in recent work to express and/or target new Ca^{2+} sensors and these techniques are outlined briefly.

AM Ester Loading Process

This technique is by far the most popular method for loading Ca^{2+} sensors, as it is generally noninvasive and technically straightforward. The carboxylate groups of sensors for Ca^{2+} are derivatized as acetoxymethyl esters, neutralizing molecular charge and rendering the sensor permeant to membranes and insensitive to Ca^{2+}. Once inside the cell, acetoxymethyl ester groups are hydrolyzed by intracellular esterases, releasing the ion-sensitive, acidic, or polyanionic sensor.

In practice, a concentrated stock solution (1–10 mM) of the sensor is prepared in anhydrous dimethyl sulfoxide (DMSO) and is divided into strategically sized aliquots that can be stored desiccated at $-20°$. We choose to size these aliquots to provide sufficient sensor for a maximum of 1 week of experimentation, thereby minimizing any spontaneous ester hydrolysis that can occur with exposure to exogenous moisture. The DMSO stock solution is diluted in a suitable experimental solution to a final concentration of about 1–10 μM (or up to 25 μM for sensors such as Fura Red that have a low fluorescence yield). After incubation at 20–37° for 15–60 min, the cells should be washed two to three times with fresh medium. As this may require gentle centrifugation or settling of cells, we prefer dilution of the extracellular (unloaded) sensor with fresh medium for cells (such as cardiac or smooth muscle cells) that may be particularly sensitive to this mechanical stimulus.

Solubility of the AM esters is a significant problem in the standardization of cell and tissue loading protocols. Lack of uniform dispersal of sensor–AM in the loading medium commonly results in a large disparity of recorded fluorescence intensity levels between individual cells in a suspension or between different regions of a tissue sample. A number of dispersing agents can be used to facilitate cell and tissue loading. Among the most common, largely through the free distribution of the compound with paid orders for sensors from Molecular Probes, is pluronic F-127. It, like a number of other viable alternatives, including fetal calf serum and bovine serum albumin,

[7] BioProbes 28, "New Products and Applications." Molecular Probes Inc., Eugene OR, 1998.

has low UV absorbance, but without posing the "contamination" or binding problems that may be exhibited by the alternatives. As the majority of our sensor loading protocols utilize freshly constructed, simple media we choose to dissolve pluronic F-127 powder (Molecular Probes) directly in the medium.

Incubation time and temperature, agitation of the incubation mixture, and initial sensor concentration (external) and cell density are the experimental variables that influence the success of the AM loading process and should be modified to suit the cell type in use and the experimental requirements. In general, we have found that experiments that require kinetic measurements of changes in cytosolic ion concentrations in single cell preparations benefit from short loading times (10–30 min), high ($>30°$) incubation temperatures, low initial sensor concentrations, and high cell densities. The latter two variables have a complex influence on internal sensor concentrations, as has been described previously in detail.[8] In short, the combination of these factors controls, in a probabilistic fashion, the maximum potential number of sensor molecules that associate with (and become internalized by) an individual cell. In some situations the ion concentrations of organelles are also of experimental interest, as may be those of deep lying cell layers in a multicellular preparation. These situations necessitate longer loading times (>60 min), lower incubation temperatures ($<30°$), higher sensor concentrations, and/or lower cell densities.

Using Temperature to Refine AM Loading Process

Temperature has an obvious influence on the loading process and the distribution of the sensor, particularly in large tissue samples and intact organs such as whole muscles. High temperatures within the physiological range (30–37°) induce elevated esterase enzyme activity. An internalized sensor is likely to be cleaved rapidly in the first compartment (usually the cytosol) or cell it enters (if in a tissue preparation) and as a result the distribution of the sensor is restricted. Lowering the temperature of the incubation mixture below this range has dramatic effects on enzymatic processes such as deesterification (i.e., high Q_{10}), but has much less influence on the physical process of diffusion (low Q_{10}). Sensors should be able to diffuse extensively before cleavage restricts their distribution. In practice we have found that it is essential that low temperature incubation (5–10°) should be followed by a period of incubation at a higher temperature. This ensures that the partially cleaved sensor does not accumulate in cells or

[8] E. D. Moore, P. L. Becker, K. E. Fogarty, D. A. Williams, and F. S. Fay, *Cell Calcium* **11,** 157 (1990).

organelles as these moieties have been shown to introduce significant error into the determination of intracellular Ca^{2+} concentrations. This loading paradigm has proven to be particularly useful in providing the uniform distribution of sensor in large multicellular structures: tumor cell spheroids,[9] intact skeletal muscles,[10] and whole-heart perfusions.[11] In addition, some studies have used a similar strategy to specifically enhance the loading of some sensors into intracellular compartments such as mitochondria.[12]

A typical loading protocol for intact heart[11] requires isolation of the heart with segments of major vessels intact. A standard Langendorff perfusion of the organ is established with the heart perfused in a standard solution such as Krebs–Henseleit buffer at 37°. The perfused heart is equilibrated for 10 min before the perfusate is changed to cold (5–8°) recirculating Krebs–Henseleit buffer containing 5–10 μM of the esterified form of the relevant sensor. During this phase the heart is also immersed externally in the sensor solution chilled in an ice bath at the desired temperature. All solutions should be gassed constantly with 95% O_2/5% CO_2 (v/v) or 100% O_2. The duration of this phase (10–60 min) should be adjusted to suit the degree of tissue invasion that is required. After the loading period, the perfusion is reverted to sensor-free Krebs–Henseleit solution to wash out the unloaded sensor and to provide for a 20-min, 37° incubation period. This postloading period ensures complete cleavage of the intracellular esterified sensor. We use a modified "bottomless" organ bath[13] in conjunction with a saline-immersible objective lens on an inverted microscope. This combination is necessary (i) to provide depth discrimination to look exclusively at individual cell layers within the intact functioning organ or tissue and (ii) to allow for reduction of the potential contraction-induced movement. It can be used with a long working distance water-immersion lens (e.g., Olympus WPlan FL 40XUV; NA 0.7, working distance 3.1 mm) to focus directly onto the epicardial surface or with higher magnification and numerical aperture water-immersion objectives (e.g., Nikon X60, NA 1.2) for higher resolution images of individual cells, especially for Ca^{2+} measurements in organelles such as mitochondria or cell nuclei.

[9] S. H. Cody, P. N. Dubbin, A. Beischer, N. D. Duncan, J. S. Hill, A. Kaye, and D. A. Williams, *Micron* **24** (1993).

[10] D. N. Bowser, S. H. Cody, P. N. Dubbin, and D. A. Williams, *in* "Fluorescent and Luminescent Probes for Biological Activity" (W. T. Mason, ed.), pp. 65–81. Academic Press, London, 1999.

[11] T. Minamikawa, S. H. Cody, and D. A. Williams, *Am. J. Physiol.* **272**, H236 (1997).

[12] D. R. Trollinger, W. E. Cascio, and J. J. Lemasters *Biochem. Biophys. Res. Commun.* **236**, 738 (1997).

[13] S. H. Cody and D. A. Williams, *J. Microsc.* **185**, 94 (1997).

Delivery of Genetically Encoded Sensors

Transfection

Transfection is the delivery of foreign molecules, DNA in particular, into eukaryotic cells. It can be transient or stable, depending on how long the DNA is retained following cell division. In transient transfections, a large amount of DNA is delivered to the cell nucleus, but does not integrate into the chromosomal DNA. Following transcription and translation the RNA or protein of interest is formed. Many copies of the plasmid DNA can be introduced into the nucleus of the cell, resulting in abundant expression of the protein of interest. This is not the case with stable transfection, which typically results in a lower level of expression as transfected DNA is either integrated into the chromosomal DNA or maintained episomally. The advantages of maintaining cell lines with the gene of interest are obvious, but this must be weighed against the attendant reduction in gene expression levels. Cells that are transfected stably can be distinguished from those that are not by the use of selectable markers that encode for resistance to neomycin, blasticidin, hygromycin or zeocin, and others. Many commercial mammalian expression vectors, such as Invitrogen's pcDNA3.1 (Carlsbad, CA), can be used for transient and stable transfections as they contain the necessary promoters and polyadenylation sequences as well as the selectable marker genes.

There have been a large number of transfection techniques reported. Among the more popular are calcium phosphate[14] electroporation, liposome-mediated, and activated dendrimer methods. Most reagents can be purchased as part of a transfection "kit" available from a number of suppliers. We have had good success using commercial transfection reagents such as lipofectin and Qiagen's Superfect and Effectene (Chatsworth, CA).

Continuous cell lines such as HEK293 and COS, which result from immortalization due to a heritable alteration, vary quite significantly from their progenitors, can be propagated, and are routinely transfected. However, for studies on the physiological role of Ca^{2+}, it is often necessary to use primary cell lines. These cells arise from the outgrowth of cells from a tissue in culture or from freshly isolated cells and are, thus, most similar to the cells of the parent tissue. The transfection of primary cells is usually more problematic than the transfection of cell lines. Qiagen's Superfect and Effectene have been used in cell lines as well as primary cells with great success and are excellent first choices. We currently use, and describe in detail, Superfect-based transfection procedures.

---

[14] C. A. Chen and H. Okayama, *Biotechniques* **6**, 632 (1988).

Because high-purity DNA is essential for transfection, we use commercial DNA purification kits for all our work. Most of our work is done with transiently transfected cells that we plate onto round coverslips that can be transferred easily into imaging chambers. On the day before we transfect we plate approximately 10^5–10^6 cells per 60-mm dish. This number is fairly empirical, as the desired end point is that on the day following seeding we wish to obtain cells that are between 50 and 70% confluent. Transfection protocols closely follow those provided by the manufacturer. We store most of our DNA stock solutions at around 1–2 $\mu g/\mu l$ in Tris–EDTA (TE) solution at $-20°$. Five micrograms of this DNA is added to serum- and antibiotic-free medium to a total volume of 150 μl. Thirty microliters of Superfect is then added, mixed, and allowed to complex with the DNA for 10 min at room temperature. Cells are washed with phosphate-buffered saline. One milliliter of normal cell growth medium is added to the 150 μl of the DNA–Superfect complex and this mixture is then added to the 60-mm plate. Cells are normally incubated for 3 hr in a CO_2 incubator after which they are washed and available for use after 24 hr for up to 3 days following transfection.

Virus-Mediated Infection of Cells

This is an alternative to transfection and provides several advantages, as it is efficient and can be used in a wide range of hosts. Viral constructs can accommodate large inserts, and by exploiting the natural tropism of some viruses, such as herpes simplex virus type 1 (HSV-1), for neural tissue, some of the difficulties encountered when trying to transfect neural cells can be overcome. Alonso et al.[15] have demonstrated successfully the use of HSV-mediated transfer of ER targeted recombinant aequorin to a number of cells, including primary cells. Alternatives such as adenovirus-mediated and baculovirus-mediated transfer are also available to infect mammalian cells and can carry either targeted cameleons or aequorins.

VP22-Mediated Protein Transfer

The discovery that VP22, a (HSV-1) structural protein, displays the remarkable property of *inter*cellular protein transport heralds yet another approach to targeting and delivering fusion proteins to cells.[16] It was found that VP22 trafficking is via a nonclassical Golgi-independent mechanism

[15] M. T. Alonso, M. J. Barrero, E. Carnicero, M. Montero, J. Garcia-Sancho, and A. Alvarez, *Cell Calcium* **24,** 87 (1998).
[16] A. Phelan, G. Elliott, and P. O. O'Hare, *Nature Biotech.* **16,** 440 (1998).

and that VP22 fusion proteins, such as p53-VP22, would also translocate between cells while maintaining fusion partner function. This unique ability could be exploited readily to use cameleons (and recombinant aequorin) on cells that are difficult to transfect or culture, such as adult cardiac muscle cells or smooth muscle cells that change their phenotype with culture. All that would be required is the transient transfection of a cameleon–VP22 fusion construct into a cell line, such as HEK293 or COS, followed by a period of a day or two to allow for sufficient expression. These cells could then be lysed and the lysate poured directly onto the cells under study. It has been shown that within 10 min of such treatment that VP22 fusion proteins have translocated to the cell nucleus. Such strategies open new avenues into the study of *freshly* dissociated cells with a combined confocal and molecular genetic technique. Construction of VP22 fusion proteins has been simplified greatly by the introduction of the Voyager VP22 vectors by Invitrogen. They come in two forms, VP22/myc-His, which possesses a number of restriction sites to enable in-frame cloning of fusion partners, and a second form, VP22/myc-His-TOPO, which possesses 3′ T overhangs and is thus suited ideally for one-step cloning of PCR products to create fusion proteins. A *myc* epitope and polyhistidine tag are incorporated and allow for the detection of recombinant protein and purification.

Targeting of Ca^{2+} Sensors to Intracellular Organelles

Most commercially available Ca^{2+} sensors possess some ability to load into cellular organelles such as nuclei and mitochondria. In most experimental studies, such organelle loading is often undesirable and adds a margin of error to cytosolic calcium measurements. However, these undesirable properties can be exploited and even enhanced to enable the measurement of calcium within cellular organelles.

Selective Loading of Organelles with Chemically Synthesized Sensors

While large organelles such as nuclei can be imaged using regions of interest (see Imaging Modalities), this is not the case for smaller organelles such as mitochondria. Modifications of the AM-ester loading procedure can be used to selectively localize the sensor into these organelles of interest. These techniques require the sensor to be modified prior to use or, alternatively, to remove a loaded sensor that enters compartments other than those of interest.

Loading of Organelles of Interest

The "cold loading" technique (described earlier) has been used to measure mitochondrial calcium using the sensor rhod-2/AM[12,17] and the following methodology is used.

The calcium fluorophore rhod-2/AM obtained from Molecular Probes is prepared as a stock solution in DMSO and refrigerated. Cells either in suspension or cultured on glass coverslips (No. 1.5) are loaded with 10 μM rhod-2/AM for at least 1 hr at temperatures below 5° (ice–water slurry in an insulated box, kept in darkness) in a HEPES-buffered medium (containing in mM: 118 NaCl, 4.8 KCl, 1.2 MgSO$_4$, 1.2 KH$_2$PO$_4$, 25.0 Na-HEPES, 11.0 glucose, and 1.0 CaCl$_2$). Following the cold loading period, the temperature is slowly (over 10 min), brought up to 37°, and cells are further incubated for 30 min to allow adequate cleavage of the AM groups by mitochondrial esterases. Depending on the preparation, cells can be incubated in a HEPES-buffered nutrient medium containing serum for 3–5 hr. The extra 3- to 5-hr incubation allows for the cytosolic fluorophore to be removed by exocytosis, further enhancing mitochondrial loading. Cells are then washed twice with medium to remove the extracellular fluorophore before imaging. Adequate excitation of rhod-2 can be obtained with the 488- or 514-nm lines of an argon ion laser, and fluorescence is collected through a 580/32-nm bandpass filter.

Care must be taken in the interpretation of mitochondrial calcium images obtained through the cold loading of rhod-2. The technique often results in residual cytosolic loading of the sensor, and therefore changes in mitochondrial fluorescence must always be compared to changes occurring in the cell cytosol. The easiest way to correct for such inaccuracies is to subtract the cytosolic (or nuclear) background fluorescence from that of the mitochondrial signal in each individual image. This is especially difficult in cells with high mitochondrial density (mitochondria are ~30% of cardiac myocyte cell volume), but can be counteracted by using 1-μm confocal optical sections, which would optically slice through, and limit data collection to, mitochondria with minimal contamination from cytosolic fluorescence.[12]

Removal of Contaminating Cytosolic Fluorescence

Some calcium sensors often preferentially load into the mitochondrial matrix. One such probe is indo-1/AM, 75% of which loads into mitochon-

---

[17] D. N. Bowser, T. Minamikawa, P. Nagley, and D. A. Williams, *Cell Calcium*, submitted (1999).

dria.[18] Indo-1/AM can be imaged using expensive UV laser scanning confocal microscopes; however, it is more common to use an epifluorescence system, which allows collection of the emitted fluorescence at two different wavelengths. The remaining cytosolic fluorescence can be removed through selective quenching of the fluorescence[19] or dialyzing the cell through a patch pipette if simultaneous electrophysiological recordings are to be made.

Indo-1/AM is prepared as a stock in DMSO and refrigerated. Cells suspended in HEPES-buffered media are exposed to 10 μM indo-1/AM for 30 min at room temperature (23°), with pluronic F-127 added to the medium to aid cellular loading. Cells are then washed twice with fresh buffer and kept at room temperature throughout the experimental protocol. Cells are then bathed/dialyzed with buffer containing 100–250 μM MnCl$_2$ for up to 1 hr to quench the cytosolic fluorescence. MnCl$_2$ is used because the dissociation constant (K_d) of indo-1 for Ca^{2+} is ~30 times higher than that for Mn^{2+}. The time and concentration depend on the cell type studied and therefore close observation of the cytosolic and mitochondrial fluorescence is required so that the Mn^{2+} quenching of mitochondrial fluorescence is kept minimal. Indo-1 has a maximum excitation of 335 nm but can be excited adequately with the 351-nm line of argon UV lasers. The emission maximum of indo-1 shifts from ~475 to ~405 nm when saturated with Ca^{2+}, allowing the fluorophore to be calibrated ratiometrically. These emission wavelengths can be split through a dichroic mirror containing a 440-nm long-pass filter, therefore allowing the 405- and 475-nm wavelengths to be detected by different photodetectors. Additionally, bandpass filters may be placed in the light path before the photodetectors to further select the wavelengths of interest. Postcollection analysis of the simultaneously acquired images can be performed by most confocal systems' software, which involves background subtraction and ratio image calculation. Additionally, newer systems can automatically calculate the Ca^{2+} concentration from previously determined R_{min} (minimal ratio in calcium free solution), R_{max} (maximal ratio in high calcium solution), and apparent K_d.[20]

Enhancing Mitochondrial Loading of Rhod-2

The sensor rhod-2/AM (discussed earlier) is the only esterified Ca^{2+} sensor with a net positive charge. This positive charge allows rhod-2 to

[18] Z. Zhou, M. A. Matlib, and D. M. Bers, *J. Physiol.* **507**, 379 (1998).

[19] H. Miyata, H. S. Silverman, S. J. Sollott, E. G. Lakatta, M. D. Stern, and R. Hansford, *Am. J. Physiol.* **261**, H1123 (1991).

[20] G. Grynkiewicz, M. Poenie, and R. Y. Tsien, *J. Biol. Chem.* **260**, 3440 (1985).

be sequestered into negatively charged organelles such as mitochondria prior to cleavage of the AM group. Further mitochondrial loading can be accomplished using the reduced form, dihydrorhod-2 (DH-rhod-2), which is oxidized to active rhod-2 within the mitochondrial matrix. A product information sheet on the preparation of DH-rhod-2 is available from Molecular Probes; however, a slightly adapted version is described briefly.

DH-rhod-2 is prepared by first dissolving a 50-μg vial of rhod-2/AM in 100 μl of DMSO. Add a small amount (2–3 μl) of NaBH$_4$ as a methanol solution (1 g NaBH$_4$ to 10 ml methanol; prepare in a fume hood due to the production of noxious gas) and shake vigorously. The rhod-2 solution (purple color) should become colorless within seconds. The DH-rhod-2 stock can be kept refrigerated for 2–3 weeks before oxidizing in air. Cell suspensions (HEPES-buffered media) are incubated in ~10 μM DH-rhod-2 at 37° for 60 min in a shaking water bath. Following incubation, cells are washed, resuspended in fresh buffer, and incubated for a further 10 min prior to imaging.

Cytosolic and nuclear fluorescence is minimal if nonexistent in DH-rhod-2-loaded cells due to low oxidant activity in the cytosol. Therefore, only DH-rhod-2 oxidized within mitochondria will fluoresce on calcium binding. An image of a DH-rhod-2-loaded myocyte is shown in Fig. 2 (Molecular Probes) exhibiting a typical mitochondrial pattern.

We have used DH-rhod-2 to examine the loss of mitochondrial calcium following the uncoupling of the mitochondrial respiratory chain. Mitochondrial uncoupling with poisons such as sodium cyanide results in loss of mitochondrial membrane potential and subsequent opening of a nonspecific pore in the mitochondrial membrane (mitochondrial permeability transi-

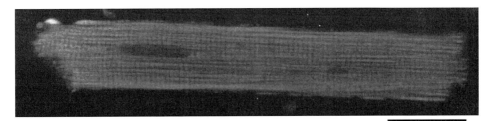

20 μm

FIG. 2. Cardiac myocyte loaded with the mitochondrially localized Ca²⁺ sensor DH-rhod-2. A typical mitochondrial (banded) distribution is observed. Note the absence of fluorescence in the in-focus cell nucleus (left) emphasizing the regional loading specificity of DH-rhod-2.

tion). We were able to visualize the loss of mitochondrial calcium using this technique.[21]

Measuring Mitochondrial Ca^{2+} with Multiple Sensors

We have already mentioned that most calcium sensors load both cytosolic and mitochondrial compartments, and in some cases loading protocols can be used to minimize cytosolic fluorescence. When this is not possible, some investigators have used single wavelength-emitting calcium sensors with mitochondrial-specific sensors and estimated mitochondrial calcium in coloaded regions. This is theoretically possible through the use of thin confocal optical sections that predominantly contain mitochondria, with minimal cytosolic contamination.

Potential problems such as spectral overlap and resonance energy transfer can be minimized through the careful selection of sensor combinations. The use of Fura Red as a calcium sensor whose fluorescence intensity falls on calcium binding and MitoTracker green as the mitochondrial marker whose fluorescence is independent of changes in mitochondrial membrane potential have shown promise in preliminary investigations. Both probes can be excited with the 488-nm line of an argon ion laser with the emission of each sensor resolved readily. The methodology is described next.

Fura Red/AM is prepared as a stock (1 mM) in DMSO and refrigerated. Typically, cardiac myocytes are loaded with relatively high concentrations of Fura Red (25 μM) at 37° for between 30 and 40 min. Suspensions are then washed with HEPES-buffered media prior to MitoTracker green loading. MitoTracker green is also available from Molecular Probes and is prepared as a stock (10 μM) in DMSO and kept refrigerated. Unlike Fura Red, only small amounts of MitoTracker green are required for adequate loading (100 nM), and cells are once again incubated at 37°, but only for 15 min. The suspensions are washed twice with buffer before imaging. As mentioned earlier, Fura Red/MitoTracker green-loaded myocytes can be excited with the 488-nm line of an argon ion laser and the emission split with a 640-nm long/short-pass dichroic mirror. Bandpass filters are then selected to minimize cross-detection of either probe: 522/32 nm for MitoTracker green and open filters for Fura Red (due to the 640-nm long-pass filter). This technique makes use of thin confocal optical sections, which ensure that predominant mitochondrial fluorescence is being collected.

Similar coloading protocols can be used to simultaneously image calcium in both cytosolic and mitochondrial compartments. Once again, there are problems with bleed-through fluorescence that can be minimized through

[21] D. N. Bowser, T. Minamikawa, P. Nagley, and D. A. Williams, *Biophys. J.* **75,** 2004 (1998).

the selection of appropriate fluorophores with distinct emission spectra, and electronic image subtraction performed by the confocal system. We have used two combinations of fluorophores: fluo-3 and DH-rhod-2 or Fura Red and DH-rhod-2. The order of loading can be very important when using two probes. We first load myocyte suspensions with DH-rhod-2 (as described earlier) and then add the cytosolic calcium sensor. Exciting both sensors at 488 nm and collecting the emission of each sensor separately will allow the imaging of myocytes coloaded with fluo-3 and DH-rhod-2. The emission can be split with a 560-nm long-pass dichroic, and bandpass filters are used to select appropriate wavelengths (522/32- and 580/32-nm filters) before reaching the photodetectors. Unfortunately, the greater increase in the fluorescence of fluo-3 on binding Ca^{2+} can bleed through to the photodetector used with DH-rhod-2, causing false reporting of mitochondrial calcium. Such bleed through can be minimized through partial electronic subtraction of the fluo-3 fluorescence image from the DH-rhod-2 image. This methodology has allowed us to examine spontaneous cytosolic calcium transients in cardiac myocytes that have lost mitochondrial calcium most probably due to induction of the mitochondrial permeability transition (Fig. 3).

Targeting of Genetically Encoded Sensors to Organelles

Protein-based sensors can be directed to their intended cellular location by "targeting sequences" that are encoded within the sequence of a protein. Studies on the targeting of single-chain antibodies to specific cellular locations have resulted in strategies that utilize well-characterized targeting sequences. For nuclear localization the targeting sequence can be placed within a protein or at the N or C terminus and directs protein translocation to the nucleus of the cell.[22] Similarly, for mitochondrial targeting, a cleavable leader sequence directs the translocation of the protein to the mitochondria.[23] In the case of endoplasmic reticulum (ER), the actions of a signal peptide and ER retention sequence act together to direct and then retain a protein within the ER.[5] Proteins lacking a targeting sequence are retained within the cytoplasm and are thus "targeted" to the cytoplasm.

In many cases, interesting changes in Ca^{2+} signals occur in close proximity to the plasma membrane and thus sensors targeted to the membrane would be highly desirable. There are no reports of a recombinant aequorin or cameleon that is targeted to the membrane. In fact, there are very few reports of any fluorescent measurement of near-membrane Ca^{2+}. A notable exception is the use of FFP18 to make measurements of Ca^{2+} near the

[22] L. Fisher-Fantuzzi and C. Vesco, *Mol. Cell. Biol.* **8,** 5495 (1988).
[23] R. Rizzuto, A. W. Simpson, M. Brini, and T. Pozzan, *Nature* **358,** 325 (1992).

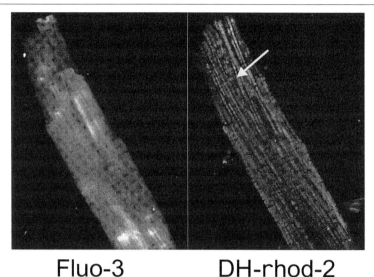

Fluo-3   DH-rhod-2

FIG. 3. Cardiac myocyte loaded with DH-rhod-2 (mitochondrial Ca^{2+}) and fluo-3 (cytosolic Ca^{2+}). Fluorescence was excited simultaneously from both sensors at 488 nm. Appropriate filter sets were used to separate the emission light of each sensor (see text). The higher intensity fluorescence regions in the fluo-3 image (left) represent areas of localized cytosolic Ca^{2+} fluctuations. The arrow in the DH-rhod-2 image (right) indicates single or small groups of mitochondria that have lost matrix Ca^{2+}, causing "black holes" in the otherwise regular-banded fluorescence pattern.

plasma membrane.[24] Two basic strategies can be used to achieve membrane targeting for a genetically encoded sensor. The first would use membrane retention signals, such as acylation consensus sequences added to the cameleon construct to target the cameleon to the lipid membrane. The second strategy would involve the construction of cameleon and membrane-bound protein fusions that would not only carry the cameleon to the membrane, but by judicious choice of fusion partner would carry the cameleon to the membrane of a particular organelle if so desired. The added advantage of this procedure would also be the ability to measure Ca^{2+} levels near a receptor or ion channel during a physiological stimulus. We are presently developing such strategies in our laboratories in an effort to examine the role of near-membrane Ca^{2+} in physiological signaling.

For construction of most fusion proteins we rely heavily on PCR, although in some cases, targetable cameleons can be made easily by direct subcloning of the cameleon construct into pShooter plasmid vectors (In-

[24] E. F. Etter, A. Minta, M. Poenie, and F. S. Fay, *Proc. Natl. Acad. Sci. U.S.A.* **93,** 5368 (1996).

vitrogen). These vectors already possess the necessary targeting sequences for the endoplasmic reticulum, nucleus, and mitochondria. They also have multiple cloning sites, which should enable the in-frame subcloning of the cameleon into the plasmid. If it is not possible to keep the targeting sequence and cameleon in frame using shared restriction sites, then it may be necessary to use PCR to make a cameleon with appropriate restriction sites at either end to enable in-frame cloning.

Calibration of Ca^{2+} Sensors

This process is of primary importance in the establishment of accurate quantitative data for any confocal imaging study employing fluorescent Ca^{2+} sensors. Calibration of chemically synthesized and genetically encoded sensors can be performed in similar ways and hence the protocols described are relevant to both classes of molecules. The calibration process requires the exposure of internalized sensor to conditions of known Ca^{2+} concentration. This process requires the mastery of techniques to prepare solutions of known or measurable Ca^{2+}, in addition to techniques that equilibrate cells or tissues with these defined solutions.

The working range of the sensor to be calibrated dictates the predominant buffer required for constructing a range of calibration solutions. As the majority of chemically synthesized sensors are derivatives of EGTA or BAPTA, these parent compounds are inherently suitable for calibration of the derived sensors. A brief description of the strategies and constraints for solution preparation is included but these have been described in detail in numerous previous publications.[25–28]

A series of solutions with different ionized Ca^{2+} concentrations can be achieved from the mixing of two stock solutions (denoted A and B) in varied proportions, maintaining the same total volume. Both solutions contain (mM): Mg^{2+} 1.0, K$^+$ 100 to 140.0; Na$^+$ 10 to 30.0; ATP 5.0 to 8.0; HEPES (pH buffer) 10.0; and creatine phosphate 10.0. Additionally, solution A contains 10 mM EGTA whereas solution B contains 10 mM nominally equimolar Ca-EGTA (with a slight inequality in favor of excess EGTA over Ca^{2+} preferred). The absolute concentration of EGTA employed in these solutions is not critical and can be varied to set the ionic strength of solutions at a defined level. However, the absolute

[25] D. M. Bers, C. W. Patton, and R. Nuccitelli, in "Methods in Cell Biology" (R. Nuccitelli, ed.), Vol. 40, p. 3. Academic Press, San Diego, 1994.
[26] D. A. Williams and F. S. Fay, *Cell Calcium* **11,** 75 (1990).
[27] A. Fabiato, *Methods Enzymol.* **157,** 378 (1998).
[28] J. A. S. McGuigan, D. Luthi, and A. Buri, *Can. J. Physiol. Pharmacol.* **69,** 1733 (1991).

level of EGTA should be high enough that the sensor to be calibrated represents an insignificant Ca^{2+}-buffering contribution in comparison (EGTA : sensor $> 100 : 1$). The exact ionic and composition should be varied to closely match the conditions expected to exist within the cell type of interest. The free Ca^{2+} concentration can be estimated from the relative proportion of Ca-EGTA to EGTA in each of the solution mixtures with an apparent binding constant K'_{Ca} and assuming the law of mass action:

$$[Ca^{2+}] = (1/K'_{Ca}) ([Ca\text{-}EGTA]/[EGTA]_{free})$$

However, such estimations contain many sources of error due to uncertainty in the actual total concentration of EGTA in a solution due to impurities and an unknown and variable degree of hydration of EGTA[28] and the existence of numerous values for the apparent affinity constant (K'_{Ca}) for EGTA. Direct experimental potentiometric measurement of the actual proportion of Ca-EGTA and EGTA in each solution can be performed to remove this source of error. This methodology, described in detail else-where,[29,30] takes advantage of the pH change that occurs during the binding of Ca^{2+} to EGTA over a wide pH range. Calibration kits available from Molecular Probes offer a way of standardizing the reporting of $[Ca^{2+}]$ in confocal studies. These kits contain a range of premade solutions that provide consistency for calibrations made on different days or with different cell types.

The ionophore for divalent cations A-23187 is used commonly for *in situ* calibrations of fluorescent Ca^{2+} sensors, to equilibrate intracellular and extracellular Ca^{2+} concentrations, and can also permit the addition of Mn^{2+} to enter the cell to quench intracellular sensor fluorescence. Although the intrinsic fluorescence of A-23187 is high, it is suitable for use with the visible light-excitable sensors, including Calcium Green, Calcium Orange, Oregon Green 488 BAPTA, fluo-3, rhod-2, and Fura Red. A brominated derivative of A-23187 (4-bromo A-23187) is essentially nonfluorescent and is the best ionophore for use with UV-excitable sensors such as indo-1. Ionomycin is also used commonly in calibration procedures, although it can be problematic because of an often overlooked pH dependence for ion transport.[31]

Equilibration of intracellular and extracellular compartments with Ca^{2+} requires initial determination of an effective concentration of ionophore. It is not suitable to simply adopt an ionophore concentration from previous literature or from previous experiments with other cell or tissue types,

[29] D. J. Miller and G. L. Smith, *Am. J. Physiol.* **246,** C160 (1984).
[30] D. G. Moisescu and R. Thieleczek, *J. Physiol.* **317,** 241 (1978).
[31] C. M. Liu and T. E. Hermann, *J. Biol. Chem.* **253,** 5892 (1978).

although such information can provide a useful starting point. A dose–response curve for ionophore should be determined by exposing the cell suspension or tissue containing the internalized sensor to a cocktail comprising a starting dose of ionophore and a calibration solution with a buffered $[Ca^{2+}]$ that is significantly different from the expected intracellular $[Ca^{2+}]$. Suitable fluorescence data are collected (dependent on the mode of sensor signaling; see next section) as the ionophore concentration is progressively increased. At some point the measured fluorescence should change markedly, indicating an effective exposure of the internal sensor to the Ca^{2+} level of the bathing solution. The ionophore concentration should be further increased to establish the saturation of this fluorescence change, indicating equilibration of intracellular with extracellular $[Ca^{2+}]$. We have found the effective dose to fall in the concentration range of 10 to 25 μM for all cell/tissue types we have used. Cell aliquots can then be subjected to a range of Ca^{2+} solutions at the effective ionophore dose. Such a procedure will allow determination of the necessary calibration constants for sensors and also effectively provides data to define the *in situ* Ca^{2+} dissociation constant (K_d) for a sensor (see Table I).

Specific calibration examples are described for some sensors that can be employed with confocal microscopy.

Single-Wavelength Methods ("Single–single")

Calibration of sensors that respond to Ca^{2+} with enhanced fluorescence emission without a shift in the spectrum of emitted light requires the determination of fluorescence limits: the maximum (F_{max}) and minimum (F_{min}) fluorescence intensity levels for the sensor within individual cells. At its simplest this procedure involves exposing sensor-containing cells to the ionophore A-23187 in the presence of Ca^{2+}-free and Ca^{2+}-saturating (see Table I) solutions. Use of a full range of Ca^{2+} solutions is not necessary where the *in situ* K_d has been determined previously for this cell and sensor combination. Alternatively, sensor-containing cells can be loaded with a second sensor to provide a $[Ca^{2+}]$ frame of reference for calibration for the sensor of interest. Several different combinations have proven useful. We have previously described a technique that employs coloading of a sensor such as fluo-3 with fura-2. The direct Ca^{2+} calibration of fura-2 ratio images for individual cells provided a defined baseline $[Ca^{2+}]$ on which changes in fluo-3 fluorescence intensity could be expressed as $[Ca^{2+}]$ changes. This calibration paradigm for coloaded cells requires prior knowledge of the fluorescence intensity enhancement of fluo-3 following Ca^{2+} binding (i.e., F_{max}/F_{min}) and the dissociation constant of Ca^{2+}–fluo-3 (see Table I).

In several cell types, the simultaneous labeling with Fura Red and fluo-3 has enabled researchers to use pseudo-ratiometric measurements for estimating intracellular Ca^{2+} levels using confocal laser scanning microscopy.[32] Both sensors can be coexcited efficiently with the 488-nm line of an argon ion laser, with the emission light from each sensor, an increased fluorescence of fluo-3, and a decrease in Fura Red emission with an intracellular elevation of Ca^{2+} resolved readily with separate detectors. While useful, the combination has a number of assumptions that govern its ability to provide accurate determinations of $[Ca^{2+}]$. These are largely the result of having two sensor pools providing components of a single reporter signal. It is essential that the proportion of the two sensors is the same in each intracellular compartment and that any change in the apparent concentrations of each sensor (i.e., photobleaching, leakage) is also equivalent. These uncertainties have restricted the widespread application of this combination in confocal Ca^{2+}-imaging studies. Some of these problems can be reduced through the conjugation of a Ca^{2+} sensor (e.g., Calcium Green or Oregon Green 488-BAPTA) with bioinert dextran polysaccharides, representing a single complex of sensor and fluorescence reference. However, these complexes are large, requiring microinjection or similar loading procedures, and the degree of conjugation is variable, enforcing the need for a detailed signal calibration of new production batches.

Dual-Wavelength Methods ("Single–dual")

Fluorescent sensors that show a shift in excitation or emission spectra on ion binding can be calibrated using a ratio of the fluorescence intensities measured at two different excitation or emission wavelengths. This derives from the ability to describe the spectrum at each Ca^{2+} concentration uniquely by the ratio of any two values on the curve. This greatly simplifies, and increases the accuracy of, the collection of experimental data. Ratiometric data result in the cancellation of artifactual variations in the fluorescence signal that might otherwise be misinterpreted as changes in ion concentration. These artifactual influences include photobleaching and leakage of the sensor, variable cell thickness and hence excitation and emission path lengths, nonuniform sensor distribution within cells (due to compartmentalization) or among populations of cells (due to loading efficacy variations), and fluctuations of detection efficiency during an experiment. Calibration procedures are identical to those employed for single wavelength indicators, although the calibration constants that are measured are limiting ratios (R_{max}, ratio of intensities collected at the two measurement wavelengths

[32] R. A. Floto, M. P. Mahaut-Smith, B. Somasunduran, and J. M. Allen, *Cell Calcium* **18,** 377 (1995).

for saturating Ca^{2+}; R_{min}, ratio of intensities for Ca^{2+}-free conditions) rather than simple intensities.

The choice of true ratiometric Ca^{2+} sensors for confocal microscopy is limited, mainly because of the limited excitation wavelengths that can be provided by present laser light sources. Fura Red is an analog of fura-2, i.e., visible light excitable, and hence offers possibilities for the ratiometric measurement of Ca^{2+} in single cells by confocal microscopy that have yet to be realized. We have explored this possibility of dual excitation by sequentially using the 457- and 488-nm lines of a 100-mW argon ion laser to excite this sensor in cardiac cells, collecting emitted light in the red spectrum (>600 nm)—a "dual–single" confocal imaging mode.[33] While both wavelengths are on the same side of the isosbestic point on the spectrum of this sensor, and emission decreases with the binding of Ca^{2+} in both cases, the ratio of the two emission intensities still provides a valid ratiometric indication of $[Ca^{2+}]$.[20] The long-wavelength emission maximum (~660 nm) of this sensor is also well removed from the autofluorescence wavelengths of most cells. One drawback of this sensor is the weaker fluorescence evident even in the absence of Ca^{2+}, necessitating the use of higher concentrations of the sensor in cells to produce equivalent fluorescence to sensors such as fluo-3. The ability to provide dual excitation wavelengths for Fura Red with conventional higher power argon ion lasers should also provide for an increased future use of this sensor for a large variety of physiological questions with low to medium time resolution requirements. The same confocal imaging strategy should also be possible for the coumarin benzothiazole-based Ca^{2+} sensor BTC,[3] although to date we have not tested this possibility directly.

In addition, we suggest that the fluo-3/Fura Red coloading procedure be modified to make use of this feature of Fura Red. All researchers that consider the use, or presently use, this combination of sensors should be encouraged to intersperse the collection of single excitation wavelength (488 nm), pseudo-ratiometric emission data with a number of measurements at 457 nm excitation. The strategy for collection of initial ratiometric and subsequent single wavelength measurements interspersed with regular fluorescence ratio data has been described elegantly for fura-2 as a means of enhancing the time resolution of nonconfocal imaging protocols.[34] The combination of this type of approach for Fura Red, but with the additional advantage of simultaneous excitation of a second Ca^{2+} sensor (fluo-3),

[33] S. H. Cody, A. M. Reilly, D. N. Bowser, G. S. Lynch, and D. A. Williams, *in* 11th International Conference on 3D Image Processing in Microscopy. 10th International Conference on Confocal Microscopy, 1998.
[34] L. Leybaert, J. Sneyd, and M. Sanderson, *Biophys. J.* **75,** 2025 (1998).

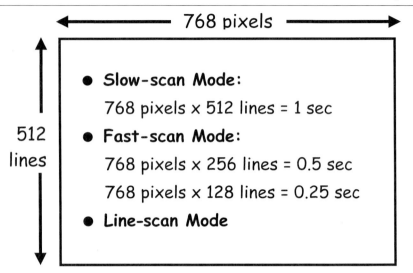

FIG. 4. Resolutions of the most common image acquisition modes for a Bio-Rad (600, 1000) confocal microscope.

offers to significantly increase the quantitative certainty of $[Ca^{2+}]$ determinations with the existing fluo-3/Fura Red, dual sensor approach.

Imaging Modalities

The majority of confocal microscopes are point-scanning systems that irradiate a biological sample with a focused point of light scanned in a raster pattern across the specimen to form an image. This creates limitations in terms of image acquisition speed when compared to video- or CCD-based imaging systems. Tradeoffs are inevitably made between spatial resolution and time resolution to reach the most acceptable compromise for both requirements. We predominantly use Bio-Rad confocal systems and accumulate images in a variety of laser scanning modalities (see Fig. 4). As most of our present and previous confocal Ca^{2+} studies have been on contracting muscle cells,[21,34–36] we commonly acquire "fast-scan" images (768 pixel $\times$ 128 lines) at 250-msec intervals or scan regions of interest (ROI) that occupy less than the full image spatial resolution of a Bio-Red image (768 pixel $\times$ 512 lines). A useful ROI is a defined single raster line that can be orientated along one axis of an individual polarized cell (see Fig.

[35] D. A. Williams, L. M. Delbridge, S. H. Cody, P. J. Harris, and T. O. Morgan, *Am. J. Physiol.* **262,** C731 (1992).

[36] D. A. Williams, *Cell Calcium* **14,** 724 (1993).

5A). Multiple consecutive scans of this line can be displayed to represent an image, commonly called a "line-scan" image. Scanning of a single raster line takes 1.5 to 2 msec (approximately 2 μsec per pixel), and both the time between acquisition of each line and the total number of acquired lines scans can be defined. In addition, collection of a series of line-scan images at predefined time points can be preprogrammed into an image collection protocol. We commonly collect 512 lines (so as to fill out the display of each "line-scan" image), but vary the timing so that this collection can take from 3 to 12 sec, depending on the phenomenon that is imaged. Orientation of this single line array along the long axis of a cardiac cell produces an image representing cell length and fluorescence intensity (or [Ca²⁺]) at a relatively high time resolution. Line-scan and fast-scan image acquisition modes have been employed successfully to monitor the changes in cell length and fluorescence that occurred during stimulated or spontaneous contractions of isolated cardiac cells.[11,21,35,37,38]

Using Confocal Regions of Interest

The combination of selected ROIs within an optical section and constrained optical sections define three-dimensional (x, y, z) volumes of interest within the biological preparation. The absolute fluorescence intensity emanating from these regions can be measured at defined time points, with the time resolution dependent on the size of (number of pixels in) the region and the scanning characteristics of the particular confocal imaging system used.

Myocytes loaded with fluo-3/AM generally exhibit diffuse fluorescence, with the nucleus the major discernible intracellular organelle (Fig. 6A). We commonly define an ROI that encompasses one of the two nuclei of a cardiac cell and record and present graphically the change in fluorescence intensity within the selected region as a function of time. While such a methodology is adequate for imaging the slower nuclear calcium transients in nonexcitable cells, calcium transients in cardiac myocytes are faster and often require higher time resolution image acquisition, such as line scanning. A typical line scan image is shown in Fig. 6B and is derived from a line array that bisects the nucleus of a cardiac myocyte that has been stimulated electrically. Vertical intensity profiles in these images represent time-dependent changes in fluorescence ([Ca²⁺]) intensity at defined cell locations. Intensity changes of two such profiles are shown in Fig. 6C, one for a section of cytosol and one in the nucleus. This form of ROI intensity collection and display allows investigation of the properties of nuclear and

[37] D. A. Williams and S. H. Cody, *Micron* **24**, 567 (1993).
[38] D. A. Williams, *Cell Calcium* **11**, 589 (1990).

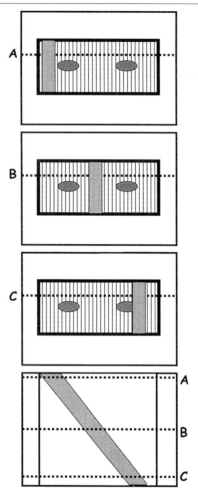

FIG. 5. Diagrammatic representations of a line-scan image. (A) A single raster line of interest (dotted line) is defined within the standard $(X-Y)$ image. For a cardiac cell (depicted), we commonly align this along the long axis of the cell. In cells containing a Ca^{2+} sensor (such as fluo-3) initiated spontaneously, localized elevations of cytosolic Ca^{2+} (Ca^{2+} waves) propagating along the cell (left to right) produce bands (gray) of intensity in standard images (A, B, C) that translate to a diagonal band in the line-scan image (lower panel). These line-scan images thereby produce a cell length (horizontal distance) and Ca^{2+} plot as a function of time (vertical direction). Propagation rates for Ca^{2+} waves (slope of band) and Ca^{2+} levels (intensity at any cell location) are determined readily from these images.

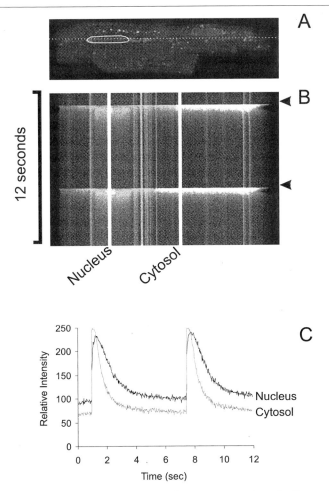

FIG. 6. Cardiac myocyte loaded with the Ca^{2+} sensor fluo-3 that distributes throughout the cytosol and nuclear compartments. (A) X–Y slow-scan confocal image of the myocyte with the nucleus defined to represent a region of interest. A raster line (dotted) that passes through the nucleus is defined and scanned repeatedly to form the line-scan image (512 lines, 12 sec total) that is represented in (B). During acquisition of the line-scan image the cell was stimulated electrically, resulting in two regions of horizontal brightness representing the rapid elevation of Ca^{2+} throughout the cell length at these times. The vertical lines in (B) represent the time transects (one each in the cytosol and nucleus), and the profiles of intensity changes occurring at these cell locations are represented graphically in (C). Note the significantly different Ca^{2+} transients that occur in these locations following electrical stimulation.

cytosolic calcium transients in cardiac myocytes, effectively highlighting differences in the time course of these events.

Image Acquisition Requirements for New Genetically Encoded Ca^{2+} Sensors

The majority of confocal microscopes are already equipped to perform single-excitation, dual-emission wavelength detection ("single–dual") and are thus ideally suited for use with some of the cameleons. The choice of GFP pairs for a cameleon will be governed to a large extent by the properties of the existing laser source and fortunately there is some flexibility in the choice of GFP pairs. The original cameleons described[5] were composed of either enhanced blue fluorescent protein (EBFP), enhanced green fluorescent protein (EGFP) pairs or enhanced cyan fluorescent protein (ECFP), enhanced yellow fluorescent protein pairs (EYFP). The EBFP–EGFP pair requires excitation light at 370 nm (adequately provided by the 363-nm argon ion line) and monitoring of emission light at both 440 and 510 nm, whereas ECFP–EYFP pairs require excitation light at 440 nm (457-nm argon ion laser line would suffice) and monitoring of emission light at 480 and 535 nm. If these combinations of GFP prove unsuitable for the existing laser excitation sources, they can be substituted with different GFP pairs, provided that spectral properties allow for RET to occur. However, other factors, such as the expression efficiency of the various GFP mutants, need to be considered when making choices of RET pairs. We have used Stratagene's seamless cloning kit to create deletions and substitutions of large fragments with great success. The type IIS restriction enzyme *Eam*1104 I does not demand shared or introduced restriction sites and thus gives complete freedom of choice in the selection of potential splice sites.

Multiphoton Confocal Microscopy: The Future

Multiphoton confocal microscopy is a relatively new, laser-based microscopy technique that utilizes longer wavelength (infrared) light to bring about the excitation of commonly used fluorophores such as Ca^{2+} sensors that normally need higher energy (lower wavelengths) for activation. Multiphoton fluorophore excitation occurs as the energy of multiple coincident photons of lower energy summate to provide the necessary excitation energy. Techniques for dual-[39] and three-photon excitation have been described and there is already some indication of the potential success of a six-photon variation of this principal.

---

[39] W. Denk, J. H. Strickler, and W. W. Webb, *Science* **248,** 73 (1990).

Multiphoton excitation is provided by illuminating tissue samples with short, intense bursts of light from mode-locked Ti:sapphire (e.g., Spectra-Physics Tsunami laser) or Nd–YLF lasers. The probability of multiphoton excitation is proportional to the square of the photon density[39,40] and hence effective excitation falls off rapidly above and below a defined point of focus. Out-of-focus fluorescence is simply not produced and hence confocal pinholes are not required to produce optical sectioning performance. This also allows all of the light that is collected by the imaging lens or detector directly to contribute to an image. Infrared light used for multiphoton excitation is scattered much less than visible wavelengths used to excite most commonly used Ca^{2+} sensors for confocal microscopy, and hence the focused beam will penetrate deeper into thick biological specimens. As multiphoton excitation localizes light energy in the focal plane, there is less photodamage than in conventional confocal microscopy, which bombards specimens with high-intensity, shorter wavelength light. All of these features offer significant benefit for confocal Ca^{2+} imaging. Apart from preliminary studies,[41] this potential has been largely untapped due primarily to the high costs of lasers required to provide the power necessary for this technique. However, it is not too far-fetched to predict that the predominance of confocal Ca^{2+} imaging studies will soon employ further evolutionary stages of the current multiphoton developments.

[40] V. E. Centonze and J. G. White, *Biophys. J.* **75,** 2015 (1998).
[41] D. W. Piston and W. W. Webb, *Biophys. J.* **59,** 156 (1991).

[26] Using Confocal Microscopy and the Fluorescent Indicator, 6-Methoxy-*N*-ethylquinolinium Iodide, to Measure Changes in Intracellular Chloride

By Jon R. Inglefield and Rochelle D. Schwartz-Bloom

Introduction[1]

Dynamic intracellular parameters have been measured extensively by combining fluorescent dyes with video-imaging techniques. By adapting

[1] Research described in this article has been reviewed and approved by the National Health and Environmental Effects Research Laboratory, U.S. Environmental Protection Agency. Approval does not signify that the contents necessarily reflect the views and policies of the agency nor does mention of trade names or commercial products constitute endorsement or recommendation for use.

physiological imaging for use with the confocal microscope, we are rapidly expanding our knowledge of neuronal activity with subcellular resolution, whether the cells are cultured,[1a] reside in intact tissues,[2-4] or even reside *in vivo*![5] While the power of conventional epifluorescence microscopy has been strengthened by continued developments in microscope optics, digital image processing and image display, there remain limitations in spatial resolution and the contribution of out-of-focus fluorescence to degrade the image. As a result, the laser scanning confocal microscope (LSCM) is widely accepted as an alternative for achieving quality high-resolution images. Although the concept of confocal microscopy is a few decades old, the emergence of suitable electronic and computer components to support it have made it a realistic tool for biologists only in the last decade or so.[6] In general, the advantage of a confocal microscope over conventional methods comes from raster scanning the sample with a focused point of laser light, enabling the elimination of stray, out-of-focus emitted light at the detector. Emitted image data from thin optical sections of the specimen can be collected, whether taken at a depth up to $\sim$100 μm within an isolated tissue or brain slice or simply from a cell on a slide. An additional feature is that subjacent optical sections can be collected and then reconstructed using special software to give even greater structural information/spatial resolution. The exceptional view of the complex three-dimensional structure of static (or fixed) biological samples afforded by the laser scanning confocal microscope has given fruit to many biomedical applications; for reviews that also include discourses on the problems (and potential solutions), see Paddock,[6] Laurent *et al.*,[7] and Pawley.[8]

Function in the living cell, including dynamic processes such as ion regulation, can be studied with imaging techniques through the application of special fluorescent dyes. Because of the signal-to-noise ratio enhancement, the confocal microscope is used increasingly in combination with fluorescent dyes as a tool for studying ion regulation. Ion-sensitive dyes have spectral properties that are sensitive and selective (although not always) for the local chemical environment and are available thanks to important

[1a] C. Luscher, P. Lipp, H.-R. Luscher, and E. Niggli, *J. Physiol.* **490.2,** 319 (1996).

[2] W. G. Regehr, J. A. Connor, and D. W. Tank, *Nature* **341,** 533 (1989).

[3] H. Miyakawa, W. N. Ross, D. Jaffe, J. C. Callaway, N. Lasser-Ross, J. E. Lisman, and D. Johnston, *Neuron* **9,** 1163 (1992).

[4] S. S.-H. Wang and G. J. Augustine, *Neuron* **15,** 755 (1995).

[5] J. R. Fetcho and D. M. O'Malley, *Curr. Opin. Neurobiol.* **7,** 832 (1997).

[6] S. W. Paddock, *Proc. Soc. Exp. Biol. Med.* **213,** 24 (1996).

[7] M. Laurent, G. Johannin, N. Gilbert, L. Lucas, D. Cassio, P. X. Petit, and A. Fleury, *Biol. Cell* **80,** 229 (1994).

[8] J. B. Pawley, "Handbook of Biological Confocal Microscopy" Plenum Press, New York, 1995.

advances made in optical probe chemistry.[9,10] Changes in the anion, chloride (Cl⁻), are necessary in cellular processes such as neuronal synaptic inhibition and regulation of water movement and pH.[11-15] Until recently, investigations of intracellular Cl⁻ regulation relied mostly on indirect measurements of Cl⁻ movement, such as [36]Cl⁻ tracer methods or Cl⁻-selective microelectrodes. [36]Cl⁻ tracer methods are done in cell-free preparations and have poor sensitivity because of the low specific activity of this isotope, whereas the use of Cl⁻-selective microelectrodes is invasive, restricted to larger cells (>15–20 μm diameter), and has selectivity problems. Physiological imaging with a Cl⁻-sensitive fluorescent indicator gives significant improvements over earlier methods and, in combination with a confocal microscope, allows Cl⁻ movement in various types of cells to be studied in relatively thick specimens where the native environment is largely preserved. This article documents approaches of studying dynamic changes in intracellular Cl⁻ in living neurons with the LSCM and a fluorescent Cl⁻ indicator, 6-methoxy-N-ethylquinolinium iodide (MEQ).

Method

Practical Considerations of Fluorescent Indicators for Chloride

Verkman and colleagues[16-18] have developed and characterized several different UV-excited Cl⁻-sensitive dyes. Unlike Ca²⁺-sensitive dyes where emission is proportional to increased Ca²⁺ levels, the fluorescence of Cl⁻-sensitive dyes is quenched collisionally by increases in Cl⁻. The Cl⁻ indicators, 6-methoxy-N-[3-sulfopropyl]-quinolinium (SPQ) and N-(6-methoxyquinolyl)acetoacetyl ester (MQAE), have been used frequently in nonneuronal cells,[19,20] synaptoneurosomes,[21,22] and brain-cultured neu-

[9] R. Y. Tsien and A. Waggoner, in "Handbook of Biological Confocal Microscopy" (J. B. Pawley, ed.), p. 169. Plenum, Press, New York, 1995.

[10] R. P. Haugland "Handbook of Fluorescent Probes and Research Chemicals." Molecular Probes, Eugene, OR, 1996.

[11] K. Krnjevic and S. Schwartz, *Exp. Brain Res.* **3,** 320 (1967).

[12] G. L. Collingridge, P. W. Gage, and B. Robertson, *J. Physiol.* **356,** 551 (1984).

[13] C. J. Schwiening and W. F. Boron, *J. Physiol.* **475,** 59 (1994).

[14] M. A. Valverde, S. P. Hardy, and F. V. Sepulveda, *FASEB* **9,** 509 (1995).

[15] J. L. Leaney, S. J. Marsh, and D. A. Brown, *J. Physiol.* **501,** 555 (1997).

[16] A. S. Verkman, *Am. J. Physiol.* **259,** C375 (1990).

[17] J. Biwersi and A. S. Verkman, *Biochemistry* **30,** 7879 (1991).

[18] J. Biwersi, B. Tulk, and A. S. Verkman, *Anal. Biochem.* **219,** 139 (1994).

[19] R. Krapf, C. A. Berry, and A. S. Verkman, *Biophys. J.* **53,** 955 (1988).

[20] C. Koncz and J. Daugirdas, *Am. J. Physiol.* **267,** H2114 (1994).

[21] A. C. Engblom and K. E. O. Akerman, *J. Neurochem.* **57,** 384 (1991).

[22] A. C. Engblom and K. E. O. Akerman, *Biochim. Biophys. Acta* **1153,** 262 (1993).

rons.[23,24] Despite MQAE's advantage of greater Cl⁻ sensitivity relative to SPQ or MEQ, MQAE and SPQ are limited in their usefulness because either they require an invasive loading procedure (hypotonic shock for SPQ) or they exhibit a rate of leakage from loaded cells that must be accounted for in the Cl⁻ measurement (MQAE).[20] ABQ-dextran is another Cl⁻-sensitive dye with the advantage that it will remain trapped intracellularly for days once it is dialyzed intracellularly by micropipette.[25] However, the dye MEQ, similar to SPQ and MQAE, is particularly useful in physiological and pharmacological studies because it permeates rapidly and noninvasively into cells in culture or in slices (once converted to a reduced state), it is retained for longer periods than MQAE, and it has resistance to photobleaching caused by laser illumination.[10,17,26–28] For these reasons, MEQ appears to be the Cl⁻-sensitive dye of choice for use with pharmacologic studies using microscopy or spectroscopy at the present time.

Reducing MEQ to a Cell Membrane-Permeable Form

Prior to an experiment, the positively charged and cell-impermeable MEQ (Molecular Probes, Eugene, OR) is first reduced with sodium borohydride (NaBH$_4$; Sigma, St. Louis, MO) to a cell-permeable form, 6-methoxy-N-ethyl-1,2-dihydroquinoline (dihydro-MEQ).[17] Dissolved MEQ (5 mg/0.1 ml distilled H$_2$O, 16 μmol) in a small glass test tube is placed under a slow stream of N$_2$, 10 μl of a 12% (w/v) NaBH$_4$ solution is added, and the solution is kept under the N$_2$ stream for 0.5 hr. The resulting yellow oil can be transferred immediately for cell loading,[29] but we find it easier to extract the oil from the reaction product by adding distilled H$_2$O and ethyl acetate (both 0.5 ml) to the reaction test tube, vortex, and allow the aqueous and organic layers to separate. The organic layer (top) having dihydro-MEQ is removed and placed in a fresh test tube. The aqueous dihydro-MEQ can be extracted a second time with another 0.5 ml of ethyl acetate. Combined organic extracts are dehydrated for 5 min with ~100 mg anhydrous MgSO$_4$. Finally, this organic layer is transferred using a pipette to

[23] A. C. Engblom, I. Holopainen, and K. E. O. Akerman, *Brain Res.* **568,** 55 (1991).
[24] M. Hara, M. Inoue, T. Yasukura, S. Ohnishi, Y. Mikami, and C. Inagaki, *Neurosci. Lett.* **143,** 135 (1992).
[25] J. Biwersi, N. Farah, Y.-X. Wang, R. Ketcham, and A. S. Verkman, *Am. J. Physiol.* **262,** C243 (1992).
[26] L. J. MacVinish, T. Reancharoen, and A. W. Cuthbert, *Br. J. Pharmacol.* **108,** 469 (1993).
[27] R. D. Schwartz and X. Yu, *J. Neurosci. Methods* **62,** 185 (1995).
[28] J. R. Inglefield and R. D. Schwartz-Bloom, *J. Neurosci. Methods* **75,** 127 (1997).
[29] E. Woll, M. Gschwentner, J. Furst, S. Hofer, G. Buemberger, A. Jungwirth, J. Frick, P. Deetjen, and M. Paulmichl, *Pflug. Arch.* **432,** 486 (1996).

a glass microvial and the ethyl acetate is evaporated with a stream of N_2. The dried dihydro-MEQ is sealed under N_2 and then stored at $-20°$ unless used shortly thereafter.

Sample Preparation and Dye Loading

The preparation to be used will depend on the experiment; in our past studies we have used brain slices,[27,28,30,31] clonal cell lines,[32] and primary neuronal cultures.[33] This article focuses on the brain slice preparation. Brain slices of the hippocampus (containing somatosensory cortex) or cerebellum are prepared from rats between 8 and 21 days old. (Younger tissue loads well, but suboptimal bath loading of cell soma occurs if older tissue is used.) Transverse (cortical) or sagittal (cerebellar) slices (200–400 μm) are obtained using a Vibratome in cold, oxygenated [95% O_2/5% CO_2 mixture (v/v)] Ringer's physiological buffer ($4°$) containing (in mM) 119 NaCl, 2.5 KCl, 1.0 NaH_2PO_4, 1.3 $MgCl_2$, 2.5 $CaCl_2$, 26 $NaHCO_3$, and 11 glucose, pH 7.4. Slices are placed on a net submerged in a beaker containing 30 ml oxygenated Ringer's buffer (room temperature) with a blunt end of a Pasteur pipette (rubber bulb placed over end that was cut and fire polished). Inclusion of 1 mM kyneurinic acid in the buffer (modified Ringer's solution) to block NMDA receptors improves slice viability considerably.

Slices (or cells plated on glass coverslips) are incubated for 0.5 hr in modified Ringer's buffer containing $\sim$300 μM diH-MEQ solution at room temperature (resuspend dye using 15 μl ethyl acetate and add to 30 ml buffer containing the slices) [more dilute diH-MEQ (e.g., $\sim$50 μM) is used for cells in culture]. Intracellular oxidation to the Cl⁻-sensitive MEQ traps the polar form of the dye inside the cells (analogous to the hydrolysis of the AM-ester forms of Ca^{2+}-sensitive dyes). Dye loading does not affect cell morphology or viability. MEQ will remain loaded for a few hours, although for improved cellular fluorescence in later specimens, these are maintained in the dye-containing beaker until needed.

Practical Considerations of Confocal Microscope Design

With physiological imaging, a series of collected images are necessary to track the fluorescence response to a drug or stimulus over time. A limitation of using the laser to do quantitative, physiological imaging is

[30] J. R. Inglefield and R. D. Schwartz-Bloom, *J. Neurochem.* **70,** 2500 (1998).

[31] J. R. Inglefield and R. D. Schwartz-Bloom, *J. Neurochem.* **71,** 1396 (1998).

[32] A. C. Grobin, J. R. Inglefield, R. D. Schwartz-Bloom, L. L. Devaud, and A. L. Morrow, *Brain Res.,* in press (1999).

[33] J. R. Inglefield and T. J. Shafer, *Toxicologist* **48**(1S), 1357 (1999).

that coherent UV laser light will cause photodamage to the fluorophore (photobleach). Slit-scanning variations of confocal microscopes (e.g., Noran Odyssey, Noran Instruments Inc., Middleton, WI) are often used because they have the capability to collect real-time, full-frame (e.g., 512 × 512 pixel) images at video rates (e.g., 32 frames/sec) and therefore an adequate image is collected with only 0.5 to 1 sec of laser illumination. Lower image quality and detector efficiency have previously plagued the slit-scanning variation,[8] but current designs are aimed at offering improvements in these areas. While confocal images with the highest clarity are obtained with pinhole or point scanning confocal microscopes, the acquisition of high signal-to-noise images with this type is relatively slow, often requiring seconds per image. Of note, if using a true point scanning confocal design (such as the Leica TCS NT) and faster frame rates are required (to reduce the likelihood of photobleach), one can reduce the pixel array for the imaged area to, e.g., 125 × 125 pixels to significantly enhance the speed of the scans (~4 frames/sec or 1 per ~250 msec) at the expense of a reduced frame size.

Confocal Imaging

For confocal imaging of MEQ fluorescence, individual samples that have been rinsed of unloaded dye for at least 10 min in fresh Ringer's buffer are transferred one at a time (using blunt end of Pasteur pipette) to a plexiglass imaging chamber (custom-made or obtained from a commercial source, e.g., Warner Instrument Co.) having inlet and outlet tubing where the buffer is perfused at 1.5–2.0 ml/min saturated in a 95% O_2/5% CO_2 mixture. Cellular MEQ fluorescence is excited with the 364-nm line of a UV laser (e.g., 80 mW; Enterprise 653, Coherent Laser Group, Santa Clara, CA). Laser light is transmitted through a UV water-immersible objective lens (40, NA 0.7, Olympus) and emission (em_{max} = 440 nm) is imaged using a 400-nm barrier filter of the confocal microscope. Photomultipliers (PMT) receive the signal through a confocal slit (widths of 25–100 μm) at 1× electronic zoom; higher zooms will lead to a faster rate of photobleach (see later). The digitized images are stored using a computer system running imaging software (e.g., MetaMorph, Universal Imaging Co.) and a Noran plug-in module. Figure 1 shows digitized confocal images of MEQ fluorescent cells in slices of somatosensory cortex layer IV/V and cerebellum. (Any remaining dihydro-MEQ within the slice has emission spectra different from MEQ and does not interfere with MEQ fluorescence detection.) The power of this technique is evident by the ability to simultaneously study living, neighboring cells with their circuitry intact. Note the varying intensity of the fluorescence between neighboring neurons. Also, the labeling of proximal

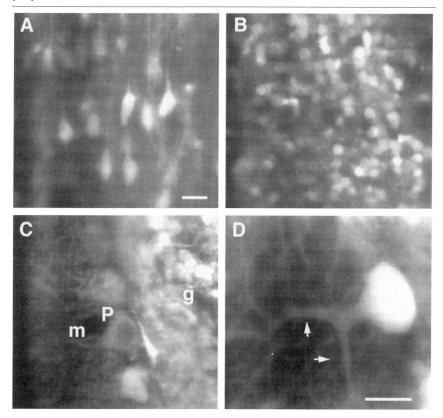

FIG. 1. Confocal digitized images of single optical sections of MEQ-loaded neurons within slice preparations of the neocortex and cerebellum. Images were acquired below the cut surface of the slices with the Noran Odyssey UV laser scanning confocal microscope. (A) Layer V of somatosensory cortex; (B) granule cells in cerebellum; (C) Purkinje (P) and granule cells (g) in cerebellum clearly distinguished from the molecular layer (m); and (D) increased magnification of a Purkinje cell (680×). In this neuron, note the distinct dendritic fluorescence (arrows) as well as the heterogeneous levels of fluorescence between the somata and dendrites. (A–C) are at the same magnification. Bars: 20 μm. Reprinted from J. R. Inglefield and R. D. Schwartz-Bloom, *J. Neurosci. Methods* **62,** 127 (1997) with permission from Elsevier Science.

dendrites with MEQ (as in Fig. 1D) not only provides morphological data, but also allows the possibility to compare changes in Cl⁻ between soma and dendrite.

As mentioned earlier, with coherent UV laser light it is important to be aware of the potential for photodamage to the fluorophore. Intracellular MEQ fluorescence is no exception, as continuous UV illumination focused on loaded cells in a hippocampal slice (with a 40× objective and 1× elec-

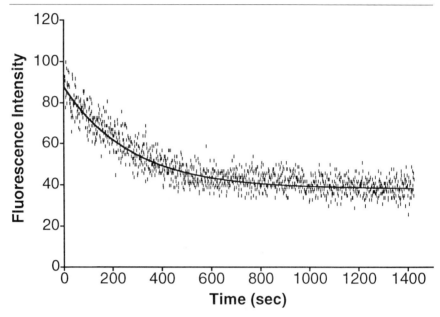

FIG. 2. Time course of fluorescence quenching in a hippocampal neuron within a brain slice that was illuminated continuously with ~33 μW from a UV laser (364-nm laser line) through a 40× objective for 20 min. Ticks represent individual data points (collected 1/sec) and an exponential curve was fit to the data. The exponential decay of the fluorescence of the cell is characteristic and demonstrates an important biophysical issue to be considered in the design of experiments using a laser scanning confocal microscope. See text for additional discussion of this issue.

tronic zoom) for 20 min gave an exponential decay (e.g., Fig. 2) with loss at 20 min equal to >50% of the initial fluorescence. The photobleaching is first order with an average rate constant of decay (k) of 0.00418 ± 0.00068 sec^{-1} (n = 3) that corresponds to a half-life ($t_{1/2}$) of 173.8 ± 24.6 sec.[28] From this rate of photobleach we estimated that an overall 20 sec of illumination during an experiment corresponds to a ~6 ± 3% decrease from the beginning intensity[28]; therefore it is desirable to restrict cumulative illumination during the course of an entire experiment to less than 20 sec. In addition, the considerable photobleaching of the fluorescence by repeated laser illumination can be overcome partially by a combination of low laser power, proper photomultiplier setting, and increasing the inter-image duration. For dynamic imaging of changes in fluorescence, we take advantage of the video frame rate (32 images per second) of the Noran Odyssey confocal microscope and acquire rapidly the average image of 8 or 16 frames (shutter opened for ~0.25 or 0.5 sec, respectively) every 1–2 min. The images are recorded on a computer hard drive or Panasonic optical memory disk recorder (OMDR, Panasonic Model LQ-3031) for off-line playback and analysis (the OMDR is able to save successive images much faster than a hard drive). Prior to switching to Ringer's buffer containing a drug, a minimum of 10 min of baseline is recorded to allow determination of the stable fluorescence of the cells. A stable baseline recording (i.e., one that does not deviate more than 10%) is desirable when measuring drug- or stimulus-induced changes in MEQ.

Data Analysis

Using previously saved images opened in image analysis software, fluorescence data are recorded for several morphologically distinct somata within the imaged field. Unlike Ca^{2+}-sensitive dyes, MEQ and other quinoline compounds are collisionally quenched by increases in intracellular Cl^-, so increases in cellular Cl^- are reflected by a corresponding decrease in fluorescence. Background signals, obtained from a neighboring area devoid of fluorescent cells, can be subtracted from each cell's measurement before any calculations. The change in fluorescence for a cell after drug addition (or stimulation by some other means) (ΔF) can be expressed as a percentage of the average of two to three frames preceding drug addition (F_b). Expression of data as $\Delta F/F_b$ permits each cell to serve as its own control. If performing experiments above room temperature or with a rate of illumination that will cause unknown amounts of photobleach, we image parallel control slices (unexposed) identically to collect the rate of change in the baseline ("baseline drift"). Then, a corrected ΔF (ΔF_s) can be calculated by subtraction of data from regression-generated values for baseline drift.

Calibration of Intracellular MEQ Fluorescence

In contrast to the ratioable dyes available for studying changes in intracellular Ca^{2+}, H^+, and Na^+,[10] commercially available Cl^--sensitive dyes are restricted to simple intensity measurements because the dyes currently lack either an excitation or an emission shift. Quenching of MEQ fluorescence by Cl^- is calibrated inside the dye-loaded cells by superfusing them with buffers containing an ionophore to equilibrate the Cl^- concentration inside the cell to the controlled external Cl^- concentration. In the presence of tributyltin (used to equilibrate Cl_o and Cl_i) and nigericin (a K^+/H^+ iono-phore to maintain constant intracellular pH during changes of Cl_o), changes in extracellular Cl^- (0–60 mM) cause incremental (or step) decreases in intracellular MEQ fluorescence. We used sodium sulfate as the Cl^- substitute for NaCl and small amounts (2.5 mM) of potassium gluconate to replace KCl (see next section). KSCN and valinomycin (a K^+ ionophore) are used at the end to determine the minimal fluorescence (background) as KSCN completely quenches MEQ fluorescence.[17] From these data, estimates of the change in intracellular [Cl⁻] are then made according to the Stern–Volmer equation

$$F_0/F_{Cl^-} = 1 + K_{Cl^-}[Cl],^{16,19}$$

where F_0 is fluorescence intensity in the absence of Cl⁻ and F_{Cl} is fluorescence intensity in the presence of varying concentrations of Cl⁻. K_{Cl} is the Stern–Volmer constant, equal to the slope of the line from the plot of F_0/F_{Cl} versus [Cl⁻]. Calibration using the confocal microscope should be done with restricted illumination (refer to the previous section on confocal microscopy and Fig. 2). Using the confocal microscope, the calibration was carried out on hippocampal slices at 26–28° to obtain the Stern–Volmer constant (K_q) from the Stern–Volmer plot (Fig. 3). Using these conditions, K_q (the slope of the plotted line) was calculated as 52 ± 3 M^{-1}. Other Stern–Volmer values from different preparations using MEQ fluorescence include 19 M^{-1} (fibroblasts),[17] 62.5 M^{-1} (cell free preparation of synaptic vesicles),[22] and 32 M^{-1} (brain slices).[30] This range in values points out the

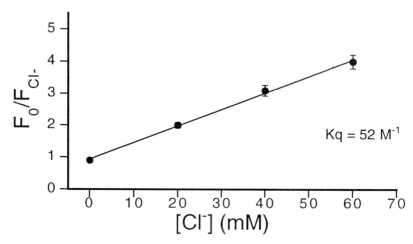

FIG. 3. The relationship of MEQ fluorescence versus Cl⁻ concentration in neurons of the hippocampal slice (maintained at 27°) expressed in the form of a Stern–Volmer plot. Calibration of the MEQ fluorescence was performed inside the neurons ($n = 6$) by perfusing the slice with buffers containing set Cl⁻ concentrations (0–60 mM) and the ionophores, nigericin (100 μM) and tributylin (50 μM), for equilibration of extra- and intracellular Cl⁻ (lower concentrations of these drugs are used in cell preparations). At the conclusion, KSCN (150 mM) and valinomycin (25 μM) were added to obtain the minimum fluorescent signal (background). Data were plotted as F_0/F_{Cl} vs [Cl⁻] (mM), and a regression line was fit ($R = 0.991$) to give the Stern–Volmer constant (K_q; slope of the line) of 52 $M.^{-1}$

need to do a calibration in each preparation and under identical conditions to the experiment.

Cl⁻-Selectivity of MEQ Fluorescence

MEQ fluorescence in cells is very selective for changes in Cl⁻ with some caveats. There is more efficient quenching by halides other than Cl⁻, such as Br⁻ and I⁻, and by other anions, such as thiocyanate (SCN⁻); however, nonphysiological concentrations of nonchloride ions and nonphysiological pH are needed to significantly affect the fluorescence of the Cl⁻ indicator.[17] One word of caution, the chloride substitute, gluconate, is able to quench MEQ fluorescence[29] (see Fig. 4). The addition of 50 mM sodium gluconate to the perfusion buffer for 7 min reduced the MEQ fluorescence within CA1 pyramidal cells by 26 ± 1% (P <0.05 relative to control, n = 4). Because use of sodium sulfate as a Cl⁻ substitute in the buffer instead of sodium gluconate does not quench MEQ fluorescence,[17,30] the quench of MEQ fluorescence by gluconate probably occurs as a result of its entry into neurons via anion transport mechanisms, as has been reported for NIH 3T3 fibroblasts.[29] If gluconate is used at all in this technique, it should only be used in low concentrations (e.g., 6 mM quenches MEQ only 11 ± 4%.[29] This important caveat should be taken into account for MEQ

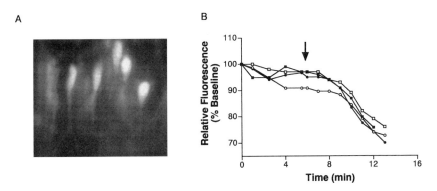

FIG. 4. (A) Confocal video image of MEQ fluorescence within a rat hippocampal slice containing area CA1 pyramidal cells. (B) Time-dependent change of MEQ fluorescence in individual hippocampal CA1 pyramidal cells following the addition to the Cl⁻ substitute gluconate (50 mM; at arrow). Following baseline recordings in which the slice was superfused with standard Ringer's buffer, the buffer was changed to a sodium gluconate-containing one. Measurements were made from a series of confocal images saved over time and analyzed off-line with Image-1 image analysis software. Parallel experiments using a different Cl⁻ substitute (Na₂SO₄) showed no quench. The result indicated that gluconate quenches MEQ fluorescence, probably after its entry into neurons via anion transport mechanisms.

fluorescence data collected using buffer solutions containing high concentrations of gluconate salts.

Discussion

The goal of this article was to provide details and issues associated with the imaging of dynamic changes in intracellular Cl⁻ with the LSCM. All laser scanning confocal microscopes have an improved signal-to-noise ratio and the ability to penetrate into tissues, although slit variations, e.g., Noran Odyssey, also convey an ability to capture the image more rapidly. Although we have employed a slit- based confocal microscope for these studies, much of the information can be generalized to nonconfocal (epifluorescence) systems and other confocal microscope systems as long as the problem of photobleaching is addressed. Protocols are becoming available to standardize the operation of the confocal microscope across experiments and users.[34] For increased image quality, it is important to balance the laser power (we use 20–25% intensity), slit width, gain, and brightness intensity level (the latter two are collectively a function of the photomultiplier tube and contribute greatly to the image "noise"). Photobleaching can be reduced if one minimizes the PMT level to a range where less noise is introduced, performs image averaging, and restricts the length of laser illumination to pulses of 1–2 sec every minute or so.

We have used this methodology to study changes in intracellular Cl⁻ within neurons of both juvenile cortical (including the hippocampus) and cerebellar slices. The Cl⁻ channels in these areas include the recently discovered ClC-2 family[35] as well as GABA$_A$ (γ-aminobutyric acid) receptors[36]; activation of the latter in adult animals leads to neuronal inhibition via a stimulated Cl⁻ conductance. Applying MEQ and confocal microscopy to the brain slice preparation, we demonstrated an appropriate GABA$_A$ receptor pharmacology.[27,28] In addition, a comparison of simulated ischemia versus activation of excitatory amino acid receptor subtypes identified different mechanisms for Cl⁻ influx and reduced GABAergic function.[30,31] In those experimental conditions where cell swelling is present, the estimation of changes in intracellular [Cl⁻] by MEQ fluorescence (or any other nonratiable dye) is inaccurate due to intracellular dilution of the dye. The presence of cell swelling can be investigated with fluorescence microscopy

[34] R. M. Zucker and O. Price, *Methods* **18**(4), in press (1999).

[35] R. L. Smith, G. H. Clayton, C. L. Wilcox, K. W. Escudero, and K. J. Staley, *J. Neurosci.* **15**, 4057 (1995).

[36] E. Costa, *Annu. Rev. Pharmacol. Toxicol.* **38**, 321 (1998).

or microfluorimetry with the volume-sensitive fluorescent dye, calcein-AM.[30,31,37,38]

Future Directions

Although MEQ is presently the Cl⁻ indicator of choice for physiological imaging, development of a ratioable dye would be highly valuable for this technique, especially for confocal imaging. The ratio method allows direct determination of ion concentration while canceling out differences in dye concentration or intensity as may occur from photobleach or leak of the fluorophore. The ideal Cl⁻ ratioable dye for use with a laser scanning confocal system would be one where a single excitation leads to a dual emission that is ratioed (as is possible with Indo-1) after their separation with a proper dichroic mirror. Employing fluorescent dyes that have two different excitations (e.g., fura-2 and SBFI) is problematic for use with confocal microscopy because the limited number of available laser lines generated by argon (or krypton) lasers probably will not coincide with the requirements of the dyes. This limitation would probably preclude the use of a dual excitation dye for Cl⁻.

Until a ratiometric indicator for Cl⁻ becomes available, perhaps the most stringent method for the measurement of Cl⁻ movement might be to conduct pseudo-ratio analysis to rule out such factors as bleaching or between cell differences in dye concentration. In this approach, the ratio image is made from two separate indicators (separate in their excitation/emission spectra and so that one is not Cl⁻ sensitive) imaged alternately. Calcein-AM (with excitation/emission = 488/530 nm after deesterification) is not sensitive to ions and thus, when dual loaded with MEQ into brain slices, gives complementary information such as viability or degree of the swelling of the cells (Inglefield and Schwartz-Bloom, unpublished data).

Results reported thus far with imaging of changes in intracellular Cl⁻-sensitive fluorescence have been at the resolution of several hundred milliseconds[28]; in contrast, Ca²⁺-imaging studies in slices have been performed with a temporal resolution of 1 msec or less.[2-4] Improvements are possible because MEQ can respond to Cl⁻ changes in <1 msec as measured with a stopped-flow technique.[17]

[37] W. E. Crowe, J. Altamirano, L. Huerto, and F. J. Alvarez-Leefmans, *Neuroscience* **69,** 283 (1995).
[38] J. Farinas, V. Simanek, and A. S. Verkman, *Biophys. J.* **68,** 1613 (1995).

Section VI

Imaging of Specialized Tissues

[27] Measurement of Mineral Gain and Loss on Dental Enamel and Dentin Using Confocal Microscopy

By CARLOS GONZÁLEZ-CABEZAS, MARGHERITA FONTANA, and GEORGE K. STOOKEY

Introduction

The health of teeth is dependent on the equilibrium of the mineral exchange between the teeth's outer surface and saliva and dental plaque. Tooth minerals are lost and regained constantly in a normal, human oral environment. If this equilibrium is disrupted in favor of demineralization for a lengthy period of time it will initiate a carious lesion. When this process continues without a mineral exchange balance, the lesion progresses until it cavitates and become irreversible. Consequently, it is very important to detect caries progress as early as possible to reverse the carious lesions prior to cavitation. It is also necessary to monitor the carious lesions process to evaluate the performance of different therapies.

Techniques such as microhardness measurement of the enamel surface,[1] microhardness measurements of enamel cross sections,[2,3] polarized light microscopy,[4] different transversal microradiography techniques (TMR),[5–7] iodine absorptiometry,[8] and light scattering[9] have been used to measure mineral changes in human enamel *in vitro*. Although all of them have advantages and disadvantages,[10,11] TMR is considered to be the most practical technique for the direct and quantitative measurement of mineral content, mineral changes, and mineral distribution.[12]

Confocal laser scanning microscopy (CLSM) has been used to measure the tooth mineral changes in enamel and dentin as an alternative to TMR.

[1] J. Arends, J. Schuthof, and W. L. Jongebloed, *Car. Res.* **14,** 190 (1980).
[2] C. L. Davidson, I. J. Hoekstra, and J. Arends, *Car. Res.* **8,** 135 (1974).
[3] J. D. B. Featherstone, J. M. ten Cate, M. Shariati, and J. Arends, *Car. Res.* **17,** 385 (1983).
[4] L. M. Silverstone, *Oral Sci. Rev.* **3,** 100 (1973).
[5] B. Angmar and D. Carlström, *J. Ultrastruct. Res.* **8,** 12 (1963).
[6] E. De Josselin de Jong, A. H. I. M. van der Linden, and J. J. ten Bosch, *Phys. Med. Biol.* **32,** 1209 (1987).
[7] F. M. Herkströter, J. Noordmans, and J. J. ten Bosch, *Car. Res.* **24,** 399 (1990).
[8] H. Almqvist, J. S. Wefel, F. Lagerlöf, F. Ekstrand, and C. O. Hendrikson, *J. Dent. Res.* **67,** 1217 (1988).
[9] J. J. ten Bosch, H. C. van der Mei, and P. C. F. Borsboom, *Car. Res.* **18,** 540 (1984).
[10] J. J. ten Bosch and B. Angmar-Månsson, *J. Dent. Res.* **70,** 2 (1991).
[11] J. D. B. Featherstone, *J. Dent. Res.* **71**(SI), 955 (1992).
[12] J. Arends and J. J. ten Bosch, *J. Dent. Res.* **71**(SI), 924 (1992).

Studies have focused on the measurement of fluorescent dye penetration in the porosities of demineralized tissue, on the measurement of changes in autofluorescence of the tissue, or on the changes in the reflection of light from the lesion. Analyses have been done on intact surfaces and cross sections, with the latter being studied most frequently at the moment. This article describes briefly the up-to-date methodologies used to measure mineral gain and loss in dental enamel and dentin using confocal microscopy.

Measurement of Enamel Mineral Loss

Enamel is more mineralized than bone or dentin. It is approximately 99% mineral by weight and averages 87% mineral by volume.[13] Jones and Boyde[14] used different fluorescent dyes, brilliant sulfaflavine, tetracycline, and alizarin red to analyze small carious lesions in extracted teeth and artificially demineralized enamel specimens. Specimens were sectioned longitudinally through the lesion, soaked in one of the fluorescent dyes, and analyzed by confocal microscopy using oil-immersion objective lenses. The working hypothesis was that imbibition of a fluorescent dye into porosities of demineralized enamel enables rapid quantitative analysis of thick samples using CLSM. This study focused on the analysis of the characteristics of the advancing front of the demineralization process. The investigators observed gross variations in the fluorescence intensity within a demineralized lesion. Benn and Watson[15] used rhodamine B to measure the depth (from 170 to 900 μm) of natural carious lesions. They found a good visual correlation in the lesion size when the confocal images were compared to backscattered electron images.

To determine the possibility of using confocal measurements of fluorescent dye penetration to determine the size and the mineral content of very early demineralization of enamel, a study was designed in our laboratories to compare results of CLSM and TMR, which is the current gold standard for this type of analysis. The following methodology is a brief description of the already fully published methodology.[16]

Sample Preparation. Enamel specimens (3 mm in diameter) are drilled from extracted, human teeth that had been obtained from oral surgeons and sterilized by soaking in 10% (v/v) buffered formalin (pH. 6.7) for at

---

[13] N. O. Harris and R. P. Suddick, *in* "Primary Preventive Dentistry" (N. O. Harris and A. G. Christen, eds.). Prentice Hall, Englewood Cliffs, NJ, 1991.

[14] S. J. Jones and A. Boyde, *Scan. Microsc.* **1**(4), 1991 (1987).

[15] D. K. Benn and T. F. Watson, *Quintessence Int.* **20**, 131 (1989).

[16] M. Fontana, Y. Li, A. J. Dunipace, T. W. Noblitt, G. M. Fischer, B. P. Katz, and G. K. Stookey, *Caries Res.* **30**, 317 (1996).

least 3 months. Each specimen is mounted in a polyacrylic rod using methyl methacrylate resin. The specimens are ground using 600 grade silicon carbide paper to remove approximately 50 μm of the surface and then polished to a high luster with Gamma Alumina (Buhler, Chicago, IL; 0.05 μm) using standard methods.

Lesion Formation. The specimens are numbered and assigned randomly to four groups. Each specimen is placed in 11 ml of a 50% saturated hydroxyapatite/0.1 M lactic acid–carbopol solution (pH 5.0) at 37° for 0, 24, 64, or 96 hr in order to create lesions with different degrees of demineralization. Following each demineralization period, the specimens are rinsed in deionized water for 2 min and sectioned using a water-cooled, Silverstone–Taylor hard tissue microtome (Scifab, Lafayette, CO) to obtain 100-μm-thin sections, 1.5–2.0 mm in length.

TMR Examination. Thin sections are mounted on microradiographic, X-ray plates [Kodak (Rochester, NY) IMTEC, poly-edged, H.R.P.] along with an aluminum step wedge. Microradiographs are made using Ni-filtered Cu Kα radiation excited at 40 kV and 35 mA. The mineral content of the lesions (ΔZ) is analyzed using densitometry and transversal microscopy software (Inspektor Research Systems B.V., The Netherlands). Depths of the lesions are measured from the outer demineralized surface of the specimen to the point where the mineral content becomes 83%, which is approximately 95% of that in sound enamel.[17] Two areas, each 208 μm in length, are analyzed per section. Measurements are made approximately 208 μm from each end of the enamel section in order to obtain two ΔZ measurements per specimen. If the selected area has artifacts or cracks, another specified area of 208 μm is analyzed. The two ΔZ values for each section are averaged to obtain a single value per specimen.

Confocal Microscopy Examination. Following microradiographic analysis, thin sections are stained with freshly prepared rhodamine B solution (Aldrich Chem. Co., Milwaukee, WI). The stained surfaces are analyzed using a laser scanning confocal microscope (Odyssey, Noran Instruments, Inc., Middleton, WI). Measurements are made in the same two 208-μm areas that are analyzed by TMR. Areas are scanned planoparallel to the transversal cut surface of the specimen and perpendicular to the natural surface of the tooth. However, the areas are not scanned directly at the cut surface of the specimen because of concern regarding the smear layer created during the cutting procedure. Areas are scanned below the cut surface. Samples are analyzed using Image-1 software (version 4.14 C; Universal Images Corp., West Chester, PA). After being brought into focus (using a 10× Nikon objective, NA 0.25), the specimens are illuminated

[17] J. Arends, T. Dijkman, and J. Christoffersen, *Caries Res.* **21,** 249 (1987).

using an argon laser with a 488-nm excitation wavelength. Confocal slits are set at 15 μm with a 515 nm long pass barrier filter, and the argon laser intensity is set at 100% (1.23 mW per scanned point). After examination of all the specimens, confocal settings (contrast and brightness) are maximized and held constant. For the collection of the images, the sample is frame averaged using 128 frames per image. Three parameters are measured: area of the fluorescent lesion (determined as the area comprised by all the pixels that fluoresced above a fluorescent threshold in the region of 208 μm in length), total fluorescence of the lesion (a measurement that results from adding all the pixel intensities in the measured area), and average fluorescence of the lesion (a measurement that represents the average fluorescent pixel value in the measured area). Duplicate measurements obtained for the three parameters from each section are averaged to obtain a single measurement for each parameter per specimen. We also measure one of the two remnant specimen pieces produced after sectioning to obtain the thin sections. These specimens are more than 1 mm in thickness. CLSM images are obtained from the two areas that are located adjacent to those areas analyzed by TMR before sectioning, ensuring that the same or equivalent areas of the carious lesions are used for the analysis.

Statistical Analysis. Data are analyzed using a single factor analysis of variance model. Where a significant effect is detected (α: 0.05), multiple comparisons are conducted using the Student–Newman–Keuls procedure. TMR ΔZ and the three confocal microscopy parameters (lesion area, total fluorescence, and average fluorescence) are correlated by calculating Pearson's correlation coefficients.

Results. The means and standard errors for the four groups analyzed in the first study are shown in Table I. As demineralization time increased, both TMR (ΔZ) and confocal parameters (lesion area and total lesion fluorescence) increased numerically. The Student–Newman–Keuls test failed to show a significant difference ($p > 0.05$) between 64- and 96-hr data for total lesion fluorescence and average lesion fluorescence.

Pearson's correlation coefficients calculated for all four groups, between TMR and confocal microscopy parameters measured in the same thin sections, were all statistically significant ($p < 0.01$): ΔZ vs area, $r = 0.97$; ΔZ vs total fluorescence, $r = 0.72$; and ΔZ vs average fluorescence, $r = 0.70$. Pearson's correlation coefficients demonstrated strong significant correlations between TMR ΔZ measurements and confocal microscopy lesion area measurements. Total lesion fluorescence correlated well with ΔZ. Average lesion fluorescence was the confocal parameter that correlated the least with ΔZ.

In the confocal analysis of thick sections, as demineralization time increased, confocal parameters increased significantly ($p < 0.05$), except for

TABLE I

MEASUREMENT OF DEMINERALIZED ENAMEL LESIONS USING TMR AND CLSM

| | | CLSM | | | | | |
| | | Thin sections | | | Thick sections | | |
Time (hr)	TMR ΔZ (×10²)	A^a	TF^b	AF^c	A	TF	AF
0	2.9 ± 0.5^d	0 ± 0	0 ± 0	0 ± 0	0 ± 0	0 ± 0	0 ± 0
24	8.6 ± 0.7	35.8 ± 3.1	14.8 ± 7.3	16.8 ± 3.7	58.2 ± 5.1	$28.3 \pm 19.0 \,]^e$	$33.1 \pm 18.6 \,]^e$
64	19.8 ± 0.8	109.9 ± 4.6	$79.1 \pm 11.3\,]^e$	$61.5 \pm 6.8\,]^e$	117.4 ± 4.3	$54.4 \pm 7.4\,\}$	$40.7 \pm 5.1\,\}$
96	26.6 ± 0.8	147.7 ± 5.2	$94.4 \pm 18.2\,\}$	$56.1 \pm 10.2\,\}$	162.9 ± 7.4	231.1 ± 27.9	123.3 ± 11.9

[a] Area of the fluorescent lesion ($\mu m^2 \times 10^2$).
[b] Total lesion fluorescence is the sum of fluorescent intensity values in all pixels in the measured area of the lesion ($\times 10^5$).
[c] Average lesion fluorescence is the average fluorescence intensity value of all pixels in the measured area of the lesion.
[d] Mean ± SEM.
[e] Groups within braces were not significantly different ($p > 0.05$).

total fluorescence and average fluorescence between 24- and 64-hr groups. However, all confocal values increased numerically as demineralization time increased. Data from all four time groups were pooled and Pearson's correlation coefficients were calculated for ΔZ and the confocal parameters: ΔZ vs lesion area, $r = 0.95$; ΔZ vs total fluorescence, $r = 0.80$; and ΔZ vs average fluorescence, $r = 0.74$. Although all three confocal parameters correlated strongly with ΔZ, the lesion area showed the highest correlation coefficient. Total lesion fluorescence correlated very well with ΔZ, whereas average lesion fluorescence correlated the least.

Discussion. Strong statistically significant correlations occurred between TMR ΔZ and fluorescent lesion area determined by the confocal analysis of both thin sections and thick specimens. Both TMR and confocal measurements (lesion area and total lesion fluorescence) increased over demineralization challenge time. However, not all confocal parameters increased significantly over time. A small sample size and large variations in lesion size could be responsible for the lack of statistical significance.

The decision regarding which fluorescent dye to be used in this study was based on the results of Benn and Watson[15] and many preliminary pilot studies (not published). One important concern in the selection of the dye was to find a substance that could penetrate into the smaller pores of the carious lesions, a concern similar to that for immersion media in polarized light microscopy. The fact that the lesion area measured with confocal microscopy correlated so strongly with ΔZ indicates that the rhodamine B fluorescent dye penetrated well, at least in the same part of the lesion that was analyzed by the well-accepted method of TMR. We also tested this dye at different concentrations, from 0.01 to 100 mM, and found that 0.1 mM showed the strongest correlations, whereas 0.01 mM was found not suitable for this kind of experiments.

The success of quantitative confocal microscopy using fluorescent dyes is dependent, therefore, on the choice of a dye concentration that will provide the best correlation coefficients with a well-validated method for quantifying mineral loss, such as TMR. It was of no surprise that lesion area and total lesion fluorescence (which is a value dependent on area) were the confocal parameters best correlated with ΔZ, which is a parameter that also proportionally relates to lesion area.[17,18] It must be made clear that the lesion area, the confocal parameter that showed the strongest correlation coefficient with ΔZ, does not measure mineral density, which is what is measured by TMR ΔZ. However, if the hypothesis is true, total lesion fluorescence, the parameter that showed the second strongest correlation coefficients in all studies, depends on the amount of dye filling spaces in enamel where mineral has been lost and therefore is the confocal parameter that more closely represents mineral density measured by ΔZ. In enamel demineralization studies, it seems appropriate to conclude from these studies that lesion area data obtained with confocal microscopy from thick specimens as well as from thin sections is the parameter that best correlates with ΔZ obtained from thin sections using TMR.

It is well known that greater amounts of fluoride accumulate in the surface of the enamel and because teeth collected for our studies come from different places in the United States, the enamel surfaces that we use have very different fluoride levels. Therefore, in this investigation we used polished enamel surfaces to reduce the variability among our specimens. However, to test if this CLSM methodology works to analyze clinical lesions, we needed to be able to develop early lesions in natural enamel surfaces (nonpolished). To solve this problem, an artificial mouth microbial model[19] was modified to develop early lesions on natural enamel surfaces.[20] Sixty specimens were divided into five groups of different demineralization times: 0, 24, 48, 72, and 96 hr. When lesions were analyzed using the same methodology described previously, we found that the microbial model created uniform carious lesions in all groups but the baseline. Mean lesion depths (in micrometers) for baseline, 24-, 48-, 72-, or 96-hr groups were 0.5 ± 1.7,

[18] J. R. Mellberg, W. G. Chomicki, D. E. Mallon, and L. A. Castrovince, *Caries Res.* **19**, 126 (1985).

[19] M. Fontana, A. J. Dunipace, R. L. Gregory, T. W. Noblitt, Y. Li, K. K. Park, and G. K. Stookey, *Caries Res.* **30**, 112 (1996).

[20] C. González-Cabezas, M. Fontana, M. E. Wilson, and G. K. Stookey, *J. Dent. Res.* **77**(SI-B), 782 (1998).

8.3 $\pm$ 3.7, 35.9 $\pm$ 8.3, 49.3 $\pm$ 5.5, and 59.2 $\pm$ 10.9, respectively. All groups had significantly different lesion depths ($p < 0.05$). These results support the hypothesis that this methodology is able to distinguish small differences in the mineral loss of natural enamel surfaces.

Currently, this confocal microscopy methodology is being used to validate clinical methodologies for the detection and quantification of early carious lesions, from lesions on smooth surfaces[21] to lesions in pits and fissures.[22] This technique has also been used to measure the effectiveness of dentifrices in preventing caries formation[23] and to evaluate the preventive properties of different restorative materials.[24]

Measurement of the penetration of fluorescent dyes in the demineralized enamel is not the only method used in confocal microscopy to analyze enamel mineral loss. Analyses of the changes in ligth reflected from the lesions of transversally cut specimens using an oil-immersion 40× objective (NA 1.3) have found that changes in reflection follow the shape of the lesion.[14,25,26] Apparently, the increase in reflection is because of the increase of "space" of the carious lesions.[14] Using the same principles, Duschner and co-workers[27] studied early carious lesions taking subsequent images from the surface of the lesion to a depth of 100 μm. Those images were used to reconstruct and analyze the carious lesion in three dimensions. These techniques have been used to describe the carious lesion. However, quantitative measurement of the lesions using this technique has not been correlated to a gold standard.

Measurement of Enamel Mineral Gain

Based on the results of the study described briefly previously[16] and using the same working hypothesis, we have developed a new protocol to measure the mineral gain in enamel after remineralizing early carious

[21] M. Ando, A. F. Hall, G. J. Eckert, B. R. Schemehorn, M. Analoui, and G. K. Stookey, *Car. Res.* **31,** 125 (1997).

[22] A. G. Ferreira, Zandoná, M. Analoui, B. B. Beiswanger, R. L. Isaacs, A. H. Kafrawy, G. J. Eckert, and G. K. Stookey, *Car. Res.* **32,** 210 (1998).

[23] M. Fontana, C. González-Cabezas, M. E. Wilson, and G. K. Stookey, *J. Dent. Res.* **77**(SI-B), 782 (1998).

[24] M. Fontana, A. J. Dunipace, R. L. Gregory, C. González-Cabezas, M. E. Wilson, and G. K. Stookey, *J. Dent. Res.* **77**(SI-A), 282 (1998).

[25] B. Øgaard, H. Duschner, J. Ruben, and J. Arends, *Eur. J. Oral Sci.* **104,** 378 (1996).

[26] A. B. Sønju Clasen, B. Øgaard, H. Duschner, J. Ruben, J. Arends, and T. Sönju, *Adv. Dent. Res.* **11**(4), 442 (1997).

[27] H. Duschner, A. B. Sønju Clasen, and B. Øgaard, *in* "Early Detection of Dental Caries" (G. K. Stookey, ed.). Indiana University School of Dentistry, Indianapolis, 1996.

lesions.[28] Remineralization would reduce the porosities of demineralized enamel, reducing the imbibition of the fluorescent dye into enamel. Therefore, the aim this study is to determine if mineral changes during remineralization, measured in a thin section by TMR, could also be monitored reliably by measuring lesion parameters (area, total, and average dye fluorescence) on the same thin section or on thick specimens with CLSM.

Remineralization Regimen. Specimens are prepared and demineralized in a manner similar to that described in the previous section, and half of each specimen is covered with an acid-resistant varnish to maintain the baseline lesion. The demineralization time is 96 hr for all the specimens. The specimens are then divided randomly (12 specimens/group) into three treatment groups: (A) placebo dentifrice (0 ppm fluoride); (B) 250 ppm fluoride dentifrice (NaF), and (C) 1100 ppm fluoride dentifrice (NaF). The cyclic remineralization regimen[29] consists of a 4-hr/day acid challenge with four 1-min dentifrice treatment periods (1:3, dentifrice:water) per day at room temperature. For the remaining time (approximately 20 hr/day), the specimens are placed in pooled, human saliva. The regimen is repeated for 20 days.

TMR Examination. Specimens are sectioned perpendicularly to the varnished area so that each section includes the varnish-covered, baseline lesion area and the uncovered, posttreatment lesion area of the specimen. The thin sections are microradiographed and ΔZ values are calculated as described earlier. For each section, measurements are made in two areas, each 340 μm in length: one measurement is made in the unvarnished-treated lesion area and the second is made in the varnished-baseline lesion area. Measurements are made approximately 510 μm from each end of the enamel specimen. $\Delta M,$ the change in mineral content of a lesion after it is exposed to the remineralization treatment, is calculated for each specimen as follows: $\Delta M =$ posttreatment ΔZ-baseline ΔZ.

Confocal Microscopy Examination. Following TMR analysis, the same thin section analyzed by TMR and the remaining half of each specimen are stained as explained earlier. The stained baseline and posttreatment lesions are then analyzed for area of the lesion, total lesion fluorescence, and average fluorescence of the lesion as described previously.

Statistical Analysis. For each group, the mean, standard deviation, and standard error of the mean (SEM) are calculated for all measured parameters. Multiple group comparisons are made using a single factor analysis of variance (ANOVA). Because of the exploratory nature of this research,

[28] C. González-Cabezas, M. Fontana, A. J. Dunipace, Y. Li, G. M. Fischer, H. M. Proskin, and G. K. Stookey, *Car. Res.* **32**, 385 (1998).
[29] A. J. Dunipace, W. Zhang, A. J. Beiswanger, and G. K. Stookey, *Car. Res.* **28**, 315 (1994).

post-ANOVA pairwise comparisons are performed using one-sided t tests. Pearson correlation coefficients are calculated to investigate the relationship between TMR scores (ΔM) and the three confocal microscopy parameters (lesion area, total fluorescence, and average fluorescence).

Results. The means and standard errors of the mean for the three treatment groups analyzed with TMR and confocal microscopy are shown in Table II. All data presented in the tables were calculated by subtracting the varnish-covered, baseline lesion values from the values obtained from the uncovered, post, treatment lesion. A positive value indicated that the lesion had demineralized, further, whereas a negative value indicated that remineralization had occurred. As fluoride concentration increased, both TMR (ΔM) and confocal parameters (lesion area and total lesion fluorescence) decreased numerically and eventually became negative for both thin and thick samples, indicating remineralization. Specimens treated with the placebo dentifrice (0 ppm) became more demineralized because the remineralization protocol included a 4-hr acid challenge per day. Multiple t tests found a significant difference in the TMR and confocal parameters studied on specimens halves (average and total fluorescence) between 250- and 1100-ppm dentifrice-treated groups and marginally significant results for thin section-area data ($p = 0.06$). However, in all cases, the score associated with the 1100-ppm dentifrice was numerically greater for ΔM, confocal area, and total fluorescence measurements than the 250-ppm dentifrice. Average lesion fluorescence measured in thin sections was the only confocal parameter that did not show a numerically greater amount of remineralization with the 1100-ppm dentifrice than with the 250-ppm dentifrice.

TABLE II

DIFFERENCES BETWEEN TREATED AND CONTROL LESIONS DETERMINED BY TMR AND CLSM

		CLSM					
		Thin sections			Thick sections		
F(ppm)	TMR ΔM ($\times 10^2$)	A^a	TF^b	AF^c	A	TF	AF
0	5.5 ± 1.1^d	39.1 ± 9.7	23.0 ± 2.6	75.3 ± 5.4	32.8 ± 16.8	20.8 ± 5.7	64.9 ± 20.1
250	-1.7 ± 1.6	-9.8 ± 8.4]^e	-3.1 ± 3.5]^e	-6.9 ± 13.7]^e	-20.0 ± 8.8]^e	-0.3 ± 3.8	-0.4 ± 17.4
1100	-5.3 ± 1.0	-33.6 ± 13.4]	-4.8 ± 4.9]	5.6 ± 14.9]	-43.9 ± 16.7]	-12.5 ± 4.8	-42.5 ± 17.5

[a] Area of the fluorescent lesion ($\mu m^2 \times 10^2$).
[b] Total lesion fluorescence is the sum of fluorescent intensity values in all pixels in the measured area of the lesion ($\times 10^5$).
[c] Average lesion fluorescence is the average fluorescence intensity value of all pixels in the measured area of the lesion.
[d] Mean $\pm$ SEM.
[e] Groups within braces were not significantly different ($p > 0.05$).

Statistical significant Pearson correlation coefficients were presented between the TMR (ΔM) and confocal microscopy parameters measured across treatment groups. For the thin section parameters, the correlations were 0.88 for ΔM vs area, 0.63 for ΔM vs total fluorescence, and 0.40 for ΔM vs average fluorescence. Pearson correlation coefficients demonstrated strong positive significant correlations between TMR ΔM measurements and confocal microscopy lesion area measurements. Average lesion fluorescence was the confocal parameter that correlated the least with ΔM. For the specimen halves parameters the correlations were 0.71 for ΔM vs area, 0.70 for ΔM vs total fluorescence, and 0.61 for ΔM vs average fluorescence.

Results from this remineralization study are similar to those described for enamel demineralization using CLSM.[16] Confocal microscopy can be used on thick specimens or thin sections of enamel stained with a 0.1 mM solution of rhodamine B as a method to replace or augment TMR when quantitating demineralization and remineralization of enamel specimens. Therefore, in enamel remineralization studies it is concluded that a statistically significant correlation exists between mineral changes (ΔM) measured using TMR and the changes in lesion parameters (area and total fluorescence) analyzed by confocal microscopy.

Measurement of the penetration of fluorescent dyes in the lesion is not the only method used in confocal microscopy to analyze changes in the mineral content of enamel. Measurement of changes in the autofluorescence of enamel has been used with some success to determine the size of the lesions.[30] To determine if early enamel mineral changes can be quantified by measuring differences in autofluorescence (AF) using CLSM, 30 human polished enamel specimens (3 mm in diameter) were demineralized for 96 hr and then half of each specimen was covered with an acid-resistant nail varnish. Specimens were divided into three groups and subjected for 20 days to a cyclic remineralization regimen with 0, 250, or 1100 ppm F dentifrice slurries. Specimens were cut and analyzed by TMR, analyzed with CLSM for AF, and stained with a fluorescent dye (FD) for 1 hr and analyzed again with CLSM. The three different techniques detected significantly greater remineralization ($p < 0.05$) in the specimens treated with the fluoride-containing dentifrices than in the 0-ppm F-treated specimens. Pearson correlation coefficients were calculated for AF-, TMR-, and FD-stained samples. In demineralized samples, TMR ΔZ vs AF depth = 0.72, TMR ΔZ vs FD depth = 0.76, and FD depth vs AF depth = 0.74. In remineralized samples, TMR ΔZ vs AF depth = 0.84, TMR ΔZ vs FD depth = 0.80, and FD depth vs AF depth = 0.79. In lesion change from demineralization to

---

[30] C. González-Cabezas, M. Fontana, M. E. Wilson, and G. K. Stookey, *J. Dent. Res.* **77**(SI-A), 116 (1998).

remineralization, TMR ΔM vs AF depth difference = 0.71, TMR ΔM vs FD depth difference = 0.71, and FD depth difference vs AF depth difference = 0.76. It was concluded that early enamel mineral changes can be monitored by measuring the changes in autofluorescence of the enamel with CLSM. However, the change in autofluorescence of enamel was not always in the same direction: sometimes the lesioned area was more fluorescent, sometimes less fluorescent. Nonetheless, the change in fluorescence marked the depth of the lesion.

Measurement Mineral Loss of Dentin

Dentin is more mineralized than bone but less than enamel. It is composed (volume) of 46% inorganic material, 33% organic material, and 21% water. van der Veen and ten Bosch[31] found that the autofluorescence of demineralized dentin is up to 10 times greater than that of normal dentin when measured with a Raman spectroscope. They also observed the same phenomenon when using CLSM. Based on the results of their study and on some of the previously discussed studies on enamel, we developed an investigation to determine if dentin demineralization could be quantified by measuring differences in autofluorescence using CLSM.[32] Sixty-six human root surface dentin specimens were collected and assigned randomly to four different groups (A, B, C, D). Group D was the baseline whereas groups A, B, and C were subjected to a HAP/0.1 M lactic acid–carbopol solution (pH 5.0) at 37° for 12, 24, and 40 hr, respectively, to create lesions with different degrees of demineralization. Thin specimens (100 μm thick) were prepared and analyzed using TMR and then analyzed with CLSM for autofluorescence. Pearson correlation coefficients were calculated for confocal parameters and TMR ΔZ. CLSM parameters that were correlated were area (A), total fluorescence (TF), average fluorescence (AF), and average depth (AD). Correlation results indicated ΔZ vs A = 0.79, ΔZ vs TF = 0.80, ΔZ vs TF = 0.47, and ΔZ vs AF = 0.42. Data suggested that certain parameters of CLSM can be used to measure the dentin demineralization of root surfaces using autofluorescence. We are currently studying the use of CLSM to measure mineral gain in dentin.

Inaba et al.[33] measured the diameter of dentin tubules to determine mineral changes. Most of the demineralization and remineralization process

[31] M. H. van der Veen and J. J. ten Bosch, *Car. Res.* **30,** 93 (1996).

[32] S. Hart, C. González-Cabezas, M. Fontana, M. E. Wilson, and G. K. Stookey, *J. Dent. Res.* **77**(SI-A), 247 (1998).

[33] D. Inaba, H. Duschner, W. Jongebloed, H. Odelius, O. Takagi, and J. Arends, *Eur. J. Oral Sci.* **103,** 368 (1995).

occurs within the dentin tubules. However, this technique has not been compared with any other validated methodology. Büyükylmaz et al.[34] analyzed the changes in light reflected from dentin in specimens exposed to tetrafluoride. They observed differences in reflection of the surface of the lesion when treated with the remineralizing solution.

Conclusion

The studies described in the article support the use of confocal laser scanning microscopy as an effective technique for measuring in vitro mineral changes in dental tissues. However, most of the research up to date has focused on the two-dimensional analysis of affected areas, not taking into account the potential for optical sectioning of the teeth that CLSM offers. Future studies need to focus on the three-dimensional analysis of carious lesions, as new information could be gained by analyzing volume changes and mineral distributions.

[34] T. Büyükylmaz, B. Øgaard, H. Duschner, J. Ruben, and J. Arends, Adv. Dent. Res. 11(4), 448 (1997).

[28] Characterization of Ocular Cellular and Extracellular Structures Using Confocal Microscopy and Computerized Three-Dimensional Reconstruction

By DANIEL BROTCHIE, NEIL ROBERTS, MIKE BIRCH, PENNY HOGG, C. VYVYAN HOWARD, and IAN GRIERSON

Introduction

Visualizing biological specimens in two dimensions (2D) using techniques such as bright-field microscopy or transmission electron microscopy can provide invaluable data on tissue morphology and morphogenesis. However, before these techniques can be used it is often necessary to cut the specimens into thin 5- to 10-μm sections. By analyzing serial tissue sections using these techniques it is possible to appreciate the 3D structure of specimens. However, when specimens have a more complex 3D architecture and exhibit subtle structural changes, this approach can result in the collection of data and the formation of conclusions that are not representative of the specimen as a whole. This problem may be compounded by structural artifacts such as tearing of the specimen at the time of sectioning.

Conceptions of 3D architecture can be made using scanning electron microscopy (SEM). However, although SEM allows 3D observation, the image is only of the surface of a specimen. The SEM processing procedure may also introduce structural artifacts into a specimen as a result of either enzymatic/physical forces used to remove overlying tissue[1] or the process of critical point drying used in conventional SEM to dehydrate the specimens prior to imaging within the SEM vacuum chamber.[2]

Confocal microscopy combined with computerized 3D reconstruction overcomes the problems associated with 2D and SEM techniques and allows the entire specimen to be observed in 3D with only minimal tissue disruption. This is achieved through the confocal principle of restricting the collection of an image to the plane of the specimen located within the optical focal plane of the confocal microscope system.[3,4] The image is hardly degraded as a result of detection of either stray light or out-of-focus structures located either above or below the focal plane. By using a motorized stage to sequentially move the specimen vertically up through the focal plane of the system, the technique allows a series of images to be collected from adjacent optical sections through the depth or z axis of the specimen. This 3D data set of images is collectively known as a z series.

Images collected by confocal microscopy systems are usually stored in an 8-bit digital form. Each image in a z series is displayed as a 2D array of square picture elements or pixels, with each pixel having both a specific x, y (column, row) coordinate and an intensity value. The x, y coordinate refers to the spatial position of the pixel within the array, whereas intensity values are derived from a range of 256 gray scales[5] with pixels appearing black having a gray-scale value of 0 and those appearing white having a value of 255. As the images in a z series are obtained with only translation of the specimen in the z direction, the images in each series are in perfect x, y registration with each other. This article uses a series of ocular tissues to demonstrate how intact or thick specimens can be prepared and analyzed using confocal microscopy and computerized 3D reconstruction.

Tissue Preparation

Although it is possible to analyze some specimens by confocal microscopy utilizing either their natural reflectance or autofluorescence, the major-

[1] M. Birch, D. Brotchie, N. Roberts, and I. Grierson, *Ophthalmologica* **211,** 183 (1997).

[2] A. Boyde, E. Bailey, S. J. Jones, and A. Tamarin, *Scan. Electr. Microsc.* **1,** 507 (1977).

[3] A. Hall, M. Browne, and V. Howard, *Proc. R. Microsc. Soc.* **26,** 63 (1991).

[4] R. H. Webb, *Methods Enzymol.* **307,** [1] (1999) (this volume).

[5] F. Morgan, E. Barbarese, and J. H. Carson, *Scann. Microsc.* **6,** 345 (1992).

ity of specimens are first processed with a fluorescent marker directed against a structural component of interest. The localization of this fluorochrome is then detected by exciting it using light of a suitable wavelength emitted by a laser connected to the confocal microscopy system.

Processing thick specimens with fluorochrome markers provides a greater technical challenge compared to their use on thin sections analyzed by conventional fluorescence microscopy. However, even with highly dense neurological tissues, depths of up to 50 μm can be labeled routinely. In contrast, failure of light penetration limits the depth through which a specimen can be imaged by confocal microscopy.[6] This problem can be partially overcome by the use of longer excitation wavelengths in multiphoton confocal microscopy.[7] When intact specimens are still too thick to be analyzed intact by confocal microscopy, they can be sectioned into 50-μm-plus-thick volumes using either a cryostat[8] or a vibratome.[9]

Although penetration of reagents into thick specimens may be promoted by prior incubation with reagents such as Triton X-100 or Tween 20, potential disruption induced by these reagents may be avoided by incubating the free-floating specimen within the appropriate solution.[10] The wells of cell culture multiwell plates provide an ideal environment for these incubations as, in addition to their ordered arrangement, their dimensions allow economic use of reagents, which is particularly important when immunocytochemicals are being used. More economic and even less problematic are the use of fluorescent histochemical reagents. When 3D imaging of all tissue components is either required or acceptable, the specimen can be incubated with the standard solution of 1% eosin Y that is normally used as one of the components of the hematoxylin and eosin histology stain. Eosin has similar absorption and emission spectra to fluorescein and rhodamine allowing imaging using standard FITC or rhodamine detection filter sets.[11,12] When specific analysis of the 3D arrangement of a specimen's connective tissue framework is required, the specimen can be incubated with a 0.1% solution of sirius red diluted in saturated picric acid.[8,13,14] After prior incuba-

[6] J. B. Pawley and V. E. Centonze, in "Cell Biology: A Laboratory Handbook" (J. E. Celis, ed.), 2nd Ed., Vol. 3, p. 149. Academic Press, San Diego, 1998.

[7] W. Denk, J. H. Strickler, and W. W. Webb, Science 248, 73 (1990).

[8] D. Brotchie, M. Birch, N. Roberts, C. V. Howard, V. A. Smith, and I. Grierson, J. Neurosci. Methods 87, 77 (1999).

[9] J. J. Camp, C. R. Hann, D. H. Johnson, J. E. Tarara, and R. A. Robb, Scanning 19, 258 (1997).

[10] A. Herzog and C. Brösamle, J. Neurosci. Methods 72, 57 (1997).

[11] H. F. Carvalho and S. R. Taboga, Histochem. Cell. Biol. 106, 587 (1996).

[12] J. M. Apgar, A. Juarranz, J. Espada, A. Villanueva, M. Cañete, and J. C. Stockert, J. Microsc. 191, 20 (1998).

[13] F. Sweat, H. Puchtler, and S. I. Rosenthal, Arch. Pathol. 78, 69 (1964).

[14] L. C. U. Junqueira, G. Bignolas, and R. R. Brentani, Histochem. J. 11, 447 (1979).

tion of the specimen in a 0.2% solution of phosphomolybdic acid (PMA), the picrosirius red (PSR) solution binds to both interstitial and basement membrane collagens and, as with eosin, this localization can be observed using filter sets optimized for detection of either FITC or rhodamine. In addition, the PMA solution quenches autofluorescence of various tissue components, allowing more accurate localization of labeled structures and aiding subsequent 3D analysis.[8]

Tables I and II describe protocols for the use of histochemical and immunocytochemical reagents for analysis of specimens by confocal micros- copy. The descriptions given in Tables I and II relate to analysis of ophthal- mic tissue, a particular area of research interest for our group. However, by appropriate variation in incubation times the protocols may be adapted for the analysis of any other tissue.

TABLE I

HISTOCHEMICAL LOCALIZATION OF CONNECTIVE TISSUE IN THICK OPTIC NERVE HEAD SECTIONS USING PICOSIRIUS RED

1. Obtain human adult eyes within 48 hr postmortem
2. Dissect out region of each eye containing the optic nerve head and attached stalk of retrobulbar optic nerve and fix in 10% neutral buffered formalin for up to 6 months
3. Cryoprotect specimens in 10% sucrose/distilled H_2O for 48–72 hr/4°
4. Snap freeze specimens in liquid nitrogen-cooled methyl-2-butane and cryosection specimens trans- versely at 50–60 μm (e.g., with 2800 Frigocut-E cryostat, Reichert-Jung/Leica, Milton Keynes, UK)
5. Transfer sections into individual wells of 24-well cell culture plates (Costar, High Wycombe, UK) containing 1 ml distilled H_2O. Store plates at 4°
6. Remove sections from storage and incubate[a] them for 30 min in 1 ml 0.2% phosphomolybdic acid[b] (PMA)/distilled H_2O (B.D.H., Poole, UK). Preadjust PMA solution to pH 1.8–2.2
7. Wash sections by incubating them for 3× 4 min in fresh distilled H_2O
8. Incubate sections for 90 min in 0.1% sirius red F3B[c] (Color Index No. 35780) (H.D. Supplies, Aylesbury, UK) diluted in saturated aqueous picric acid (Sigma, Poole, UK). Preadjust PSR solu- tion to pH 1.8–2.2
9. Wash sections by incubating them for 3× 4 min in fresh distilled H_2O[d]
10. Transfer each section to a glass microscope slide and mount in glycerol beneath a number 1½ glass coverslip.[e] Store sections at 4° prior to imaging

[a] Perform each incubation of the procedure with the sections free-floating within fresh wells of cell culture plates. Turn sections over within their wells halfway through each incubation. Protect cell culture plates from light during each incubation.
[b] PMA is also known as dodecamolybdophosphoric acid.
[c] Sirius red is also known as direct red 80 or solophenyl red 3BL.
[d] When required, an optional final 4-min incubation in 0.01 M HCl/pH 1.8–2.2 may be used to remove background staining.
[e] Adjusting the glycerol mountant to pH 2 reduces elution of the dye from the sections after long- term storage.

TABLE II
IMMUNOCYTOCHEMICAL LOCALIZATION OF VIMENTIN IN MIGRATING CULTURED TRABECULAR
MESHWORK CELLS

1. Culture trabecular meshwork cells in Dulbecco's modified Eagle's medium supplemented with 10% fetal calf serum
2. Place chemoattractant[a] in lower wells of a modified 48-well Boyden microchemotaxis chamber (Neuroprobe Inc., Cabin John, MD)
3. Cover wells with gelatin-coated polycarbonate membrane (pore diameter 10 μm) (Neuroprobe), gasket, and upper section of chamber
4. Pipette 50 μl aliquot of cells[b] (20,000 cells per aliquot) into each of the upper wells
5. Incubate chamber in a 37° humidified incubator in 95% air/5% CO_2 (v/v) for 2.5 hr
6. Disassemble chamber, remove membrane, and fix in cold (−20°) methanol for 7 min. Cut membrane into 6-well sections and store at −20° until required
7. Localize intermediate filament vimentin using a standard FITC antibody-labeling procedure.[c] Mount membrane in aqueous antifade mounting medium
8. Store at 4° in the dark prior to imaging

[a] Chemoattractant 20 μg/ml soluble fibronectin (Sigma, Poole, UK).
[b] Cells in suspension in serum-free culture medium.
[c] Primary antibody; monoclonal antivimentin (Dako, Denmark) dilution 1:100. Time 2 hr (1 hr per side). Secondary antibody, FITC-conjugated goat antimouse (Sigma) dilution 1:20. Time 2 hr (1 hr per side).

Table I describes the use of PSR for the analysis of connective tissue in thick sections of the optic nerve. Connective tissue such as collagen and elastin is known to be present along the length of the optic nerve where it provides structural support to the axon bundles that transmit visual impulses from the retina to the brain. Of particular functional importance is the relationship between connective tissue and axon bundles in the optic nerve head, i.e., the anterior segment of the nerve into which retinal axons exit from the eye. Using 2D and SEM observations, it has been suggested that the axon bundles are structurally supported by approximately 11 connective tissue sheets that extend across the face of the nerve, perpendicular to the direction of the axon bundles. A change in the shape and therefore support provided by this framework occurs prior to the onset of the eye disease glaucoma.[15] Improved analysis of the 3D structure of the optic nerve head using confocal microscopy and computerized 3D reconstruction may help identify why some people are more susceptible to developing glaucoma and hence aid clinical diagnosis and treatment of this disease that can cause blindness.

Table II refers to our use of vimentin antibodies to localize the cytoplasm of intact cultured trabecular meshwork cells migrating through 10-μm pores

[15] H. A. Quigley, R. M. Hohman, E. M. Addicks, R. W. Massof, and W. R. Green, *Am. J. Ophthalmol.* **95,** 673 (1983).

in a polycarbonate membrane. The trabecular meshwork is a branching network of connective tissue located at the base of the iris in the anterior chamber of the eye. It forms the drainage channel for the continuously secreted aqueous humor that fills that chamber. The cells that line the meshwork trabeculae are thought to perform a vital purpose in maintaining a healthy functioning drainage system. These cells are gradually lost with age and this loss is noticeably higher in forms of glaucoma where increased pressure in the eye is followed by retinal damage and loss of sight. As the aqueous fluid contains many chemoattractants it is possible that the cells are stimulated to detach and migrate away from the trabecular beams, leaving behind a potentially compromised drainage system. A gelatin-coated polycarbonate membrane has been used as an *in vitro* model in order to investigate the effect of fibronectin, a major component of aqueous humor, on the attachment and motility of trabecular meshwork cells[16] and to also analyze their 3D cytoskeletal organization at different stages of the migratory process.

Optimizing Three-Dimensional Reconstruction

Accurate analysis of the 3D structure of biological specimens using confocal microscopy and computerized 3D reconstruction first requires that the original 3D data sets are optimally collected and processed.

Integral to the optimal collection of a 3D data set is knowledge of the confocal system's lateral (x or y) and axial (z) resolution. Lateral resolution is often set, not by optical theory, but by the x or y dimension of each pixel. This value can be determined for different objectives (and electronic zooms) by imaging a field of view of a known calibrated width and dividing this width by the number of pixels in the width of a 2D image.

The axial resolution of a standard epiconfiguration confocal microscope system is typically two to three times poorer than the optical lateral resolution due to the ellipsoidal nature of the point-spread function at the system's focal plane.[17] In practice, this axial resolution is poorer with low magnification objectives due to them usually having a lower numerical aperture (NA) than higher magnification ones. The axial resolution is also reduced further if the diameter/width of the confocal detection pinhole is increased, as the resulting image will contain signals detected from out-of-focus planes. The axial resolution of a system is also a measure of optical section thickness and therefore influences the appearance of 3D reconstructions. Changes

[16] P. Hogg, M. Calthorpe, S. Ward, and I. Grierson, *Invest. Ophthalmol. Vis. Sci.* **36,** 2449 (1995).
[17] P. J. Shaw and D. J. Rawlins, *J. Microsc.* **163,** 151 (1991).

occurring over the depth of a specimen will only be visible if this depth is thicker than the optical section thickness.

The axial resolution of a confocal microscopy system can be measured by analyzing a specimen that simulates an infinitely thin surface. Such specimens include a mirrored coverslip analyzed by reflectance confocal microscopy or coverslip-mounted 200-nm-diameter microspheres or mono-molecular films analyzed by fluorescence confocal microscopy.[18,19] Instead of scanning the specimen along each x and y direction of the field of view, the specimen is imaged by scanning the light source down x for a constant value of y. A series of these images is obtained through the specimen, starting and stopping at positions just above and below its surface. This so-called x, z sectioning results in the production of an image showing a side view of the specimen. A graph of pixel gray-scale intensity versus z can be plotted and should produce a bell-shaped plot with the intensity rising from a low level when the specimen is positioned above or below the focal plane to a maximum when its center is located within this plane. The plot represents the axial point-spread function of the confocal system, a measure of how far the point light source of the system extends axially from the focal plane. The width of the plot at the point where the pixel intensity is half the sum of the minimum and maximum intensities detected can then be measured, and this so-called full-width at half-maximum (FWHM) measure is also a measure of the axial resolution of the particular imaging configuration (wavelength, objective, pinhole).

Setting the vertical increments of the motorized stage of the confocal microscope system to match the value of the current axial resolution should ensure efficient collection of z series images. However, ideally the mode of image collection should be modified to satisfy the Nyquist sampling theorem. This states that if a specimen is to be optimally reconstructed in a digital image, the specimen has to be sampled at intervals of approximately double the axial resolution of the imaging configuration used. Therefore, for a system configuration with an axial resolution of 6 μm, the specimen should be optically sectioned at ~3-μm intervals. This situation is further complicated by the fact that the vertical movement of the microscope stage only results in the same incremental change in the plane of the specimen imaged if the entire medium between the objective lens and the plane of focus has the same refractive index (RI).[20] A constant RI may be present when viewing a glycerol-mounted specimen using a glycerol immersion objective, but not when viewing a glycerol (RI 1.47)-mounted specimen

[18] C. J. Cogswell, C. J. R. Sheppard, M. C. Moss, and C. V. Howard, *J. Microsc.* **158,** 177 (1990).
[19] M. Scharder, U. G. Hofmann, and S. W. Hell, *J. Microsc.* **191,** 135 (1998).
[20] H. Jacobsen and S. W. Hell, *Bioimaging* **3,** 39 (1995).

using a dry objective (RI of air = 1). In this case the vertical movement of the specimen plane imaged can be approximated by Eq. (1) based on Snell's law[21] in which

$$\text{Movement}_{\text{imaged plane}} = \text{Movement}_{\text{stage}} \times (RI_{\text{mountant}}/RI_{\text{immersion medium}}) \quad (1)$$

In the glycerol/air example, a vertical stage movement of 3 μm would therefore result in a specimen plane movement of ~4.5 μm. Therefore, to achieve 3-μm vertical movements of the imaged plane using this configuration requires sequential vertical stage movements of ~2 μm. With high NA objectives, in particular, RI mismatches can also lead to a reduction in both signal intensity and axial resolution over the depth of thick specimens.[20] When feasible, it is therefore advantageous if the RI of the immersion medium is matched to that of the specimen.

If the spatial characteristics of the original specimen are to be correctly reconstructed and measured in 3D, then it is essential that the reconstructed z axis is scaled properly. With advanced 3D reconstruction of a confocal z series, the computer interprets data displayed in each of the pixels forming each of the 2D images as belonging to a single cubic 3D pixel or voxel. In effect, when the z dimension of a specimen is being observed in a 3D reconstruction, you are observing the z dimensions of these voxels. A potential problem is that as the voxel is cubic, its z length does not represent the z depth of tissue over which signals were collected, but rather the lateral (i.e., x, y) distance in the original specimen represented by each pixel. Therefore, consider an imaging configuration in which lateral resolution is 2 μm and axial resolution is 6 μm. If 10 images are collected from a specimen at 6-μm-depth intervals, the z dimension of the 3D reconstruction will extend a distance of approximately 20 μm rather than the 60 μm that was actually imaged and the reconstruction therefore appears compressed. Similarly, if the z interval used is less than the original x, y pixel resolution, the reconstructed z axis would appear elongated. Consequently, to produce correctly scaled 3D reconstructions, specimens should be optically sectioned at (RI corrected) depth intervals equal to the x, y pixel resolution of the 2D image. Taking into account this factor happens to also help satisfy the Nyquist sampling theorem described earlier.

Although collecting images at lateral resolution-dependent depth intervals is optimal, a combination of a thick specimen and a high lateral resolution requires the collection of a large number of sections. Apart from the financial implications of the extended use of the confocal microscopy system, the prolonged illumination required for this collection may result in signals from deeper regions of the specimen fading before these regions

---

[21] E. M. Glaser, *J. Neurosci. Methods* **5**, 201 (1982).

are imaged. Although the use of antifading agents will help minimize this problem,[22] it is possible to avoid this situation and still obtain spatially correct 3D reconstructions by choosing a motor step size based solely on determined axial resolution. In the example where lateral resolution is 2 μm and axial resolution is 6 μm, this results in the collection of a data set of 10 images (60 μm depth/6 μm increments) unsuitable for correct spatial 3D reconstruction. However, this can be compensated for by computerized addition of 20 duplicate images into the data set so that the final number of 30 images equals the number that would have been obtained at lateral resolution-dependent, 2-μm intervals. The 20 duplicate images are added sequentially so that the first two are added between original images 1 and 3 and the final 2 after original image number 10. Whichever detection/processing regime is chosen, it is possible to check the validity of the reconstruction procedure by imaging and reconstructing fluorochrome-coated microspheres of known dimensions.[9]

To improve the appearance of 3D reconstructions and allow more accurate interpretation, preprocessing is often required. Histogram equalization algorithms equalize the spread of pixel intensities over the entire 256 gray-scale intensity range. The overall effect is to increase the contrast of an individual image. In addition, a global histogram algorithm acts to match the contrast distribution of images obtained from near the base of a specimen to those obtained near its surface which the user considers to have a more optimal contrast distribution. Histogram equalization generally results in images within 3D data sets having a more uniform appearance. In addition, median filtering is particularly effective in removing noise as it adjusts the gray-scale intensity value of each image pixel to match the median value of its eight neighbors.

Basic Three-Dimensional Reconstruction

A basic technique for extracting 3D information from a confocal data set is to project the entire series into one single image. A commonly applied algorithm is one that calculates the maximum gray-scale intensity value detected in each x, y position over all the z levels imaged. This results in the production of a so-called maximum intensity projection image in which unstained areas appear transparent. Structures localized near the base of the specimen are therefore observed in the same 2D image as ones localized at the surface of the specimen. More accurate interpretations of 3D architecture can be made by generating stereoscopic views of the 3D data set. This

[22] M. Berrios, K. A. Conlox, and D. E. Colflesh, *Methods Enzymol.* **307,** [4] (1999) (this volume).

can be achieved in a number of ways.[23] A common approach is to obtain two projections in which data have been progressively translated to the right or left with depth. After pseudo coloring, merging the projections results in the formation of a 2D red/green anaglyph. When viewed through red/green spectacles the viewer is able to appreciate depth by stereopsis.

Advanced Three-Dimensional Reconstruction

Z Dimension

Advanced processing features are often beyond the capabilities of the standard computer software and/or hardware associated with confocal microscopy systems and require the use of more advanced computer workstation-operated software. Fortunately, the production of 3D data sets by clinical imaging techniques such as magnetic resonance has driven the development of numerous software packages that are also potentially suitable for the advanced processing of confocal data sets. These include the public domain NIH Image (http://rsb.info.nih.gov/nih-image/) and commercially available packages such as ANALYZE (http://www.mayo.edu/bir/analyze/ANALYZE_Main.html) and 3DVIEWNIX (http://mipgsun.mipg.upenn.edu:80/~Vnews/). An up-to-date list of such software with details of their capabilities can be found within or via Lance Ladic's WWW site at http://www.cs.ubc.ca/spider/ladic/software.html.

Using advanced software packages it is possible to manipulate voxel-based 3D data sets to form a single image that allows all three dimensions of the original specimen to be viewed simultaneously, i.e., in addition to the imaged x, y surface plane, the image also shows a reconstructed view of at least one z plane edge. The direction of view can be adjusted interactively so that it appears the reconstruction is being animated on the computer monitor. This rendering of the data set therefore allows appreciation of 3D structural variation. These procedures can be applied to interactively chosen subsets of the 3D data by translation of a cut plane in a direction orthogonal (i.e., at 90°) to the other faces of the data set. This gives the appearance of the reconstructed specimen being interactively microtomed and allows 3D analysis of interior volumes of the specimen whose presence in the intact specimen was masked by surrounding tissue. In addition, data can be reformatted along any user defined oblique plane through the data array.

[23] C. J. R. Sheppard and D. M. Shotton, "Confocal Laser Scanning Microscopy." BIOS Scientific Publishers, Oxford, 1997.

Some of these effects are demonstrated in Fig. 1. Figure 1A shows four 2D confocal images collected 10 μm apart from within a PSR-stained, transverse 60-μm section of a human optic nerve head. A framework of "white" connective tissue surrounds the unstained axon bundles. Figure 1B shows a 3D reconstruction obtained using ANALYZE software to process the entire z series containing these four images. Reconstructing the original imaged volume in 3D allows an enhanced appreciation of the relationship between the connective tissue and the axon bundles. The connective tissue can be seen to both envelop the axon bundles and extend continuously through the depth of tissue imaged. Although the ~5-μm axial resolution of the imaging configuration may have masked the presence of finer spatial aspects, these observations partially contradict previous observations in which, perhaps due to the use of 2D imaging and/or disruptive processing procedures, the connective tissue arrangement appeared more segmented.[1] The 3D reconstruction shown in Fig. 1C was obtained from the same 3D data set as in Fig. 1B, except that only a subvolume of data was reconstructed. It is apparent from this "digitally microtomed" reconstruction that the arrangement of connective tissue identified in Fig. 1B is not restricted to the edge of the imaged volume.

A surface rendering process can be applied when more extensive information on the shape and surface of a specimen is required. Surface rendering first requires prior separation of structures of interest from a user determined background. This is achieved through a process of segmentation in which each image pixel is classified according to its gray-scale intensity value as representing either part of a structure of interest or part of the background. This classification typically involves computerized thresholding using a window of gray-scale intensity values chosen by the user. The next step is to convert the original 8-bit data set into a 1-bit binary data set in which all pixels representing a structure of interest appear white (binary value 1) whereas all other pixels appear black (binary value 0). The surface rendering algorithm is applied to the 3D binary data set and the surface of the reconstructed specimen becomes represented by a mesh composed of aligned geometric structures, e.g., triangular facets.[24] The mesh corresponds to the surface contours of the specimen and assignment to it of textural, light, and shading properties allows an enhanced appreciation of the surface's morphology. It is also possible to apply the surface rendering algorithm to two binary data sets corresponding to two features of interest extracted from the same original 8-bit data set, but using different threshold

[24] A. Martinez-Nistal, M. Alonso, F. González-Rio, A. Sampedro, and R. Astorga, *Microsc. Anal.* **Issue 63,** 19 (1998).

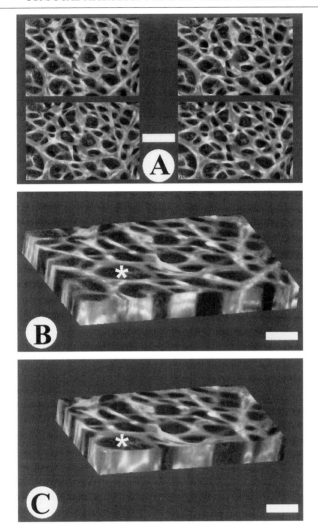

FIG. 1. (A) Four 2D confocal images collected 10 μm apart from within a PSR-stained, transverse 60-μm section of a human optic nerve head. (B) Three dimensional reconstruction obtained by advanced processing of the data set containing images shown in (A). (C) Three dimensional reconstruction obtained by "digitally microtoming" the reconstruction shown in (B). One nerve bundle whose depth can be observed is marked with an asterisk. The same nerve bundle is similarly marked in (B) where its 3D presence is masked by surrounding tissue. Axial resolution of imaging configuration ~ 5 μm. Bars: 162 μm (A) and 86 μm (B, C).

windows. Differentially pseudo coloring and then merging the binary data sets further aids analysis.

An alternative rendering procedure is volume rendering, which is useful if information on internal as well as surface structures is of interest. In contrast to surface rendering, volume rendering does not require prior segmentation and binarization of the original data. Instead the entire data set is pseudo-illuminated with rays from a chosen viewpoint. The degree to which each voxel contributes to the volume rendered reconstruction is determined by its original gray-scale value as well as both a given transparency attribute and its depth beneath the plane displayed.[24,25] The surfaces of the reconstructed specimen can be made to appear translucent and, as the nonsegmented data sets still contain all of the originally collected information, internal structures can be appreciated while surface perspective is retained. Alternatively, as shown in Fig. 2A, the surface voxels can be made less and less transparent until, finally, they appear opaque, allowing just the surface of the specimen to be viewed in 3D.

A maximum intensity projection algorithm can also be applied to data forming a 3D reconstruction. Figure 2B shows a maximum projection through the z plane of a x, y imaged trabecular meshwork cell. The cell was incubated with a vimentin–FITC antibody and is migrating through a nonfluorescent polycarbonate membrane. There is a concentration of vimentin cytoskeleton at both poles of the migrating cell. The size of the vimentin-rich area in the nonmigrated pole is indicative of a vimentin nuclear basket, suggesting that the portion of the cell containing the nucleus is the last to migrate. The concentration of vimentin at the migrated pole mirrors previous observations of cytokeratin distribution in epitheloid cells.[26]

X, Y Dimensions

Although confocal microscopy allows thick tissue specimens to be sectioned optically, this ability is dependent on the use of an objective lens with a suitably high NA. Although a 1.4 NA, 60× oil immersion objective can be used to image optical sections as thin as 0.7 μm,[27] this is achieved with a restricted field of view. Reducing objective magnification at the expense of optical section thickness improves the field of view, but in practice the lowest magnification objective suitable for analyzing thick

[25] W. C. Loftus, M. J. Tramo, and M. S. Gazzaniga, *Hum. Brain Map.* **3**, 257 (1995).
[26] H. L. Robey, P. S. Hiscott, and I. Grierson, *J. Cell Sci.* **102**, 329 (1992).
[27] D. L. Gard, in "Methods in Cell Biology" (B. Matsumoto, ed.), Vol. 38, p. 241. Academic Press, New York, 1993.

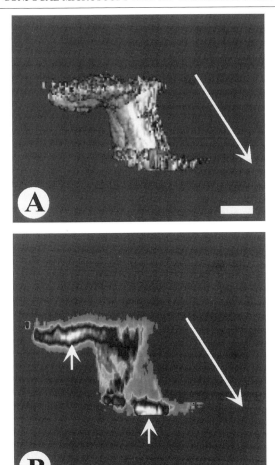

FIG. 2. Three-dimensional views of a cultured trabecular meshwork cell migrating through a polycarbonate membrane. The cytoskeleton of the cell has been labeled with a vimentin–FITC antibody. Long arrows show the direction of migration. (A) Volume rendered reconstruction showing the surface of cell. (B) Maximum intensity projection. Vimentin is concentrated in both migrated and nonmigrated poles of the cell (short arrows). Bars: 6 μm.

specimens is usually 10×. With a suitably high NA of approximately 0.5, a 10× objective would optimally allow imaging of approximately 7-μm optical sections. However, even with a 10× objective, only a small proportion of the x, y cross section of many specimens will be imaged within the field of view. Consequently, it would be difficult to analyze 3D changes in the structure of these specimens through their entire volume.

It is possible to overcome this problem as the reconstruction of specimens using confocal imaging and advanced computer software need not be restricted to their z dimension. Many advanced software packages also have the ability to reconstruct series collected from adjacent, but overlapping, fields of view. One way in which this can be achieved is to first obtain a z series through the depth of a central x, y region of a specimen. Similar z series can then be obtained from adjacent, but minimally overlapping x, y fields. In our analysis of the optic nerve, these procedures result in the collection of five z series, which between them typically cover ~90% of the volume of the optic nerve section. Only small isolated volumes around the edge of the section are left unimaged. The surface image from the centrally collected z series is first selectively processed along with the corresponding image from one of the four adjacent series. The central region image is interactively edited so that all areas are deleted except for the two to three structures common to both it and the second image. This process is repeated with the second image and a second edited image is obtained containing the same two to three structures as retained in the first. Using an algorithm designed to match common surfaces, a transformation matrix is then produced that is capable of translating the second edited image so that when combined with the first, the two images are superimposed. Applying this same matrix to all of the images in the original second z series will then similarly translate these data and when combined with the entire central z series, the two originally separate sets of data are displayed as a single series with their overlapping regions in register. If these procedures are repeated with the central z series and each adjacent z series in turn and all four translated data sets are combined with the nontranslated central one, the end result is a reconstruction of the surface of the entire specimen. This reconstruction will be of superior quality to the equivalent field of view image that could be obtained by conventional microscopy using a low magnification objective. As the reconstruction also contains depth information it can be viewed along any arbitrary plane and structures of interest can be highlighted in 3D by surface or volume rendering algorithms. Figure 3 shows the end result of applying a surface matching algorithm to a series of confocal data sets to reconstruct a cross section through a human optic nerve head. High signal regions denote the positions of overlap between the four outer fields and the central field of view.

If it is not feasible to image an intact specimen by confocal microscopy, z series can be collected from sequentially obtained, optimally thick sections and data from different sections combined using surface matching. Because of the nature of confocal microscopy image collection, images within each individual z series are aligned with one another with respect to the x, y field of view. However, the z series images obtained from one physical

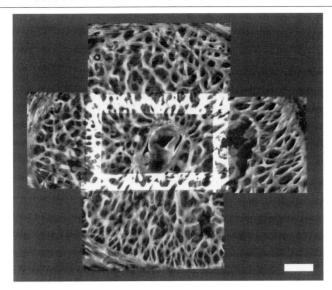

FIG. 3. Reconstruction of a low magnification x, y plane halfway between top and bottom surfaces of a PSR-stained, transverse 60-μm section of a human optic nerve head. Reconstruction was prepared by surface matching high magnification images obtained at that depth from adjacent fields of view. Bar: 222 μm.

section will not be x, y aligned with the z series images obtained from the underlying section. By surface matching the image collected at the base of one section with the image collected at the top of the deeper section, a transformation matrix can be obtained that will allow the two z series to be aligned and then combined with one another.

Future Directions: Incorporation of Stereological Methods

Between 1990 and 1998, over 200 scientific articles have been published concerned with the 3D reconstruction of features of interest in 3D image data sets obtained using confocal microscopy (source of information BIDS ISI Service, Bath, UK). The motivation for the majority of these studies is to reveal to the investigator the morphology of structures (cells, tissue, compartments, particles, etc.) that have hitherto only been imagined. However, inevitably, after having seen a glimpse of this new world, the scientist is immediately keen to measure these structures so as to be able to objectively test hypotheses about them.

It is surprising, therefore, that over the same 8-year period there have been only 25 scientific articles that describe the application of stereological methods in conjunction with confocal microscopy. Stereology, which may be defined as the statistical inference of geometric parameters from sampled information, provides a set of methods by which, for example, the volume, surface area, length, and number of structures of interest may be estimated efficiently and without bias from appropriately sampled sections through, or projections of, the 3D image data set.[28] Indeed, it is an important point that prior 3D reconstruction is not required in order that these measurements can be made.

There are likely to be a variety of reasons why so few measurements of geometric parameters have been made using confocal microscopy. A confocal microscope system can provide the appropriate images through a specimen noninvasively. Specimen preparation, however, remains important. Not all specimens can be studied and for many, as described earlier special staining techniques have had to be developed. In addition, there is concern over the thickness of the tissue section represented in the focal plane,[29,30] and algorithms and approaches are continually being developed aimed at narrowing this. As these problems are solved, a more lively interaction between the methods of 3D reconstruction and the methods of modern design stereology will surely evolve, with the interactive stereological measurement probes inevitably becoming an inherent feature of the computer software supplied with the microscope system.

Always, one will be intrigued to see a 3D computer rendering of a structure of interest. However, with this inquisitiveness appeased, one can go on to apply convenient and efficient stereological methods to obtain measurements of, for example, size distributions,[31] surface area using the vertical spatial grid[32] or isotropic fakir methods,[33] length from vertical slices or total vertical projections,[34] number,[35,36] and second-order properties such as spatial distribution of point processes[37,38] and curve length,[39] for an

[28] C. V. Howard and M. G. Reed, "Unbiased Stereology: Three-dimensional Measurement in Microscopy." BIOS Scientific Publishers, Oxford, 1998.
[29] H. Brismar, A. Patwardhan, G. Jaremko, and J. Nyengaard, J. Microsc. 184, 106 (1996).
[30] T. Masuda, J. Kawaguchi, H. Oikawa, A. Yashima, K. Suziuki, S. Sato, and R. Satodate, Path. Int. 48, 179 (1998).
[31] L. M. Karlsson and L. M. Cruz-Orive, J. Microsc. 165, 391 (1992).
[32] L. M. Cruz-Orive and C. V. Howard, J. Microsc. 178, 146 (1995).
[33] L. Kubinova and J. Janacek, J. Microsc. 191, 201 (1998).
[34] C. V. Howard, L. M. Cruz-Orive, and H. Yaegashi, Acta Neurol. Scand. 85, S14 (1992).
[35] D. A. Peterson, C. A. Lucidi-Phillipi, D. P. Murphy, J. Ray, and F. H. Gage, J. Neurosci. 16, 886 (1996).
[36] G. Kempermann, H. G. Kuhn, and F. H. Gage, J. Neurosci. 18, 3206 (1998).
[37] L. M. Karlsson and A. Liljeborg, J. Microsc. 175, 186 (1994).
[38] M. G. Reed, C. V. Howard, and C. G. Shelton, J. Microsc. 185, 313 (1997).
[39] N. Roberts and L. M. Cruz-Orive, J. Microsc. 172, 23 (1993).

unbiased sample of the same structures. In addition, the connectivity of a network may be measured using the conneulor technique.[40] It is likely that the combined use of these stereological tools with confocal microscopy and computerized 3D reconstruction will improve our understanding of the structural features and relationships that occur within normal and diseased tissues.

---

[40] H. J. G. Gundersen, R. W. Boyce, J. R. Nyengaard, and A. Odgaard, *Bone* **14,** 217 (1993).

[29] Multiphoton Excitation Microscopy, Confocal Microscopy, and Spectroscopy of Living Cells and Tissues; Functional Metabolic Imaging of Human Skin *in Vivo*

By Barry R. Masters, Peter T. C. So, Ki Hean Kim, Christof Buehler, and Enrico Gratton

Introduction

Multiphoton excitation microscopy combined with functional metabolic imaging based on NAD(P)H is an important development in cellular imaging. Nonlinear multiphoton excitation processes were predicted by Maria Göppert-Mayer in 1931.[1] An example of this nonlinear process is two-photon-induced fluorescence in which the simultaneous absorption of two photons of red light (800 nm) causes the subsequent emission of blue light (Fig. 1).

In practical terms this new technology has been implemented in nonlinear optical microscopes that use pulsed near-infrared light to induce the fluorescence of chromophores (ultraviolet and visible) in cells, tissues, and organs. The rate of a two-photon absorption process is a function of the square of the instantaneous intensity of the excitation light. The microscope objective focuses the light pulses into a diffraction-limited volume. At sufficiently high light intensity absorption and fluorescence can be induced from the fluorophores at a rate suitable for microscopy. Photobleaching and phototoxicity are limited to the same focal region. Outside the focal volume, the nonlinear multiphoton process occur with a much lower probability due to the reduced photon flux.

---

[1] M. Göppert-Mayer, *Ann. Phys. Lpz.* **9,** 273 (1931).

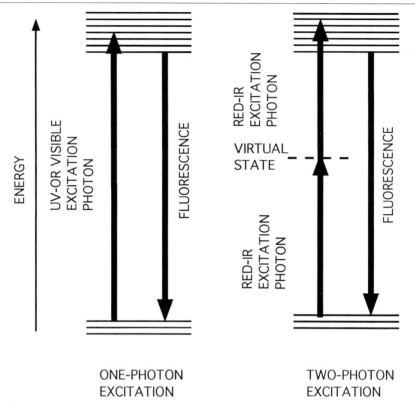

FIG. 1. Jablonski diagram of two-photon process versus single-photon process. For the two-photon excitation process there is a short-lived virtual state shown by a dashed line.

Denk and colleagues[2,3] first demonstrated this new type of microscope based on two-photon excitation of molecules by integrating a laser scanning microscope (scanning mirrors, photomultiplier tube detection system) and a mode-locked laser that generates femtosecond pulses of near-infrared light. The pulses of red or near-infrared light (700 nm) were less than 100 fsec in duration and the laser repetition rate was about 100 Hz. The high intensity, short pulses of near-infrared light cause significant multiphoton excitations; however, the low average power incident on the sample minimizes cell and tissue damage. The capabilities of these nonlinear microscopes include improved spatial resolution without the use of pinholes or

[2] W. J. Denk, J. H. Strickler, and W. W. Webb, *Science* **248,** 73 (1990).
[3] W. Denk, J. P. Strickler, and W. W. Webb, U.S. Patent 5,034,613 (1991).

slits for spatial filtering, improved signal strength, deeper penetration into thick, highly scattering tissues, confinement of photobeaching and photodamaging to the focal volume, and improved background discrimination.

In order to correctly interpret the results of applying multiphoton excitation microscopy to three-dimensional functional imaging of thick, highly scattering *in vivo* tissue such as human skin it is necessary to demonstrate the following. First, images of the cells based on the cellular autofluorescence are obtained and these images are correlated with similar images of the same tissue obtained with reflected light confocal microscopy. This use of correlative microscopy is important to demonstrate the histology of the tissue based on images acquired with multiphoton excitation microscopy. Second, the correct interpretation of skin physiology using multiphoton microscopy and spectroscopy requires that the tissue fluorophores responsible for the fluorescence contrast be identified correctly. Because only a few fluorophores in tissues have been studied, the correct identification of the fluorescent species requires a knowledge of its excitation and emission spectra, which requires knowing whether the fluorophores are excited by either two or higher photon processes. The characterization of the multiphoton process can be performed in the following manner. For a two-photon excitation process, it is necessary to show a plot of the quadratic dependence of the fluorescence intensity on the excitation power. A plot of the logarithm of the fluorescence intensity versus the logarithm of the excitation power should have a slope of 2 for a two-photon excitation process. Similarly, a slope of 3 indicates a three-photon process. Finally, for each type of cell or tissue studied, the emission spectra and the lifetime of the fluorescence should be determined to confirm the nature and biochemical state of the fluorophore, e.g., NAD(P)H or flavoprotein.

To demonstrate a multiphoton excitation process it is necessary to show the nonlinear dependence of the fluorescence on the laser excitation power. In the human skin system, we have measured the fluorescence intensity as a function of excitation power to determine whether the excitation at 960 nm is due to two-photon excitation or involves higher photon processes (Fig. 2). A 0.5-mm-thick slice of human skin from the second digit of the right hand was excised. The sample was sandwiched between a piece of cover glass and a microscope slide. The fresh sample was imaged immediately. The incident power was controlled by a polarizer. Fluorescence images of the same area were collected at power levels ranging from 1 to 10 mW with 960 nm excitation. Equivalent portions (10 × 10 pixel areas) of these images were averaged and compared. The slopes of the plots verify multiphoton excitation processes.

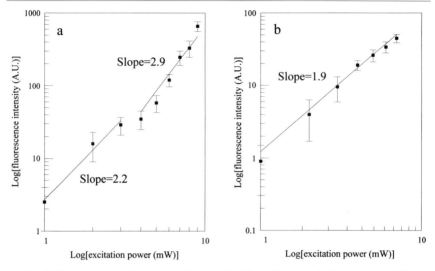

FIG. 2. Fluorescence emission intensity as a function of laser excitation power at 960 nm. The log–log plot is the experimental verification of a multiphoton excitation process; the slopes of 3 and 2 indicate three-photon and two-photon excitation processes, respectively. From B. R. Masters, P. T. C. So, and E. Gratton, *Biophys. J.* **72,** 2405 (1997).

How does multiphoton excitation microscopy compare with confocal laser scanning microscopy?[4] The main limitations of two-photon excitation microscopy are (1) multiphoton excitation microscopy is only suitable for fluorescent imaging; reflected light imaging is not currently available, and (2) the technique is not suitable for imaging highly pigmented cells and tissues that absorb near-infrared light. Multiphoton excitation microscopy has the following important advantages: (1) reduced phototoxicity, (2) reduced photobleaching, (3) increased penetration depth, (4) ability to perform uncaging or photobleaching in a diffraction limited volume, (5) ability to excite fluorophores in the ultraviolet without a ultraviolet laser, (6) the excitation and the fluorescence wavelengths are well separated, and (7) no spatial filter is required.

Functional Metabolic Imaging of Cellular Metabolism Based on NAD(P)H Fluorescence

Redox fluorometry is a noninvasive optical method used to monitor the metabolic oxidation–reduction (redox) states of cells, tissues, and organs.

[4] B. R. Masters, "Selected Papers on Confocal Microscopy" (B. R. Masters, ed.). SPIE, The International Society for Optical Engineering, Bellingham, WA, 1996.

It is based on measuring the intrinsic fluorescence of the reduced pyridine nucleotides, NAD(P)H, and the oxidized flavoproteins of cells and tissues.[5–12] Both the reduced nicotinamide adenine dinucleotide, NADH, and the reduced nicotinamide adenine dinucleotide phosphate, NADPH, are denoted as NAD(P)H. The fluorescence of NAD(P)H is in the range of 400–500 nm with a single photon absorption in the region of 364 nm. Redox fluorometry is based on the fact that the quantum yield of the fluorescence, and hence the fluorescence intensity, is greater for the reduced form of NAD(P)H and lower for the oxidized form. For flavoproteins, the quantum yield, and hence the fluorescence intensity, is higher for the oxidized form and lower for the reduced form. The reduced pyridine nucleotides are located in both the mitochondria and the cytoplasm. The flavoproteins are uniquely localized in the mitochondria. Fluorescence from oxidized flavoproteins occurs in the region from 520 to 590 nm with a single photon absorption in the region of 430 to 500 nm. Fluorescence from the reduced pyridine nucleotides is usually measured in tissue investigations as the measured fluorescence is higher than in the case of the flavoprotein fluorescence. Redox fluorometry has been applied to many physiological studies of cells, tissues, and organs.

Functional imaging of cellular metabolism and oxygen utilization using the intrinsic fluorescence has been studied extensively in various types of cells and tissues. Comprehensive reviews of instrumentation, techniques, and experimental results based on fluorescence measurements of the autofluorescence of cells have been published previously. Specific studies based on redox fluorometry include redox measurements of *in vivo* rabbit cornea based on flavoprotein fluorescence (Fig. 3), monitoring of oxygen

[5] B. Chance, B. Schoener, R. Oshino, F. Itshak, and Y. Nakase, *J. Biol. Chem.* **254**, 4764 (1979).
[6] B. R. Masters, S. Falk, and B. Chance, *Curr. Eye Res.* **1**, 623 (1981).
[7] B. R. Masters and B. Chance, *in* "Current Topics in Eye Research" (J. A. Zadunaisky and H. Davson, eds.), Noninvasive Corneal Redox Fluorometry, p. 140. Academic Press, London, 1984.
[8] B. R. Masters, *in* "The Cornea: Transactions of the World Congress of the Cornea III" (H. D. Cavanagh, ed.), p. 281. Raven Press, New York, 1988.
[9] B. R. Masters, A. K. Ghosh, J. Wilson, and F. M. Matschinsky, *Invest. Ophthalmol. Vis. Sci.* **30**, 861 (1989).
[10] B. R. Masters, *in* "Noninvasive Diagnostic Techniques in Ophthalmology" (B. R. Masters, ed.), p. 223. Springer-Verlag, New York, 1990.
[11] B. R. Masters and B. Chance, *in* "Fluorescent and Luminescent Probes for Biological Activity" (W. T. Mason, ed.), p. 44. Academic Press, London, 1993.
[12] B. R. Masters, *in* "Medical Optical Tomography: Functional Imaging and Monitoring" (G. Müller, B. Chance, R. Alfano, S. Arridge, J. Beuthan, E. Gratton, M. Kaschke, B. R. Masters, S. Svanberg, and P. van der Zee, eds.), p. 555. The International Society for Optical Engineering, Bellingham, WA, 1993.

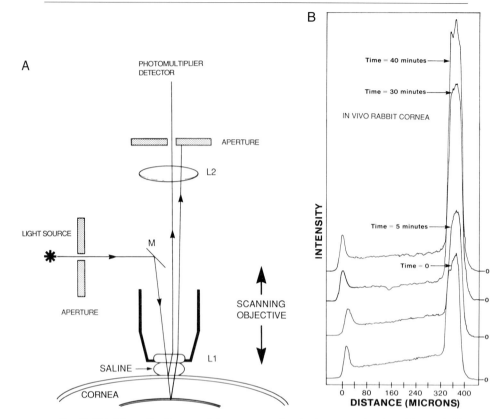

FIG. 3. (A) Schematic diagram of the z-scan confocal microscope developed for optically sectioning the cornea *in vivo*. The light source is connected to the confocal microscope with a quartz fiber-optic light guide. Two conjugate slits make the instrument a confocal microscope. One slit is imaged onto the cornea; the second slit is in front of the photon detector. The z-scan confocal microscope has a scanning microscope objective that moves on the optical axis under computer control via a piezoelectric driver. This device scans the focal plane through the full thickness of the cornea under computer control. (B) *In vivo* redox metabolic functional imaging of the rabbit cornea. The z-scan confocal microscope shown in (A) repeatedly scans the 400-μm-thick cornea and measures the NAD(P)H fluorescence intensity as a function of position in the cornea and shows the time dependence of the intensity of NAD(P)H fluorescence (across the full thickness of the cornea *in vivo*) after the application of a thick PMMA contact lens on the cornea. The small peak on the left side is the corneal endothelium. The larger peak on the right side is from the corneal epithelium. Time is the duration that the contact lens was on the cornea. Successive z scans of the cornea are displaced vertically for clarity. Note that during the 40-min period of contact lens wear (blocks oxygen flow into the cornea from the air) the NAD(P)H fluorescence intensity of the epithelium doubled. This effect is reversed completely when the contact lens is removed.

tension under the contact lens in live rabbits, chemical analysis of nucleotides and high energy phosphorous compounds in the various layers of the
rabbit cornea, and redox fluorescence imaging of the *in vitro* cornea with
ultraviolet confocal fluorescence microscopy.

One important aspect of this technique, which is often overlooked, is
the experimental verification of the chemical identities of the fluorophores
that contribute to the fluorescence signal. Several experimental techniques
are used to verify that the predominate contribution to the measured fluorescence is from the reduced pyridine nucleotides, NAD(P)H. Methods of
microanalysis and enzyme cycling have been used to measure the concentrations of NADH and NADPH, the reduced pyridine nucleotides, in the
normoxic state and in the anoxic states of the rabbit cornea. This analysis
was performed for the epithelial, the stromal region, and the endothelium.
Another method (Fig. 3B.) measures the time course of the increase of
fluorescence following full reduction of the oxidized pyridine nucleotide,
NAD^+, following the application of cyanide, which blocks cytochrome oxi-
dase in the mitochondria. An alternative technique to verify NAD(P)H
as the source of fluorescence, which is discussed later, is to measure the
fluorescence spectrum and the fluorescence lifetime.

Ultraviolet Confocal Fluorescence Microscopy Compared with Multiphoton Excitation Microscopy

It is instructive to compare redox imaging of NAD(P)H fluorescence
from the cornea with both single-photon confocal microscopy and two-
photon excitation microscopy. The cornea is the 400-μm-thick, transparent
tissue located on the front surface of the eye.

In confocal microscopy studies, an ultraviolet confocal laser scanning
fluorescence microscope was used to investigate the *ex vivo* rabbit cornea.
The laser scanning confocal microscope used an argon–ion laser with a
light output at 364 nm. The excitation wavelength of 364 nm and the
emission at 400–500 nm were used to image the fluorescence of the reduced
pyridine nucleotides, NAD(P)H. A Zeiss water-immersion microscope objective 25× with a numerical aperture of 0.8 and corrected for the ultraviolet, was used for imaging the fluorescence.

Rabbit eyes were obtained from New Zealand White rabbits. The rabbits
were euthanized with an injection into the marginal vein of the ear with
ketamine hydrochloride (40 mg/kg) and xylazine (5 mg/kg). The eyes were
freed immediately of adhering tissue and swiftly enucleated. A specimen
chamber was designed and developed to maintain the eye in a physiological
state during the data acquisition period and to gently immobilize it during
microscopic imaging. The intact eye was placed in the specimen chamber,

perfused with aerated, bicarbonate Ringer's solution, which contained 5 mM glucose and 2 mM calcium, and kept at 37°. The specimen chamber preserved the normal corneal morphology for several hours; however, in the absence of the special specimen chamber, cells in the cornea of the *in situ* eye showed time-dependent morphological alterations.[13]

Analysis of the images obtained with ultraviolet confocal laser scanning fluorescence microscopy showed photobleaching of the NAD(P)H during the course of the measurements. The microscope objective was designed for the refractive index of water. However, there are significant variations of the index of refraction across the full 400 μm of corneal thickness. While the average refractive index of the cornea is known, the profile of refractive index is unknown. The microscope objective had chromatic aberrations for excitation and emission wavelengths. These effects resulted in reduced image quality in the ultraviolet confocal microscope.

Two-photon excitation microscopy was used to image cells of the freshly excised rabbit cornea.[14–16] The initial experiments used a modified confocal scanning laser microscope (Bio-Rad, Richmond, CA, MRC-600) that was coupled to a Zeiss Universal upright microscope. The epi-illumination collimator lenses were removed. This is an example of how to modify an existing Bio-Rad laser scanning confocal microscope in order to convert it to an instrument for multiphoton excitation microscopy. It is suggested that in order to maximize the sensitivity of the modified instrument, the fluorescence light from the specimen should not be descanned prior to photon detection. Instead, the fluorescence light should be detected with external photon detection, which bypasses the descanning unit. This follows from the physics of the multiphoton excitation process; the optical sectioning capability is on the excitation side and it is derived from the nonlinear optical absorption of the fluorophore.

A Satori ultra fast dye laser from Coherent Corporation was used to generate 150-fsec pulses of 705-nm light at a repetition rate of 76 MHz. A dichroic mirror (Omega Optical 550DCSP) was used to direct the excitation light to the specimen. The microscope objective was a Leitz water-immersion objective lens (50×/1.0) with a free working distance of 1.7 mm. The resulting fluorescence from the NAD(P)H in the cornea was collected by the microscope objective and descanned by the confocal microscope scanning mirrors. The confocal aperture was open in order to maximize the fluorescence signal. Other microscope modifications included replacing the inter-

[13] B. R. Masters, *Scan. Microsc.* **7**(2), 645 (1993).
[14] B. R. Masters, A. Kriete, and J. Kukulies, *Appl. Opti.* **32**(4), 592 (1993).
[15] D. W. Piston, B. R. Masters, and W. W. Webb, *J. Micros.* **178,** 20 (1995).
[16] B. R. Masters, P. T. C. So, and E. Gratton, *Cell Vision* **4**(2), 130 (1997).

nal mirrors with those optimized for 380–650 nm and replacing the existing photomultiplier tube (PMT) with a Thorn-EMI 9924B. The average excitation power at the specimen was 10–15 mW.

NAD(P)H fluorescence was imaged from the superficial cells of the corneal epithelium and from the corneal endothelium (Fig. 4). In comparison to single photon confocal microscopy there were several advantages. There was minimal photobleaching of the NAD(P)H. Images from the basal epithelial cells showed improved contrast and image quality.

More recently, a specially designed and constructed multiphoton excitation microscope was used to investigate the functional imaging of the *in situ* cornea based on NAD(P)H imaging. The use of emission spectroscopy was used to further validate that the fluorophore was NAD(P)H. The use of external photon detection in this design (described later) permitted multiphoton imaging of the stromal keratocytes in the normoxic cornea. In comparison, the modified Bio-Rad instrument with descanning of the fluorescence prior to photon detection was not capable of imaging the weakly fluorescent stromal keratocyte cells in the absence of complete reduction of the pyridine nucleotides with cyanide treatment, which results

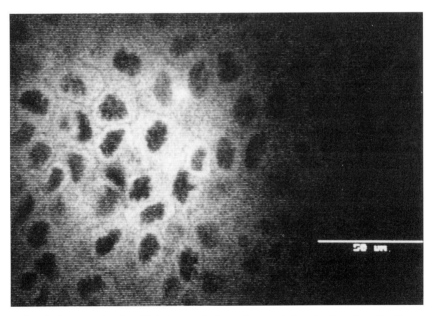

FIG. 4. An example of multiphoton excitation microscopic imaging based on the fluorescence of NAD(P)H. The dark oval regions are cell nuclei of the basal epithelial cells in a freshly excised rabbit cornea. The cell borders are shown as dark lines between adjacent cells. Scale bar: 50 μm. From D. W. Piston, B. R. Masters, and W. W. Webb, *J. Micros.* **178**, 20 (1995).

in a doubling of the concentration of NAD(P)H. The two-dimensional imaging of the corneal endothelial cells based on NAD(P)H functional redox imaging with multiphoton excitation microscopy and spectroscopy presents a new technique for investigating the heterogeneity of cellular respiration in the corneal cell layers.

Laser Sources for Multiphoton Excitation Microscopy

The appropriate selection of a laser light source is critical in designing a multiphoton excitation microscope. For a one-photon absorption process, the rate of absorption is the product of the one-photon absorption cross section and the average of the photon flux density. For a two-photon absorption process, in which two photons are absorbed simultaneously by the fluorophore, the rate of absorption is given by the product of the two-photon absorption cross section and the average squared photon flux density. In respect for the work of Maria Göppert-Mayer, the units of two-photon absorption cross section are measured in GM (Göppert-Mayer) units. One GM unit is equal to $10–50$ cm^4 s/photon.

To achieve a reasonable probability of the occurrence of a multiphoton excitation process, a pulsed laser source of high peak power is required. A mode-locked dye laser can provide sufficient excitation intensity at the focal volume of the microscope objective. Typical laser pulses are in the red region of the spectrum (630 nm) with a duration of less than 100 fsec and a repetition ration of 80 MHz. These pulses are of sufficient intensity to generate fluorescence in a fluorophore that usually absorbs at 315 nm. Alternatively, two photons in the infrared region at 1070 nm could excite a fluorophore that usually absorbs at 535 nm in the visible region. The two photons must interact simultaneously with the molecule. The following quantities are useful for rough calculations. Typically the laser pulse width is 10^{-13} sec, the fluorescence decay time is 10^{-9} sec, and the pulse separation time is 10^{-8} sec. At a laser pulse repetition rate of 100 MHz, there is a laser pulse every 10^{-8} sec, or once every 10 nsec. Two-photon excitation processes do not require that the two photons absorbed simultaneously be of the same wavelength. Two different wavelengths can be combined by superimposing pulsed light beams of high peak powers. The two wavelengths can be chosen using the following equation:

$$1/\lambda_{ab} = 1/\lambda_1 + 1/\lambda_2 \tag{1}$$

where λ_{ab} is the short wavelength of the absorber and λ_1 and λ_2 are the incident beam wavelengths.

The most critical component of a multiphoton microscope is the laser light source. The fluorescence excitation rate is insufficient for scanning microscopy in deep tissue applications unless femtosecond or picosecond pulsed lasers are used. Although the use of a CW infrared laser for two-photon imaging of cell cultures has been achieved, its applicability in deep tissue work is limited. The choice between femtosecond and picosecond light sources for multiphoton excitation remains controversial. As discussed, multiphoton excitation can be achieved with both types of lasers. To generate the same level of fluorescence signal, picosecond laser systems will need a significantly high average input power whereas femtosecond laser systems have a much higher peak power.

Several laser sources are available for multiphoton excitation microscopy.[17-21] Important considerations in laser selection for multiphoton excitation microscopy include peak power, pulse width, range of wavelength tunability, cooling requirement, power requirement, and cost. The range of tunability should cover the region of interest for multiphoton excitation processes of typical fluorophores. It is important to note that single-photon and two-photon excitation spectra for many molecules are different. Published two-photon excitation spectra (plots of two-photon excitation cross sections as a function of wavelength) are extremely useful in optimizing the selection of a laser wavelength for a specific molecule. This article focuses on multiphoton excitation microscopy based on NAD(P)H functional imaging; however, progress is being made in synthesizing molecules with very large two-photon absorption cross sections.

Titanium–sapphire laser systems are good choices. These systems provide a high average power (1–2 W), high repetition rate (80–100 MHz), and short pulse width (80–150 fsec). The titanium–sapphire laser provides a wide tuning range from below 700 nm to above 1000 nm. Titanium–sapphire lasers require pump lasers. The older systems use argon ion lasers, but the new ones use solid-state diode-pumped Nd : YAG lasers. The newer lasers are basically turnkey systems. Ultracompact (the size of a shoe box), single wavelength femtosecond lasers combining diode, Nd : YAG, and titanium–sapphire lasers are becoming available commercially. Furthermore, other single wavelength solid-state systems, such as diode-pumped Nd : YLF lasers and diode-pumped erbium-doped fiber laser systems, have also become available.

[17] J. A. Valdmansis and R. L. Fork, *IEEE J. Quant. Electron.* QE-**22**, 112 (1986).

[18] E. Gratton and M. J. vandeVen, *in* "Handbook of Biological Confocal Microscopy" (J. B. Pawley, ed.), Laser Sources for Confocal Microscopy, p. 69. Plenum Press, New York, 1995.

[19] D. L. Wokosin, V. E. Centonze, J. White, D. Armstrong, G. Robertson, and A. I. Ferguson, *IEEE J. Select. Top. Quant. Electr.* **2**(4), 1051 (1996).

[20] C. Xu, R. M. Williams, W. Zipfel, and W. W. Webb, *Bioimaging* **4**, 198, (1996).

[21] C. Xu and W. W. Webb, *in* "Nonlinear and Two-Photon-Induced Fluorescence" (J. Lakowicz, ed.), p. 471. Plenum Press, New York, 1997.

Pulse Compensation Techniques

For multiphoton excitation processes the number of photon pairs absorbed for each laser pulse is related inversely to the pulse width. The passage of ultrashort pulses from the mode-locked laser through a dielectric medium results in pulse broadening due to group velocity dispersion. This pulse broadening, called pulse dispersion, is a serious problem in multiphoton excitation microscopy because it results in a reduction in the probability of multiphoton excitation. This is observed as a strong reduction in the intensity of the fluorescence signal from the fluorophores in the specimen. An experimental technique called "prechirping" the laser pulses can be used to compensate the laser pulse dispersion. A "prechirp" unit can be constructed with two prisms and mirrors. This system compensates for the group velocity dispersion and causes all of the different wavelengths in each pulse to arrive simultaneously at the specimen after propagating through the microscope optics.

An important paper describes how to measure the group velocity dispersion for a microscope with a variety of microscope objectives and how to "prechirp" the pulses to achieve pulse compensation.[22] A Michelson-type interferometric autocorrelator was attached to the microscope, which was used to measure the pulse length of the laser pulses at the specimen. For example, a Coherent MIRA 900-F titanium–sapphire femtosecond laser system (780 nm, 72 MHz, 92 fsec) was adapted to the Zeiss confocal laser scanning microscope LSM 410 with a C-APOCHROMAT 40×/1.2 water-immersion microscope objective. The measured pulse length at the specimen was 355 fsec. Recompression of the pulse width to 135 fsec at the specimen was achieved with the use of a "prechirp" unit consisting of a pair of prisms.

Multiphoton Excitation Microscope

The instrumentation and design of a basic multiphoton microscope have been described in a number of previous publications.[23–31] In addition to

[22] R. Wolleschensky, T. Feurer, and R. Sauerbrey, *Scanning* **19**(3), 150 (1997).
[23] P. T. C. So, T. French, W. M. Yu, K. M. Berland, C. Y. Dong, and E. Gratton, *in* "Fluorescence Imaging and Microscopy" (X. F. Wang and B. Herman, eds.), p. 351. Wiley, New York, 1996.
[24] C. Y. Dong, P. T. C. So, T. French, and E. Gratton, *Biophys. J.* **69**, 2234 (1995).
[25] P. T. C. So, T. French, W. M. Yu, K. M. Berland, C. Y. Dong, and E. Gratton, *Bioimaging* **3**, 49 (1995).
[26] J. Art, *in* "Handbook of Biological Confocal Microscopy" (J. B. Pawley, ed.), Photon Detectors for Confocal Microscopy, p. 183. Plenum Press, New York, 1995.
[27] B. R. Masters, P. T. C. So, and E. Gratton, *Biophys. J.* **72**, 2405 (1997).
[28] B. R. Masters, P. T. C. So, and E. Gratton, *Ann. N.Y. Acad. Sci.* **838**, 58 1998.

the laser light source, a critical element is a high throughput microscope system with beam-scanning electronics. High numerical aperture microscope objectives are critical for efficient two-photon excitation and the detection of low level signals. As compared to a confocal system, chromatic aberration is not a crucial factor as the different color emission light does not have to be descanned. Maximizing infrared transmission decreases the scattering of the high power excitation light, which interferes in detecting the low-level fluorescence signal. Most implementations of multiphoton scanning microscopes incorporate infrared femtosecond light sources with an existing scanning confocal microscope (Fig. 5). This implementation is fairly straightforward where only mirrors in the scanning head have to be modified to efficiently reflect infrared laser light. It is also critical to modify the emission beam path such that high loss descanning optics are not used. Because the mechano-optics involved in multiphoton microscopes is much simpler than the confocal system, modifications needed to convert a high throughput fluorescence microscope for multiphoton scanning are actually quite simple. It mainly involves modification of the excitation light path to incorporate a scan lens and to provide an electronic interface to synchronize the scanner mirror system and the data acquisition electronics.

Properly designed detection electronics is essential for a high-performance multiphoton scanning microscope. Typical scanning systems use photomultiplier tubes with analog detection circuitry. Photomultiplier tubes have good quantum efficiency (10–25%) in the blue-green region, but poor (less than 1%) quantum efficiency in the red. Although analog circuity does not provide very sensitive detection at very low signal levels, it can handle fairly high intensity signals without saturation. It is also critical that analog conversion circuitry have sufficient dynamic range in cellular imaging applications where signal strength can vary greatly. Typically, a dynamic range of at least 12 bits is desirable. Another common detection method uses a single photon counting scheme that is very sensitive at extremely low light situations and has excellent dynamic range, but suffers from the difficulty of easy saturation at high intensity. Other detectors that are promising for multiphoton microscopes are high sensitivity single photon-counting avalanche photodiode detectors, which have excellent quantum efficiency

[29] B. R. Masters, P. T. C. So, and E. Gratton, *Lasers Med. Sci.* **13**(3), 196 (1998).
[30] K. H. Kim, P. T. C. So, I. E. Kochevar, B. R. Masters, and E. Gratton, *in* "Functional Imaging and Optical Manipulation of Living Cells" (D. L. Farkas and B. J. Tromberg, eds.). SPIE, Bellingham, WA, 1998.
[31] P. T. C. So, T. French, W. M. Yu, K. M. Berland, C. Y. Dong, and E. Gratton, *Bioimaging* **3**, 49 (1995).

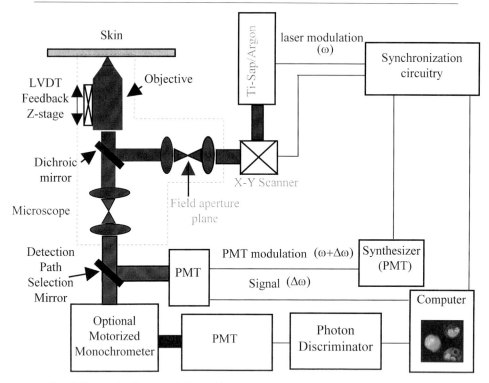

FIG. 5. Schematic diagram of the multiphoton excitation scanning microscope. Two detection beam paths are shown: one for measuring fluorescence lifetimes and the other for measuring the fluorescence emission intensity/spectrum. These two beam paths are selected by rotating a single mirror. PMT, photomultiplier tube; LVDT, linear variable differential transformer; titanium–sapphire/argon, argon ion laser pumped titanium–sapphire laser. From B. R. Masters, P. T. C. So, and E. Gratton, *Biophys. J.* **72,** 2405 (1997).

(over 70%) at the red spectra range and very good efficiency (30–50%) in the green range.

The laser used in the microscope constructed at the Laboratory for Fluorescence Dynamics, University of Illinois at Urbana-Champaign, and at the Department of Mechanical Engineering, M.I.T., is a mode-locked titanium–sapphire laser (Mira 900, Coherent Inc., Palo Alto, CA). Two-photon excitation microscopy at 730 and 960 nm was used. A Glan-Thomson polarizer and a laser pulse picker are used either alone or in combination to control the excitation laser pulse train profile. The beam-expanded laser light is directed into the microscope via a galvanometer-driven x–y scanner (Cambridge Technology, Watertown, MA). Images are generated by raster scanning the x–y mirrors. The excitation light enters the Zeiss Axioskopt microscope (Zeiss Inc., Thornwood, NY) via a modified epiluminescence light path. The scan lens is positioned such that the x–y scanner is at its eye point, whereas the field aperture plane is at its focal point. Because the objectives are infinity corrected, a tube lens is positioned to recollimate the excitation light. The scan lens and the tube lens function together as a beam expander, which overfills the back aperture of the objective lens. The excitation light is reflected by the dichroic mirror to the objective. The dichroic mirrors are custom-made, short-pass filters (Chroma Technology Inc., Brattleboro, VT) that maximize reflection in the infrared and transmission in the blue-green region of the spectrum. The water-immersion objective used in most of these studies is a Zeiss c-Apochromat 40× with numerical aperture of 1.2. The objective axial position is driven by a stepper motor interfaced to a computer. The microscope field of view is about 50–100 mm on a side. The typical image acquisition time is about 2 sec.

The fluorescence emission is collected by the same objective and is transmitted through the dichroic mirror along the emission path. An additional barrier filter is needed to further attenuate the scattered excitation light because of the high excitation intensity used. Because two-photon excitation has the advantage that the excitation and emission wavelengths are well separated (by 300–400 nm), suitable short-pass filters such as 2 mm of the BG39 Schott glass filter (CVI Laser, Livermore, CA) eliminate most of the residual scatter with a minimum attenuation of the fluorescence. A descan lens is inserted between the tube lens and the photomultiplier tube. The descan lens recollimates the excitation. It also ensures that the emission light strikes the PMT at the same position independent of scanner motion. A single photon-counting signal detection system is implemented. The fluorescence signal at each pixel is detected by a R5600-P PMT (Hamamatsu, Bridgewater, NJ), which is a compact single photon-counting module with high quantum efficiency.

The instrumentation for wavelength-resolved spectroscopy measurements in multiphoton microscopy is essentially similar to the technology used in one-photon systems with two exceptions. First, because the two-photon excitation wavelength is red shifted to twice the normal spectra region, there is always good separation between two-photon excitation and emission spectral regions. This wide separation allows the emission intensity to be isolated easily from the high-intensity excitation. More importantly, the overlap between excitation and emission spectra in the one-photon case prevents the shorter wavelength region of the emission spectra to be studied, but this is not a problem in the two-photon case. Second, two-photon excitation at the properly chosen wavelength region can simultaneously excite fluorophores in the near-ultraviolet, blue, green, and red ranges. In the multiple labeling experiment, this capability allows all the fluorescent probes to be excited simultaneously and their individual distributions mapped using wavelength-resolved spectroscopy. For a rough wavelength separation, as in applications where a number of chromophores need to be resolved, different wavelength channels can be constructed using a number of dichroic filters. When finer wavelength resolution is needed to study spectral features, single-point measurement using monochromameters or spectrographs can be performed.

The lifetime-resolved method is another powerful spectroscopy technique that has been integrated with multiphoton microscopy. Lifetime imaging can resolve multiple structures in cells and tissues similar to wavelength-resolved imaging. It can also be used to quantitatively measure cellular metabolite concentration such as calcium using nonratiometric probes such as Calcium Green. The instrumentation involved in lifetime imaging inside a multiphoton microscope is very similar to that of a confocal system. In the confocal system, lifetime measurements have been performed using both time-domain and frequency-domain techniques. Because multiphoton microscopy is still a relatively new approach, only the incorporation of frequency domain techniques have been attempted. Lifetime measurements were performed using frequency-domain heterodyning techniques, which have been described previously.[25]

Simultaneous Multiphoton Excitation Microscopy and Reflected Light Confocal Microscopy

There are many experimental studies when it would be of great advantage to have the capability of simultaneous multiphoton excitation microscopic and reflected light confocal imaging. Because the multiphoton excitation microscope is limited to fluorescence imaging, it is very important to have a simultaneous capability to study the morphology of cells and tissues based on reflected light confocal microscopy. In the original two-photon excitation microscope the infrared photons that are reflected by the sample are rejected at the dichroic mirror. If an additional photon detection path is incorporated into the standard multiphoton excitation microscope, these infrared photons scattered from the sample could be collected and detected to form a confocal image.[30] The reflected infrared photons are collected by the microscope objective and then collimated by the excitation tube lens and scan lens combination. This collimated beam is then descanned by being directed backward through the x–y scanning mirrors. This descanned beam is then focused by a lens through a confocal pinhole aperture onto a photodetector. A polarizer can be used to reduce the specular reflection of the infrared light beam back into the confocal detector path (Fig. 6). Choices of the photodetector include a fast silicon photodiode (PDA50, Thorn Labs, Newton, NJ) or a high-sensitivity avalanche photodiode (Ad-

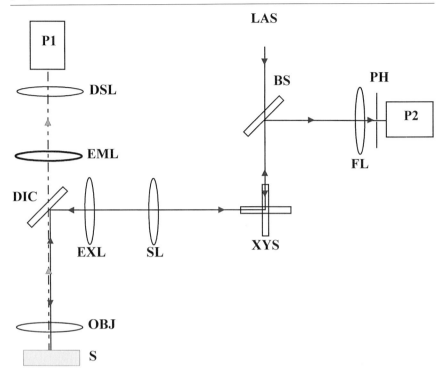

FIG. 6. A schematic of a new multiphoton excitation microscope design showing the new reflected light confocal beam path. S, sample; OBJ, objective; DIC, dichroic mirror; EXL, excitation tube lens; SL, scan lens; XYS, x–y scanner; BS, beam splitter; FL, focusing lens; PH, pinhole aperture; LAS, laser light source; EML, emission tube lens; DSL, descan lens; P1, fluorescence PMT; P2, confocal PMT.

vanced Photonics, Camarillo, CA). Data acquisition of the single photon-counting circuit for the multiphoton excitation fluorescence signal and the reflected light signal are synchronized by a custom interface circuit. The only effect on the multiphoton excitation microscope is a factor of two loss in the excitation power at the input beam splitter. This loss of power can be compensated easily by increasing the initial power of the laser source.

Video Rate Multiphoton Excitation Microscopy

There are a number of studies on living cells, tissues, and organisms in which a video rate of data acquisition is required.[32–35] For example, in order

[32] M. Rajadhyaksha, M. Grossman, D. Esterowitz, R. H. Webb, and R. Anderson, J. Invest. Dermatol. **104,** 946 (1995).
[33] B. Guild and W. W. Webb, Biophys. J. **68,** 290a (1995).

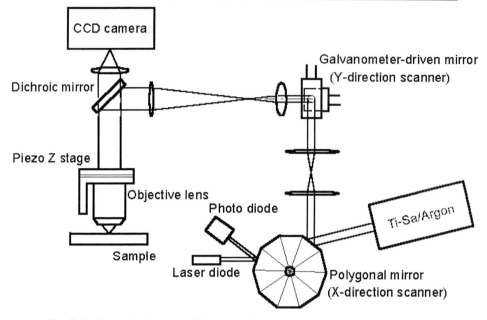

FIG. 7. A schematic of a new multiphoton excitation microscope design showing the system components critical to video rate capability. The 50 facet polygonal mirror can run at a maximum speed of 3000 rpm. A pair of relay lens project the laser spot from the polygonal mirror facets onto the galvanometer-driven x–y scanner. Only the y axis of the galvanometer scanner is used for raster scanning because the scanning in the x direction is provided by the polygonal mirror. The x-axis galvanometer-driven mirror is used for easy positioning of the scan region.

to acquire a stack of images of thick human skin *in vivo,* it is necessary to minimize the motion of the skin. This problem is substantially easier if the time of data acquisition can be reduced to the range of tens of milliseconds per frame. One technique to achieve video rate data acquisition is to use a rapidly rotating polygon mirror for scanning on one axis and to use a galvanometer-driven mirror for scanning on the orthogonal axis (Fig. 7). Other promising approaches include line scanning and multifocal scanning microscopes.

[34] J. Brakenhoff, J. Squier, T. Norris, A. C. Bliton, W. H. Wade, and B. Athey, *J. Microsc.* **181**(Pt 3), 253 (1996).

[35] J. Bewersdorf, R. Pick, and S. W. Hell, *Opt. Lett.* **23,** 655 (1998).

Multiphoton Excitation Microscopy of *in Vivo* Human Skin: Emission
Spectroscopy and Fluorescent Lifetime Measurements

Multiphoton excitation microscopy at 730 and 960 nm was used to image
in vivo human skin autofluorescence (Fig. 8).[27] The lower surface of the
right forearm (of one of the authors) was placed on the microscope stage
where an aluminum plate with a 1-cm hole is mounted. The hole is covered
by a standard cover glass. The skin was in contact with the cover glass to
maintain a mechanically stable surface. The upper portion of the arm rested
on a stable platform prevented motion of the arm during the measurements.
The measurement time was always less than 10 min. The estimated power
incident on the skin was 10–15 mW. The photon flux incident on a diffraction-limited spot on the skin is on the order of 10 MW/cm^2. We observed
individual cells within the thickness of the skin at depths from 25 to 75 μm
below the skin surface. No cells were observed in the stratum corneum.

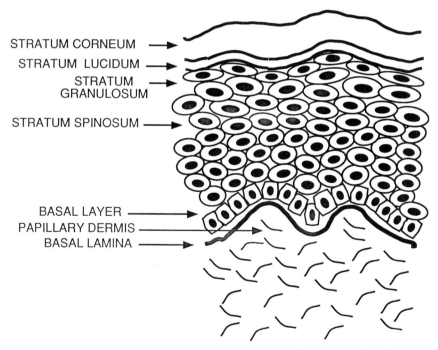

STRATUM CORNEUM →
STRATUM LUCIDUM →
STRATUM →
GRANULOSUM
STRATUM SPINOSUM →
BASAL LAYER →
PAPILLARY DERMIS →
BASAL LAMINA →

FIG. 8. A vertical section of skin showing the following cell layers from the skin surface
to the dermis: stratum corneum, stratum lucidum, stratum granulosum, stratum spinosum,
basal cell layer, and the papillary dermis. Individual cells forming the stratum corneum and
the thin stratum lucidum are not shown. From B. R. Masters, P. T. C. So, and E. Gratton,
Biophys. J. **72,** 2405 (1997).

These results are consistent with studies using reflected light confocal microscopy

In order to show the three-dimensional distribution of the autofluorescence, we acquired optical sections with the two-photon excitation microscope and formed a three-dimensional visualization across the thickness of the *in vivo* human skin [Figs. 9 (color insert), 10 (color insert), and 11][27, 36–38]

It is important to characterize the source of the fluorescence that is imaged with multiphoton excitation microscopy. Two types of measurements are useful in the characterization of the fluorophore: emission spectroscopy and lifetime measurements. We measured these characteristics at selected points on the skin. Fluorescent spectra were obtained close to the stratum corneum (0–50 μm) and deep inside the dermis (100–150 μm). Measurements were made for both 730- and 960-nm excitation wavelengths corresponding to one-photon excitation wavelengths of about 365 and 480 nm, respectively. The fluorescent lifetimes were measured at selected points on the skin to complement the fluorescent spectral data obtained. The lifetime results support NAD(P)H as the primary source of the autofluorescence at 730 nm excitation. The chromophore composition responsible for the 960-nm excitation is more complicated.

Use of Laser Pulse Picker to Mitigate Two-Photon Excitation Damage to Living Specimens

For a single-photon process, fluorescence excitation typically uses continuous wave lasers and requires an average power of about 100 μW. For a two-photon excitation microscope, lasers with 100-MHz and 100-fsec pulses are often used and require an average power of about 10 mW. This corresponds to a peak power on the order of 1 kW. For typical cells and tissues, the two-photon photodamage pathways still need to be better determined.[39,40] Because typical two-photon excitation microscopy requires milliwatt average power and laser pulse trains consisting of nanojoule pulses, severe thermal damage can occur resulting from one-photon excitation. This situation is particularly severe for highly pigmented cells such as melanophores. Furthermore, the high peak power required for two-photon excitation may produce cell and tissue damage through dielectric break-

[36] B. R. Masters, G. Gonnord, and P. Corcuff, *J. Microsc.* **185**(3), 329 (1996).

[37] B. R. Masters, *Bioimages* **4**(1), 13 (1996).

[38] B. R. Masters, D. J. Aziz, A. F. Gmitro, J. H. Kerr, T. C. O'Grady, and L. Goldman, *J. Biomed. Opt.* **2**(4), 437 (1997).

[39] K. König, P. T. C. So, W. W. Mantulin, B. J. Tromberg, and E. Gratton, *J. Microsc.* **183**(3), 197 (1996).

[40] K. König, P. T. C. So, W. W. Mantulin, and E. Gratton, *Opt. Lett.* **22**, 135 (1997).

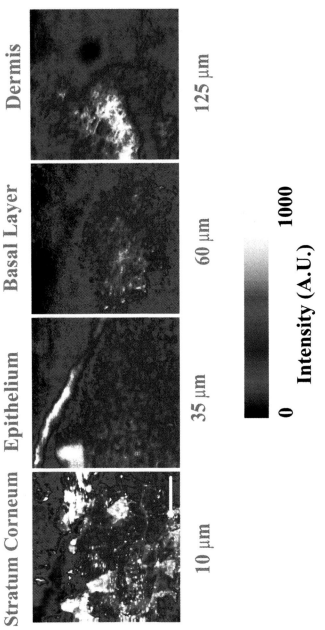

Stratum Corneum   **Epithelium**   **Basal Layer**   **Dermis**

10 μm   35 μm   60 μm   125 μm

0   Intensity (A.U.)   1000

FIG. 9. Multiphoton excitation microscopy of human skin with an excitation wavelength of 780 nm. The stratum corneum is clearly seen on the cell surface. Keratinocytes 15–20 μm in diameter were imaged at an approximate depth of 40–50 μm in the epidermis. Basal cells of about 10 μm were observed at a depth of 60 μm below the skin surface. Punctated fluorescence was observed in the cytoplasm of the larger cells. Similar findings were reported previously. These fluorescent organelles are likely to be mitochondria with a high concentration of NAD(P)H. In the dermis layer between 80 and 150 μm, collagen/elastin fibers can be clearly discerned. Bar: 32 μm.

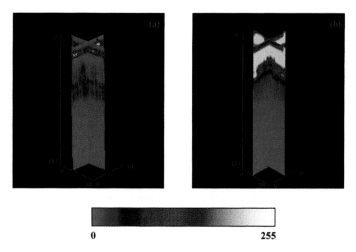

0   255

FIG. 10. Three-dimensional images of *in vivo* human skin: (a) 730 nm excitation and (b) 960 nm excitation. *x* and *y* orthogonal slices are shown. The axis dimensions are in micrometers. (a) The top bright layer corresponds to the stratum corneum. The second bright band at a depth of 80–100 μm is the basal cell layer at the top of the papillary dermis. (b) The top bright band extends throughout the stratum corneum. From B. R. Masters, P. T. C. So, and E. Gratton, *Biophys. J.* **72,** 2405 (1997).

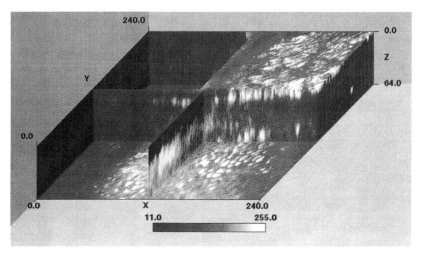

FIG. 12. Three-dimensional image of human skin *in vivo*. A tandem scanning reflected light confocal mioscope was used to acquire the optical sections. The human skin was visualized with Dicer software (Spyglass, Inc., Champaign, IL), which permits the display of three-dimensional volume data as a set of orthogonal slices. From B. R. Masters, G. Gonnord, and P. Corcuff, *J. Microsc.* **185**(3), 329 (1996).

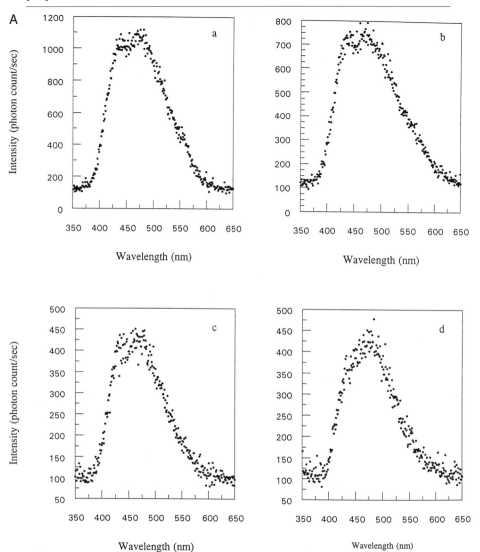

FIG. 11. Point emission spectrum of *in vivo* human skin at 730 (A) and 960 (B) nm excitation. Intensity is given in photon counts per second. (a and b) Two distinct points on the surface of *in vivo* human skin between 0 and 50 μm deep. (c and d) Two distinct points at a depth between 100 and 150 μm. From B. R. Masters, P. T. C. So, and E. Gratton, *Biophys. J.* **72,** 2405 (1997).

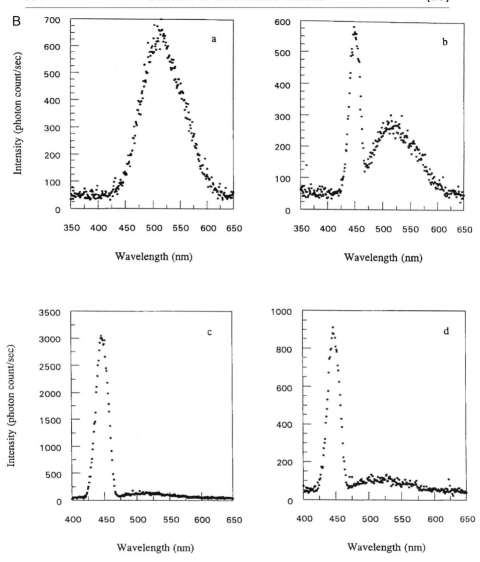

FIG. 11. (*continued*)

down mechanisms. For multiphoton excitation microscopy, oxidative photodamage is also generated. The processes that result in photodamage
to the specimen are complex. As a comparison between two-photon and
confocal methods, less photodamage occurs in the two-photon case as the

excitation volume is less than a femtoliter, whereas photodamage in the confocal case occurs throughout the optical path in the specimen.

During a two-photon excitation process, it is important to realize that three-photon excitation can also be generated. Although the emission filters may be selected for fluorescence detection from the two-photon process, there may still be photodamage from the unintentional three-photon excitation process. When intentionally using three-photon excitation for tissue imaging, it is important to realize that three-photon excitation microscopy often requires an order of magnitude higher incident laser power than two-photon excitation in order to achieve similar rates of excitation. It is important to study the associated three-photon cellular damage mechanisms and to identify the damage threshold. Nevertheless, three-photon excitation microscopy provides a useful technique to excite those fluorophores with single photon absorption bands in ultraviolet and short-ultraviolet wavelengths.

For the successful study of highly pigmented cells such as melanophores, it is critical to reduce thermal damage by reducing power deposition from individual laser pulses as well as reducing heat accumulation from the pulse train. The power management of laser pulses is typically accomplished by using an attenuator to decrease laser pulse energy. With energy sufficiently attenuated such that thermal damage from individual pulses is insignificant, heat accumulation from the pulse train remains a major problem. It is possible to reduce the laser pulse energy further to minimize the effect of heat accumulation. However, given a constant pulse width, reducing the pulse energy decreases fluorescence excitation efficiency quadratically. It is preferable to reduce the laser pulse repetition rate by using a laser pulse picker that maximizes the ratio between the peak power and the average power. This strikes the best compromise between minimizing thermal effects and maintaining fluorescence excitation efficiency. We have accomplished this goal using a laser pulse picker.[41] If a Glan-Thompson polarizer is used to reduce the light intensity of the laser pulse train, then the laser average power, as well as the laser peak power, is reduced. With an ideal pulse picker, the average light intensity of the laser pulse train can be decreased, not affecting the peak intensity. This is because the pulse picker operates by removing a fraction of the laser pulses, but does not affect the peak power of each laser pulse. Real pulse pickers, which are based on acousto-optical deflectors, result in about 50% instantaneous power loss. We have demonstrated that when the laser power is attenuated with a polarizer, the fluorescence intensity of the fluorophores decreases quadrati-

[41] B. R. Masters, C. Y. Dong, P. T. C. So, C. Buehler, W. M. Mantulin, and E. Gratton, *J. Micros.,* in preparation (1999).

cally with the excitation power. When laser power is attenuated with a laser pulse picker, the fluorescence intensity decreases only linearly with the excitation power. It is concluded that decreasing the pulse repetition rate with a laser pulse picker is the preferred method of choice for mitigating tissue thermal damage when studying living cells and tissues with a multiphoton excitation microscope. In the future three-dimensional optical biopsy will provide a new diagnostic tool (Fig. 12).

[30] Video-Rate, Scanning Slit Confocal Microscopy of Living Human Cornea in Vivo: Three-Dimensional Confocal Microscopy of the Eye

By Barry R. Masters and Matthias Böhnke

Biomicroscopy of the Living Eye from Slit Lamp to Confocal Microscope

The optimal application of confocal microscopy to ocular tissue requires specially designed and optimized instrumentation and an examination technique that takes into account the unique structural and optical properties of the human eye.[1] This article describes the unique capabilities of a video-rate, scanning slit confocal microscope and its application in the living human cornea. A confocal microscope provides two enhancements compared to a standard light microscope: enhanced lateral resolution and enhanced axial resolution. The latter property is the basis of its capability to optically section a thick, highly scattering specimen.

The cornea is an avascular, transparent, living optical element in the front portion of the eye. Investigation of the physical basis for the transparency of the cornea and the structural alterations that affect corneal transparency, such as corneal wounds and corneal refractive surgery, are topics of active study. For the designer of a video-rate, clinical confocal microscope the optical problem is how to image a moving, transparent cornea with very little intrinsic contrast.

There is a direct and interesting lineage from the confocal microscope developed by Goldmann, to the development of the specular microscope by Maurice, Koester, and others, to the various types of clinical confocal

[1] R. L. McCally and R. A. Farrell, in "Noninvasive Diagnostic Techniques in Ophthalmology" (B. R. Masters, ed.), p. 189. Springer-Verlag, New York, 1990.

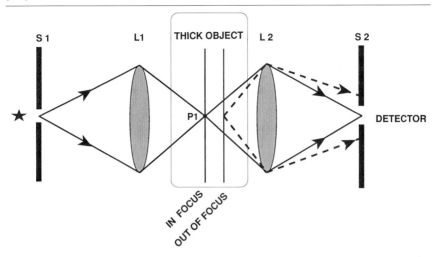

Fɪɢ. 1. Diagram illustrating the principle of a confocal imaging system. From M. Böhnke and B. R. Masters, *Progr. Retina Eye Res.* **18**(3), 1 (1999).

microscopes.[2] These instruments represent partial solutions to the problem of how to image thin optical sections from a 500-μm-thick, transparent, moving object: the human cornea *in vivo*.

Optical Principles of Confocal Microscope

It is useful to illustrate optical principles of the confocal microscope with a simple diagram. A confocal microscope is a type of microscope in which a thick object, such as the cornea, is illuminated with a focused spot of light. The same microscope objective used to illuminate a point, P1, in the object is used to collect the scattered and reflected light from the same point P1. This is illustrated in Fig. 1. For simplicity, we show two separate microscope objectives placed on opposite sides of a thick translucent object. The microscope objective on the left side is used to illuminate a point (P1) in the object, whereas the microscope objective on the right side is used to detect the illuminated point within the object. A point source of light at aperture S1 is focused by lens L1 onto a small spot in the focal plane within the object. The second lens L1 is positioned so that its focus is also in the same focal plane and coincident (confocal) with the spot of illumination. Lens L2 forms an image of the illuminated spot at aperture

[2] C. J. Koester and C. W. Roberts, *in* "Noninvasive Diagnostic Techniques in Ophthalmology" (B. R. Masters, ed.), p. 99. Springer-Verlag, New York, 1990.

S2. All of the light that lens L2 collects from the illuminated spot enters the aperture at S2 and reaches the detector.

How does a confocal microscope discriminate against light that is not in the focal plane? For real thick scattering objects the point of light imaged by lens L1 onto the thick object will focus the light as a spot in the focal plane; however, there will be light of lower intensity in the double cone on both sides of the focal plane. In Fig. 1 the light paths that are from an out-of-focus plane are shown as dotted lines. In this case, some of the scatted light from an out-of-focus plane is collected by lens L2. In this case the collected light is spread out at aperture S2. Only a small amount of the light that is spread out enters aperture S2. Therefore, the detector will detect only a small amount of the light from the out-of-focus planes. This is the physical basis for strong discrimination against out-of-focus light in a confocal microscope. In Fig. 1 the diameter of the apertures is enhanced greatly for visual clarity. Clinical confocal microscopes have a single microscope objective. The single objective is typically used for point illumination of the object and to simultaneously image the point of illumination on the aperture of the light detector.[3] The optical principle of the confocal microscope has been simply illustrated for a single point of the object being illuminated and the same point being imaged simultaneously on the detector. Both aperture S1 and aperture S2 are cofocused (confocal) on the same point in the focal plane. This is the origin of the word confocal.

Objective Scanning Confocal Microscope

Figure 2 illustrates the mechanical and optical components of a confocal microscope with computer-controlled objective scanning on the z axis. The device is a confocal microscope because it contains two conjugate slits. The confocal microscope has been used in the vertical mode for work on tissue culture and for studies of *in vitro* eyes. When the confocal microscope is shifted into the horizontal mode, it is used for *in vivo* studies on animals or human subjects. This confocal microscope can be used to measure the following signals: z-axis profiles of fluorescence from NAD(P)H, fluorescence from oxidized flavoproteins, fluorescence from extrinsic probes, i.e., mitochondrial or nuclear stains, and simultaneous measurements of backscattered light. This confocal microscope is used to investigate the metabolism of the cells in the cornea *in vivo*. The unique feature of the confocal microscope is a z-scanning objective. A piezoelectric driver, under computer

[3] B. R. Masters, "Selected Papers on Confocal Microscopy," v. MS131. The International Society for Optical Engineering, Bellingham, 1996.

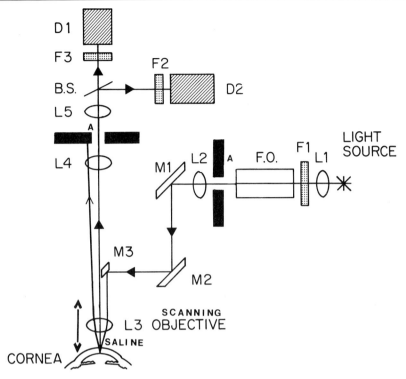

Fig. 2. Schematic diagram of the scanning one-dimensional confocal microscope showing a light ray path. The light source is either a laser or a mercury arc lamp connected to the microscope by a fiber optic. F1 and F2 are narrow band interference filters to isolate the excitation wavelengths. F3 is a narrow band interference filter to isolate the emission light. There are two conjugate slits. M1, M2, and M3 are front surface mirrors, and B.S. is a quartz beam splitter. L3 is the scanning objective 50×, NA 1.00. The piezoelectric driver scans the microscope objective along the optic axis of the eye. This confocal microscope is suitable for use with tissue culture or in the horizontal mode for use with living animals or human subjects. The depth resolution is 6 μm with a 100× microscope objective and 18 μm with a 50× microscope objective. From B. R. Masters, *in* "The Cornea: Transactions of the World Congress on the Cornea III" (H. D. Cavanagh, ed.), p. 281. Raven Press, New York, 1988.

control, scans the microscope objective along the z axis; this motion of the objective scans the focal plane across the thickness of the cornea. The z axis spatial resolution of this instrument system is shown in Fig. 3.

Several design features of this instrument were incorporated into more recent designs of clinical confocal microscopes.[4] The slit confocal microscope has *two adjustable slits* that can be used to optimize the signal-to-

[4] M. Böhnke and B. R. Masters, *Progr. Retina Eye Res.* **18**(5), 1 (1999).

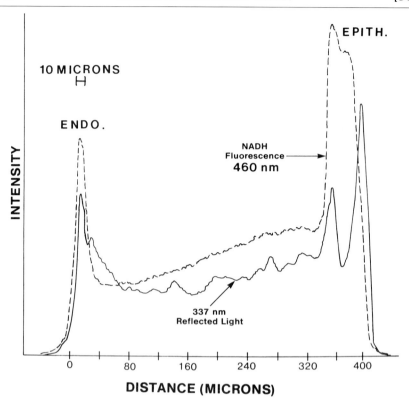

F IG. 3. One-dimensional *in vivo* confocal microscopy of the living eye. An optical section through a rabbit cornea illustrating the range resolution for back-scattered light (solid line) and NAD(P)H fluorescence emission (broken line). The intensity of the back-scattered light is 10 times that of the fluorescence. The tear film is on the right-hand side of the scan and the aqueous humor is on the left-hand side. The ordinate is relative intensity and the abscissa is distance into the cornea. From B. R. Masters, *in* "The Cornea: Transactions of the World Congress on the Cornea III" (H. D. Cavanagh, ed.), p. 281. Raven Press, New York, 1988.

noise ratio from the cells in the cornea and the image contrast. At the early stages of its development, it was decided to eliminate the applanating microscope objectives, which were designed for specular microscopes, and to use nonapplanating, high numerical aperture water-immersion microscope objectives (Leitz, 50 and 100× water-immersion microscope objectives). These microscope objectives are used routinely with the more recently developed video-rate, scanning slit confocal microscopes.

Confocal Microscopy of Cornea

Tandem Scanning Nipkow Disk Confocal Microscopes

A video-rate tandem scanning confocal microscope was developed by Petran and co-workers to optically section living neural tissue. Petran acknowledged the contribution of the Nipkow disk, which was invented over 100 years ago by Nipkow.[3] The principle of the tandem scanning confocal microscope is as follows. Sets of conjugate pinholes (20–60 μm in diameter in various instruments) are arranged in several sets of Archimedes spirals. Each pinhole on one side of the disk has an equivalent and conjugate pinhole on the other side of the disk. The light transmission of the different disks varies from 0.25 to 1%. The illumination light passes through a set of pinholes (about 100 at a time) and the reflected light passes through a conjugate set (about 100 at a time) on the other side of the disk. As the disk rotates it causes the illumination to focus on the focal plane of the cornea, and the detected light passes through sets of conjugate pinholes. Both the illumination and the reflected light are scanned in parallel over the sample to generate the two-dimensional image of the focal plane. Because the ratio of the area of the holes to the area of the disk is usually only about 0.25%, only a small fraction of the illumination reaches the sample, and a similar small fraction of the light reflected from the sample passes the disk and reaches the detector.[5] The microscope has a very low light throughput. Therefore, the illumination must be very bright (a xenon or mercury arc lamp is usually required). The potential advantages of the tandem scanning confocal microscope include (1) video-rate operation, (2) true color confocal imaging, and (3) the image can be viewed with the eye.

Because of their low light throughput, confocal microscopes based on a tandem scanning Nipkow disk are best suited for reflected light confocal imaging of highly reflecting specimens (i.e., hard tissue such as bone).[6–9] Therefore, although the resolution of a pinhole-based confocal microscope is higher than a scanning slit-based confocal microscope (using the same microscope objectives and light source), it is a poor choice for weakly

[5] M. A. Lemp, P. N. Dilly, and A. Boyde, *Cornea* **4**, 205 (1986).
[6] B. R. Masters and S. Paddock, *J. Microsc.* **158**(2), 267 (1990).
[7] B. R. Masters, *in* "Confocal Microscopy" (T. Wilson, ed.), p. 305. Academic Press, London, 1990.
[8] B. R. Masters and G. S. Kino, *in* "Noninvasive Diagnostic Techniques in Ophthalmology" (B. R. Masters, ed.), p. 152. Springer-Verlag, New York, 1990.
[9] B. R. Masters, *Comments Mol. Cell. Biophys.* **8**(5), 243 (1995).

reflecting specimens such as the human cornea *in vivo*. The low intensity of light, reflected from the cornea and passed through the tandem scanning disk with a 0.25% transmission, that reaches the intensified video camera results in single video images of marginal image quality and low contrast. This result is predominately due to noise and low signal level. Thus, there is a need for digital processing (frame averaging) after image acquisition in order to obtain acceptable images.

Scanning Slit Confocal Microscopes

An alternative to point scanning is to use a slit of illumination that is scanned over the back focal plane of the microscope objective.[9] The advantage of this optical arrangement is that because many points on the axis of the slit are scanned in parallel, the scanning time is markedly decreased. The microscope operates at video rates. Another very important advantage is that scanning slit confocal microscopes have superior light throughput as compared to point scanning Nipkow disk systems. The disadvantage is that the microscope is truly confocal only in the axis perpendicular to the slit height. In comparison to a pinhole-based confocal microscope, a slit-based confocal microscope provides lower lateral and axial resolution. This comparison is for the same wavelength of illumination and reflected light and the same microscope objective in each case. However, for confocal imaging of weakly reflecting living biological specimens such as the cornea, the trade-off of lower resolution with higher light throughput is acceptable and preferable. Several arrangements have been developed to provide the scanning of the slit of illumination over the specimen and the synchronous descanning of the reflected light from the object.

There are several advantages to scanning slit confocal microscopes. The slit height can be adjusted, which allows the user to vary the thickness of the optical section. What is more important, the user can vary the slit height and therefore control the amount of light that reaches the sample and the amount of reflected light that reaches the detector. This is important for samples that are very transparent and therefore can be imaged with the slit height very small; more opaque samples require that the slit height is increased. The light throughput is much greater for a slit scanning confocal microscope than for a confocal microscope based on the Nipkow disk containing sets of conjugate pinholes.

A confocal microscope has enhanced transverse (x and y coordinates in the plane of the specimen) and axial resolution (z coordinate, which is orthogonal to the plane of the specimen) in comparison with a nonconfocal microscope using the same wavelength of light and the same microscope objective. It is the enhanced axial resolution in a confocal microscope that permits the improved optical sectioning of thick specimens and their three-dimensional reconstruction. The transverse resolution of a confocal microscope is proportional to the numerical aperture (NA) of the microscope objective. However, the axial resolution is more sensitive to the NA of the microscope objective. Therefore, to obtain the maximum axial resolution, and hence the best degree of optical sectioning, it is necessary to use a microscope objective with a large numerical aperture. For work with living cells and tissues, we recommend long working distance, water-immersion microscope objectives with a high numerical aperture such as the Leitz 50×, NA 1.0.

The advantage of a slit scanning confocal microscope over those based on Nipkow disks containing pinholes is shown in the following example. For cases of weakly reflecting specimens, such as living, unstained cells and tissues, the advantage of the much higher light throughput from the slit scanning systems is crucial for observation. The basal epithelial cells of the normal, *in vivo* human cornea cannot be observed with a tandem scanning confocal microscope. However, corneal basal epithelial cells are always observed *in vivo* in normal human subjects when they are examined with a video-rate, scanning slit *in vivo* confocal microscope. The reason for this discrepancy is that although the tandem scanning confocal microscope has higher axial and transverse resolution, the very low light throughput of the disk does not pass enough reflected light from the specimen to form an image on the detector (in a single video frame) that has a sufficient signal to noise and, therefore, contrast to show an image of the cells.

A video-rate, direct view tandem scanning confocal microscope was developed in the mid-1960s by Petran and co-workers based on a spinning Nipkow disk. The disk contains many sets of conjugate pinholes, one pinhole is on one side of the disk and its conjugate pinhole is on the other side of the disk. A number of confocal microscopes based on the tandem scanning disk have been developed for use in imaging the eye.

Lemp and co-workers[5] produced a series of studies on the rabbit eye and the *in vivo* human cornea based on the tandem scanning confocal microscope. They used a low numerical aperture applanating microscope objective developed for specular microscopy. The *in vivo* cornea was flattened by the applanating microscope objective. Disadvantages of the system are high noise in the intensified video camera and scan lines on the single images. Postprocessing and frame averaging reduced the noise and removed the scan lines; however, the instruments were no longer video rate. In general, single frames acquired with low light-intensified video cameras are noisy; frame averaging is usually required to increase the signal-to-noise ratio. In addition, many of the acquired video frames are blurred because of eye motion; postprocessing is usually required to align and average sequential video frames and to preserve the best nonblurred images. We define video rate as the acquisition and display of usable video frames. If data acquisition is video rate (video) but postprocessing, such as frame averaging, is required to enhance the quality of the images, then the system cannot be called video rate.

A clinical confocal microscope based on a Nipkow disk with an intensified video camera as a detector was developed by the Tandem Scanning Corporation, Inc. It used a higher numerical objective than was used in the first system that Lemp used at Georgetown University. Their later version of the instrument contained an internal focusing lens that varied the depth of focus while the applanating microscope objective was held stationary on the surface of the deformed cornea. The design of an internal focusing lens was first proposed by Masters.[7]

There are several disadvantages to using a tandem scanning confocal microscope in imaging the living cornea. The most important problem is with the tandem scanning disk itself. In order to reduce the cross torque between sets of adjacent pinholes, the pinholes are configured with a separation distance. The result is that only about 1% or less of the area of the disk is used for the pinholes. On the illumination side the incident light is very bright as observed by the patient; however, only about 1% or less of the light passes through the tandem scanning disk and reaches the cornea. The reflected light from the cornea is collected by the microscope objective and only 1% or less of this light (the signal containing the image) is passed through the disk toward the photon detector. Therefore, most of the signal from the cornea is lost and never detected. In the real instrument there are significant additional light losses from optical elements within the microscope and at the photocathode of the video camera. The result is that individual video frames need to be postprocessed after image acquisition, averaged, and digitally enhanced in order to reduce the noise in the final image. The video-rate property of the tandem scanning confocal microscope is lost due to the necessity for digital postprocessing of the images.

The advantage of an applanating microscope objective is that it helps stabilize the cornea against motion along the optical axis of the eye. The eye still undergoes rotatory motion, although it is reduced.

Although in 1990 we advocated the use of applanating objectives, based on our experience in using the wide-field specular microscope with its applanating microscope objective, we have since decided against the use of applanating objectives. Our experience is that applanating microscope objectives, which press against and deform the shape of the cornea in the applanating process, induce artificial alternations in the structure of the cornea. This observation has been confirmed by other groups using a modified, Koester wide-field confocal microscope with applanating objectives. They also observed that the instrument caused flattening-induced corneal bands and ridges.[10] Therefore, in the development of a video-rate, scanning slit clinical confocal microscope, we introduced the use of nonapplanating, high numerical aperture, water-immersion microscope objectives. The front surface of these microscope objectives does not contact the corneal surface directly; between the front surface of the objective and the surface of the cornea is a layer of an index matching gel. This feature is an important distinction between the scanning slit clinical confocal microscope described in this article and the alternative confocal microscopes used by other investigators.

Video-Rate, Scanning Slit Clinical Confocal Microscope

A new, video-rate, scanning slit confocal microscope was developed by Thaer for the observation of the *in vivo* human cornea. The image of a slit is scanned over the back focal plane of the microscope objective. The slit width can be varied in order to optimize the balance of optical section thickness and image brightness. The instrument is based on the double-sided mirror, which is used for scanning and descanning. This confocal microscope used a halogen lamp for illuminating the slit. The detector is a video camera that acquires images at video rates. This confocal microscope can image basal epithelial cells and the adjacent wing cells in the living human cornea due to its high light throughput. This design was first developed into a video-rate confocal microscope over 20 years ago. Svishchev designed and constructed a video-rate confocal microscope based on an oscillating two-sided mirror (bilateral scanning) and used this microscope to observe living neural tissue in the reflected light mode.[11,12] Figure 4 illustrates the optical design of the video-rate, scanning slit *in vivo* confocal microscope developed by Dr. A. Thaer. The design consists of two adjustable slits placed in conjugate planes of the confocal microscope. Both scanning of the illumination slit over the back focal plane of the microscope objective and descanning of the reflected light from the object are accomplished with an oscillating two-sided mirror. The optical design of the video-rate, scanning slit *in vivo* confocal microscope developed by Dr. A. Thaer incorporates the double-sided mirror slit scanning system first developed

[10] J. D. Auran, C. J. Koester, R. Rapaport, and G. J. Florakis, *Scanning* **16,** 182 (1994).
[11] G. M. Svishchev, *Opt. Spectrosc.* **26**(2), 171 (1969).
[12] G. M. Svishchev, *Opt. Spectrosc.* **30,** 188 (1971).

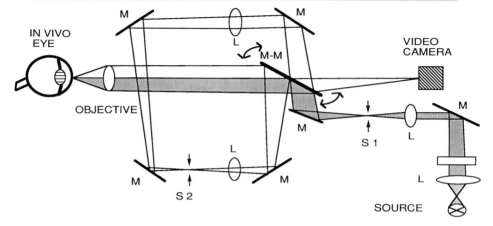

FIG. 4. Optical system of a clinical video-rate, scanning slit confocal microscope. The light source is a halogen lamp. S1, S2, adjustable confocal slits; L, lens. The objective shows the position of the microscope objective. One-half of the objective is used for illumination of the cornea and one-half of the objective is used to collect the light from the cornea. The illumination light path is drawn with black dots, and the collection light path is shown in white. M, fixed front surface mirror; M-M, oscillating double-sided mirror that is used for both scanning and descanning. The video camera is an intensified video camera. From M. Böhnke and B. R. Masters, *Progr. Retina Eye Res.* **18**(5), 1 (1999).

by Svishchev.[13–15] The double-sided mirror slit scanning system of Svishchev has also been incorporated into other "bilateral scanning" confocal microscopes.

For corneal imaging, a 25×/0.65 NA, a 40×/0.75 NA, or a 50×/1.0 NA water-immersion lens (Leitz, Germany) can be used. The resolution of the different objectives and the thickness of the optical section to some extent are influenced by the light levels used and the geometry and the reflectivity of the structures studied. As a practical guideline, the lateral resolution of the 25× objective is about 1.4 μm, whereas the thickness of the optical sections is about 16 μm. The lateral resolution of the 50× objective is better than 0.8 μm, whereas the thickness of optical section is about 10 μm.

The following design parameters were incorporated into the video-rate,

[13] B. R. Masters and A. A. Thaer, *Appl. Opt.* **33**(4), 695 (1994).
[14] W. Wiegand, A. A. Thaer, P. Kroll, O.-C. Geyer, and A. J. Garcia, *Ophthalmology* **102**, 568 (1995).
[15] B. R. Masters, A. A. Thaer, and O.-C. Geyer, *in* "Contact Lens Practice" (M. Ruben and M. Guillon, eds.), p. 390. Chapman & Hall, London, 1994.

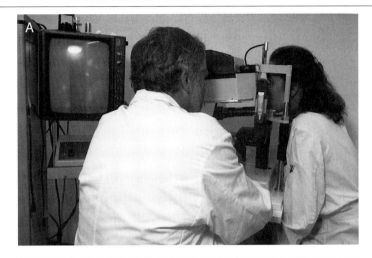

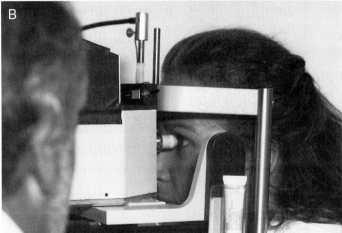

FIG. 5. (A) The video-rate, scanning slit confocal microscope in a clinical setting. (B) Close-up of the instrument showing the nonapplanating water-immersion microscope objective with a drop of index matching gel that contacts the cornea.

scanning slit confocal microscope (Fig. 5). (1) The use of nonapplanating, high numerical aperture, water-immersion microscope objectives, Leitz 50× and Leitz 100× microscope objectives. (2) The microscope objective would use a methylcellulose gel to optically couple the tip of the microscope objective to the cornea. There was no applanation or direct physical contact, which deforms the cornea and introduces artificial folds and ridges in the cornea, between the objective and the surface of the cornea. (3) One-half of the numerical aperture was used for illumination, and one-half of the numerical aperture was used for collection of the reflected and fluorescent light. (4) Optical sectioning in the plane of the cornea was obtained with two sets of conjugate slits. The slit heights are variable and adjustable. (5) An oscillating, two-sided mirror (bilateral scanning) was used for scanning the image of the slit over the back focal plane of the microscope objective and for descanning the reflected and back-scattered light collected by the microscope objective from the focal plane in the specimen. (6) The light source is a 12-V halogen lamp. For fluorescence studies, a mercury arc lamp or a xenon arc lamp can be used. (7) The scanning was synchronized with the readout of an interline CCD camera in order that the full vertical resolution of the intensified CCD camera could be utilized.

The microscope used standard nonapplanating microscope objectives with RMS threads that are interchangeable. Several different microscope objectives can be used, which permits the use of various magnifications and fields of view. Typically a Leitz 50×, 1.0 NA, water-immersion objective is used. When a larger field of view is required, a Leitz 25×, 0.6-NA water-immersion microscope objective is used.

An intensified video camera (Proxitronic) with video output to a Sony S-VHS tape recorder is used. The synchronization of the bilateral scanning and the readout of the intensified video camera have been described previously. The PAL video format provides 625 lines. The S-VHS video recorder provides high band width. In parallel with the video recording, there is a video monitor so that the operator can observe the confocal images of the subject's eye in real time. This video-rate, scanning slit confocal microscope does not require any frame averaging for producing the image quality and contrast shown in this article.

What are the advantages of using a scanning slit confocal microscope such as is described and demonstrated in this article? Slit scanning confocal microscopes have a much higher light throughput than confocal microscopes based on a Nipkow disk. This has two consequences. First, the illumination incident on the patient's eye can be much less. This allows for a much longer duration of the use of the confocal microscope on the patient's eye without the severe patient discomfort and high light intensity that is necessary with the use of the confocal microscope based on the Nipkow disk. Second, it is possible to image the low reflecting layer of wing cells that are immediately adjacent to the basal epithelial cells in the normal human cornea. This layer of wing cells has been imaged, at video rates, as single video frames without the need for any analog or digital image processing using the video-rate, scanning slit confocal microscope. No other video-rate confocal microscope has been able to image these wing cells in normal, *in vivo* human cornea. This new confocal instrument has unique advantages over other confocal systems. The bright, high-contrast confocal images of the wing cells and the basal epithelial cells demonstrate its unique optical characteristics. The low reflectivity of wing and basal cells in normal human cornea presents a low-contrast benchmark test specimen for various types of confocal microscopy. The high rejection of stray light and the narrow depth of field, coupled with the high numerical aperture microscope objective (1.0 NA), result in the ability of the instrument to image these cell layers clearly in live normal human cornea. The clear advantage of slit scanning confocal microscopes for ophthalmic diagnostics and basic eye research is best appreciated when the basal epithelium in the anterior cornea is imaged. The video-rate scanning slit confocal microscope provides high-contrast, high-resolution images of both wing cell and basal epithelial cells in the normal *in vivo* human eye.

In 1990, it was suggested that an internal lens system would permit focusing at different depths within the cornea; however, in order to keep light losses to a minimum, and therefore maximize the light sensitivity of the z-scanning instrument, it was decided to use a z-scanning microscope objective.[7] This capability of a z-scanning confocal has been implemented into the clinical video-rate, scanning slit confocal microscope. In addition to microscopic pictures with a lateral movement of the scanning slit, a confocal measurement of the tissue reflectivity can be performed by an automated z scan through all layers of the cornea. For this procedure, the lateral scan is switched off. Reflectivity, as recorded by a photomultiplier, can be used to measure corneal haze, which may be of special interest in keratorefractive procedures.

Optical Low Coherence Reflectometry to Measure Position within Cornea

Optical low coherence reflectometry instruments can noninvasively measure distances within the *in vivo* cornea with a precision of less than 1 μm.[16–18] In order to relate the distance that the focal plane of the microscope objective has moved into the thickness of the cornea, it is necessary to know the profile of the refractive index across the thickness of the cornea. This requirement exists for any form of optical pachometry and includes optical low coherence reflectometry.

[16] B. R. Masters, "Selected Papers on Optical Low Coherence Tomography," SPIE Press, Bellingham, 1999.
[17] M. Böhnke, P. Chavanne, R. Gianotti, and R. P. Salathe, *J. Refract. Surg.* **14,** 140 (1998).
[18] M. Böhnke, R. Wölti, F. Lindgren, R. Gianotti, P. Bonvin, and R. P. Salathe, *Klin. Monatsbl. Augenheilkd.* **212,** 367 (1998).

The technique is rapid, noncontact, and noninvasive and can easily be incorporated into a variety of ocular imaging instruments. Optical low coherence reflectometry is an optical imaging technique based on a Michelson interferometer and a low coherence light source. It can obtain micrometer resolution cross-sectional imaging of biological tissue. The technique is analogous to ultrasound B-mode imaging; the difference is that with optical low coherence reflectometry, light is used instead of acoustic waves. Low coherence optical reflectometry is noninvasive in that is operated in a noncontact fashion. It has been used to measure corneal thickness and retinal thickness and for ocular biometry. Optical low coherence reflectometry has promise as a useful noninvasive technique for biometry of the anterior segment. It has been applied to measure corneal thickness in a clinical setting. This new optical pachometer has been implemented on a slit lamp headrest as well as on an eximer laser system for refractive surgery. Optical low coherence reflectometry is an ideal noninvasive technique to measure position within the full thickness of the cornea at a precision of 1 μm.

Clinical Examination Technique with Confocal Microscope

The technique of video-rate, scanning slit confocal microscopy of the cornea has been applied at the University Eye Clinic, University of Bern, in about 1000 clinical sessions since 1994. This accumulated clinical experience is factored into the current clinical examination technique, which is presented here in sufficient detail for others to follow.[4] Care must be taken not to mechanically damage the cornea.

The instrument used for our investigations is a video-rate, scanning slit confocal microscope. The microscope is equipped with a halogen lamp for illumination and a slit scan-synchronized, high sensitivity video camera with adjustable black level suppression. Phototoxicity from the halogen lamp is negligible. Patients have reported this examination to be less disturbing than contact corneal endothelial cell photography with a specular microscope. Complications have not been observed in our patients. The healthy cornea with its low reflectivity requires a high primary light output from the halogen lamp. In cases of corneal opacities such as scar tissue or other deposits, the light source has to be dimmed down to a much lower level.

In clinically normal corneas, the halogen lamp will have to be set to full power to supply enough reflected light from the corneal structures. In pathological corneas with scar tissue or other highly reflective contents, the lamp power has to be reduced by about 50%. With selected filters, which can be inserted into the optical path of the microscope, the spectrum emitted from the halogen lamp can be confined to selected spectra, either improving optical penetration by selecting longer wavelengths or improving image contrast by blocking out longer wavelengths. For studies using fluorescent dyes, a 300-W mercury lamp with an appropriate fluorescence excitation and emission filter set is supplied. Depending on the experience of the investigator, the instrument should be readjusted periodically to give an optimum and homogeneous illumination of the optical section, a perfect alignment of the scanning slits, and the lowest possible levels of stray light, which may degrade the image contrast. The projected confocal slit width, which can be adjusted in some instruments, was selected to be 10 μm, which is a compromise between the best resolution and the best illumination and contrast.

In order to check the homogeneity of the field illumination and also to practice coordination of the microscope movements, a piece of black paper is placed vertically in front of the microscope. The microscopy is equipped with a 10× objective and advanced until the surface structure of the object is visible. The halogen lamp has to be dimmed for this procedure. With this overall weakly reflective object, the technical status of the microscope optics can be checked easily and, if required, also adjusted. By mounting the black paper not strictly *en face* before the objective, the investigator can practice to judge and establish perpendicularly by using extra drives tilting the frontal plane of the microscope.

To investigate the patient's cornea, the instrument objective is brought from its most backward position into optical contact with the corneal tissue by a high viscosity acrylic ocular gel. The S-VHS tape recorder is started when an optimum centration has been achieved, as judged from a well-centrated light reflex on passing the epithelial or endothelial layer. The video-rate *en face* sections from all layers of the cornea are then recorded on videotape. The position of the optical plane in the z axis is controlled with a manual micrometer drive. Bowman's layer and the corneal endothelium are used as additional reference structures for the z-axis position.

For a detailed analysis, the recorded video sequences are reviewed frame by frame. For the patients' record, selected frames can be printed with a video printer. Pictures for publication can be photographed directly off the monitor. Alternatively, if the appropriate equipment is available, selected frames can be digitized and entered directly into an image file server to be available either online or be exposed to photographic film with a laser film printer. We recommend that the examiner routinely digitize images from all corneal layers, plus specific findings for a given case.

Patients are evaluated routinely by clinical slit lamp biomicroscopy before and after the confocal scanning procedure. Before the examination with the confocal microscope, the patients are informed about the nature of the confocal scanning procedure. The investigator should decide beforehand which objectives are going to be used. For a quick routine examination, the 40× objective offers the best overview type of characteristics. However, this objective will miss some details visible with the 50× objective and does not give a generous overview (and a better link to slit lamp morphology) like the 25× objective. In most of our patients, we try to work with all three objectives and select the one or two most appropriate to image the specific pathology in further follow-up sessions.

One drop of acrylic eye gel is placed on the microscope objective and the instrument is moved to a full backward position. After topical anesthesia with one drop of 0.4% Novesine or any other topical anesthetic, the patient's head is positioned on an adjustable headrest. The confocal microscope with the 40× objective is then placed 1–1.5 mm above the apex of the corneal center. The patient is asked to look into the light so that the optical center of the cornea is aligned with a lateral accuracy of probably less than 1 mm. The microscope is then brought into optical contact with the cornea by manual advancement of the micrometer-controlled z drive. From this point on, all further $x–y–z$ movements of the instrument are controlled from the video-rate picture displayed on the monitor. The image is recorded on a S-VHS tape recorder and can be reviewed in slow motion. Simultaneously, the observer comments on the videotape sound track on position and other findings to supplement the information recorded. Care must be taken not to mechanically damage the cornea.

For manual imaging with the 40 and 50× objective setup, the following pattern is usually employed.

1. Establish optical contact with cornea, focus on basal epithelial layer, centrate on the surface to image parallel sections, and start videotape.
2. Move backward to image superficial epithelial cell layers.
3. Proceed to basal cell layer.
4. Proceed to Bowman's layer and look for subepithelial nerve plexus.
5. When first stromal keratocytes are visible, return to step 2.
6. Repeat steps 2–5 until enough frames of epithelium have been sampled.
7. From Bowman's layer proceed to endothelium at about 0.1–0.5 mm/ sec with manual micrometer advancement.
8. When endothelium reflex just above the anterior chamber becomes visible, recentrate the microscope if required and then return to the kerato-cyte layer just above Descemet's membrane.
9. Return to endothelium and anterior chamber, making sure that some frames of the endothelium have been captured.
10. Repeat steps 8 and 9.
11. Return to epithelium and reverse movement as in step 7.

12. Repeat steps 2–11 at least once or more times if required.

If required, the same procedure (steps 1–12) can be carried out with the 50 and 25× microscope objectives. For every eye and microscope objective, a minimum of 0.5- to 1-min good quality tape recording should be collected. Extended recording times are usually due to a nontrained investigator, which lead to decreased patient comfort and later on to extended tape reviewing times. With sensitive or less cooperative patients, the recording time, however, may be extended to obtain at least 1000 useful optical sections from all corneal layers. Included in this number is usually a minimum of at least optical sections from all z positions of the central corneal stroma. If a specific finding is observed during the examination, e.g., in the epithelium, extra recording time is spent on the region of interest.

To investigate noncentral locations on the cornea, the limbus, or even conjunctival areas, the patient's direction of gaze is aided with a fixation light for the contralateral eye. For these special locations, the angle of the frontal plane has to be tilted to achieve perpendicularity to the surface of the tissue studied. Occasionally, a somewhat oblique section may also be interesting, as it gives more information on the thickness of some structures imaged in one optical section. The videotapes should be evaluated in slow motion or single frame mode of the video recorder.

From patient examinations stored on S-VHS tapes, we usually digitize a standard set of frames (from 40 or 50× objective recordings) for every patient, which includes (1) epithelial surface cells, (2) epithelial wing (intermediate) cells, (3) epithelial basal cells, (4) subepithelial nerve plexus/ Bowman's layer, (5) first keratocyte layer, (6) anterior stroma keratocytes (10–100 μm below/behind Bowman's layer), (7) intermediate stroma (100– 350 μm depth), (8) posterior stroma (350 to >500 μm depth), (9) most posterior keratocyte layers, just anterior to Descemet's membrane, and (10) endothelium (see Figs. 6–11).

In addition, we digitize a variable number of frames that show specific findings for a given case. The digitized files are given identifying numbers, which clearly link them to a record of the database carrying information about the identity of the patient, the location of the recording, and relevant technical information regarding the microscope. When video images are digitized, we do not apply image enhancement techniques except for a median filter function, which basically makes the image appear smoother and reduces pixel noise. With the advent of digital video recording technology, the video images can now be stored in a digitized format with a unique identifier of every image on a given tape. Thus, the need to digitize selected frames within due time to prevent deterioration of image quality may be

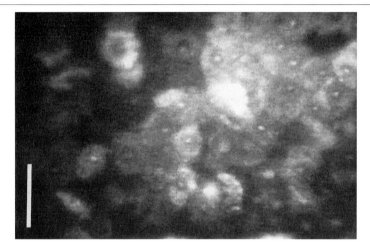

FIG. 6. Confocal section of surface cells in a normal human cornea. Dark and light cells are visible with dark nuclei. In some cells the borders are bright, possibly indicating a loss of contact and the process of desquamation. Bar: 50 μm. From M. Böhnke and B. R. Masters, *Progr. Retina Eye Res.* **18**(5), 1 (1999).

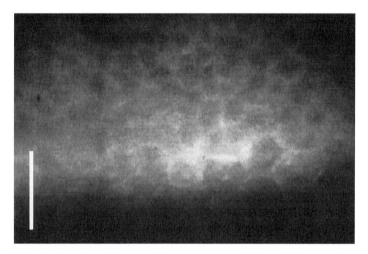

FIG. 7. Confocal section of a normal human cornea at the level of intermediate or wing cells. The orientation of the section is slightly oblique, showing maturation and enlargement of the cells from top to bottom. Bar: 50 μm. From M. Böhnke and B. R. Masters, *Progr. Retina Eye Res.* **18**(5), 1 (1999).

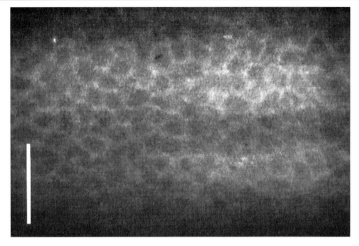

FIG. 8. Confocal section of a normal human cornea at the level of basal cells. Dimly reflective cell borders are shown. Bar: 50 μm. From M. Böhnke and B. R. Masters, *Progr. Retina Eye Res.* **18**(5), 1 (1999).

reduced; however, a critical review of a recorded sequence after the recording session is still mandatory.

In contrast to the extended sessions in the early days of the procedure, a central corneal examination of all layers now typically consists of 40–100

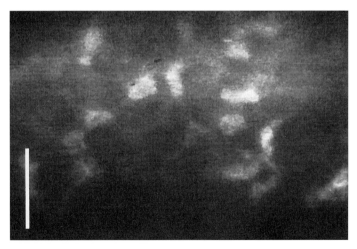

FIG. 9. Confocal section of a normal human cornea at the level of the most anterior keratocyte layer. Bar: 50 μm. From M. Böhnke and B. R. Masters, *Progr. Retina Eye Res.* **18**(5), 1 (1999).

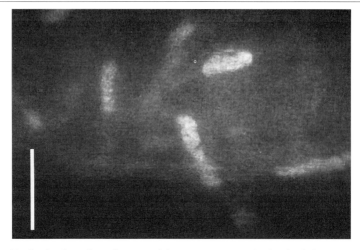

FIG. 10. Confocal section of a normal human cornea at the level of the most posterior keratocyte layer. Keratocyte nuclei with oval shape and cytoplasmic invaginations, which are the most predominant type throughout all tissue layers. A second type of elongated nuclei in the last or the second to last keratocyte layer just before Descemet's membrane is visible. Bar: 50 μm. From M. Böhnke and B. R. Masters, *Progr. Retina Eye Rese.* **18**(5), 1 (1999).

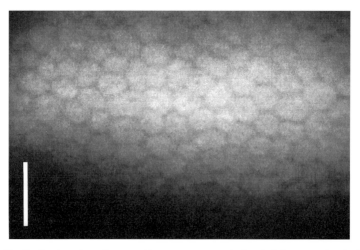

FIG. 11. Confocal section of a normal human cornea at the level of the corneal endothelium. Bar: 50 μm From M. Böhnke and B. R. Masters, *Progr. Retina Eye Res.* **18**(5), 1 (1999).

sec of recording time for every microscope objective used. Considering patient and instrument handling, the total examination time does not exceed 10 min. It should be kept in mind, however, that the observer has to review the tape frame by frame later on and digitize and store selected images. The total time for review, selection, processing and storage of the confocal images can be much more than 1 hr per patient. If, however, only one specific aspect has to be answered, the total investigation time can be considerably shorter.

Clinical Findings with Scanning Slit Confocal Microscope

The scanning slit confocal microscope provides *in vivo* details of the normal and pathologic structure of the cornea with a resolution of less than 1 μm.[19–24] The confocal microscope provides *en face* images of the cornea; the plane of the image is orthogonal to the thickness of the cornea. These images are very different and are oriented perpendicular from the typical sections obtained in histopathology in which the tissue is cut along the thickness of the cornea (vertical sections). Some examples from various clinical conditions are given.[25–31]

1. Contact lens-associated changes.[25,26] This is an important example of a new corneal degeneration associated with microdot deposits in the corneal stoma. This clinical study is a seminal example of the efficacy of the scanning slit confocal microscope to image submicrometer cellular alterations in the living human cornea. In patients with long-term contact lens wear, all layers of the cornea stroma show multiple fine particles with high reflectivity. These changes persist for

[19] B. R. Masters and A. A. Thaer, *Microsc. Res. Techn.* **29,** 350 (1994).
[20] M. Böhnke and A. A. Thaer, *in* "Bildgebende Verfahren in der Augenheilkunde" (O.-E. Lund and T. N. Waubke, eds.), p. 47. Ferdinand Enke Verlag, Stuttgart, 1994.
[21] B. R. Masters and A. A. Thaer, *Bioimages* **3**(1), 7 (1995).
[22] B. R. Masters and A. A. Thaer, *Bioimages* **4**(3), 129 (1996).
[23] B. R. Masters, *Scann. Microsc.* **7**(2), 645 (1993).
[24] B. R. Masters, *Opt. Eng.* **34**(3), 684 (1995).
[25] M. Böhnke and B. R. Masters, *Ophthalmology* **104,** 1887 (1997).
[26] R. Cadez, B. Früh, and M. Böhnke, *Klin. Monatsbl. Augenheilkd.* **212,** 257 (1998).
[27] M. Böhnke, I. Schipper, and A. Thaer, *Klin. Monatsbl. Augenheilkd.* **211,** 159 (1997).
[28] R. Cadez, B. Früh, and M. Böhnke, *in* "Kongress der Deutschsprachigen Gesellschaft für Intraokularlinsen: Implantation und refraktive Chirurgie" (D. Vörösmarthy, ed.), p. 403. Springer, Berlin, 1997.
[29] B. E. Früh, U. Korner, and M. Böhnke, *Klin. Monatsbl. Augenheilkd.* **206**(5), 317 (1995).
[30] A. G. Chiou, R. Cadez, and M. Böhnke, *Br. J. Ophthalmol.* **81**(2), 168 (1997).
[31] B. R. Masters and M. Böhnke, *in* "Corneal Disorders: Clinical Diagnosis and Management" (H. Leibowitz and G. Waring, eds.), p. 123. Saunders, Philadelphia, 1998.

years after discontinuing contact lens wear. These stromal deposits of submicrometer size have not been reported previously. This may be due to the superior resolution and contrast that is achieved routinely with the video-rate, scanning slit confocal microscope. These microdots were only visible when the $50\times/1.0$ NA water-immersion microscope objective was used (Fig. 12).

2. Corneal trauma. After a plant injury from *Dieffenbachia*, confocal microscopy was used to locate the oxalate crystals in all layers of the cornea down to Descemet's membrane. Over time the crystals were found to fragment and dissolve without signs of cell toxicity or inflammatory infiltrate (Fig. 13).

3. Corneal infection. Confocal microscopy is an excellent tool used to easily identify a fungal infection in the cornea (Fig. 14).

4. Corneal surgery. After lamellar cornea surgery, the viability of the keratocyte population, as well as the progress of reinnervation, can be studied. One year postoperatively, normal patterns of keratocytes and nerve fibers are found in the donor lenticule. Following photorefractive keratectomy, various changes even in the deep corneal layers can be found. A new finding is the occurrence of a highly reflective

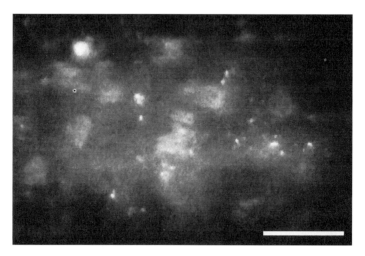

Fig. 12. Confocal section of a normal human cornea after 15 years of soft contact lens wear. Numerous small structures, called microdots, are present in all regions and layers of the corneal stroma. These structures are associated frequently with keratocytes, giving rise to the theory that they consist of lipofuscin granules that are of intracellular origin. Some of these granules may be deposited in the extracellular compartment, where they stay and accumulate over time. Bar: 50 μm. From M. Böhnke and B. R. Masters, *Progr. Retina Eye Res.* **18**(3), 1 (1999).

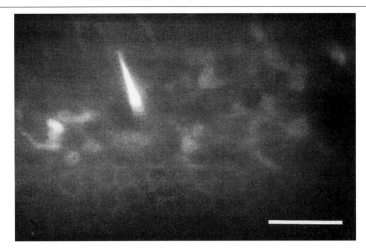

FIG. 13. A corneal injury with oxalate crystals derived from the plant sap of *Dieffenbachia*. Four weeks after the trauma, needle-like highly reflective crystals are visible in all layers of the corneal stroma down to Descemet's membrane. Confocal section of highly reflective oxalate crystal just below the basal epithelial cells, located in Bowman's layer. Bar: 50 μm. From M. Böhnke and B. R. Masters, *Progr. Retina Eye Res.* **18**(5), 1 (1999).

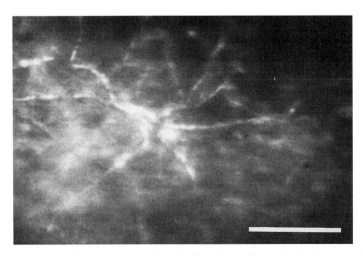

FIG. 14. Confocal microscopic image showing fungal mycelia growing in the stroma. Fungal infection after minor corneal trauma in a contact lens wearer. Bar: 50 μm. From M. Böhnke and B. R. Masters, *Progr. Retina Eye Res.* **18**(5), 1 (1999).

spindle-shaped structure, which in size may correspond to some extra-cellular deposit oriented along the collagen fibers. These changes persist for years after photorefractive keratectomy and possibly correlate with the amount of tissue ablated and the type of instrument used (Fig. 15).

Three-Dimensional Confocal Microscopy and Visualization
of the Living Eye

Three-dimensional visualization is an important application of confocal microscopy based on the enhanced z-axis resolution. This feature permits the acquisition of a stack of z sections (optical sections) and the subsequent three-dimensional visualization. The three-dimensional visualization of the living *in situ* eye and studies of the structural changes of the three-dimensional structures over time (four-dimensional visualization) are important applications of multidimensional confocal microscopy (see Figs. 16 and 17). There are many diverse applications of confocal imaging and three-dimensional reconstruction of the living eye. Each application involves its special techniques for both data acquisition and three-dimensional visual-

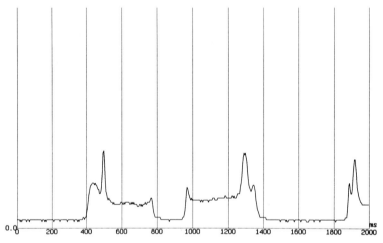

FIG. 15. Repetitive forward–backward confocal z scan of a human cornea with grade 1 haze 1 year after photorefractive keratectomy. The small amount of subepithelial scar tissue causes a significant peak in the stromal light scatter. In each z scan the intensity of scattered light is plotted on the ordinate in arbitrary units, and the distance within the cornea is plotted on the abscissa. The largest peak of scattered light in each scan is due to the reflectivity from the subepithelial region of the cornea. From M. Böhnke and B. R. Masters, *Progr. Retina Eye Res.* **18**(5), 1 (1999).

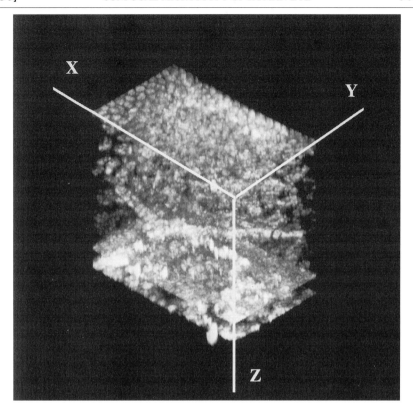

FIG. 16. Three-dimensional visualization of an *in situ* rabbit cornea based on the maximum intensity technique. The single layer of endothelial cells is shown on the top. The epithelium is shown at the bottom. A large bifurcating nerve fiber is shown in the anterior stroma. A stack of 365 confocal reflected light images were used to generate projections. The step size, which is the distance on the z axis between consecutive optical sections, was set equal to the pixel size. In a maximum intensity projection the three-dimensional image is formed from the maximum intensity or the brightest pixel along each line of sight. The visualization was performed with software from Molecular Dynamics, Inc.

ization. Examples of the authors' prior work in the three-dimensional visualization of stacked optical sections acquired with a confocal microscope include the three-dimensional visualization of the full thickness of the *ex vivo* rabbit cornea.[32–34] This application required the use of a specially developed specimen chamber that is required to maintain the freshly excised

[32] B. R. Masters and S. W. Paddock, *Appl. Opt.* **29,** 3816 (1990).
[33] B. R. Masters, *Mach. Vis. Appl.* **4,** 227 (1991).
[34] B. R. Masters and M. A. Farmer, *Comp. Med. Imag. Graph.* **17**(3), 211 (1993).

Fig. 17. Three-dimensional visualization of an *in situ* rabbit cornea based on the volume rendering technique. The reconstruction is shown in an isometric view. The single layer of endothelial cells is at the top. The epithelium is shown at the bottom. The full thickness of the central cornea is 400 μm. The volume rendering technique was performed with Voxel View software (Vital Images, Fairfield, IA).

eye in a normal physiological state and with minimal stress on the cornea during data acquisition with a confocal microscope.[23] The *in vivo* human fundus and optic nerve were visualized in three dimensions from a stack of optical sections acquired with a modified scanning laser ophthalmoscope.[35] The *in vivo* human lens has been visualized in three dimensions from a series of optical slices acquired with a rotating slit confocal microscope.[36]

Special techniques of multidimensional confocal microscopy and visualization of the living eye have been described. As in all techniques of microscopy, the fidelity of the images and their correct interpretation require appropriate instrumentation as well as a proper microscopic technique. Correlative microscopy with both confocal microscopy and electron microscopy has been shown to be critical in the analysis of confocal reflected light

[35] F. W. Fitzke and B. R. Masters, *Curr. Eye Res.* **12,** 1015 (1993).
[36] B. R. Masters and S. L. Senft, *Comp. Med. Imag. Graph.* **21**(3), 145 (1997).

images.[37-40] For *ex vivo* studies of living ocular tissue, the maintenance of the normal physiological state and cell and tissue morphology required a specially developed specimen chamber.[23] For clinical confocal microscopy of the living human eye, the correct examination techniques are required to achieve the full potential of this important new optical technique. Video-rate, scanning slit confocal microscopes offer many advantages over Nipkow disk (pinhole-based), tandem scanning confocal microscopes for imaging the *in vivo* human cornea. This noninvasive, nonapplanating confocal micro-scope provides a new tool to image cellular details of the *in vivo* living human eye.

Acknowledgments

This work was supported by a grant (B.R.M.) from the NIH, National Eye Institute, EY-06958. The authors thank Dr. Andreas A. Thaer for collaboration in the development of the clinical confocal microscope. We thank Dr. David Hanzel of Molecular Dynamics for helping us process the confocal images.

[37] B. R. Masters, J. Vrensen, B. Willenkens, and J. Van Marle, *Exp. Eye Res.* **64**(3), 371 (1997).
[38] B. R. Masters, *Opt. Express* **3**(9), 332 (1998). URL = http://www.opticsexpress.org
[39] B. R. Masters, *Opt. Express* **3**(9), 351 (1998). URL = http://www.opticsexpress.org
[40] B. R. Masters, *Opt. Express* **3**(10), 356 (1998). URL = http://www.opticsexpress.org

[31] *In Vivo* Imaging of Mammalian Central Nervous System Neurons with the *in Vivo* Confocal Neuroimaging (ICON) Method

By RALF ENGELMANN and BERNHARD A. SABEL

Introduction

Anatomical investigations of animals usually require that the animal is sacrificed and the tissue of interest, such as the brain, is prepared with special procedures. However, these anatomical studies are only "snapshots" in time, with the one-time preparation of tissue precluding the possibility for repeated observations. For many reasons, it would be advantageous to study microscopic detail in the living organism, particularly in the brain. The use of confocal techniques for such *in vivo* imaging of neurons has been considered an advantageous approach since the early days of confocal microscopy.[1] The optical sectioning ability and the rising availability of

[1] A. Villringer, U. Dirnagl, A. Them, L. Schurer, F. Krombach, and K. M. Einhäupl, *Microvasc. Res.* **42**, 305 (1991).

cell-specific markers especially provided opportunities for successful *in vivo* imaging in the nervous system.[2] The first attempts for such *in vivo* imaging had to overcome challenges such as choosing the right animal models, assuring the stability of the markers, and overcoming the photodamaging effects following fluorescence excitation. Imaging with cranial window preparations or on the exposed spinal cord provided novel approaches for *in vivo* observation, but these kinds of experiments were highly invasive and often difficult to repeat. New two-photon excitation techniques[3] might change this now, but tissue penetration ability and resolution are still limited because the observed tissue itself acts as an optical barrier.

An elegant way to overcome this problem is to find animal models in which it is possible to visualize nerve cells in a transparent environment. This is the case, for example, in amphibia during early development or in the adult zebrafish.[4] Here, high resolution imaging of nerve cells is possible in a noninvasive manner, allowing repeated observation of changes related to development or degeneration. While these models of *in vivo* imaging help to elucidate some fundamental neurobiological problems, they do not allow the observation of mammalian nerve tissue, as these animals are nonmammals. Therefore, there is still the need to find a mammalian tissue model for *in vivo* imaging of neurons. Two models have been proposed: one in the peripheral and one in the central nervous system (CNS). Imaging through the skin of mammals with fiber-optic confocal imaging (FOCI)[5] allows subsurface observation of fluorescing nerve fibers around hair follicles, sometimes also related to nerve regeneration. The "*in vivo* confocal neuroimaging" (ICON) method allows observation of retinal ganglion cells in the eye.[6] Because nerve cells of the retina derive directly from the brain during early development, the ICON procedure is the first method to observe mammalian CNS neurons in a noninvasive manner and without damaging effects.

Investigators interested in *in vivo* imaging have long been interested in the mammalian eye as a model, but early attempts using scanning laser ophthalmoscopes or modified tandem scanning microscopes had problems with the high refractive power that characterizes the eyes of a typical rodent laboratory animal.[7] However, small rodents are the most frequently used

[2] W. M. Petroll, J. V. Jester, and H. D. Cavanagh, *Scanning* **16,** 131 (1994).
[3] J. R. Fetcho and D. M. O'Malley, *Curr. Opin. Neurobiol.* **7,** 832 (1997).
[4] F. Zimprich, R. Ashworth, and S. Bolsover, *Pflug. Arch.* **436,** 489 (1998).
[5] L. J. Bussau, L. T. Vo, P. M. Delaney, G. D. Papworth, D. H. Barkla, and R. G. King, *J. Anat.* **192,** 187 (1998).
[6] B. A., Sabel, R. Engelmann, and M. F. Humphrey, *Nature Med.* **3,** 244 (1997).
[7] H. D. Cavanagh, J. V. Jester, J. Essepian, W. Shields, and M. A. Lemp, *CLAO* **16,** 65 (1990).

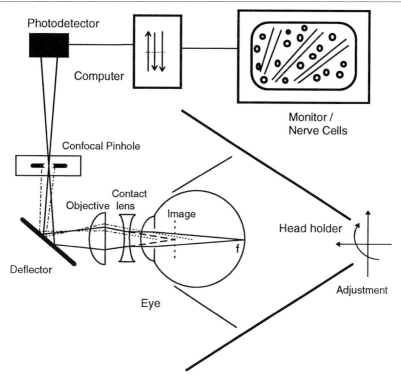

FIG. 1. Schematic drawing of an ICON setup. The contact lens projects the observed image plane onto the retina of the eye (f). Without such a lens, the image would be located in the vitreous, even with extremely long working distance objectives. The contact lens does not fully compensate the refractive power of the eye, which is useful in achieving further magnification. The confocal pinhole effectively eliminates out-of-focus information, even with the low-power optics used.

animals in studies of development or degeneration in the nervous system.[8] ICON now provides both the instrumental adaptation for cellular resolution imaging of embedded nerve cells and the use of a mammalian animal model for degeneration and plasticity studies. This article describes the instrumentation and experimental details of a typical ICON setup.

Equipment

In general, ICON requires versatile laser scanning microscopes and a mechanical and optical microscope basis (Fig. 1) that fits the physical

[8] S. Sautter and B. A. Sabel, *Eur. J. Neurosci.* **5,** 680 (1993).

requirements of the chosen animal model, in our case the rat. Similar sized animals may also be used. In our initial experiments, ICON was developed using a standard confocal laser scanning microscope (CLSM) with an acousto-optical deflector, providing very flexible scan speeds and image averaging with an argon laser (488/546 nm). This was used in conjunction with a modified light microscope to which a special head holder for the rat was attached. The best instrumentation might vary depending on the commercial availability of the models, but as of 1998, the following components can be recommended.

Equipment List:

1. Zeiss Axiotech 100 microscope: The Axiotech microscope is a versatile technical inspection microscope with sufficient probe space and an incident light illumination. It fits the Noran Instruments scan head very well and may also be used for general fluorescence microscopy.
2. Animal size specimen holder plate: The standard table and object guide are too small for a rodent, so the object guide has to be replaced by a simple plate with the approximate size of the animal.
3. Narishige nontraumatic rat head holder: For fixation and adjustment of the eye in the optical axis of the microscope, a head holder is needed. Narishige provides a nontraumatic version using the teeth and forehead for fixation. Some small ball heads and clamps will mount the head holder on the stage.
4. Noran OZ scan head with an argon laser: The Noran scan head is recommended because of the very large range of different scan speeds.
5. Noran Intervision software with a slow scan mode: In general, slow scan modes are required, but because the animal will move slightly under the anesthesia due to heartbeat and breathing, it is recommended using moderate short scanning times and averaging a number of pictures.
6. SGI O2 workstation computer: The Noran system comes with a powerful SGI workstation and the acquisition software "Intervision," which offers all options needed for image analysis and editing with optional deconvolution modules.
7. Zeiss Epiplan 5×/10× ICS-lens: The optical adaptation uses a low-power microscope lens (4–5×) with sufficient working space (10 mm or more), which will be combined with the animals eye optics using a contact lens by immersion with optical gel.
8. Newport f = −12.5 (KPC-013) contact lens: The best contact lens to be used is a concave contact lens of −160dpt. When combined

with the objective described earlier, the resulting overall magnification power will be 20–25× in case of a rat eye. This may be adapted
to study other species.

9. Molecular Probes FluoSpheres 40 nm/5%: The choice of the fluorescence label will depend on the topic to be studied, but as an
example, FluoSpheres (Molecular Probes Eugene, OR) are particularly suitable for retrogradely labeling retinal ganglion cells, to be
able to analyze their morphology (such as cell number and size).
FluoSpheres are nontoxic to cells and provide a strong retrograde
label of about 90–95% of the retinal ganglion cells. The label will
not fade for several months, even with repeated illumination.

10. Stoelting Lab Standard stereotactic instrument: Any stereotaxic
instrument is suitable for the standard intracranial injection of the
retrograde label.

Animal Preparation

In vivo imaging in general requires proper animal handling, otherwise
tissue damage and bad image quality will result. However, in control experiments it was assured that typical laser illumination in order to excite fluorescent dyes will not cause any damage to the retinal cells. Anesthesia is
the most critical step with ICON. For use with the hardware mentioned
earlier, the following protocol can be followed (see Fig. 2).

Experiment CheckList

1. Labeling: stereotactic injection of retrograde markers into the brain
under standard anesthesia. The injections should be performed several days before the first imaging session to allow sufficient time for
retrograde transport. Five to 7 days are sufficient for the Fluoro-
sphere transport.

2. Anesthesia for imaging session: Imaging requires a very stable anesthesia for about 20 min with a quick recovery. This anesthesia can
be repeated every fifth day or so, preferably not less than 2–3 days
to avoid anesthesia-related problems.

3. Nontraumatic fixation: The head of the animal has to be held in a
fixed position such that the laser illumination of the retina is stable.
Preferably, the head holder can be rotated such that larger retinal
areas (typically 0–45° eccentricity) can be visualized. The head holder
should not use ear bars; rather a head holder such as the one from
Narishige should be used, which uses the forehead and teeth for fixation.

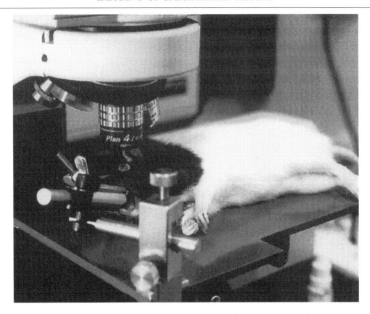

Fig. 2. Experimental setup with an animal. An anesthetized rat is lying on the modified microscope table and is adjusted in the optical path of the microscope with a nontraumatic head holder. A Hruby-style 160dpt contact lens is attached to the surface of the cornea with the help of immersion gel, and a standard low-power microscope lens is focused on the retina.

4. Protection of the eyes: To increase the field of view and image brightness, iris relaxation drugs are very useful. We have tested different ones and found that standard Neosynephrin-POS eye drops (e.g., Ursapharm, Germany) are most useful. It is also necessary to protect the cornea during the imaging session.
5. Contact lens: The eye-sided surface should be immerged with optical gel, which will also protect the cornea. Any normal contact lens gel will work, e.g., Vidisic optical gel (Dr. Mann Pharma, Germany).
6. Illumination wavelength: Use visible range (488/546 nm) excitation, with moderate laser power. Fluorescence will not be much stronger with high laser power.
7. Scan speed: A submillisecond scanning, with averaging of several pictures/sec, provides the best image quality and sensitivity.

Typical Substances and Settings

Green FluoSpheres 40 nm, 2.0 μ per brain injection site, 5–7 days transport time
Injection sites for the rat sup. colliculus: A/P: -6.5 mm, lat.: 1–1.7 mm, 3–3.5 mm below dura

Ketamine/Rompune anesthesia, 50 mg/10 mg per kg body weight intra-
peritoneally

Neosynephrin-POS 5% eye drops for iris relaxation, wait 5–10 min for
maximum effect

Vidisic optical gel for protection of the cornea and contact lens immersion

200–400 nsec scan acquisition time, 7–15p/sec scan rate, averaging of
16–32 pictures

Applications and Future of ICON

The use of confocal *in vivo* imaging techniques provides advantages
especially for the observation of developmental and degenerative changes
in the nervous system. Cells can be observed repeatedly in a noninvasive
manner and in a totally natural environment. Even with cell culture approaches and long-term two-photon imaging this cannot be achieved. As
Fig. 3 shows, both the cell number and the cell size can be followed over

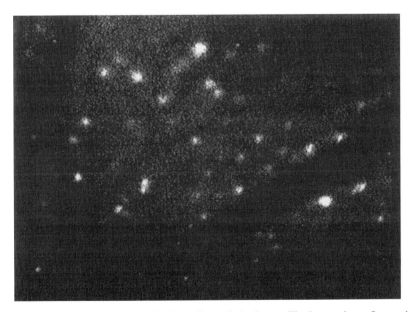

FIG. 3. Observation of living retinal ganglion cells in the rat. The image shows fluorescing
ganglion cells at 30° eccentricity. The observed field is about 0.08 mm² in size. The two darker
radial stripes are blood vessels of the retina, being about 20 μm in diameter. These blood
vessels can be used as natural scale bars and will also help relocate the observed region in
later imaging sessions. The ganglion cells visible are of different size and fluorescence intensity
and belong to different types. This image was taken after an optic nerve damage (controlled
crush), and here only the surviving or unaffected ganglion cells are labeled.

time with ICON in the same animal. Because of this ability to image cells with microscopic resolution repeatedly over time, very small changes (e.g., in size and number) can be observed. Therefore, we hope that over time ICON will become an important tool for neurobiological research, as a number of questions can be studied *in vivo* that hitherto could only be investigated *in vitro*.

In the future, the use of ICON imaging will be influenced both by the available markers and by technical advances in imaging resolution. Some examples of new ICON applications might include the following. A new fluorescent marker may be found that, beyond retrograde labeling, may be useful to show other cell types or particular cell functions, including molecular alterations. It is conceivable that glial cells could be labeled via the blood system, and with GFP and viral vectors, even certain nerve cell types may be labeled specifically. Furthermore, while anatomical markers work well for describing entire cell populations, many markers exist that show specific cellular functions as well, e.g., cell death markers or Ca^{2+} dyes for the monitoring of nerve cell activity.

The current drawback of ICON is that the light scattering properties of the surrounding tissue are not optimal. With deconvolution techniques and improved optical corrections, enhanced resolution may be possible. This is currently under development with the goal to visualize structural details such as nerve cell processes or even dendritic details.

ICON can be combined with other microscopic evaluation techniques as well. For example, after ICON has been applied in the living animal repeatedly, it is still possible to prepare the tissue and perform physiological or anatomical studies with this tissue. With the setup and procedures described, tissue slices or retinal whole mounts can be analyzed with conventional microscopic techniques, and the observed areas can be relocated without problems.

ICON thus provides the opportunity to visualize neurons of the mammalian CNS in a noninvasive manner repeatedly and without any damages. This opens up the opportunity to image neurons already in their natural environment and may be, at least for some questions, a very good alterative to *in vitro* experiments.

Suppliers

The addresses of local distributors can be found at the following sites: light microscope, www.zeiss.de; confocal scanner, www.noran.com; head holder, www.narishige.co.jp; contact lens, www.newport.com; and fluorescence markers, www.probes.com.

Section VII

Imaging Viruses and Fungi

[32] Identification of Viral Infection by Confocal Microscopy

By DAVID N. HOWELL and SARA E. MILLER

Introduction

Confocal microscopes have found increasing use as diagnostic tools.[1] In the clinic, confocal instruments have been used to perform *in vivo* biomicroscopy of tissues such as the cornea and skin. Surgical pathologists have employed confocal imaging to examine thick or difficult to section biopsy specimens. In the laboratory, confocal optics have been used to scan solid-phase arrays of oligonucleotide probes hybridized with DNA samples.

As an isolated technique, confocal laser scanning microscopy (CLSM) is of limited utility in the field of diagnostic virology. Confocal imaging has been employed in a few studies to detect *in situ* hybridization of virus-specific probes,[2,3] but in most diagnostic settings offers little advantage over standard methods such as routine histology, enzyme immunohistochemistry, and conventional immunofluorescence microscopy. Used in concert with electron microscopy (EM), however, confocal microscopy provides a powerful survey tool, allowing ready identification of focal areas of pathologic alteration in large tissue samples for subsequent ultrastructural analysis.

Ultrastructural pathologists are handicapped by severe limitations on the size of tissue specimens that can be accommodated by electron microscopes. This is a particular problem for ultrastructural diagnosis of viral infections in tissues, which are often quite focal. Although a variety of survey methods have been developed to allow selection of tissue samples for EM,[4] many of these are cumbersome and time-consuming. The method employed most frequently, light microscopic examination of toluidine blue-stained survey sections produced from small epoxy-embedded blocks of

[1] J. L. Caruso, R. M. Levenson, and D. N. Howell, *in* "Biomedical Applications of Microprobe Analysis" (P. Ingram, J. D. Shelburne, V. L. Roggli, and A. LeFurgey, eds.). Academic Press, San Diego, in press.

[2] A. J. C. van den Brule, F. V. Cromme, P. J. F. Snijders, L. Smit, C. B. M. Oudejans, J. P. A. Baak, C. J. L. M. Meijer, and J. M. M. Walboomers, *Am. J. Pathol.* **139**, 1037 (1991).

[3] S. E. Mahoney, M. Duvic, B. J. Nickoloff, M. Minshall, L. C. Smith, C. E. M. Griffiths, S. W. Paddock, and D. E. Lewis, *J. Clin. Invest.* **88**, 174 (1991).

[4] S. E. Miller, R. M. Levenson, C. Aldridge, S. Hester, D. J. Kenan, and D. N. Howell, *Ultrastruct. Pathol.* **21**, 183 (1997).

randomly selected tissue, is labor-intensive and occasionally fruitless. Another popular alternative, light microscopic survey of sections of paraffin-embedded tissue with subsequent reprocessing of selected areas for EM, affords ultrastructural preservation insufficient for identification of any but the largest viral particles.

With their ability to produce optical sections of relatively thick tissues, confocal microscopes can be used to scan large tissue specimens for focal areas of interest, including foci of viral cytopathology. These foci can then be excised and examined by EM. Confocal microscopic examination can be performed either before[4–6] or after[7,8] tissue embedment. Each approach has advantages and disadvantages. Preembedment scanning facilitates examination of a large amount of tissue for features of interest, but isolation and processing of the selected areas for EM are somewhat cumbersome. Postembedment scanning facilitates the subsequent production of sections for EM, but is best suited for relatively small specimens, as production of numerous large embedment blocks is tedious and time-consuming. Some fluorescent labels useful for localizing viral infections withstand EM embedment poorly, limiting their use in postembedment scanning techniques.

Because the search for foci of viral infection often requires examination of a considerable amount of tissue, we have concentrated our efforts on preembedment survey techniques. A standard method developed in our laboratories[4–6] is illustrated in Fig. 1. Fixed, unembedded tissue specimens are sliced using a vibrating microtome. The slices are stained by a variety of methods and examined with a confocal microscope for areas of pathologic alteration. These areas are dissected out, embedded, and examined by EM. An example of the application of this method, in which a focal adenovirus infection was identified in liver tissue, is provided in Fig. 2. With rapid processing techniques, the entire procedure can be accomplished in a day or less.

This article provides a step-by-step discussion of the processing and examination of tissue specimens by confocal microscopy for subsequent ultrastructural study. Although the method was originally developed for examination of focal viral infections, it is equally applicable to the analysis of other focal pathologic features (e.g., foci of differentiation within poorly differentiated tumors).

[5] S. E. Miller and D. N. Howell, *Immunol. Invest.* **26,** 29 (1997).
[6] C. T. Chu, D. N. Howell, J. C. Morgenlander, C. M. Hulette, R. E. McLendon, and S. E. Miller, *Am. J. Surg. Pathol.,* in press.
[7] J. S. Deitch, K. L. Smith, J. W. Swann, and J. N. Turner, *J. Electron Microsc. Technol.* **18,** 82 (1991).
[8] X. J. Sun, L. P. Tolbert, and J. G. Hildebrand, *J. Histochem. Cytochem.* **43,** 329 (1995).

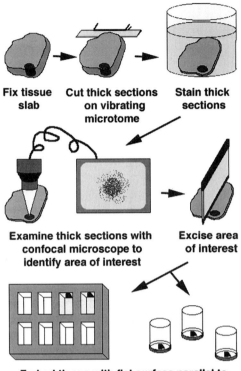

**Fix tissue   Cut thick sections   Stain thick
slab   on vibrating   sections
microtome**

**Examine thick sections with   Excise area
confocal microscope to   of interest
identify area of interest**

**Embed tissue with flat surface parallel to
eventual plane of section**

Fɪɢ. 1. Schematic diagram of the technique for confocal microscopic examination of unembedded tissue, with subsequent embedment of focal area of interest for EM. Reprinted, with permission and with minor modifications, from S. E. Miller, R. M. Levenson, C. Aldridge, S. Hester, D. J. Kenan, and D. N. Howell, *Ultrastruct. Pathol.* **21,** 183 (1997).

Confocal Microscopy

Confocal microscopes are a relatively recent addition to the armamentarium of diagnostic pathologists and virologists. Many of the techniques for preparing specimens for CLSM, however, are similar to methods used in conventional histology laboratories; a vast majority of the necessary reagents and supplies are readily available in such laboratories. The optical sections produced by confocal microscopes will look familiar to pathologists accustomed to performing conventional immunofluorescence microscopy on tissue sections produced with a microtome. Some confocal optical systems allow the microscopist to examine a specimen in real time, in much the same manner that a section would be viewed with a conventional microscope.

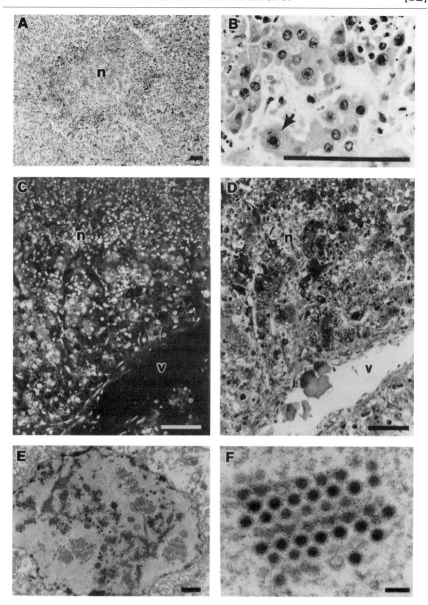

Specimen Selection and Fixation

Confocal microscopic survey is most useful for specimens too large to be embedded in a limited number of conventional EM blocks. (Using conventional methods, this limit is reached with fairly small specimens; complete embedment of a cube of tissue 3 mm on a side in 1-mm^3 tissue segments, a typical size for EM specimens, would require 27 blocks.) Although tissues with a variety of shapes can be accommodated, ideal specimens are rectangular solids measuring approximately $1 \times 1 \times 0.3$ cm. Specimens of this shape and size can be fixed rapidly and are well suited for subsequent sectioning and examination (see later).

Rapid fixation of specimens is of paramount importance for examination by EM. Prompt fixation is necessary to achieve optimal ultrastructural preservation, particularly for diagnostic virology, where many of the potential pathogens are quite small and easily confused with endogenous cellular structures (e.g., ribosomes). Fixation has the added benefit of rendering tissues somewhat less pliable, facilitating the subsequent production of thick sections by vibrating microtomy.

Choice of fixative depends partially on the type of staining to be used for confocal microscopic examination. For many general histochemical stains, any good EM fixative can be used.[9,10] A 2–4% (v/v) solution of glutaraldehyde in cacodylate buffer is an excellent choice, but other fixatives, such as 2–4% (w/v) paraformaldehyde, also work well. Ultrastructural preservation is usually quite acceptable in tissue fixed promptly with the 10% neutral-buffered formalin used by most general pathology services for routine specimen fixation. In cases where confocal examination is contemplated, a portion of a biopsy specimen can be reserved in buffered formalin

[9] A. M. Glauert, "Fixation, Dehydration, and Embedding of Biological Specimens." Elsevier, New York, 1975.
[10] M. A. Hayat, "Fixation for Electron Microscopy." Academic Press, New York, 1981.

FIG. 2. Detection of adenovirus infection in liver tissue using CLSM and EM. (A) Light micrograph of an H&E-stained section of paraffin-embedded tissue showing a focus of necrosis (n). (B) Higher magnification view of the edge of necrotic focus shown in (A), demonstrating intranuclear inclusion in a hepatocyte (arrow). (C) Confocal micrograph of a glutaraldehyde-fixed, propidium iodide-stained tissue showing an area of necrosis (n) adjacent to a hepatic vein (v). (D) Toluidine blue-stained semithin section of focus shown in (C) following epoxy embedment, showing same area of necrosis (n) and hepatic vein (v). (E and F) Electron micrographs of infected cell from the edge of necrotic focus, showing paracrystalline arrays of 75-nm adenovirus particles within nucleus. Bars: (A–D), 100 μm; (E), 1 μm; (F), 100 nm. Reprinted, with permission, from S. E. Miller, R. M. Levenson, C. Aldridge, S. Hester, D. J. Kenan, and D. N. Howell, *Ultrastruct. Pathol.* **21**, 183 (1997).

and subsequently apportioned for CLSM/EM or processed for routine histologic sectioning as needed.

Glutaraldehyde-fixed tissues autofluoresce, a property that can actually be used as a basis for confocal imaging (see later), although it represents a severe impediment to subsequent immunofluorescence microscopy. Paraformaldehyde fixation is often a good choice for specimens being prepared for immunofluorescent staining. As with any form of immunohistochemical staining, many antigenic epitopes are affected adversely by fixation, and determination of optimal fixation times and fixative concentrations for preservation of antigenicity as well as ultrastructural detail may be required.

Preembedment Sectioning

Although confocal microscopes allow imaging beneath the surfaces of tissue specimens, the maximal depth for optical sectioning is dependent on a number of factors, including penetration of stain and inherent opacity of the tissue under study. With the exception of unusually clear tissues such as cornea, the practical limit is generally a few dozen micrometers at best. As a result, most of the interior of typical biopsy specimens is inaccessible to confocal imaging.

This problem can be circumvented by producing slices of the specimens with a vibrating microtome. Several versions of this device are available, including the Vibratome Tissue Sectioning System (Ted Pella, Redding, CA) and the Oscillating Tissue Slicer (Electron Microscopy Sciences, Ft. Washington, PA). All employ a rapidly oscillating blade (generally a conventional razor blade) to produce slices of wet, unembedded tissues. Slices produced in this manner are mounted easily on microscope slides and have flat surfaces that are ideally suited for rapid confocal scanning.

Specimens for vibrating microtomy are affixed to a glass slide or Teflon support; a fast-drying cyanoacrylate glue works well for this purpose. Handling and sectioning of small specimens can sometimes be facilitated by encasing them in molten gelatin and allowing it to solidify, forming a larger block. For this purpose, a 3–7% aqueous solution of agarose or gelatin with a hardness of 275-300 Bloom, melted and cooled to 42°, can be employed.

The support with attached tissue is immersed in a chamber in the vibrating microtome containing buffer such as phosphate-buffered saline. Slices float off into the surrounding buffer as they are produced and are picked up and stored (usually in their original fixative) pending staining and examination. The speed of the oscillating blade, rate of specimen advance, and thickness of the sections produced must be adjusted for the tissue type under study. With insufficient oscillating speed and/or excessive specimen advance speed, the tissue may be pushed aside or compressed rather than sliced by the advancing blade. In general, the thinner the slices produced, the greater the percentage of tissue accessible for confocal scanning. However, friability of the tissue generally imposes a lower limit of approximately 50 μm on slice thickness. Slices less than 100 μm thick, particularly tissues with a paucity of fibrous stroma such as brain and liver, are often quite fragile. Grasping such tissues with forceps frequently causes them to tear and crumble; they are most easily picked up and transferred by draping them over a plastic pipette tip, blunt probe, or camel-hair brush.

Small, irregular tissue fragments can also be examined by CLSM, although the results are less satisfactory than those achieved with vibrating microtome slices. Nonplanar specimens are difficult to mount with coverslips, and much of their volume is often inaccessible to confocal scanning. When examining such specimens with higher magnification objectives with short working distances, care must be taken to avoid racking the objective into the specimen while attempting to visualize features at deeper focal planes.

Staining

A wide variety of staining methods can be utilized to prepare tissues for examination by CLSM. These methods have variable applicability to the use of confocal microscopy as a survey tool for subsequent EM. Occasional tissue components (e.g., lipofuscin) are autofluorescent and can serve as landmarks for confocal imaging. Fixation with glutaraldehyde induces a more generalized autofluorescence, which in some cases can be employed to produce very detailed confocal images. This method has been used, for example, to study whole-mount preparations of human nerves.[11] An obvious additional benefit of this "staining technique" is the optimal ultrastructural preservation afforded by glutaraldehyde fixation.

Numerous tissue stains employed in conventional fluorescence microscopy can be applied to confocal imaging. The fluorescent nucleic acid-binding dye propidium iodide works well for visualizing many forms of viral cytopathology[4,5]; similar results should be achievable with bisbenzimide (Hoechst 33258) and 4,6-diamidino-2-phenylindole (DAPI). Tissues stained with propidium iodide and viewed with a green excitation filter exhibit strong nuclear staining as well as weaker cytoplasmic staining, the latter presumably due to binding of the dye to cytoplasmic RNA.[4,5] Erythrocytes, which contain low levels of residual cytoplasmic nucleic acids, frequently exhibit moderate fluorescence in propidium iodide-stained sections; care

---

[11] R. J. Reynolds, G. J. Little, M. Lin, and J. W. Heath, *J. Neurocytol.* **23,** 555 (1994).

must be taken not to confuse them with nucleated cells. Several stains used in routine histochemistry (e.g., eosin, Congo red) also fluoresce when excited at appropriate wavelengths and can be used to stain tissues for CLSM.

Tissues can also be stained by immunohistochemical and *in situ* hybridization methods for confocal microscopy. Direct or indirect immunofluorescent staining with a variety of fluorochromes can be employed, for example, to pinpoint focal infections with known viruses for subsequent ultrastructural study.[4] Reflectance-mode confocal imaging has been used to study the localization of colloidal gold-labeled *in situ* hybridization probes specific for human papillomavirus[2] and human immunodeficiency virus.[3] With proper embedment and sectioning, examination of probe localization in preparations stained in this manner by EM should be possible. Labeling can also be performed with a mixture of fluorescent and colloidal gold-tagged secondary reagents to allow sequential imaging of the same labeled feature by CLSM and EM.[8]

The development of techniques for the genomic incorporation of genes encoding green fluorescent protein (GFP) has allowed endogenous fluorescent tagging of a wide variety of tissue components. Experimental infections with viruses genetically engineered to express GFP can be monitored by confocal microscopy, and infected foci can be selected for subsequent examination by EM (Fig. 3).[12]

Choice of stains for confocal examination of tissues depends on a variety of factors. If examination of portions of tissue beneath the most superficial layer is desired, penetration of the stain is an important consideration. This is obviously not a concern for endogenous fluorescent features (including tissue components expressing GFP). Small molecules such as glutaraldehyde and many histochemical stains also penetrate multiple cell layers with relative ease. Staining with large molecule probes such as antibodies, however, is generally confined to the most superficial portion of the tissue. Techniques for enhancing the penetration of such reagents exist, including treatment of tissue with dilute detergent solutions or graded series of ethanol,[8] but their employment invariably sacrifices ultrastructural preservation to some extent.

In general, the identification of putative infectious foci within tissues for subsequent diagnostic EM is best performed using dyes, such as propidium iodide, which are capable of staining a broad range of cell types and tissue features.[4,5] Use of more specific labeling techniques such as immunofluo-

---

[12] S. C. Henry, K. Schmader, T. T. Brown, S. E. Miller, D. N. Howell, G. G. Daley, and J. D. Hamilton, submitted for publication.

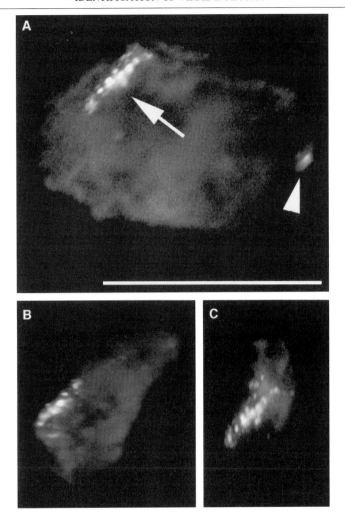

Fig. 3. Isolation of small focus of cells infected with recombinant murine cytomegalovirus expressing green fluorescent protein (GFP) from a small, irregularly shaped fragment of splenic tissue. (A) The position of the infected focus (arrow) in the tissue fragment is shown; a second, smaller focus of infection is also present (arrowhead). The larger GFP-expressing focus was excised in two successive steps, as illustrated in (B) and (C), and snap-frozen for ultracryomicrotomy. Tissue courtesy of Mr. Stanley C. Henry and Dr. John D. Hamilton, Veterans Affairs Medical Center, Durham, NC. Bar: 1 mm; magnification same in all panels.

rescent staining presupposes a knowledge of the infecting organism and is useful primarily in research settings.[4]

For staining, tissues are immersed in the stain of choice and incubated with gentle agitation. Care should be taken that both sides of tissue slices produced by vibrating microtomy are exposed to the stain, allowing staining and subsequent examination of both faces. The exact procedure followed depends on the stain employed. A 3-min incubation at 25° with a 50-μg/ml aqueous solution of propidium iodide followed by a brief rinse in distilled water provides strong nuclear staining of superficial cell layers; deeper penetration of the stain can be achieved with longer incubations. Free propidium iodide fluoresces minimally, so exhaustive washing to remove unbound stain is unnecessary. Direct or indirect immunofluorescent staining can be achieved using staining protocols virtually identical to those used for routine histologic sections. In this setting, thorough washing to remove unbound reagents is advisable.

Mounting

Following staining, tissue samples are transferred to glass slides for examination. Specimens can be mounted in any suitable aqueous buffer; the solution used originally for fixation is a reasonable choice. If possible, a coverslip is applied; this provides an optically flat surface for viewing, protects the objective from inadvertent immersion in the buffer, and prevents desiccation of the tissue. Judicious blotting or aspiration of excess mounting buffer from around the edges of the coverslip minimizes slippage and facilitates the marking of areas of interest for subsequent excision (see later). Further stability can be achieved by sealing the edges of the coverslip with a melted mixture of equal volumes of paraffin wax, petroleum jelly, and lanolin. Vibrating microtome slices up to 200 μm in thickness can generally be coverslipped without difficulty. Thicker slices and irregular tissue fragments may be too large to allow the coverslip to rest flat on the slide. Such specimens can be scanned without a coverslip, but care must be taken to avoid drying of the tissue and contamination of the objective.

Thin tissue slices can also be mounted between two coverslips. The resulting "sandwich" is placed on a slide for confocal scanning and can be flipped, allowing easy examination of both faces of the slice.

Scanning

The goal of confocal scanning as a survey tool is to identify features of potential interest for subsequent ultrastructural study. For viral infections, the features in question are often quite focal, but can usually be identified at relatively modest magnification and resolution. As such, optimal results are obtained by scanning in a manner that allows rapid survey of large tissue areas.

Any available confocal laser scanning microscope can be used for tissue survey. Point-scanning microscopes provide images with optimal resolution, but generally can generate no more than one image per second. These images must be collected using a computer frame buffer and displayed on a monitor. Systematic scanning of a large piece of tissue is difficult using such an instrument; moving the stage to allow sequential viewing of overlapping fields covering the entire specimen surface requires considerable practice.

This problem can be circumvented by using a slit-scanning confocal microscope (e.g., Meridian Insight Plus, Genomic Solutions, Inc., Ann Arbor, MI). Such instruments, although not capable of the resolution achieved with point scanners, scan with sufficient rapidity to allow generation of real-time images, either on a computer monitor or directly through oculars. This capability greatly simplifies the systematic survey of an entire tissue surface and also facilitates focusing to identify features of interest in deeper tissue layers.

Confocal survey of large tissues is achieved most readily with a low-to-medium power objective (10–20×). The modest resolution afforded by such objectives is sufficient for the identification of most forms of viral cytopathology; their ample field of view and the relatively thick optical sections they produce are well suited to survey applications. The identity of features detected at scanning power can be confirmed by observation at higher magnification. A lower magnification objective (e.g., 4×) is also useful for capturing images of trimmed tissue blocks prior to embedment for EM (see later).

Given sufficient stain penetration and tissue translucency, serial images at progressive depths (z series) can be collected and used to produce three-dimensional reconstructions. Using this method, the exact location and depth of a feature within the tissue can be determined.[8] This process is fairly time-consuming, however, and is primarily useful for small specimens or selected areas within larger specimens.

While viruses themselves are generally beyond the resolution limits of optical microscopic methods, they produce a number of direct and indirect alterations in cell and tissue structure that can be detected. Large clusters of replicating virus particles can often be seen as inclusion bodies. With some exceptions, the inclusions of DNA viruses are generally intranuclear, whereas those of RNA viruses reside in the cytoplasm. Viral inclusions are typically rich in viral nucleic acids and, as a result, are stained easily with nucleic acid-binding dyes such as propidium iodide.[4]

Infection with viruses can induce a wide variety of other cell and tissue alterations, including the formation of multinucleate syncytial giant cells, cellular enlargement ("cytomegaly"), inflammation, hemorrhage, necrosis, and, in some cases, neoplastic transformation. While none of these features is specific for viral infection, they can all serve as keys in the search for areas harboring virus particles. Infected cells with optimal preservation for ultrastructural examination are found frequently at the peripheries of areas of necrosis, inflammation, or hemorrhage. Cytopathic alterations associated with specific virus infections have been catalogued.[13]

Although our initial applications of confocal microscopic survey have been in the area of diagnostic virology, the method has potential value as a survey tool for other forms of ultrastructural diagnosis. In particular, it can be used to identify focal areas of interest within tumors (e.g., foci of differentiation within poorly differentiated tumors, foci of viable tissue within largely necrotic masses) for subsequent examination by EM. An example is shown in Fig. 4 in which a small area of neuroblastoma-like rosette formation was isolated from an astrocytic brain tumor for ultrastructural study.

Trimming of Selected Tissue Specimens

Once a feature of interest has been identified within a tissue slice, the next goal is to trim away excess tissue, leaving the area of interest and a small amount of surrounding tissue. Because trimming of the specimen under direct confocal microscopic guidance is not practical, the selected area must be marked for subsequent trimming under a dissecting microscope. For specimens mounted with a coverslip, this can be accomplished using an inked object marker mounted on the microscope turret in place of an objective. This device marks the coverslip with an ink circle, approximately 2 mm in diameter, centered on the area of interest. Care must be taken to avoid slippage of the coverslip during this maneuver; this can be monitored by a second confocal imaging of the specimen after the circle has been marked.

The slide is next transferred to the stage of a dissecting microscope, and the portion of tissue inside the marked circle is examined for identifying landmarks (e.g., areas of visible hemorrhage, other discolorations, small blood vessels). The relationship of the circled area to the edges of the tissue slice is also noted. It is often helpful to sketch a diagram showing the

---

[13] J. L. Caruso and D. N. Howell, *in* "Laboratory Diagnosis of Viral Infections" (E. H. Lennette and T. F. Smith, eds.), 3rd ed., p. 21. Dekker, New York, 1999.

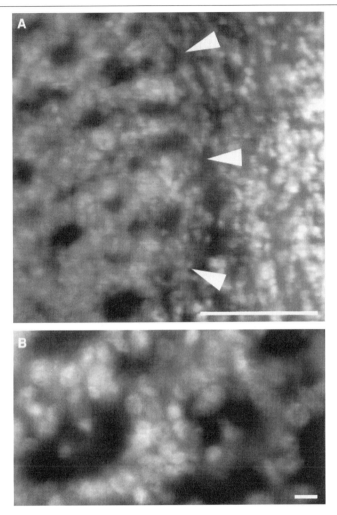

FIG. 4. Identification of a focal area of neuroblastoma-like differentiation within a vibrating microtome slice of an astrocytic neoplasm stained with propidium iodide. (A) A focus of tumor cell rosettes (arrowheads) is seen adjacent to tumor tissue without obvious differentiation; rosettes are shown at higher magnification in (B). Rosette-forming areas constituted less that 5% of the total tissue volume. Tissue courtesy of Dr. Christine M. Hulette, Duke University Medical Center. Bars: (A), 100 μm; (B), 10 μm.

approximate location of the area to be excised. The coverslip is removed carefully and the area of interest is reidentified under the dissecting microscope. A clean, fresh razor blade is then used to dissect away the surrounding tissue, leaving the area of interest in a small polygonal tissue fragment approximately 1–1.5 mm in maximal dimension. To allow distinction of the upper and lower faces of the tissue fragment (the former associated with the feature of interest) during subsequent steps, the fragment should be trimmed in an asymmetric shape. A trapezoid with two nonparallel sides and no axis of symmetry (e.g., the outline of the state of Nevada, U.S.A.) works well (Fig. 5).

After trimming, the specimen is returned to the confocal microscope and reexamined to ensure that the feature of interest has been retained. Although it is desirable to reapply a coverslip prior to scanning, small pieces of fragile tissue can be crushed by the weight of the glass. This problem can be circumvented by supporting the coverslip with the edges of additional coverslips or by mounting some of the larger, trimmed-off tissue fragments along with the selected fragment. If necessary, a second round of trimming can be performed to remove more excess tissue. Properly trimmed specimens will usually fit within the field of a 4× objective. An image captured with this objective is a useful guide for subsequent processing for EM; although the resolution of the image is insufficient for identifying most forms of viral injury, the shape of the block and the approximate location of the area of suspected infection can be documented (Fig. 5).

Trimming of small, irregular tissue fragments produced without the aid of a vibrating microtome poses a difficult challenge. Often, however, the irregular shape of the tissue provides landmarks to guide trimming. An example is shown in Fig. 3 in which a small patch of cells expressing GFP was dissected from a tissue fragment measuring approximately 1 mm in greatest dimension under confocal microscopic guidance.

When trimming is complete, the specimen is transferred to 2–4% glutaraldehyde in cacodylate buffer. At this stage, the tissue is generally too small and delicate to grasp with forceps, but can often be picked up and manipulated with the end of a wooden applicator stick sliced obliquely with a razor blade to produce a sharp, flat point. This implement is also useful during subsequent processing steps.

Electron Microscopy

An exhaustive discussion of preparative techniques for EM and a review of viral ultrastructure are beyond the scope of this article. Thorough

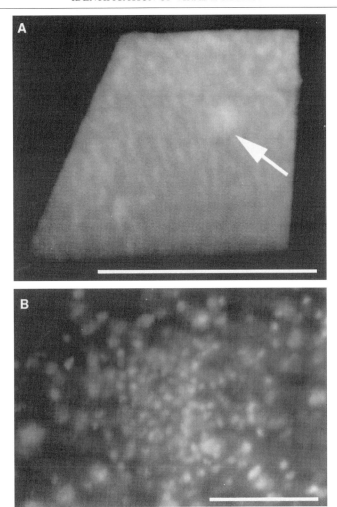

FIG. 5. Isolation of an inflamed focus in cerebral cortex from a patient with viral encephalitis. A trimmed block from a vibrating microtome section stained with propidium iodide is shown in (A); the inflamed area, visible as an indistinct white patch (arrow), is shown at higher magnification in (B). Bars: (A), 100 μm; (B), 10 μm.

guides to preparative methods are available in several texts,[9,10,14–16] general principles of viral electron microscopy are provided in a variety of sources,[17–22] and detailed descriptions of the ultrastructural features of individual viruses are available in several atlases.[23–26] A few guidelines will be offered, however, with specific reference to the processing and analysis of specimens selected by confocal microscopy.

Processing and Embedment

The method of processing and embedment selected depends on the type of EM examination to be performed. For routine ultrastructural study, specimens are postfixed with osmium tetroxide, stained *en bloc* with uranyl acetate, dehydrated, and infiltrated with epoxy resin by standard methods.[9,14–16] Spurr resin, available from most EM supply companies, is a good choice because it requires a relatively short baking time (8 hr); other epoxy resins [e.g., EMbed-812 (Electron Microscopy Sciences), Polybed 812 (Polysciences, Warrington, PA)] are also acceptable, but require 16–24 hr of baking.

[14] B. Mackay, ed., "Introduction to Diagnostic Electron Microscopy." Appleton-Century-Crofts, New York, 1981.

[15] M. A. Hayat, "Principles and Techniques of Electron Microscopy: Biological Applications." CRC Press, Boca Raton, FL, 1989.

[16] J. J. Bozzola and L. D. Russell, "Electron Microscopy: Principles and Techniques for Biologists." Jones and Bartlett, Boston, 1992.

[17] S. E. Miller, *J. Electron Microsc. Technol.* **4,** 265 (1986).

[18] S. E. Miller and D. N. Howell, *J. Electron Microsc. Technol.* **8,** 41 (1988).

[19] S. E. Miller, in "Medical Virology" (L. M. de la Maza and E. M. Peterson, eds.), Vol. 10, p. 21. Plenum Press, New York, 1991.

[20] S. E. Miller, in "Diagnostic Procedures for Viral, Rickettsial and Chlamydial Infections" (E. H. Lennette, D. A. Lennette, and E. T. Lennette, eds.), p. 37. American Public Health Association, Washington, DC, 1995.

[21] C. M. Payne, in "Pathology of Infectious Diseases" (D. H. Connor, F. W. Chandler, D. A. Schwartz, H. J. Manz, and E. E. Lack, eds.), Vol. 1, p. 9. Appleton & Lange, Stamford, CT, 1997.

[22] S. E. Miller, in "Laboratory Diagnosis of Viral Infections" (E. H. Lennette and T. F. Smith, eds.), 3rd ed., p. 45. Dekker, New York, 1999.

[23] F. W. Doane and N. Anderson, "Electron Microscopy in Diagnostic Virology." Cambridge Univ. Press, New York, 1987.

[24] E. L. Palmer and M. L. Martin, "Electron Microscopy in Viral Diagnosis." CRC Press, Boca Raton, FL, 1988.

[25] F. A. Murphy, C. M. Fauquet, D. H. L. Bishop, S. A. Ghabrial, A. W. Jarvis, G. P. Martelli, M. A. Mayo, and M. D. Summers, eds., "Virus Taxonomy: Sixth Report of the International Committee on Taxonomy of Viruses." Springer-Verlag, New York, 1995.

[26] D. H. Connor, F. W. Chandler, D. A. Schwartz, H. J. Manz, and E. E. Lack, eds., "Pathology of Infectious Diseases." Appleton & Lange, Stamford, CT, 1997.

During embedment, care must be taken to orient the tissue fragment so that the side containing the area of interest is parallel to and facing the eventual sectioning face of the embedment block. In initial experiments, we embedded specimens "on edge" in the ends of flat embedding molds. Specimens embedded in this manner frequently fell over and came to rest perpendicular to the block face before the resin hardened. To circumvent this problem, we now embed specimens in inverted cylindrical embedding capsules ("BEEM" capsules). To prepare the capsule for this form of embedment, the cap is sealed to the top of the capsule with paraffin film, and the conical tip is amputated with a razor blade. A small amount of liquid resin is placed in the inverted capsule, and the specimen is inserted with the surface containing the area of interest (the original "up" face) oriented face down. Positioning of the specimen in the capsule is best accomplished under a dissecting microscope with trans illumination from a light box, using a trimmed applicator stick. When the specimen has been positioned adequately, additional resin is added, with care being taken not to disturb the specimen, and the capsule is baked.

Specimens destined for immunoelectron microscopy can be embedded in a hydrophilic resin or snap-frozen for ultracryomicrotomy as indicated. To ensure rapid and uniform freezing, tissue samples for ultracryomicrotomy are of necessity quite small (<0.5 mm in greatest dimension). Maintaining the orientation of such samples during freezing is generally not possible; in many cases, however, this problem is minimized by the fact that the area of pathologic alteration occupies most of the specimen (see Fig. 3).

Sectioning

In sectioning tissue specimens selected by confocal microscopy, extreme care must be taken to avoid cutting through the area of interest in the process of facing and orienting the block. Careful alignment of the block so that the plane of sectioning is parallel to the embedded tissue surface is crucial. Initial sectioning of epoxy blocks is performed using a glass knife. Serial 0.5-μm survey sections are produced, stained with toluidine blue, and examined by light microscopy. When the area of interest is identified, the glass knife is replaced with a diamond knife and thin sections are produced. A final toluidine blue-stained survey section can be cut after thin sectioning to ensure that the area of interest is still represented.

If the area of interest is not on the surface of the specimen, but at a known depth within the tissue, the microtome can be used to remove tissue to an appropriate depth before sections are collected. This should be done conservatively, however; one should err on the side of removing too little tissue rather than too much, for obvious reasons. This is particularly true as tissues can shrink during processing, diminishing the distance from the tissue surface to the desired area.

Sections should be mounted on fine-bar grids to minimize the chance of the area of interest lying on a grid bar. Alternatively, slotted grids with a Formvar support membrane can be used. However, they are technically more difficult to manipulate and are less stable in the electron beam. Stability can be considerably enhanced by carbon coating. It is preferable to place one section in the center of the grid rather than randomly positioning single sections or a ribbon, particularly if the block face is large ($\geq$1 mm).

Electron Microscopy

Locating the area of interest in the electron microscope is facilitated greatly by a variable low-magnification setting for observing the whole section at once. A print of a low-magnification confocal micrograph or drawing of a toluidine blue-stained survey section can serve as a useful map of the thin section. Depending on how the section was picked up from the water (from underneath or above) and how the microscope specimen holder is inserted (flipping the grid 180° or not), the low magnification image may be inverted from that observed by CLSM or light microscopy. Crossovers at various magnifications in the electron microscope will also invert the image. If this poses a problem, the confocal print or drawing can be viewed in strong light from the reverse side of the paper to provide an inverted image; computer inversion of confocal image files can also be accomplished easily with several software packages.

After the section has been oriented in the electron microscope, the structural landmarks identified by CLSM should be located. It is frequently possible to reidentify individual virus-infected cells initially detected by confocal imaging.[4] Once identified, areas suspected of harboring virus infection should be examined at a magnification of at least ×40,000. Keys for the ultrastructural detection and identification of virions are provided in the references cited previously.

Summary

Confocal microscopy is a valuable adjunct to electron microscopy in the fields of diagnostic and investigative virology. Confocal imaging can be used to examine large amounts of tissue stained by a variety of methods for evidence of viral infection. Areas thus identified can then be processed for ultrastructural study, allowing a highly focused search for viral pathogens. With the possible exception of the vibrating microtome, all of the equipment and reagents necessary for the preparation of specimens for confocal scanning are available in any well-stocked histology laboratory. Although originally developed to facilitate viral diagnosis by EM, the methods described herein can be applied to the ultrastructural study of any focal pathologic process.

Acknowledgments

The authors are indebted to Dr. Emilie Morphew for critical review of the manuscript and to Dr. Charleen Chu for helpful discussions.

[33] Using Confocal Microscopy to Study Virus Binding and Entry into Cells

By Alain Vanderplasschen and Geoffrey L. Smith

Introduction

For decades, the quantitative study of virus binding and entry relied on assays requiring the use of purified virus preparations that were labeled in some way. One approach was to incorporate radioactive tracers into the culture medium during virus growth and then to purify radioactive virus. Other approaches involved labeling the virus after purification by nonradioactive methods such as biotinylation[1,2] or incorporation of fluorescent label.[3] The results generated were relative numbers representing the mean number of virions bound to the cell population[4] or individual cells.[1] We have developed new approaches to study vaccinia virus (VV) binding and entry based on confocal microscopy. These techniques do not require virus purification or labeling and generate data that reveal the absolute numbers of virus particles that have bound to or have entered into individual cells.

In this article, these techniques are described and then illustrated with some of the results obtained. Although the utility of the techniques reviewed here have been demonstrated with VV, given the resolution of the confocal microscope, these methods should be applicable to any virus larger than

[1] G. Inghirami, M. Nakamura, J. E. Balow, A. L. Notkins, and P. Casali, *J. Virol.* **62,** 2453 (1988).

[2] P. Borrow and M. B. Oldstone, *J. Virol.* **66,** 7270 (1992).

[3] R. W. Doms, R. Blumenthal, and B. Moss, *J. Virol.* **64,** 4884 (1990).

[4] A. Vanderplasschen, M. Bublot, J. Dubuisson, P.-P. Pastoret, and E. Thiry, *Virology* **196,** 232 (1993).

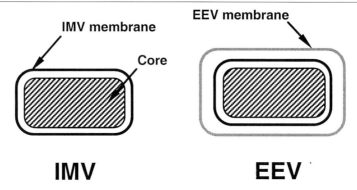

FIG. 1. Schematic representation of IMV and EEV structure.

50 nm. Finally, we will discuss how these techniques have generated data that cannot be obtained with classical binding or entry assays.

Vaccina virus is the prototype of the poxvirus family (for review, see Moss[5]). These are DNA viruses that replicate in the cell cytoplasm and have genomes between 150 and 300 kbp. The virions are large and complex, and in the case of VV, have dimensions of 250×350 nm and contain more than 100 proteins.[6] The study of VV binding and entry has been complicated by the fact that there are two morphologically distinct infectious virions, termed intracellular mature virus (IMV) and extracellular enveloped virus (EEV).[7,8] Structurally, IMV consists of a core surrounded by one membrane,[9,10] whereas EEV consists of an IMV with an additional outer membrane containing proteins, which is absent from IMV (Fig. 1). IMV remains within the cytoplasm until cell lysis and represents the majority of infectious progeny; for instance, with the Western Reserve (WR) strain of virus, IMV represents 99% of infectivity. In contrast, EEV is released actively from infected cells before cell lysis and is the form of the virus that is important for virus dissemination *in vitro* and *in vivo*.[7,11–13]

Extracellular enveloped virus is formed by the wrapping of IMV with

[5] B. Moss, *in* "Virology" (B. N. Fields, D. M. Knipe, P. M. Howley, R. M. Chanock, J. Melnick, T. P. Monath, B. Roizman, and S. E. Straus, eds.), p. 2637. Lippincott-Raven, Philadelphia, 1996.
[6] K. Essani and S. Dales, *Virology* **95,** 385 (1979).
[7] G. Appleyard, A. J. Hapel, and E. A. Boulter, *J. Gen. Virol.* **13,** 9 (1971).
[8] Y. Ichihashi, S. Matsumoto, and S. Dales, *Virology* **46,** 507 (1971).
[9] S. Dales and E. H. Mosbach, *Virology* **35,** 564 (1968).
[10] M. Hollinshead, A. Vanderplasschen, G. L. Smith, and D. J. Vaux, *J. Virol.,* in press (1999).
[11] E. A. Boulter and G. Appleyard, *Prog. Med. Virol.* **16,** 86 (1973).
[12] L. G. Payne and K. Kristensson, *J. Gen. Virol.* **66,** 643 (1985).
[13] L. G. Payne, *J. Gen. Virol.* **50,** 89 (1980).

two membranes derived from the trans-Golgi network or early tubular endosomes forming an intracellular enveloped particle (IEV) that has three membranes. This particle moves to the cell surface where a fusion event between the plasma membrane and the outer membrane of IEV forms a virus with two membranes. This particle may remain attached to the cell surface as cell-associated enveloped virus (CEV) or be released from the cell as EEV. The proportion of enveloped virus that is either retained on the cell surface or released into the medium varies with different strains of virus. At least 12 proteins that are absent from IMV are associated with the outer envelope of EEV.[14–16] Six VV genes are known to encode these proteins (for review, see Smith and Vanderplasschen[17]) and, in addition, several cellular proteins are present in the EEV envelope.[16] These proteins endow EEV with different biological and immunological properties.[7,11,16,18]

   The presence of these two infectious forms of virus complicates studies investigating the binding of virions to cells or entry into cells. In order to obtain data relating to only one form of the virus, it is necessary to either purify the different viruses from each other and to study each in isolation or devise a method to study both forms simultaneously within a mixed preparation and to distinguish one from the other. The extra membrane of EEV gives the virus different physical properties such that it may be separated from IMV by sedimentation through CsCl density gradients.[14] In these gradients, IMV and EEV sediment at 1.27 and 1.23 g/ml, respectively.[14] However, the EEV outer membrane is a loose and extremely fragile structure[19] that is damaged by virus purification,[17,20–22] and once it is ruptured, the particle retains full infectivity as an IMV.[23,24] Consequently, the mechanism of binding and penetration of intact EEV may be different from that of damaged EEV having IMV proteins exposed on its surface.[17,20–22] Therefore, binding and penetration studies should not be performed using purified EEV. To overcome this problem, we developed

[14] L. G. Payne, *J. Virol.* **27,** 28 (1978).

[15] L. G. Payne, *J. Virol.* **31,** 147 (1979).

[16] A. Vanderplasschen, E. Mathew, M. Hollinshead, R. B. Sim, and G. L. Smith, *Proc. Natl. Acad. Sci. U.S.A.* **95,** 7544 (1998).

[17] G. L. Smith and A. Vanderplasschen, in "Coronaviruses and Arteriviruses" (L. Enjuanes, S. G. Siddell, and W. Spaan, eds.), p. 395. Plenum Press, London, 1998.

[18] G. S. Turner and E. J. Squires, *J. Gen. Virol.* **13,** 19 (1971).

[19] N. Roos, M. Cyrklaff, S. Cudmore, R. Blasco, J. Krijnse-Locker, and G. Griffiths, *EMBO J.* **15,** 2343 (1996).

[20] A. Vanderplasschen and G. L. Smith, *J. Virol.* **71,** 4032 (1997).

[21] A. Vanderplasschen, M. Hollinshead, and G. L. Smith, *J. Gen. Virol.* **78,** 2041 (1997).

[22] Y. Ichihashi, *Virology* **217,** 478 (1996).

[23] A. A. G. McIntosh and G. L. Smith, *J. Virol.* **70,** 272 (1996).

[24] E. J. Wolffe, E. Katz, A. Weisberg, and B. Moss, *J. Virol.* **71,** 3904 (1997).

binding and entry assays that allow the use of fresh supernatants of infected cells as the source of EEV. These assays are based on confocal microscopy and the use of monoclonal antibodies that are specific to each form of the virus.

Methods

Cells and Virus

RK$_{13}$ cells are grown in minimum essential medium (MEM) (GIBCO) containing 10% heat-inactivated fetal bovine serum (HFBS). HeLa cells are grown as suspension cultures as described elsewhere.[25] The IHD-J VV strain is used throughout. Fresh EEV and purified IMV are prepared as described previously.[20] Briefly, for fresh EEV, cells are infected at 1 plaque-forming unit (pfu)/cell, and the culture supernatant is harvested 24 hr postinfection (hpi) and clarified by centrifugation at 2000g. After appropriate dilution, any contaminating IMV infectivity is neutralized by the addition of monoclonal antibody (MAb) 5B4/2F2 (final dilution of 1/2560) against the 14-kDa fusion protein (A27L gene product) of IMV.[26] In the conditions used in this study, MAb 5B4/2F2 neutralizes >93% of purified IMV. A fresh EEV preparation is produced for each experiment and, except where stated otherwise, MAb 5B4/2F2 is added to these EEV samples.

MAbs and Rabbit Antiserum

The culture supernatant from murine hybridoma secreting the IgM MAb B2 was kindly provided by Dr. W. Chang.[27] MAb B2 bound to the cell surface is shown to prevent IMV binding, but its effect on EEV binding or infectivity is unknown. Murine MAb AB1.1 (α-D8L)[28] and rat MAb 19C2 (α-B5R)[29] are raised against the D8L and B5R surface proteins of IMV and EEV, respectively. A rabbit antiserum, hereafter called anticore (α-core), is raised against VV cores isolated as described.[30] This serum recognizes the core proteins encoded by genes A10L, A3L, F18R, L4R, and A4L and was kindly provided by Gareth Griffiths (EMBL, Germany).

[25] P. R. Cook and I. A. Brazell, *J. Cell Sci.* **19,** 261 (1975).
[26] C. P. Czerny and H. Mahnel, *J. Gen. Virol.* **71,** 2341 (1990).
[27] W. Chang, J.-C. Hsiao, C.-S. Chung, and C.-H. Bair, *J. Virol.* **69,** 517 (1995).
[28] J. E. Parkinson and G. L. Smith, *Virology* **204,** 376 (1994).
[29] M. Schmelz, B. Sodeik, M. Ericsson, E. J. Wolffe, H. Shida, and G. Griffiths, *J. Virol.* **68,** 130 (1994).
[30] S. Cudmore, R. Blasco, R. Vincentelli, M. Esteban, B. Sodeik, G. Griffiths, and J. Krijnse Locker, *J. Virol.* **70,** 6909 (1996).

Indirect Immunofluorescent Staining

Samples are fixed in phosphate-buffered saline (PBS) containing 4% paraformaldehyde (PFA) (w/v) for 20 min on ice and 40 min at 20°. Immunofluorescent staining (incubation and washes) of fixed samples is performed in PBS containing 10% HFBS (v/v) (PBSF). When permeabilization is required after fixation, the staining is performed in PBSF containing 0.1% (w/v) saponin (Sigma, Poole, Dorset, UK). The samples are incubated at 37° for 45 min with α-D8L (diluted 1/300), biotinylated α-D8L (diluted 1/100), α-B5R (diluted 1/16), or rabbit α-core serum (diluted 1/1000) as the primary antibody. After three washes, the samples are incubated at 37° for 30 min with fluorescein isothiocyanate (FITC)-conjugated F(ab')$_2$ goat antimouse IgG (FITC-GAM) (8 μg/ml) (Sigma), rhodamine-conjugated streptavidin (Rd-Strep) (3.3 μg/ml) (Serotec, Kidlington, Oxon, UK), FITC (10 μg/ml) (Serotec), R-phycoerythrin (PE) (10 μg/ml) (Serotec)-conjugated F(ab')$_2$ rabbit antirat IgG (FITC-RAR, PE-RAR), FITC (6 μg/ml) (Sigma), or Rd (10 μg/ml) (ICN, Costa Mesa, CA)-conjugated goat IgG antirabbit IgG (FITC-GARb, Rd-GARb) as secondary conjugates. Samples are mounted in Mowiol mounting medium as described.[21]

Capping of MAb B2 Reactive Epitope

Capping of MAb B2 reactive epitope is induced in HeLa cells grown in suspension as described.[20] Undiluted hybridoma supernatant containing MAb B2 is added to HeLa cells for 1 hr on ice, and unbound antibody is removed by washing the cells with PBSF. Rabbit antimouse IgM is added for 30 min on ice. After further washing with PBSF, the cells are warmed to 37° for 10 min to permit capping and then recooled on ice.

Staining of Cell Plasma Membrane

The plasma membrane is stained with FITC-labeled wheat germ agglutinin (FITC-WGA) (5 μg/ml) (Vector Laboratories Ltd., Peterborough, UK) for 20 min on ice as described previously.[31] After extensive washing with PBSF and a final wash with PBS, the cells are fixed in PBS containing 4% paraformaldehyde (PFA) (w/v) for 20 min on ice and for 40 min at 20°.

Staining of Lysosomes

Lysosomes are visualized by labeling living cells with lysine PFA-fixable Rd-conjugated dextran (Rd-dextran) (final concentration 100 μg/ml, Mo-

[31] A. Vanderplasschen, M. Hollinshead, and G. L. Smith, *J. Gen. Virol.* **79,** 877 (1998).

lecular Probes, Eugene, OR) as described.[31] After extensive washing with PBSF, the cells are incubated with normal culture medium for 6 hr at 37°.

Confocal Microscopy Analysis

Cells are analyzed with a Bio-Rad (Richmond, CA) MRC 1000 confocal microscope (running under Comos software) using appropriate filters, a full dynamic range of gray scale, and Kalman filtration. Optical sections perpendicular to the Z axis are performed throughout the sample. Except where stated otherwise, the confocal pictures are reconstructed by the projection of sections so that the total number of virions bound on the cell surface (for binding assay) or the total number of cores inside the cell (for entry assay) could be determined by examination of the reconstructed picture.

Statistical Analysis

Student's t test is used to test for the significance of the results ($P < 0.05$).

Results

Our first approach to study IMV and EEV binding to cells was to use IMV and EEV that had been purified from infected cells (IMV) or the culture supernatant (EEV) and then detect these particles on the cell surface by fluorescent microscopy. IMV was purified using sucrose density gradient centrifugation as described by Joklik.[32] EEV was purified from the culture supernatant by first removing detached cells and large cell debris by low-speed centrifugation and then pelleting the virus by ultracentrifugation. The pelleted virus was then resuspended by vortexing and sonication and was purified on CsCl or sucrose density gradients. While CsCl gradients in particular give good separation of IMV and EEV, it was discovered that the majority of EEV particles purified from density gradients had a damaged outer envelope. This was shown by the neutralization of the majority of infectivity in these preparations by a monoclonal antibody specific for IMV (methods). The ability of an IMV-specific antibody to neutralize "EEV" indicated that the integrity of the EEV outer envelope was broken and IMV antigens were exposed, and consequently it would be unknown if the virions were binding to cells via antigens present on the IMV or EEV

[32] W. K. Joklik, *Virology* **18,** 9 (1962).

surface. Thus data obtained with such preparations would be uninterpretable.

Attempts to purify EEV by other less vigorous methods also failed to retain the integrity of the EEV envelope and even pelleting the virus and resuspension was found to damage a significant proportion of the virions. Therefore this approach was abandoned and instead we devised a method that used a preparation of virus that contained both IMV and EEV and simultaneously identified and distinguished these particles from each other.

Virus-Binding Assay Using Confocal Microscopy

Vaccinia virus is a very large virus (approximately 250 × 350 nm) and individual virions had been detected by fluorescent microscopy.[33] Therefore we investigated whether indirect immunofluorescent staining of virions and confocal microscopy could be used to develop a novel virus-binding assay. The idea was to quantify the number of individual virions bound on the cell surface at 4° (a temperature that prevents virus entry). The infectivity of fresh culture supernatant was found to contain mostly EEV, but was contaminated with a significant proportion (15–25%) of IMV, and therefore its use for binding studies required an assay that permitted the differentiation of IMV and EEV. With that goal in mind, we used a double immunofluorescent staining with MAb AB 1.1 and MAb 19C2 against the D8L and B5R gene products, respectively, on fixed and permeabilized samples (Fig. 2). D8L encodes a 32-kDa protein present on the IMV surface[28] and 19C2 detects the EEV-specific 42-kDa glycoprotein B5R.[29] Biotinylated MAb AB1.1 and MAb 19C2 were used as primary MAbs and were subsequently revealed by Rd-Strep and FITC-RAR, respectively. Following this staining procedure, EEV virions appeared as double positive (red and green) foci, whereas IMV virions were single (red) fluorescent foci on the cell surface. Each fluorescent focus was shown to represent an isolated particle and not a cluster of virions by measuring the size of the spots. The mean diameter of Rd fluorescent foci was 409.3 nm (SD = 13.1, $n = 20$), which is similar to the dimension of a single vaccinia virion.

The use of confocal microscopy rather than conventional epifluorescent microscopy allowed the determination of the total number of virions bound on the cell surface by examining the single image reconstructed by projection of the horizontal sections performed throughout the sample. To obtain the same data by conventional microscopy would require careful examination of a series of pictures taken at different focal planes throughout the sample.

---

[33] S. Cudmore, P. Cossart, G. Griffiths, and M. Way, *Nature* **378,** 636 (1995).

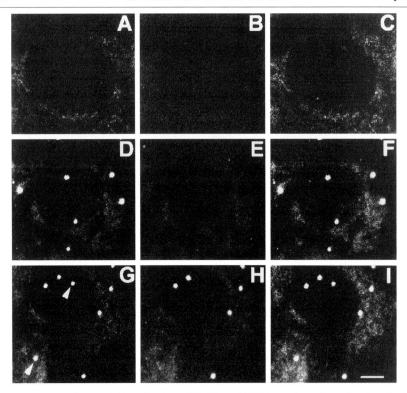

FIG. 2. Detection and identification of IMV and EEV on the cell surface by double immunofluorescent staining and confocal microscopy. RK$_{13}$ cells were mock infected (A–C) or infected on ice with purified IMV (D–F) or fresh EEV (G–I). The cells were then treated as described in Methods for the simultaneous detection of D8L and B5R gene products by double indirect immunofluorescent staining. Biotinylated MAb AB1.1 and MAb 19C2 were used as primary MAbs and were revealed by Rd-Strep and FITC-RAR, respectively. Horizontally, the three panels represent analysis of the same cells. The first (A, D, G) and second (B, E, H) columns of panels represent the analysis for Rd and FITC fluorescent emissions, respectively. The third (C, F, I) column represents the merged Rd and FITC signals. The arrows in (G) identify particles (IMV) that were detected by MAb AB1.1 but not by MAb 19C2. The bar in (I) represents 2 μm. Reproduced from A. Vanderplasschen and G. L. Smith, *J. Virol.* **71**, 4032 (1997), with permission from the American Society for Microbiology.

Particle/Plaque-Forming Unit Ratios for IMV and EEV

The examination of virus preparations by double immunofluorescent staining and microscopy enabled the total number of physical particles to be determined and the proportion of these that were EEV and IMV. The number of infectious particles in a preparation was determined by plaque assay. For purified IMV the great majority (97%, $n = 400$) of physical particles was IMV and therefore the ratio of total number of particles to the infectious particles (particle/pfu ratio) could be determined. For fresh EEV, the proportion of the infectivity attributable to IMV and EEV was calculated by measuring the infectivity of the virus preparation in the presence or absence of a MAb that neutralized IMV infectivity. These two values enabled the particle/pfu ratios of IMV and EEV in the fresh culture supernatant to be determined. These data showed that EEV had a particle/pfu ratio of 12.7 ± 6.1 ($n = 3$), considerably lower than that of IMV in the same virus preparation (45.0 ± 11.1, $n = 3$).[20] Comparison of the particle/pfu ratios of fresh IMV (45.0 ± 11.1, $n = 3$) or purified IMV (64.6 ± 16.5, $n = 3$) showed that the proportion of IMV particles that were infectious decreased during virus purification.[20]

These data also enabled the determination of the proportion of virus particle bound to cells that gave rise to a plaque. When IMV was added to cells at 10 pfu/cell, 2104 particles bound to 200 cells; therefore each IMV particle that bound gave rise to a plaque [(2104/200)/10 = 1.052]. For EEV added to cells at 4.056 pfu/cell, 3984 particles bound to 200 cells. Therefore only approximately 1 in 5 particles gave rise to a plaque [(3984/200)/4.056 = 4.9].[20] If the proportion of EEV particles bound to a cell that gave rise to a plaque was five times lower than for IMV particles, how was the greater infectivity (lower particle/pfu ratio) of EEV particles explained? This must reflect differences in the efficiency of either IMV and EEV binding to cells or the subsequent penetration event leading to release of the virus core into the cytoplasm and initiation of the infectious cycle.

IMV and EEV Bind to Different Cellular Receptors

Using the binding assay described in this article we were able to demonstrate unequivocally that IMV and EEV bind to different cellular receptors. Three independent observations allow this conclusion: (1) the efficiency with which IMV and EEV bound to different cell lines was unrelated; (2) cell surface digestion with some enzymes affected IMV and EEV binding differently; and (3) binding of a MAb B2 to cells prevented IMV but not EEV binding.

IMV and EEV Bound to Different Cell Lines with Varying Efficiency

If IMV and EEV bound to an identical receptor, variation in expression of this receptor between different cell lines should affect IMV and EEV binding to the same extent. However, if IMV and EEV bound to separate receptors, the efficiency of IMV and EEV binding might fluctuate independently between different cell lines, reflecting differing levels of expression of the different receptors. Three cells that are used commonly to grow VV

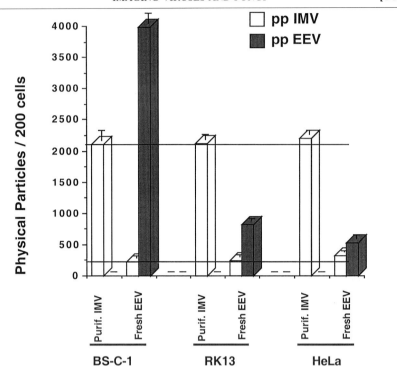

FIG. 3. IMV and EEV binding onto different cell lines. Adherent cells were grown in a 1-well culture chamber slide (Nunc, Life Technologies Ltd., Uxbridge, UK). For HeLa cells, the binding assay was performed in suspension (10^6 cells/ml). Cells were infected on ice for 1 hr (0.5 ml per 1-well chamber slide) with either purified IMV (10 pfu/cell) or fresh EEV (5.2 pfu/cell, 78% of the pfu being resistant to neutralization by MAb 5B4/2F2) diluted in PBS plus 2% FBS. After washing, the cells were treated as described in the text to reveal IMV and EEV virions by double immunofluorescent staining. Numbers of IMV and EEV virions bound on the surface of 200 cells were then determined by confocal microscopy. Data presented are the average ±SD for triplicate measurements.

were selected for study: human (HeLa) cells in suspension culture, rabbit kidney $(RK)_{13}$ cells, and African green monkey kidney (BS-C-1) cells. Two different virus preparations were added to these cells: either IMV purified from the cytoplasm of infected cells or fresh EEV from the culture supernatant. In the latter case, the presence of some contaminating IMV enabled the binding of both forms of virus to be analyzed simultaneously and provided a perfect internal control for each form of virus (Fig. 3).

The number of IMV particles bound to the surface of 200 cells was similar for the three cell lines tested. Both sources of IMV (purified IMV and fresh EEV) led to this observation. In contrast, the efficiency of EEV

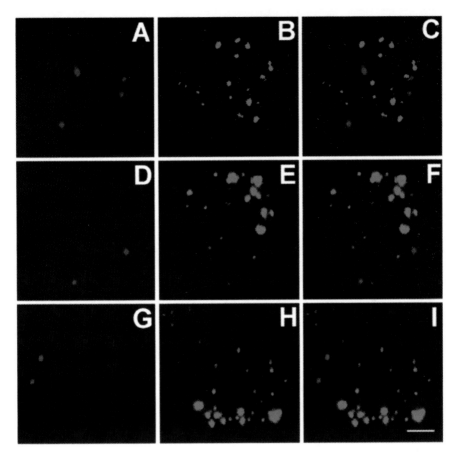

FIG. 5. Concomitant detection of EEV (red signal) and MAb B2 reactive molecule clusters (green signal) on the surface of HeLa cells. HeLa cells were grown in suspension culture and were processed as described in Methods to induce capping of the MAb B2 reactive molecule. Cells were then infected with fresh EEV at 3.5 EEV pfu/cell (10^6 cells/ml) on ice for 1 hr. After washing with cold PBS plus 2% FBS, the cells were fixed in PBS containing 4% paraformaldehyde (w/v) for 20 min on ice. MAb B2/RAM–IgM complexes were revealed by FITC-GARb. EEV were revealed by indirect immunofluorescent staining using MAb 19C2 as the primary antibody and PE-RAR as the secondary antibody. The cells were first incubated simultaneously with FITC-GARB and MAb 19C2 for 45 min at 37° and, after washing, were incubated with PE-RAR at 37° for 30 min. After final washing, the cells were mounted and examined by confocal microscopy. The three horizontal rows (A–C, D–F, G–I) of panels represent the analysis of three different cells. The first (A, D, G) and second (B, E, H) panels of each line represent the emission of red and green fluorescence, respectively, and the third (C, F, I) panel represents the merged image of the red and green emissions. Bar: (I) 2 μm. Reproduced from A. Vanderplasschen and G. L. Smith, *J. Virol.* **71**, 4032 (1997), with permission from the American Society for Microbiology.

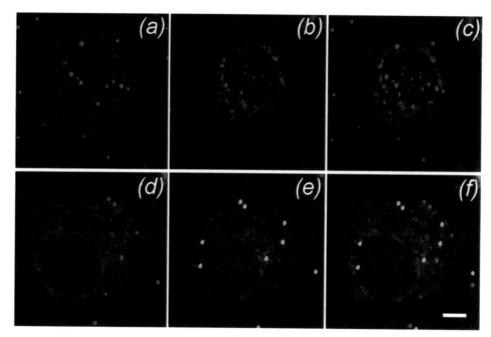

Fɪɢ. 6. Detection of IMV and virus cores in infected cells. Confluent RK_{13} cells were infected on ice with purified IMV at 10 pfu/cell, washed with ice-cold PBS, and incubated for 30 min with PBS either at 4° (a–c) or at 37° (d–f). After washing with ice-cold PBS, cells were treated as described in Methods for double indirect immunofluorescent staining of fixed and permeabilized samples. Biotinylated α-D8L and rabbit α-core were used as primary antibodies and were revealed by Rd-Strep and FITC-GARb, respectively. Sets of three horizontal panels represent analyses of the same cell. Panels a and d and b and e are analyses for Rd and FITC fluorescent emissions, respectively. Panels c and f show the merged Rd and FITC signals. Bar: 2 μm. Reproduced from A. Vanderplasschen, M. Hollinshead, and G. L. Smith, *J. Gen. Virol.* **79,** 877 (1998), with permission from the Society for General Microbiology.

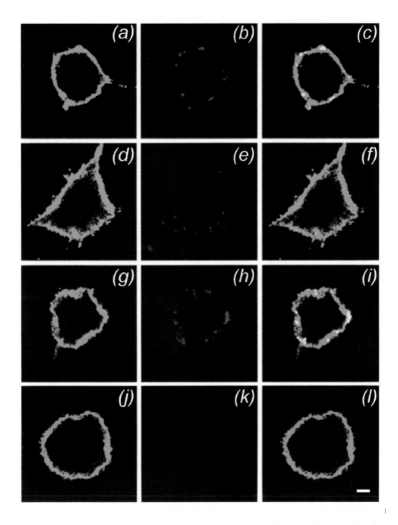

Fig. 7. Virus cores are intracellular. Subconfluent RK_{13} cells were infected on ice with purified IMV (100 pfu/cell) and then incubated for 30 min at 37°. After washing with ice-cold PBS, cells were incubated in PBSF containing FITC-WGA for staining of the plasma membrane as described in Methods. Cells were washed with ice-cold PBS and then either fixed and permeabilized (a–f) or fixed (g–l). After washing with PBS, samples were treated for single indirect immunofluorescent staining as described in Methods. Biotinylated α-D8L (b and h) or rabbit α-core (e and k) were used as primary antibodies and were revealed by Rd-Strep and Rd-GARb, respectively. The pictures represent single optical sections perpendicular to the Z axis. Sets of three horizontal panels represent analyses of the same cell. Panel a, d, g, and j and b, e, h, and k show FITC and Rd fluorescent emissions, respectively. Panel c, f, i, and l show merged FITC and Rd signals. Bar: 2 μm. Reproduced from A. Vanderplasschen, M. Hollinshead, and G. L. Smith, J. Gen. Virol. **79,** 877 (1998), with permission from the Society for General Microbiology.

binding varied between cell lines. For example, 4.9 times more EEV was detected on BS-C-1 cells than on RK_{13} cells.

The number of plaques formed by IMV or EEV on BS-C-1 and RK_{13} cells was also investigated. As for the binding assay, the number of plaques formed by IMV was equivalent on both cell types. Whereas fresh EEV, from which contaminating IMV was removed by the addition of IMV-neutralizing antibody, formed fivefold more plaques on BS-C-1 than RK_{13} cells.[20]

Cell Surface Digestion with Some Enzymes Affects IMV and EEV Binding Differently

If IMV and EEV bind to separate cellular receptors and if these receptors have different biochemical composition, cell surface treatment with some enzymes might affect the binding of IMV and EEV differently. RK_{13} cells were digested with trypsin, pronase, or neuraminidase or were mock treated as described previously.[20] Purified IMV or fresh EEV (without addition of IMV neutralizing antibody) was then added to these cells and the number of virions bound to the cell was determined (Fig. 4). Trypsin and neuraminidase treatment increased the number of IMV particles bound to cells slightly, whereas pronase treatment reduced IMV binding dramatically. This was true for purified IMV and for IMV present in fresh EEV. These enzymes had quite different effects on EEV binding: trypsin and pronase increased EEV binding approximately two- or threefold, respectively, whereas neuraminidase produced only a small increase. The most spectacular difference between IMV and EEV was observed with pronase, which reduced IMV binding drastically, while enhancing EEV binding. These data were consistent with the proposal that IMV and EEV bind to different cellular receptors.

MAb B2 Bound on the Cell Surface Does Not Affect EEV Binding

Chang *et al.*[27] showed that a mouse IgM MAb (MAb B2) bound to a trypsin-sensitive epitope on the cell surface and, when bound, inhibited the binding and consequently infectivity of IMV. This suggested that MAb B2 recognized an IMV receptor. If IMV and EEV bind to an identical receptor, MAb B2 would be expected to affect EEV binding and infectivity to the same extent as IMV. This was addressed using the binding assay described earlier.[20] The results demonstrated that when MAb B2 was bound on the cell surface, EEV binding was not inhibited. Moreover, if the MAb B2 reactive antigen was induced to aggregate (capping) by cross-linking with a secondary antibody and warming to 37°, the binding of EEV was still unaffected.[20] These results suggest that the MAb B2 reactive receptor is a receptor for IMV but not EEV binding. An alternative, but less likely,

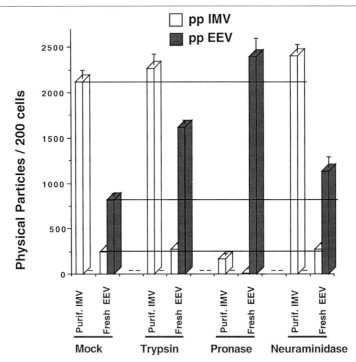

FIG. 4. Cell surface protease treatment affects IMV and EEV binding differently. Mock- or enzyme-treated RK_{13} were cells grown in a 1-well culture chamber slide (Nunc) and were infected on ice for 1 hr (0.5 ml per 1-well chamber slide) with either purified IMV (10 pfu/cell) or fresh EEV (5.2 pfu/cell, 78% of the pfu being resistant to MAb 5B4/2F2 neutralization) diluted in PBS plus 2% FBS. After washing, the cells were treated as described in the text to reveal IMV and EEV virions by double immunofluorescent staining. Numbers of IMV and EEV virions bound on the surface of 200 cells were then determined by confocal microscopy. Data presented are the average ±SD for triplicate measurements.

interpretation was that the IMV and EEV virions bound to different regions of the same receptor and that by binding to a particular epitope of the receptor, MAb B2 was able to inhibit the binding of only IMV. To exclude the possibility that EEV was able to interact with the MAb B2 reactive receptor despite its capping, we used double immunofluorescent staining to visualize simultaneously MAb B2 reactive molecule clusters and EEV (Fig. 5, color insert). One hundred cells, containing 254 bound EEV, were examined by confocal microscopy to determine the relative localization of EEV-binding sites and MAb B2 reactive molecule clusters.[20] All but six EEV particles (which were observed at the periphery of MAb B2 reactive molecule clusters) were present at positions clearly isolated from the MAb B2 clusters, and three representative cells are shown in Fig. 5. This provided definitive data that the receptors used by IMV and EEV are different and illustrated an important feature of the binding assay: namely it could provide steric information about the binding sites used by a virus in relation to a defined antigen or binding site used by another virus.

Virus Entry Assay Using Confocal Microscopy

The entry of enveloped viruses, like VV, requires the fusion of their envelope with a cellular membrane. The fusion is mediated by viral envelope proteins and can be categorized by the optimal pH required. Viruses that enter the cell in a pH-independent manner (neutral pH) fuse at the plasma membrane or eventually after being internalized within endosomes. In contrast, viruses that enter the cell in a low pH-dependent manner fuse only after internalization and exposure to an acidic environment of an intracellular vesicle.[34]

The result of the fusion is the release of the structure containing the viral genome and associated proteins into the cytosol of the infected cell. This structure is called the core for VV or capsid for other enveloped viruses. Because confocal microscopy was able to quantify individual particles bound on the cell surface, we considered that it might be possible to adapt this technique to study virus penetration by detecting virus cores inside the cell. To do this we searched for an immunofluorescent staining method that would detect uncoated cores (after virus had entered the cell) but not intact virions (virions still bound onto the cell surface or internalized whole). Suitable results were obtained with an antibody raised against VV cores using the following procedure.

IMV particles were not recognized by this rabbit anticore antiserum if they were untreated or fixed with paraformaldehyde and then permeabilized. It is likely that fixed and permeabilized IMV particles were not stained due to the formation of a lattice around the core by the cross-linking of IMV membrane proteins during fixation. This lattice may prevent access of antibodies to core epitopes despite further permeabilization. However, cores released from IMV after treatment of IMV with Nonidet P-40 (NP-40), or cores present within the cytoplasm were recognized by this antibody.[31] Samples were therefore first fixed and then permeabilized (as described in Methods) before immunofluorescent staining with a rabbit serum raised against virus cores. The ability of this procedure to detect exclusively intracellular cores resulting from virus entry but not virions adsorbed on the cell surface is illustrated in Fig. 6 (color insert). Purified IMV virions were bound to cells at 4°, and after a short incubation 4 or 37°, cells were fixed, permeabilized, and stained with α-D8L MAb and α-core serum (Fig. 6, color insert). Staining with α-D8L revealed the presence of virions on

[34] M. Marsh, *Biochem. J.* **218,** 1 (1984).

the surface of cells that had been incubated at either temperature, although fewer were present after incubation at 37° as many virions had already entered the cell (compare Figs. 6a and 6b). However, α-core serum identi-fied structures only when virus entry was allowed (37°) (compare Figs. 6b and 6e). The merged image (Fig. 6f) demonstrated the core-positive struc-tures did not stain with α-D8L.

Anticore Positive Structures Are Intracellular but Not within Lysosomes

Before using this approach as a way to quantify virus entry, it was necessary to demonstrate that the α-core positive structures were intracellu-lar and did not represent noninfectious virions being degraded within lyso-somes.

Optical sections through cells that had been treated to visualize the plasma membrane and virus cores revealed that α-core positive structures were either colocalizing with the plasma membrane or were internal of it (Fig. 7f, color insert). In contrast, staining with α-D8L revealed virions only colocalizing with the plasma membrane (Fig. 7c). To determine if the α-core positive structures colocalizing with the plasma membrane were intracellular or on the cell surface, VV-infected cells were fixed, but not permeabilized, before staining (Figs. 7g–7l). No α-core positive structures were detected without permeabilization (Fig. 7k), indicating that cores were exclusively intracellular.

The possibility that α-core positive structures represented virus being degraded in lysosomes rather than uncoated cores resulting from virus entry was excluded by the simultaneous visualization of lysosomes labeled with Rd-dextran and α-core positive structures. Merged images of lyso-somes and cores showed that these were not coincident.[31]

Weak Bases Affect EEV but Not IMV Entry

The penetration assay described in this article provided a simple way to quantify virus entry and to investigate its inhibition by chemicals. The weak bases chloroquine and ammonium chloride affect the entry of viruses that are dependent on a low pH pathway by preventing pH reduction in endosomes and phagosomes.[34] The kinetics of IMV and EEV entry in the presence or absence of weak bases was therefore investigated (Fig. 8).

The kinetics of IMV and EEV entry were measured up to only 1 hpi because at later times cores became clustered in the perinuclear region and could not be quantified accurately. Consequently, the total number of virus particles able to enter cells could not be determined. Figure 8 shows that

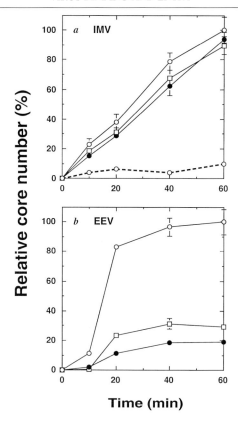

FIG. 8. Effect of weak bases on IMV and EEV entry. Confluent RK_{13} cells were grown on glass coverslips and preincubated with MEM-2% HFBS (○) or the same containing ammonium chloride (20 mM) (●) or chloroquine (0.1 mM) (□) for 1 hr at 37°. Cells were infected for 1 hr on ice with purified IMV (10 pfu/cell) (a) or fresh EEV (3.5 pfu/cell) (b) diluted in the same preincubation medium. The open circle-dotted line symbol in (a) represents cells that were treated without drugs and infected with IMV containing a neutralizing concentration of MAb 5B4/2F2. Cells were washed and incubated at 37° with preincubation medium for the indicated time and then washed with cold PBS and treated as described in the text to reveal intracellular cores by indirect immunofluorescent staining. The numbers of cores in 200 cells that were selected randomly were then determined by confocal microscopy. Data are expressed as percentages of the number of cores at time 60 min for the control and are the average ±SDs for triplicate measurements. At time 60 min, control cells infected with purified IMV and fresh EEV contained a mean of 1482 and 732 cores, respectively.

IMV and EEV enter cells with different kinetics. Whereas the number of cores deriving from IMV entry increased with linear progression and had not plateaued by 1 hpi, the number of cores deriving from EEV entry had plateaued by 20 min and reached 83 and 96% of the level at 1 hpi by 20 and 40 min, respectively (Fig. 8b). The number of cores derived from IMV entry was reduced by 90% by MAb 5B4/2F2, demonstrating that this antibody blocks IMV entry (Fig. 8a).

The weak bases chloroquine or ammonium chloride affected EEV and IMV entry differently: whereas the number of cores deriving from IMV entry was unaltered by these drugs (Fig. 8a), cores resulting from EEV entry were reduced by 71 and 81%, respectively (Fig. 8b, time 60 min). This indicated that EEV but not IMV enters cells by a low pH-dependent pathway.

Discussion

This section describes new methods for the analysis of virus binding to and entry into cells that are based on confocal microscopy. These techniques are illustrated with a study of the binding and entry of VV. The methods have been particularly useful for studying VV because this virus produces two different forms of infectious virion that are antigenically and biologically distinct and are produced in widely differing amounts. Moreover, the EEV form of VV cannot be purified from contaminating IMV without disrupting the integrity of the outer envelope. Such rupture does not destroy virus infectivity because the IMV particle within the outer envelope remains infectious, but a damaged EEV particle with both IMV and EEV antigens exposed might bind to cells via either type of antigen. Purified EEV preparations therefore are unsuitable for binding and entry assays. These problems were overcome by using fresh unpurified virus that contains both EEV and IMV and simultaneously identifying and distinguishing these forms of virus by double immunofluorescent staining followed by confocal microscopy. Data obtained with the methods described in this article have helped to establish that the IMV and EEV virions bind to different cellular receptors and enter cells by different mechanisms.

In addition to the advantages specific to the study of VV due to the two infectious forms of virus, the methods offer several other desirable features. (i) They can use fresh, unpurified virus preparations that are unlabeled. Thus there is no need to use radioactive tracers during the growth of virus or to label the virus after purification with, for instance, fluorescent markers or biotin. Potentially these latter methods might compromise the virion integrity or change its biological properties. (ii) The method generates absolute numbers of virions per cell, not mean numbers representing the average number of virions bound in the whole cell population. Therefore the binding assay could identify specific cells within a mixed cell population that are binding virus particles. For VV, the identification of cells that are unable to bind IMV and EEV, or either form of virus, would be a useful step toward identifying the cellular receptors for these viruses by expression cloning. (iii) The methods may be used to determine the particle infectivity ratios of viruses and to identify the proportion of virions that bind to cells that give rise to a productive infection. (iv) The methods provide steric information about the site on a cell to which a virion binds in relation to specific cellular antigens or the binding sites for other virions. This was illustrated by the binding of EEV to sites distinct from that identified by a MAb that blocked IMV binding. In addition, the site within cells to which uncoated virus cores or capsids migrate can be studied. (v) In studies of virus entry, the use of antibodies that recognize different virion antigens can identify which antigens are left at the cell surface and the kinetics with which different core or capsid antigens become exposed during uncoating.

In conclusion, we have described novel methods based on confocal microscopy to investigate VV binding and entry. Those methods do not require purified or labeled virus, provide absolute numbers of virus or cores/cell, give steric information concerning the position of the virus or the core, and allow the dissection of the replication cycle by investigating only binding or entry. Although the utility of those techniques has been demonstrated here with VV, they should be applicable to any virus larger than 50 nm.

Acknowledgments

Dr. Alain Vanderplasschen is a permanent senior research assistant of the Fonds National Belge de la Recherche Scientifique at the University of Liège (Belgium). This work was supported by a program grant from the UK Medical Research Council (PG8901790) and an equipment grant from the Wellcome Trust.

[34] Fluorescent Labels, Confocal Microscopy, and Quantitative Image Analysis in Study of Fungal Biology

By Russell N. Spear, Daniel Cullen, and John H. Andrews

Introduction

Microscopy in the study of fungi has been applied basically in two contexts. First, the fungus in isolation can be the subject of investigation, in which case studies typically have focused on cell dynamics, morphogenesis, cytochemistry, and organelle or spore structure and development. Within the past decade, laser scanning confocal microscopy (LSCM) has become a powerful tool in the analysis of subcellular detail such as that just noted.[1-4] It combines laser, computer, and imaging technologies to enable optical rather than mechanical sectioning of specimens. The second context pertains to the relationship between a fungus and its substratum. Examples include host–fungus interactions and the extent or spatial pattern of colonization. LSCM can reconstruct multiple, in-focus image planes, rendering them in three dimensions. Thus, it has been used to obtain sharp images of fungi within a host or colonizing highly contoured or rugged surfaces such as the phylloplane or rhizoplane of plants.

Fluorescence is the usual mode for LSCM and many natural and synthetic fluorescent probes (i.e., fluors, fluorophores, or fluorochromes) are available. Fluorescent labels have great utility because they are relatively photostable and sensitive. They are also potentially highly specific when linked to various macromolecules (e.g., proteins, nucleic acids). The main shortcoming of fluorescence-based procedures is related to noise, i.e., nonspecific fluorescence contributed by background-contaminating substances such as host cells, debris, or reagents. Operationally, the goal is usually to maximize signal-to-noise ratios.

Green fluorescent protein (GFP) from the jellyfish *Aequorea victoria* was reported[5] in 1994 as a marker and has since been expressed in its native or mutated state in prokaryotes and diverse eukaryotes, including fungi. It is a 238 amino acid protein, the fluorophore of which is formed by posttranslational cyclization and oxidation of the amino acids Ser-65, Tyr-66, and Gly-67. GFP has revolutionized cell biological research (trafficking; intracellular signaling; gene expression) and has proven to be a useful label to follow cell movement within organisms or in the environment because of its stability, versatility, and ability to function solely with the stimulus of UV or blue light and oxygen, i.e., without cofactors or substrates.[6]

Another major area of recent technical advance considered in this article is quantitative image analysis. In theory, it allows for the rapid and relatively objective enumeration and classification of fungal propagules meeting specific descriptions. Briefly, this procedure involves the manipulation of pictorial information. Its role in biological investigation is accelerating with the advent of modestly priced computers powerful enough to handle pictorial

[1] J. B. Pawley, ed., "Handbook of Biological Confocal Microscopy," 2nd Ed. Plenum, New York, 1995.
[2] Y. H. Kwon, K. S. Wells, and H. C. Hoch, *Mycologia* **85,** 721 (1993).
[3] K. J. Czymmek, J. H. Whallon, and K. L. Klomparens, *Exp. Mycol.* **18,** 275 (1994).
[4] B. Matsumoto, ed., "Methods in Cell Biology," Vol. 38. Academic Press, New York, 1993.
[5] M. Chalfie, Y. Tu, G. Euskirchen, W. W. Ward, and D. C. Prasher, *Science* **263,** 802 (1994).
[6] M. Chalfie and S. Kain, eds., "Green Fluorescent Protein: Properties, Applications, and Protocols." Wiley-Liss, New York, 1998.

data.[7] In the context of fungal biology, the information is typically microscopic and is related to the size, shape, and quantity of various cell components, cells, or cell aggregates such as hyphae. Thus, the initial image usually begins in optical form and is converted subsequently through an analog (electrical) stage by a video camera to a digital form for computer processing.[7] For example, an image of continuous gray tones is converted to a rectangular array of discrete points of brightness. Each point is known as a pixel or picture element and has an associated digital value representing its intensity. Interpretation of the digital array of quantitative information is controlled by the observer who, through different analysis operations, adjusts the image for enhanced contrast.[7] The enhanced image is then categorized by various criteria related to size, shape, brightness, and so on.

We are interested in the growth and ecology of the cosmopolitan, yeast-like fungus *Aureobasidium pullulans* on leaf surfaces. In these studies, we have developed various microscopic techniques, including quantitative fluorescence *in situ* hybridization using rRNA-specific probes, vital staining, and the tracking of and quantification of GFP-expressing transformants. Essential to this work is validation of these methods by various techniques with both confocal and conventional microscopy. The protocols described here for DNA staining and detection of actin by the probes of small (phalloidin) and large (antibody) molecular size are the outgrowth of these goals. They also demonstrate the power and versatility of the techniques.

Green Fluorescent Protein Expression in Fungi

Green fluorescent protein has proven to be a valuable aid for identifying the subcellular location of proteins and for monitoring gene expression. To date, GFP has been expressed successfully in various fungi or fungus-like organisms, including *Saccharomyces cerevisiae, Schizosaccharomyces pombe, Dictyostelium discoideum, Aspergillus nidulans, Podospora anserina, Ustilago maydis, Candida albicans,* and *A. pullulans.* In several instances, expression has been monitored *in situ.* For example, GFP can be visualized in corn plants and mouse kidney cells infected with transformants of *U. maydis*[8] and *C. albicans,*[9] respectively. In our laboratories, GFP expression has proven useful for the quantitative detection of *A. pullulans* transformants on leaf surfaces.[10] We describe later the methodology for

[7] G. A. Baxes, "Digital Image Processing: Principles and Applications." Wiley, New York, 1994.
[8] T. Spellig, A. Bottin, and R. Kahmann, *Mol. Gen. Genet.* **252,** 503 (1996).
[9] B. P. Cormack, G. Bertram, M. Egerton, N. A. R. Gow, S. Falkow, and A. J. P. Brown, *Microbiology* **143,** 303 (1997).
[10] A. J. Vanden Wymelenberg, D. Cullen, R. N. Spear, B. Schoenike, and J. H. Andrews, *BioTechniques* **23,** 686 (1997).

detection and quantification of this fungus in isolation or associated with the phylloplane. We also discuss how GFP can be combined with other fluorescent tags.

There are three basic requirements for GFP expression in fungi: a reliable transformation system; functional transcriptional control elements to drive expression; and the appropriate GFP-encoding gene. With regard to transformation, a wide range of selectable markers has been described,[11] most commonly complementation of auxotrophic markers or conferral of drug resistance. Resistance to hygromycin B is conferred by a phospho-transferase-encoding gene, *hph*, and this dominant marker has been used for several fungi expressing GFP, including *A. pullulans*.[8,12] Transformation of *A. pullulans*, like most fungal systems, involves polyethylene glycol-induced aggregation of protoplasts in the presence of plasmid DNA.[10,12] Protoplasts are prepared by treating germlings with a commercial lytic enzyme, Novozym, in an osmotically stabilized buffer, e.g., 1 *M* sorbitol. Digestion conditions often require minor alterations because enzyme lots vary.

Hygromycin-resistant colonies appear within 5 days, and at least in the case of *A. pullulans* transformed with pTEFEGFP, these may appear yellow-green without UV irradiation.[10] Transformation of *A. pullulans*, like the vast majority of filamentous fungi, involves the stable integration of plasmids within the genome. The copy number and integrative context of plasmid insertions can be ascertained by Southern blot hybridization of genomic DNA. Another common feature of fungal transformation is high frequencies of cotransformation. Thus, the GFP expression plasmid pTEFEGFP can be introduced by simultaneously transforming with a separate plasmid (pDH33)[13] carrying the selectable marker *hph*.

With respect to transcriptional control, both constitutive and regulated promoters have been used for GFP expression in fungi. Regulated expression has been obtained with promoters of the *C. albicans* maltase gene *MAL2*[9] and the pheromone-inducible gene *mfa1*[8] of *U. maydis*. For constitutive expression, we[10] and others[8] have found the translational elongation factor promoter *TEF* to be especially useful (Fig. 1). Both translational fusions[14] and direct promoter fusions[10] (Fig. 1) have produced fully active GFP.

[11] J. R. Fincham, *Microbiol. Rev.* **53,** 148 (1989).

[12] D. Cullen, V. Yang, T. Jeffries, J. Bolduc, and J. H. Andrews, *J. Biotechnol.* **21,** 238 (1991).

[13] T. Smith, J. Gaskell, R. Berka, D. Henner, and D. Cullen, *Gene* **88,** 259 (1990).

[14] C. Lehmler, D. Zickler, R. Debuchy, A. Panvier-Adoutte, C. Thompson-Coffe, and M. Picard, *EMBO J.* **16,** 12 (1997).

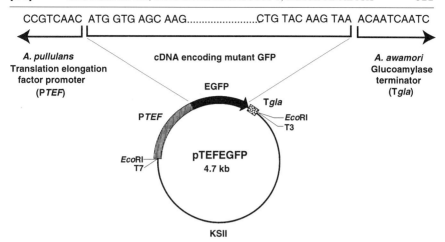

FIG. 1. The expression vector pTEFEGFP. A red-shifted, engineered mutant GFP cDNA (pEGFP) is fused to the *TEF* promoter from *Aureobasidium pullulans* and the glucoamylase terminator from *Aspergillus awamori*. The entire expression cassette is ligated into pBluescript II KS(-) and the sequence of junctions is confirmed by dideoxy sequencing. Partial sequence of the relevant regions is shown.

The choice of GFP clone is critically important to success. Several laboratories have reported that the wild-type cDNA will not yield a functional GFP in fungi.[8–10] Functional forms of GFP usually feature two mutations, (Ser-65 → Thr and Phe-64 → Leu), which dramatically increase fluorescence and frequently alter the codon bias to those codons preferred by higher vertebrates and by some plants. In the case of yEGFP, all codons are optimal for *C. albicans*.[9] The commercially available clone EGFP1[15] (CLONTECH, Inc., Palo Alto, CA) works well in *P. anserina*[16] and in *A. pullulans*[10] (Fig. 1).

Fluorescence Microscopy and GFP

Regardless of the type of fluorescence microscopy (LSCM or conventional epifluorescence), it is important to match the fluorochrome used to the illumination spectra and filters available. Several GFP variants are available and must be chosen carefully for LSCM. Few instruments are presently equipped with lasers with output lines in the UV range that can excite either wild-type GFP at its major peak of 390 nm or the UV-optimized

[15] T. T. Yang, L. Cheng, and S. R. Kain, *Nucleic Acids Res.* **24**, 4592 (1990).
[16] V. Berteaux-Lecellier, G. Steinberg, K. M. Snetselaar, M. Schliwa, R. Kahmann, and M. Bölker, *EMBO J.* **17**, 1248 (1998).

variant at 395 nm. Wild-type GFP can be excited at its secondary peak of 470 nm with the 488-nm line of either argon or argon/krypton lasers, although this is fairly inefficient. The red-shifted excitation mutants, such as EGFP, GFP-S65T, and RSGFP, possess excitation maxima of 488, 489, and 490 nm, respectively, which closely match the 488-nm line, and have emission spectra that closely match fluorescein isothiocyanate, allowing the use of configurations for that fluorochrome. Also, these red-shifted mutants possess much higher fluorescence (4- to 35-fold) than the wild-type, allowing more efficient detection.[17]

GFP is very stable after formation and can withstand fixation with either formaldehyde or glutaraldehyde, although it might be advisable to avoid the latter due to its tendency to produce autofluorescent aldehyde groups in tissues. Fixation in methanol/acetic acid (3:1) destroys fluorescence, although it is retained when methanol is used alone.[18] Fluorescence is best in a range from pH 7.2 to pH 8.0 with most common buffers, i.e., phosphate-buffered saline (PBS) 50 mM, pH 7.4, 0.8% NaCl, piperazine-N,N'-bis (2-ethananesulfonic acid) (PIPES); and N-[2-hydroxyethyl]piperazine-N'-[4-butansulfonic acid] (HEPES) 50 mM, pH 7.2. In our hands, $A.$ $pullulans$ EGFP transformants with a TEF promoter are extremely bright and do not photobleach appreciably in room light, although this should be determined with each system. Cautionary notes against the use of nail polish in sealing preparations have appeared,[17] but we have found that Revlon Epoxy 1000 has not caused any harm. It can be used to seal coverslips for periods of up to 1 month when slides are stored in the dark at 4°.

Laser Scanning Confocal Microscopy

The following section describes conditions and equipment that have performed well for us. We use a Bio-Rad (Hercules, CA) MRC-1024 confocal microscope equipped with an argon/krypton laser illumination source. The microscope and laser are controlled with Bio-Rad Laser Sharp 24-bit software. The MRC-1024 allows up to three colors of fluorochrome to be imaged either simultaneously or sequentially. In the simultaneous mode, all fluorochromes are illuminated at the same time, and collection of each channel (color) is also done concurrently. In the sequential imaging mode, each fluorochrome is illuminated only by the appropriate laser line and the

[17] S. A. Endo and D. W. Piston, in "Green Fluorescent Protein: Properties, Applications, and Protocols" (M. Chalfie and S. Kain, eds.), p. 271. Wiley-Liss, New York, 1998.

[18] W. W. Ward, in "Green Fluorescent Protein: Properties, Applications, and Protocols" (M. Chalfie and S. Kain, eds.), p. 45. Wiley-Liss, New York, 1998.

corresponding filters for that fluorochrome are changed via a filter wheel. Each image is collected, then the next laser line is used to illuminate the next fluorochrome and its image is stored, and so on. In addition, a z-series (depth) may be acquired by moving the stage focus by a step between each image in either of these collection modes.

The sequential collection mode allows for a more precise adjustment of the laser intensity and photomultiplier (PMT) gain and threshold. This enables better compensation for differences in the intensity of different fluorochromes and adjustment to compensate for spectral overlap ("cross talk") of emission signals to each PMT ("bleed through"). The disadvantage of this collection strategy is the slower speed of data collection and more exposure of the specimen to excitation illumination that could photobleach specimens. "Cross talk" compensation in the simultaneous mode can be implemented by adjusting the PMT gain and sensitivity of the different channels. This adjustment, however, can be difficult when the brightness of differing channels is great and the spectral overlap of colocalizing signals is high. Older instruments, such as the Bio-Rad MRC-600, are not capable of the consecutive detection of multicolor signals, and signals of greatly different strengths present problems in channel overlap.

Another image acquisition technique that can be applied is Kalman average filtering,[1] which is an image averaging technique that allows the removal of electronic noise. Kalman filtering is especially useful for acquiring dim images.

The image acquisition parameters for our work have been generally as follows: Objectives either 60× NA 1.40, plan-apochromatic oil, or 40× NA 1.30, plan-apochromatic oil (Nikon, Melville, NY); Kalman averaging of three or four images per focal step; z-series focal step 60× objective 0.5 μm, 40× objective 1.0 μm; consecutive collection mode with appropriate filters for EGFP (FITC set) and for propidium iodide, and rhodamine phalloidin and antibodies (TRITC set) (see following sections). For the TEFEGFP transformant on leaf surfaces, we use the fluorescein isthiocyanate (FITC) set to acquire the GFP signal, whereas the Texas Red set is used to acquire the autofluorescence signal for leaf chlorophyll.[10]

Images are saved as z-series stacks (see Fig. 2) and merged z-series projections of each color channel (green and red), allowing individual images from each signal to be displayed and analyzed. Postacquisition processing can be done with the freeware program Confocal Assistant v3.10 from (ftp://ftp.genetics.biorad.com/Public/confocal/cas/) or Adobe Photoshop 5.0 (Adobe Systems, Mountain View, CA) using Bio-Rad image conversion "plug-ins" for Photoshop (ftp://ftp.genetics.biorad.com/Public/confocal/photoshop/).

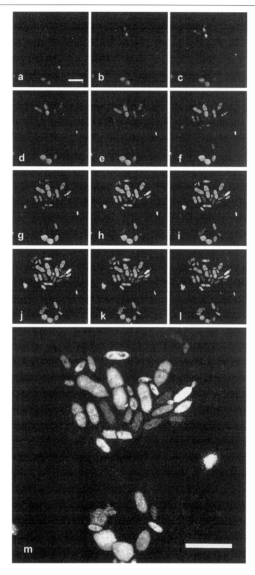

FIG. 2. GFP expression in *A. pullulans;* *z*-series image consisting of 12 0.5-μm steps (a–1); m is the composite image. Bars: 20 μm.

Multicolor LSCM: Use of GFP with Other Fluorescent Labels

Localizing GFP Concurrently with Nuclear Staining by Propidium Iodide

To a midlog phase culture grown in any standard fungal growth medium for 18–20 hr at 26–28°, with shaking, add 1/10 volume freshly prepared 30% formaldehyde in PBS, replace on shaker, and fix for 2–4 hr at room temperature.

Formaldehyde 30%. Caution: Make immediately prior to use in a fume hood. Formaldehyde is a carcinogen!

1. Place 3 g of paraformaldehyde (EM Sciences, Ft. Washington, PA) in approximately 5 ml PBS, 50 mM, pH 7.4, heat with stirring to 70° (do not allow to boil), add 0.5 ml of 5 M NaOH, allow solution to clear, and adjust pH to 7.4. Make up to 10 ml with PBS. Cool to room temperature before use.
2. Pellet cells by centrifugation at 5000 rpm for 2 min in a microfuge. Wash the cells three times with PBS. We have found that processing cells in suspension produces a lower level of background fluorescence than drying cells on slides prior to staining.
3. Suspend the cells in 1% Triton X-100 in PBS. Incubate for 3–5 min at room temperature. This permeabilizes the cell walls.
4. Wash by centrifugation three times with PBS.
5. Incubate with 10 μg/ml RNase A (Sigma, St. Louis, MO) in PBS at pH 7.2 in a 37° water bath for 1 hr. This step removes RNA that would also stain with propidium iodide.
6. Wash with PBS three times at room temperature.

Carry out the following steps under reduced light to prevent fading.

7. Incubate in 5 μg/ml propidium iodide (Sigma) in PBS for 10–15 min at room temperature. *Caution:* Propidium iodide is a carcinogen!
8. Wash in PBS, two changes, at room temperature.
9. Place drops of cell suspension on Super-Frost (Fisher Scientific, Pittsburgh, PA) microscope slides and allow to air dry at room temperature.
10. Rinse slides briefly with distilled water and mount in an antifade compound such as VectaShield (Vector Labs, Burlingame, CA), coverslip, and seal with nail polish (see earlier note). Sealing of coverslips is desirable to prevent movement when using oil-immersion objectives. Store slides prior to examination in darkness at 4° Examine as described earlier for LSCM (Fig. 3).

green channel   red channel

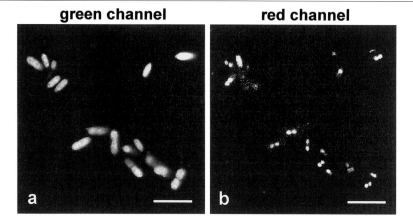

FIG. 3. GFP expression in the cytoplasm of *A. pullulans* cells in the green image channel (a) and DNA staining with propidium iodide of the nuclei in the red channel (b). These images consist of a nine image z-series of 0.5-μm steps merged into a single plane. Cells are pretreated with RNase; otherwise the red signal would have colocalized with the green in the cytoplasm. Bars: 20 μm.

Note: If a UV laser is available, the RNase treatment can be omitted, and the propidium iodide step (7) can be replaced with 5–10 μg/ml. 4′,6-diamidino-2-phenylindole (DAPI), which stains only DNA. *Caution:* DAPI is a suspected carcinogen!

Staining of Actin with Rhodamine-Conjugated Phalloidin in GFP-Expressing Cells

1. Fix cells with formaldehyde as described earlier, substituting PBS with 50 mM PIPES buffer, pH 7.4.
2. Wash three times in PIPES buffer by centrifugation at 5000 rpm for 2 min at room temperature.
3. Permeabilize with 1% Triton X-100 in PIPES for 2–5 min at room temperature.
4. Wash in three changes of PIPES.

Perform the following steps under reduced light to prevent fading.

5. Incubate in rhodamine–phalloidin conjugate (Sigma), 10 μm/ml in PIPES, for 1–2 hr at room temperature. *Caution:* Phalloidin is extremely toxic, work in fume hood, wear laboratory coat, gloves, mask, and eye protection!
6. Wash in three changes of PIPES.

7. Place a drop of cell suspension on a Super-Frost microscope slide and air dry at room temperature.
8. Rinse slide briefly in distilled water and air dry, mount in VectaShield and coverslip, and store slides in the dark at 4° prior to examination (Fig. 4).

Staining of Actin with Rhodamine-Conjugated Antibodies in GFP-Expressing Cells

If one wishes to detect actin and avoid the hazard of toxic phalloidin, immunologic detection may be employed.

1. Fix cells with 3% formaldehyde in PIPES buffer as described earlier.
2. Wash suspension by centrifugation at 5000 rpm for three times at 5 min each at room temperature.
3. Digest the cell wall using Novozym 234 (Calbiochem, San Diego, CA), 0.5 mg/ml in PIPES buffer for 2–10 min at room temperature. Monitor the cell wall digestion by phase-contrast microscopy, halting digestion by washing in ice-cold PIPES when approximately 10–20% of the cells show a decrease in refraction. Alternatively, digest several batches of cells for differing periods of time, process, and select those with the best staining.
4. Wash three times in PIPES.

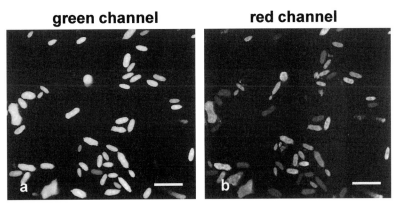

green channel   **red channel**

FIG. 4. GFP localization in the green channel (a) with phalloidin–rhodamine staining (red channel) of actin (b) in *A. pullulans*. In this case, the GFP and actin colocalize in the cytoplasm. The signals, however, differ greatly in brightness; with the sequential detection mode, laser power was varied between 1.0% for the bright signal of green GFP and 30% for the weaker red signal of the phalloidin–rhodamine. The difference of these signals would have made compensation for "bleed through" difficult if the simultaneous detection mode were used. This image is a merged z-series of nine images at 0.5-μm steps. Bars: 20 μm.

5. Incubate cells in blocking buffer consisting of PIPES buffer with 2% bovine serum albumin (Sigma) and 2% normal donkey serum (Sigma) for 1 hr at room temperature with agitation.
6. Incubate with rabbit antiactin antibody (Sigma) 1:100 in the just-described blocking buffer overnight at room temperature with agitation.
7. Wash in blocking buffer, three changes of 15 min each.

Perform the following steps in reduced light to prevent fading.

8. Incubate with goat antirabbit rhodamine-conjugated antibody (Sigma) 1:50 in blocking buffer for 6 hr to overnight at room temperature with agitation.
9. Wash with PIPES buffer, three times, for 15 min each at room temperature, dry on Super-Frost slides, rinse with distilled water, mount with VectaShield, and store in the dark at 4° prior to examination (Fig. 5).

Note: More drastic methods of permeabilization must be employed for antibody staining than for tagged phalloidin because of the higher molecular weight of antibodies. If allowed to proceed too long, Novozym digestion can completely permeabilize the cell wall and cause loss of both the GFP and the cytoplasmic antigen. Digestion is empirical, subject to considerable variability, and must be monitored carefully by microscopy.

Conventional Epifluorescence Microscopy and Digital Deconvolution

The recent increase of processor speeds of personal desktop computers and the decrease in cost of both hard drive storage devices and random access memory, coupled with computer control of image acquisition, allow an alternative to confocal microscopy.

A standard epifluorescence microscope equipped with a computer-controlled z-stepper motor and filter wheels to control excitation and emission wavelengths is used with a cooled charge-coupled device (CCD) video camera. This equipment provides a z-series of images of low light fluorescence that is stored on a hard drive. The z-series is then processed by deconvolution software to "reassign" out-of-focus light in the images.[19,20] This technique eliminates the need for point scanning and the pinhole aperture needed in the confocal system to exclude the out-of-focus light. Thus dimmer images can be collected without as much exposure to the

[19] G. Sluder and D. E. Wolf, eds., "Methods in Cell Biology," Vol. 56. Academic Press, New York, 1998.
[20] R. Rizzuto, W. Carrington, and R. A. Tuft, *Trends Cell Biol.* **8,** 288 (1998).

green channel   **red channel**

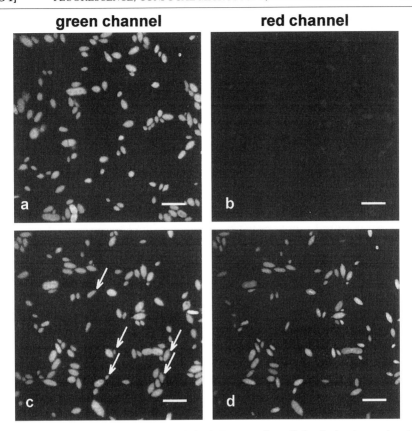

FIG. 5. Immunological detection of actin in GFP-expressing cells by rhodamine-conjugated antibodies. (a and b) Cells were not digested with cell wall-permeabilizing enzymes; note the lack of staining of actin in the rhodamine (red) channel caused by the inability of the antibodies to penetrate undigested cells walls. (c and d) Cells digested for 8 min with Novozym 234. The actin stains with the rhodamine conjugate, demonstrating that antibodies are able to penetrate permeabilized cell walls. Arrows show examples of cells staining in the green (GFP) but not in the red (rhodamine) channel, indicating that not all cells have permeabilized sufficiently to allow uptake of the labeled antibody. Merged z-series of 13 images at 0.5-μm steps. Bars: 20 μm.

excitation wavelengths, allowing more efficient imaging and less photobleaching compared with LSCM. There is no limit to the excitation of the fluorochromes due to the lack of appropriate laser lines, so fluorochromes exciting from the UV to infrared can be obtained from a single light source. With proper attention to filter selection, several GFP mutants can be monitored either simultaneously or sequentially, allowing multiparameter information to be obtained. A disadvantage is that the reconstructed "in focus" image is only obtained after postacquisition image processing. Deconvolution algorithms may be based on either measured point-spread function for the specific optical elements of the microscope or estimated from values for lens numerical aperture, mounting media refractive index, and the illumination wavelength used to acquire the image prior to deconvolution.

Our system is an Olympus BX-60 (Olympus America, Lake Success, NY) epifluorescence microscope with a mercury HBO-100 arc lamp, equipped with a cooled CCD camera DEI-470 (Optronics Engineering, Goleta, CA). A Ludl focus motor (Ludl Electronic Products, Ltd., Hawthorne, NY) controlled by Volume Scan (VayTech, Fairfield, IA) software is used to obtain the z-series image of the specimen. Deconvolution is performed by Microtome Digital Deconvolution Software (VayTech) by nearest neighbor estimation or by processing of groups of three images using a measured point-spread function, obtained by imaging fluorescent PS-Speck microspheres (Molecular Probes, Eugene, OR). In our system, filter changes are done manually, although filter wheel control of the excitation and emission wavelengths could be accomplished by software from either the filter wheel supplier or the image analysis software. Software programs can be written to integrate filter wheel excitation/emission wavelength changes, image averaging to eliminate temporal noise, z-step acquisition, and time-lapse acquisition of data.

Quantitative Image Analysis of Fungal Cells on Leaves

As discussed earlier for LSCM, we describe here the approach that has worked well in our hands. Leaves supporting naturally occurring or inoculated fungal populations should be fixed in 3% formaldehyde in PBS, pH 7.4, for 18–20 hr at 4°. If leaf disks (ca. 13 mm diameter) are used, fixation time can be reduced, if desired. The leaf material is then washed in three changes of PBS for 1–2 hr each at room temperature and mounted in VectaShield with the surface of interest uppermost. To prevent deformation of the leaf, the coverslip can be supported by placing 5-μl drops of nail polish at the corners. Slides are sealed with nail polish and stored in the dark at 4° pending examination.

We routinely process images from GFP-expressing cells by merging green channel (GFP) z-data into a single 8-bit gray-scale image consisting of pixel intensities ranging from 0 (black) to 255 (white). z-data can be acquired from either LSCM or deconvolution equipment, both as described earlier. Image analysis software (Optimas v6.2, Bothell, WA) is used to apply a gray-scale threshold to include all cells. Any touching cells are separated by applying progressive erosion filters to the image. When the cells are deemed separated, a nonmerging dilation filter is applied to restore them to their original size. The cell areas are then outlined and filled using binary filters. Finally, the processed image is analyzed for the total number of cells, individual cell areas, and total area coverage of the cells. From these data the percentage area coverage of the leaf and average cell size can be calculated; statistical information relating to certain cell population parameters (mean, SD, range, etc.) can also be calculated (Minitab v.12, State College, PA) and displayed. Figure 6 illustrates data acquired by LSCM; Figure 7 is from conventional epifluorescence microscopy combined with deconvolution.

Conclusion

As a case study of the biology of a fungus in its natural habitat, we have demonstrated that GFP-transformed *A. pullulans* may be imaged *in situ* by either confocal or conventional microscopy and by the use of single or

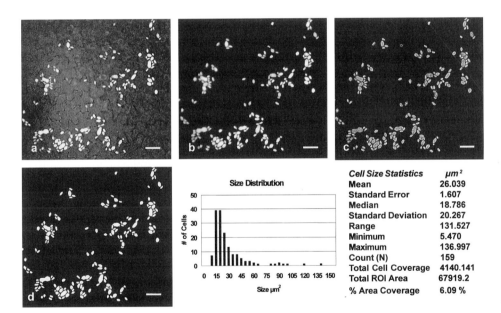

Cell Size Statistics	μm²
Mean	26.039
Standard Error	1.607
Median	18.786
Standard Deviation	20.267
Range	131.527
Minimum	5.470
Maximum	136.997
Count (N)	159
Total Cell Coverage	4140.141
Total ROI Area	67919.2
% Area Coverage	6.09 %

FIG. 6. Apple seedling leaf inoculated with GFP-expressing *A. pullulans* cells and incubated for 96 hr at high (>90%) humidity. (a) A combined green (GFP-expressing cells) and red (chlorophyll of leaf) image of 28 sections merged at 1.0-μm steps with the outline of leaf epidermal cells in the background. (b) The merged image of the green signal alone representing the fungal cell component of the image (a). Image (c) is the image of (b) following erosion, dilation, and filling of the outlined cells. Image (d) is the binary image of (c) subjected to image analysis. Quantitative and statistical output derived from the image analysis software and Minitab v12 software is shown. ROI, region of interest. Bars: 20 μm.

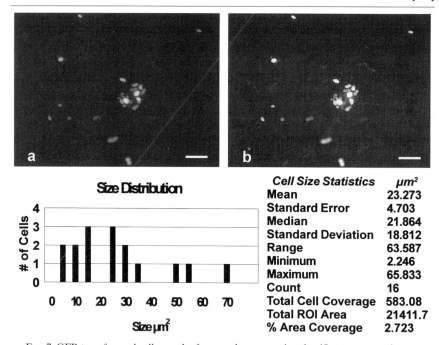

Cell Size Statistics	μm^2
Mean	23.273
Standard Error	4.703
Median	21.864
Standard Deviation	18.812
Range	63.587
Minimum	2.246
Maximum	65.833
Count	16
Total Cell Coverage	583.08
Total ROI Area	21411.7
% Area Coverage	2.723

FIG. 7. GFP-transformed cells on a leaf as seen by conventional epifluorescence microscopy with a 40×, NA 0.75 neofluor objective and a zseries of five images with 1.0-μm spacing. (a) A merged series without haze removal; (b) the same series deconvolved with Microtome software. Image analysis was performed on (b), using the technique as in Fig. 6. Bars: 20 μm.

multiple probes. For routine applications, conventional microscopy coupled with digital image collection and deconvolution of the image stacks offers an alternative to confocal microscopy and, for us, is superior in several aspects: (1) There is the ability to image by UV excitation of such probes as the vital stain CellTracker Blue (Molecular Probes), which cannot be excited without a UV laser in the confocal system. (2) Costs are lower, both in initial equipment (>$200,000 for confocal vs <$60,000 for conventional microscope, camera, computer, and software) and ongoing expenses of operation, including laser replacement. (3) Although postacquisition deconvolution is slower, the faster rate of image acquisition by wide-field microscopy reduces photobleaching of the specimen and allows imaging deeper into specimens with less light loss.

Acknowledgments

Research in our laboratories, which provided the basis for this paper, was funded in part by Grant R82-3845 to JHA from the U.S. Environmental Protection Agency–National Center for Environmental Research and Quality Assurance. We thank Drs. Victoria Centonze Froh-lich and Holly Simon for comments on the manuscript and Dr. Mark Tengowski of the W. M. Keck Neural Imaging Laboratory for assistance with the confocal microscopy.

Author Index

Numbers in parentheses are footnote reference numbers and indicate that an author's work is referred to although the name is not cited in the text.

A

Aalkjaer, C., 247
Abbe, E., 302
Abd Alla, S., 383
Abe, M., 402, 403(13), 404(13)
Acker, H., 109, 110(9), 113(7), 114(5), 115(5, 8, 9), 116(7, 8), 117(5), 128, 129(8)
Adachi, E., 333
Adam, S., 222
Adams, A. E. M., 181
Adams, C. L., 55
Adams, J. A., 351, 445, 457(5), 468(5)
Adams, M., 111
Adams, M. A., 247
Adams, S. R., 56, 111
Adamson, R. H., 418
Addicks, E. M., 500
Adolph, S., 200, 203(17)
Aebi, U., 174, 176, 178, 179(14), 180, 184, 186, 189, 208, 210
Agard, D. A., 135, 209, 210, 300, 301(21), 304, 304(19), 309(21), 313(21)
Aggarwal, S. J., 275, 418
Akerman, K. E. O., 472, 478(22)
Akey, C. W., 229
Akgoren, N., 419
Aldridge, C., 573, 574(4), 575, 577, 579(4), 580(4), 582(4), 583(4)
Alexander, J. W., 322, 324(10)
Allemenos, D. K., 321
Allen, J. M., 462
Allen, T. D., 210
Almqvist, H., 485
Alonso, M., 451, 506, 508(24)
Altamirano, J., 274, 481
Alvarez, A., 451
Alvarez-Leefmans, F. J., 274, 481
Amos, W. B., 301, 427

Amoscato, A. A., 321
Amsterdam, A., 329
Analoui, M., 491
Anderson, N., 588
Anderson, R., 529
Ando, M., 491
Andrews, J. H., 607, 610, 611(10), 613(10)
Andrews, P. M., 230(2), 231
Angmar, B., 485
Angmar-Månsson, B., 485
Angus, J. A., 247
Anson, L. F., 40
Aoki, T., 396(5), 397(5), 403(5), 404, 404(5), 405, 408(5, 21), 410, 411, 411(5, 18, 20), 412(19, 20), 413, 413(5, 18, 19), 414, 415, 416, 416(17, 21), 417
Apgar, J. M., 498
Appleyard, G., 592, 593(7, 11)
Araujo, G. D., 68, 73(29)
Archangeletti, C., 180, 184
Arends, J., 485, 487, 490(17), 491, 495, 496
Arfors, K. E., 412
Arii, T., 136
Armstrong, D., 523
Arribas, S. M., 246, 257, 259, 259(23), 261, 262, 263(28), 264, 272(21, 23)
Art, J., 301, 524
Asai, T., 419
Ashworth, R., 564
Askew, D. S., 324
Astorga, R., 506, 508(24)
Atar, D., 189
Athey, B., 529(34), 530
Aubry, J. P., 385
Augustine, G. J., 470
Auran, J. D., 545
Avilion, A. A., 77
Avinash, G. B., 41
Aziz, D. J., 532

B

Subject Index

A

Acousto-optic deflector, rapid confocal microscope scanning, 400–402

Actin, *see* Cytoskeleton, confocal microscopy

Airy disk, *see* Point-spread function

Alpha blending, rendering of confocal images, 47, 49

Angiotensin II receptor, confocal microscopy, 132–133

Antibody, fluorochrome-conjugated
secondary antibody probing, 62, 65
vendors, 65–66

Antifading agents
commercial agents, 67–68, 134
effectiveness with immunofluorescence specimens, 68–69, 73
mounting of immunofluorescence specimens, 65–67, 80
vascular casts with fluorochrome-labeled gelatin, 99–100

AOD, *see* Acousto-optic deflector

Aureobasidium pullulans, green fluorescent protein expression
confocal microscopy
image acquisition, 613
instrumentation, 612–613
propidium iodide multicolor staining, 615–616
rhodamine-conjugated antibody, actin staining in multicolor detection, 617–618
rhodamine-conjugated phalloidin, actin staining in multicolor detection, 616–617
epifluorescence microscopy and digital deconvolution, 618–620, 622
fixation, 612
imaging in living samples, 111
mutants and detection wavelengths, 611–612
quantitative analysis on leaves, 620–621
transcriptional control, 610

transformation, 610
vectors, 611

Autofluorescence, minimization in immunofluorescence specimens
buffer selection, 60
chambers and sealing, 61
reducing agent, 60–62, 185

B

BCECF, vascular remodeling studies, 257

β cell, glucose response analysis of NAD(P)H with two-photon excitation microscopy
immunostaining of enzymes and hormones, 362–365
islet cell studies, 360–362
principle, 360

Bleaching, *see* Antifading agents; Photobleaching

C

Calcium
arterial endothelial cell, *in situ* calcium imaging with confocal microscopy
acetylcholine response, 429, 431
data collection, 429
fluo-3 loading, 428
quantitative analysis, 430
signal separation from smooth muscle cells, 429
tissue preparation, 428
custom confocal microscope construction for imaging
calcium transients, 167–168
cost, 152–153
dyes, 164
electronics, 158, 161
emission path, 155, 157
excitation path, 155

advantages, 98–99
animal perfusion, 105
antifading agents, 99–100
confocal microscopy, 102–103
fetus perfusion, 105–106
fluorescence labeling of gelatin, 104–
    105
immunostaining comparison and applica-
    tion, 98, 107
mounting in polyester resin, 99–102,
    107
organ perfusion, 106
replica technique comparison, 97
GFP, *see* Green fluorescent protein
Golgi apparatus, fluorescent probes, 121
Green fluorescent protein, *see also* Calmod-
    ulin–green fluorescent protein con-
    structs
  expression in fungi
    confocal microscopy
      image acquisition, 613
      instrumentation, 612–613
      propidium iodide multicolor staining,
          615–616
      rhodamine-conjugated antibody, ac-
          tin staining in multicolor detec-
          tion, 617–618
      rhodamine-conjugated phalloidin, ac-
          tin staining in multicolor detec-
          tion, 616–617
    epifluorescence microscopy and digital
        deconvolution, 618–620, 622
    fixation, 612
    mutants and detection wavelengths,
        611–612
    quantitative analysis of *Aureobasidium
        pullulans* on leaves, 620–621
    species of fungi, 609
    transcriptional control, 610
    transformation, 610
    vectors, 611
  fluorophore formation, 608
  imaging in living samples, 111
  virus tagging, 580

H

Hoechst 33342, vascular remodeling studies,
    253, 255–256, 262–263

I

ICON, *see In vivo* confocal neuroimaging
Image data, confocal microscopy
  animation, 116–117
  Cavalieri principle in object volume esti-
      mation, 309–310
  compression formats, 40–43, 54
  cytoskeletal elements, localization and
      colocalization with software, 188–
      189
  filtering and preprocessing of images,
      51–53
  hardware selection for processing, 49–50,
      54–55
  intensity levels, 29, 377, 497
  Kalman filter averaging, 613
  limitations in processing and display,
      313–315
  morphometry in three dimensions
    line length, 54
    surface area, 53–54
    volume, 53, 116
  rendering, *see also* Cytotomography
    ion-selective dyes and volume render-
        ing, 123–124
    simple projections, 45, 47
    surface extraction, 44
    three-dimensional rendering, overview
        of stereo pairs, 43–44
    weighted projection and alpha blend-
        ing, 47, 49
  software selection for processing, 49–51,
      55, 116
  storage of data
    CD-ROM, 37–39
    costs, 33–34
    digital video disk, 38
    floppy diskettes and Zip disks, 34–35,
        39–40
    hard disks, 35–36
    magnetic tape, 30–33, 39
    magneto-optical media, 36–37
    permanence, 30
    random access memory, 29
    space requirements, 30
    speed of connection with computer, 34
    write once read many drives, 37
  three-dimensional reconstruction
    advanced reconstruction

N

ISBN 0-12-182208-7

90038